Handbook of Neurochemistry and Molecular Neurobiology

Sensory Neurochemistry

Abel Lajtha (Ed.)

Handbook of Neurochemistry and Molecular Neurobiology

Sensory Neurochemistry

Volume Editor: Dianna A. Johnson

With 68 Figures and 8 Tables

Springer

Editor
Abel Lajtha
Director
Center for Neurochemistry
Nathan S. Kline Institute for Psychiatric Research
140 Old Orangeburg Road
Orangeburg
New York, 10962
USA

Volume Editor
Dianna A. Johnson
Hiatt Professor of Ophthalmology
Associate Vice Chancellor for Academic, Faculty and Student Affairs
University of Tennessee Health Science Center
62 South Dunlap
Memphis
Tennessee, 38163
USA

Library of Congress Control Number: 2006922553

ISBN: 978-0-387-30349-9

Additionally, the whole set will be available upon completion under ISBN: 978-0-387-35443-9
The electronic version of the whole set will be available under ISBN: 978-0-387-30426-7
The print and electronic bundle of the whole set will be available under ISBN: 978-0-387-35478-1

springer.com

Printed on acid-free paper

SPIN: 11417743 2109 - 5 4 3 2 1 0

Preface

It has been a singular challenge to organize a volume on sensory systems that appropriately reflects the intent of this new edition of the *Handbook of Neurochemistry and Molecular Neurobiology.* The physiology of sensory systems relies as much on highly specialized cellular and tissue morphology as it does on unique neurochemical pathways. Hence, this volume has emerged as a series of discussions that move continually, seamlessly we hope, between molecular and anatomical views of sensory function. It is by no means comprehensive, leaving room for additional chapters and updates, as befitting on-line publications such as this one. Nevertheless, every major sensory system is represented, each with its own blend of molecular, anatomic, and genetic components. Emphasis has been placed on presenting new discoveries and insights, rather than providing a textbook review of basic principles. The "somatosensory system" is discussed by Dr. Roe in a demanding and insightful chapter, which convincingly demonstrates that positional relationships within the cortex encode for the essential quality of touch, namely the source of the incoming somatosensory signal with respect to the body plan. The "visual system" is presented by Dr. Sharma and myself from two perspectives: development and degeneration. Together, these views provide not only an appreciation of normal retinal function but also highlight the utility of the visual system as a powerful model for uncovering internal and external clues for development and death of all CNS neurons. Drs. Wu and Zuo in their chapter on the "auditory system" emphasize exciting new findings from their own work, which explains the important role of cochlear amplification as a major requirement for auditory processing. Chapters on the chemical senses, "gustation" (by Drs. Smith and Boughter) and "olfaction" (by Dr. Ennis et al.), focus more on new concepts of the neurochemistry and molecular biology of transduction events and as such perhaps fit more neatly in the mold set by other volumes in the handbook. The last two chapters address additional aspects of the chemical senses, namely "glucose sensing" (by Dr. Roth et al.) and "CO_2/H^+ homeostasis" (by Drs. Lahiri et al.). In these cases, the term "sensory system" is used in the broadest sense, and discussions include a diversity of cell types and tissue locations that function collectively to monitor the internal milieu.

I offer special thanks to Dr. Abel Lajtha for his heroic feat in bringing to fruition this treatise of amazing breadth, diversity, and potential importance. Anyone with less than Dr. Lajtha's world view and historic grasp of the paradigm shifts resulting from neurochemical research would never have succeeded. What is more, they could never have even conceived of the plan.

Dianna A. Johnson

Table of Contents

Contributors

T. G. Anthony
Department of Biochemistry and Molecular Biology, Indiana University School of Medicine, Evansville, IN 47712, USA

M. Auer-Grumbach
Department of Internal Medicine, Diabetes and Metabolism, Medical University Graz, Graz, Austria and Institute of Medical Biology and Human Genetics, Medical University Graz, Graz, Austria

P. Auer-Grumbach
Gesundheitszentrum Feldbach, Feldbach, Austria

S. M. Baby
Department of Physiology, University of Pennsylvania School of Medicine, B-400 Richards, Building, 37th & Hamilton Walk, Philadelphia, PA 19104-6085, USA

J. D. Boughter, Jr.
Department of Anatomy and Neurobiology and The Neuroscience Institute, University of Tennessee Health Science Center, Memphis, Tennessee, USA

L. M. Chen
Department of Psychology, Vanderbilt University, Nashville, TN 37212, USA

C. Di Giulio
Department of Biomedical Sciences, School of Medicine, Chieti University, 66100, Chieti, Italy

M. Ennis
Department of Anatomy and Neurobiology, University of Tennessee Health Science Center, Memphis, TN 38163, USA

R. M. Friedman
Department of Psychology, Vanderbilt University, Nashville, TN 37212, USA

D. W. Gietzen
Department of Anatomy, Physiology and Cell Biology, School of Veterinary Medicine, University of California, One Shields Ave, Davis, CA 95616, USA

K. A. Hamilton
Departments of Cellular Biology and Anatomy and Pharmacology, Toxicology, and Neuroscience, Louisiana State University Health Sciences Center, Shreveport, LA 71130, USA

S. Hao
Department of Anatomy, Physiology and Cell Biology, School of Veterinary Medicine, University of California, One Shields Ave, Davis, CA 95616, USA

A. Hayar
Department of Anatomy and Neurobiology, University of Tennessee Health Science Center, Memphis, TN 38163, USA

D. A. Johnson
Department of Ophthalmology and Hamilton Eye Institute, University of Tennessee Health Science Center, Memphis, TN 38163, USA

S. Lahiri
Department of Physiology, University of Pennsylvania School of Medicine, B-400 Richards, Building, 37th & Hamilton Walk, Philadelphia, PA 19104-6085, USA

J. J. McArdle
Department of Pharmacology & Physiology, New Jersey Medical School (UMDNJ), Newark, NJ 07101; VA Puget Sound Health Care System, Metabolism and Endocrinology Division, Seattle, Washington; Wyeth Ayerst Pharmaceutical Company, Division of Neuroscience, Princeton, NJ, USA

M. Nikinmaa
Centre of Excellence in Evolutionary Genetics and Physiology, Department of Biology, University of Turku, FI-20014, Turku, Finland

T. Pieber
Department of Internal Medicine,
Diabetes and Metabolism, Medical University Graz, Graz, Austria

G. M. C. Renshaw
Hypoxia and Ischemia Research Unit,
Heart Foundation Research Centre, School of Physiotherapy and Exercise Science, Griffith University, PMB 50 Gold Coast Mail Centre, Queensland, 9726, Australia

A. W. Roe
Department of Psychology, Vanderbilt University, Nashville, TN 37212, USA

V. H. Routh
Department of Pharmacology & Physiology, New Jersey Medical School (UMDNJ), Newark, NJ 07101; VA Puget Sound Health Care System, Metabolism and Endocrinology Division, Seattle, Washington; Wyeth Ayerst Pharmaceutical Company, Division of Neuroscience, Princeton, NJ, USA

A. Roy
Department of Physiology, University of Pennsylvania School of Medicine, B-400 Richards, Building, 37th & Hamilton Walk, Philadelphia, PA 19104-6085, USA and Department of Pediatrics/Neonatology, 600N Wolfe Street, Bldg. CMSC, Rm 6111 Johns, Hopkins Medical Institute, Baltimore, MD 21218, USA

N. M. Sanders
Department of Pharmacology & Physiology, New Jersey Medical School (UMDNJ), Newark, NJ 07101; VA Puget Sound Health Care System, Metabolism and Endocrinology Division, Seattle, Washington; Wyeth Ayerst Pharmaceutical Company, Division of Neuroscience, Princeton, NJ, USA

R. K. Sharma
Department of Ophthalmology and Hamilton Eye Institute, University of Tennessee Health Science Center, Memphis, TN 38163, USA

Assistant Professor and Director of Research,
Department of Ophthalmology,
University of Florida HSC,
Jacksonville, FL 32209, USA

D. V. Smith
Department of Anatomy and Neurobiology and The Neuroscience Institute,
University of Tennessee Health Science Center,
Memphis, Tennessee, USA

Z. Song
Department of Pharmacology & Physiology,
New Jersey Medical School (UMDNJ), Newark,
NJ 07101; VA Puget Sound Health Care System,
Metabolism and Endocrinology Division,
Seattle, Washington; Wyeth Ayerst Pharmaceutical Company, Division of Neuroscience,
Princeton, NJ, USA

R. Wang
Department of Pharmacology & Physiology,
New Jersey Medical School (UMDNJ), Newark,
NJ 07101; VA Puget Sound Health Care System,
Metabolism and Endocrinology Division,
Seattle, Washington; Wyeth Ayerst Pharmaceutical Company, Division of Neuroscience,
Princeton, NJ, USA

X. Wu
Department of Developmental Neurobiology,
St. Jude Children's Research Hospital, Memphis, Tennessee 38105, USA and Department of Anatomy and Neurobiology, University of Tennessee Health Science Center, Memphis, Tennessee 38163, USA

J. Zuo
Department of Anatomy and Neurobiology, University of Tennessee Health Science Center, Memphis, Tennessee 38163, USA

1 Multiple Representation in Primate SI: A View from a Window on the Brain

A. W. Roe · R. M. Friedman · L. M. Chen

Abstract: This chapter summarizes recent findings regarding the functional organization of primary somatosensory cortex (SI) in primates when viewed through 'windows' on the brain with optical imaging methodologies. These views have confirmed previous knowledge regarding topographic organization in SI. They have also revealed the presence of functional domains for the processing of different sensory tactile modalities (pressure, flutter, and vibration domains). Surprisingly, the representation of these tactile modalities is quite distinct in organization from that of visual modalities (form, color, and depth) in visual cortex. Rather, tactile modality maps appear similar to visual orientation maps in primate visual cortex. Implications of these findings for the relationship of cortical organization to the sensory scene are discussed.

List of Abbreviations: SI, primary somatosensory cortex; SII, second somatosensory area; PV, parietal ventral area; SA, slowly adapting; RA, rapidly adapting; PC, pacinian; VPL, ventral posterior lateral; VPI, ventroposterior inferior; CCD, charge coupled device; V1, primary visual cortex; V2, second visual area; V4, fourth visual area; IT, inferotemporal cortex

1 Functional Representation in Primary Somatosensory Cortex (SI)

1.1 Multiple Topographic Maps in SI

Primate primary somatosensory cortex (SI) in the postcentral gyrus contains four complete topographic maps of the body surface that fall within the architectonically defined Brodmann's Areas 3a, 3b, 1, and 2 (e.g., Woolsey et al., 1942; Powell and Mountcastle, 1959; Kaas et al., 1979; Nelson et al., 1980; Sur et al., 1982; Pons et al., 1985, 1987) ➋ *Figure 1-1*. Areas 3b and 1 receive input primarily from cutaneous afferents where areas 3a and 2 receive input from deep afferents (muscle spindles and joints) (e.g., Tanji and Wise, 1981). Other parietal areas, such as Areas 5 and 7, also process somatosensory information (Murray and Mishkin, 1984; Dong et al., 1994; Burton et al., 1997; Duhamel et al., 1998; Debowy et al., 2001). Somatotopic maps are also found laterally in second somatosensory area (SII) and the adjacent parietal ventral area (PV) (Burton and Fabri, 1995; Krubitzer et al., 1995) and there are other somatosensory areas in insular cortex that receive cutaneous and visceral information (Robinson and Burton, 1980; Schneider et al., 1993; Craig, 2003).

1.2 Hierarchical Relationship between Areas 3b and 1

Numerous studies suggest a hierarchical relationship between Area 3b and Area 1. Ablations of Area 3a and 3b leave Area 1 unresponsive, consistent with anatomy studies that show that Area 1 receives the bulk of its input from Area 3b. These findings suggest that direct thalamic inputs to Area 1 play either a weak or a modulatory role in cutaneous information processing (Garraghty et al., 1990). In comparison with cells of Area 1, response properties of cells in Area 3b can be described as relatively simple or closer to the physical aspects of the stimulus. Area 3b neurons (and layer 4 neurons in Area 1) have receptive fields confined to single-digit tips; in contrast, Area 1 neurons recorded in supra- or infragranular layers integrate over larger areas of skin, often spanning multiple-digit tips (Mountcastle and Powell, 1959; Hyvarinen and Poranen, 1978; Costanzo and Gardner, 1980; Iwamura et al., 1983; Sur et al., 1980, 1985). In concert with a greater degree of integration in Area 1, intrinsic connections within Area 1 are more extensive than those in Area 3b (Burton and Fabri, 1995). Both SA (slowly adapting) and RA (rapidly adapting) responsive cells are commonly found in Area 3b, whereas Area 1 is characterized by a predominance of RA cells and cells responsive to motion and orientation (Warren et al., 1986; Nelson et al., 1991). Although both mecharceptors and RA cells are responsive to textured surfaces, the firing patterns of slowly-adapting type I mechanoreceptors cells are more closely tied with roughness and texture features (Connor and Johnson, 1992; Blake et al., 1997). These findings could suggest a stronger role of Area 3b in fine spatial pattern

Figure 1-1
Somatotopic maps in SI cortex. (a) Classical somatotopic maps determined with electrophysiology mapping. Sensory maps in Areas 3b and 1 are as shown (from Sur et al., 1982). (b) Enlarged view of the hand region from a. (c) Optical window over the hand region of Areas 3b and 1 in the squirrel monkey

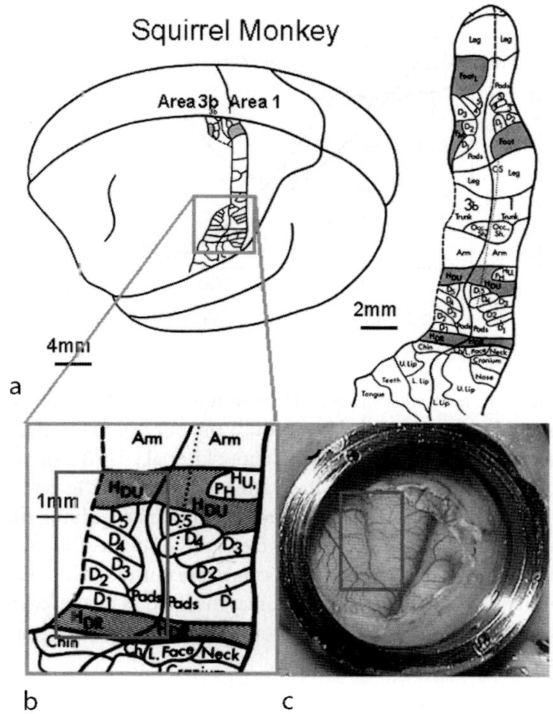

discrimination and a role of Area 1 in tactile processing during active tactile exploration. Lesions of SI have produced pronounced deficits in texture (Area 1), and size and shape (Area 2) discrimination, but have not isolated Area 3b versus Area 1 contributions (Carlson, 1981, 1984).

1.3 Representation of Multiple Cutaneous Modalities in Areas 3b and 1

1.3.1 Psychophysical and Peripheral Channels

A significant body of anatomical and physiological evidence suggests that "labeled lines" of modality-specific cutaneous information exists from the periphery to cortex. When glabrous skin is indented with vibratory stimuli, three distinct sensations can be felt. A local flutter sensation is evoked by low-frequency stimulation (2–40 Hz), a deep radiating hum of vibration is evoked by higher frequencies (40–200 Hz), and pressure is felt for stimuli below 2 Hz (Johansson et al., 1982). These pressure, flutter, and vibratory sensations are mediated by SA, RA, and pacinian (PC) receptors, respectively (➋ *Figure 1-2;* Mountcastle et al., 1972; LaMotte and Mountcastle, 1975; Cohen and Vierck, 1993).

Numerous psychophysics and microneurography studies suggest that these modalities remain separate in their central projections to somatosensory cortex (Verrillo, 1966; Talbot et al., 1968; Vallbo and Johansson, 1984). For example, direct electrical stimulation of single, identified, low-threshold mechanoreceptive afferents (SA, RA, or PC) evokes only one type of perception (pressure, flutter, or vibration,

Figure 1-2

Psychophysically determined thresholds for detection of different frequencies of vibrotactile stimulation. Peak thresholds are around 1 Hz, 30 Hz, and 200 Hz

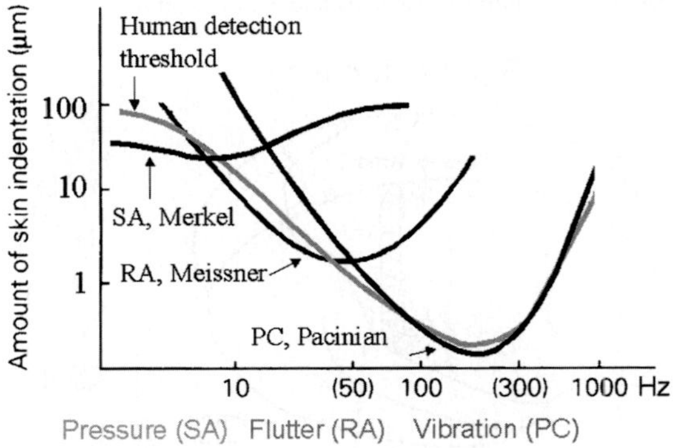

respectively) (Torebjork and Ochoa, 1980; Vallbo, 1981). Psychophysics studies fail to find vibrotactile masking and adaptation between stimulus frequencies that produce pressure (0.5 Hz), flutter (20 Hz), and vibratory sensations (200 Hz) (Gescheider et al., 1979, 1985; Bolanowski et al., 1988). In addition, frequency-specific electrical stimulation of a cortical RA-dominated site in Area 3b mimics the effect of stimulating RA receptors of the skin (Romo et al., 1998). Remarkably, even the transfer of tactile learning from one digit to another is modality-specific (Harris et al., 2001). These studies suggest a marked degree of separation in the experiences of pressure, flutter, and vibration, mediated by separate populations of receptors that remain separate in their central projections, and perhaps even to higher cortical areas involved in tactile learning and memory (see also Romo et al., 2000). Thus, both psychophysics and neurophysiology studies suggest some degree of modality-specific functional segregation in somatosensory cortex.

1.3.2 Anatomical and Physiological Pathways

Anatomical and physiological evidence also suggest parallel modality-specific pathways, from periphery through the dorsal column nuclei, to the thalamus, and into early somatosensory cortical areas. Dykes et al. (1981) have described the segregation of RA, SA, and PC responses in the VPL and VPI. Jones and colleagues (1982) have suggested that "rods" of topography and modality-specific cells project to similar modality-specific bands in Area 3b, and perhaps Area 1. Connections between Areas 3b and 1 are topographically homotopic, with feedforward projections being more robust than feedback (e.g., Jones and Powell, 1969; Jones et al., 1978; Cusick et al., 1985; Burton and Fabri, 1995). Using 2-deoxyglucose labeling methods combined with anatomical tracer injections, Juliano et al. (1990) suggested that excitatory information is transmitted from Area 3b to Area 1 in a way that connects clusters of cells with similar response properties.

1.3.3 Cortical Domains for Tactile Features

However, there is limited evidence to show whether different tactile features form multiple functional domains within each of Areas 3a, 3b, 1, and 2. Perhaps the best evidence for functional domains within SI comes from electrophysiological mapping studies describing zones of neurons with SA, RA, and PC mechanoreceptor responses within Area 3b (Paul et al., 1972; Sur et al., 1981, 1984; Sretavan and Dykes, 1983). ➋ *Figure 1-3* Based on densely spaced electrode penetrations, Sur et al. (1981, 1984) found a segregation of SA and RA cells in the middle layers of Area 3b and suggested that these are organized in

Figure 1-3
Electrophysiological evidence of modality-specific segregation in Area 3b (from Sur et al., 1984). Based on this and other studies, it was hypothesized that Area 3b contains alternating bands or zones of pressure and flutter domains

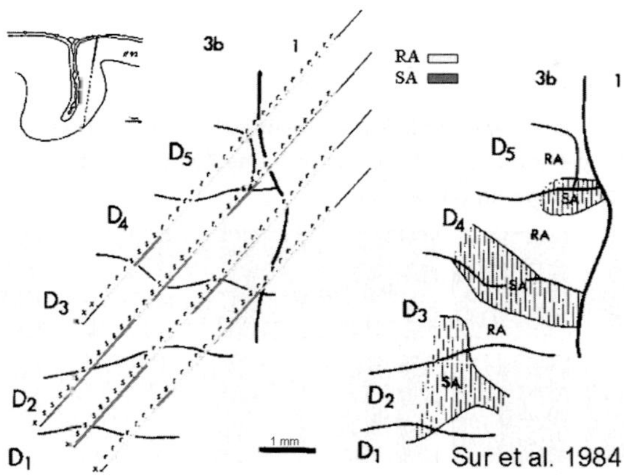

irregular antero-posterior "bands." This groundbreaking work was the first to suggest the presence of multiple maps in single cortical areas in SI. However, the limitations of electrophysiological mapping still leave open many questions regarding the degree of modality-specific segregation in Area 3b and its architecture. It is not clear whether SA/RA segregation in the middle layers implies some segregation in the superficial or deep layers of cortex. Whether a third map for PC responses exists is also unknown. It is also unknown whether Area 1, which is tightly associated with Area 3b, contains any functional organization for vibrotactile modality.

2 Intrinsic Signal Optical Imaging in Primate SI

2.1 The Methodology

2.1.1 Measuring Intrinsic Optical Signals

Optical imaging of intrinsic cortical signals is a method based on the activity-dependent reflectance changes of cortical tissue. The imaging procedure uses a charge-coupled device (CCD) camera to record the minute changes in optical absorption that accompanies cortical activity (Blasdel and Salama, 1986; Grinvald et al., 1986; Ts'o et al., 1990; Bonhoeffer and Grinvald, 1996). By presenting appropriate stimuli during optical imaging, the functional organizations of the sensory cortices can be mapped at high resolution. Although spiking response of single neurons is often predictive of the preference of the imaged domain (e.g., Bonhoeffer and Grinvald, 1991), it is known that both spiking and subthreshhold sources contribute to the intrinsic optical signal.

2.1.2 Signal Characteristics

The magnitude (typically in the 0.1–1.0% range) and timecourse of reflectance change is dependent on the illumination wavelength used (Bonhoeffer and Grinvald, 1996). In visual cortex, the typical time course under 600–630 nm illuminant and 2–3 s stimulation peaks in 2–3 s, followed by an undershoot that recovers in 4–5 s. The optical signal is similar in SI (➋ *Figure 1-4*; Chen et al., 2001; Friedman et al., 2004). Longer illumination wavelengths and longer stimulation periods produce longer signal timecourses (cf. Tommerdahl et al., 1999).

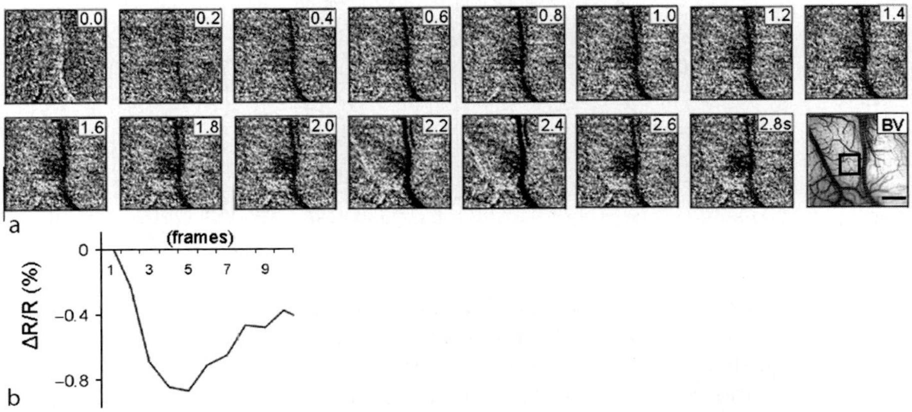

Figure 1-4
Timecourse of optical signal. (a) A series of images showing the temporal development of intrinsic cortical response (indentation of D2, pentothal anesthesia). Each frame 200 ms. BV, blood vessel map. Scale bar: 1 mm. (b) Magnitude of reflectance change (in box at BV)

2.1.3 Caveats in Interpretation

The relationship of the intrinsic signal to neural activity is by nature indirect and must be interpreted carefully. In visual cortex, due to the clustered nature of visual functional organization and microvascular relationships, the correlation of this signal with local neural excitatory response is strongly supported (e.g., Grinvald et al., 1986; Bonhoeffer et al., 1995; Roe and Ts'o, 1995, 1999; Rao et al., 1997; Issa et al., 2000). In somatosensory and auditory cortices, neurophysiological/imaging correlations also exist (cf. Bakin et al., 1996; Harel et al., 2000; Chen et al., 2001; Spitzer et al., 2001). Interpretation of optical images requires careful attention to stimulus design, multiple types of image analysis, and, if possible, accompanying electrophysiological recordings. For example, the presence of strong optical signal can indicate either a strong uniform stimulus-specific response or a diversity of response dominated by one particular component. In differential images, lack of response may be indistinguishable from equal responsiveness to opposing stimuli. Finally, it is important to bear in mind that the apparent organization of a cortical area as revealed by optical imaging is a direct function of the stimulus used to probe its organization.

2.2 Optical Imaging of Cortical Somatotopy

2.2.1 Somatotopic Maps

Optical imaging has been used in the study of somatosensory representation in rats (Masino and Frostig, 1993, 1996; Goldreich et al., 1998; Sheth et al., 1998), nonhuman primates, and humans (Cannestra et al., 1998; Schwartz et al., 2004). In the nonhuman primate, studies of somatotopy have produced images of the body map in the squirrel monkey (radial interdigital pad, D2 fingertip, and similar sites on the leg and foot)

(Tommerdahl et al., 1999), distal fingerpads of the squirrel monkey (Chen et al., 2001; Tommerdahl et al., 2002), and Area 1 of the Macaque monkey (Shoham and Grinvald, 2001). These have, in general, been consistent with the previously described somatotopic maps (e.g., Sur et al., 1982).

2.2.2 New World Monkeys

New World monkeys are an attractive species for optical imaging studies because their lissencephalic cortex allows unobstructed viewing of multiple visual cortical areas (cf. for visual cortex Malach et al., 1994; Xu et al., 2004; Roe et al., 2005). Furthermore, New World monkeys have been studied extensively in physiological, anatomical, and behavioral studies, making them prime candidates for studies of functional organization of sensation and sensory behavior. In the squirrel monkey, we have observed an orderly topographical map of the fingerpads within Area 3b (➋ *Figure 1-5*), consistent with the progression of maps determined with electrophysiological mapping (e.g., Sur et al., 1982; Merzenich et al., 1987). Activation sites were focal, measuring 0.5–1 mm in size.

In New World monkeys, it is also possible to simultaneously image multiple cortical areas, including Areas 3a, 3b, 1, and 2. As shown in ➋ *Figure 1-6*, the functional organization of SI in New World monkeys largely parallels those in Old World monkeys (cf. Sur et al., 1982). Stimulation of a single digit produces activation in Areas 3b and 1 in a topographically predictable manner (cf. Chen et al., 2002, 2005).

◘ Figure 1-5
Optical images of digit tip topography in Area 3b in squirrel monkey. (a) Five images obtained in response to indentation of digits D1 (thumb) to D5, respectively. Dark pixels indicate cortical activation (decrease in cortical reflectance). (b) Locations of D1–D5 activation zones are indicated by red circles overlaid on the blood vessel map. Green dots are locations of electrical recordings. (c) Topography consistent with published maps of digit topography (a, b from Chen et al., 2001; c from Sur et al., 1982)

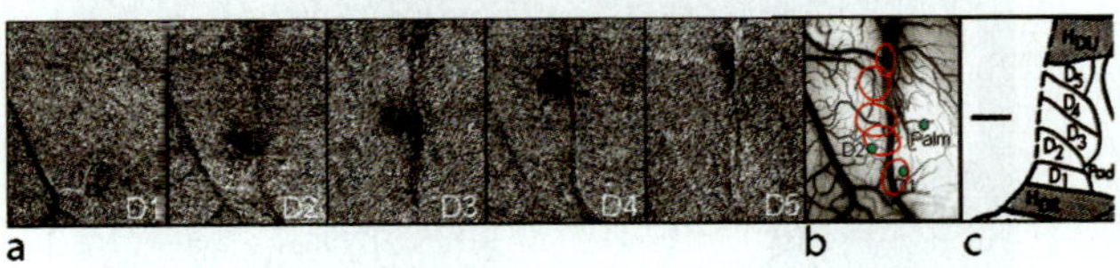

◘ Figure 1-6
Optical imaging of multiple cortical areas in SI of squirrel monkey. (a) Digit topography in Areas 3b and 1 (Sur et al., 1982). (b) Optical image of Area 3b and Area 1 to indentation of digit D5. Electrophysiological mapping of 3b (yellow dots) and 1 (red dots). Note that D5 activation zones (dark zones indicated by arrows) correlate well with electrophysiological map

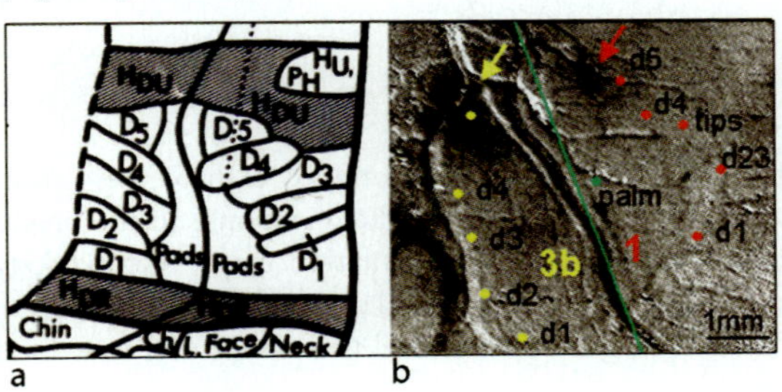

2.3 Modality Domains in SI

2.3.1 Vibrotactile Activation

As reported by Tommerdahl et al. (1996, 1998), mapping of modality-specific responses revealed vibrotactile response in Areas 3b and 1, and preferential response to skin heating in Area 3a. Comparison of flutter and vibration response revealed that extended vibration stimuli induced an initial increase in signal followed by a broad decrease, reminiscent of inhibitory influences described in neurophysiology studies (Tommerdahl et al., 1999).

2.3.2 Vibrotactile Segregation

Segregation of RA, SA, and PC responsiveness in Area 3b (Chen et al., 2001) and Area 1 (Friedman et al., 2004) of the squirrel monkey has been reported. Vibrotactile stimulation of the digit fingerpads at frequencies that produce the sensations of pressure (1 Hz), flutter (30 Hz), and vibration (200 Hz) were used in the anesthetized squirrel monkey (➋ *Figure 1-7*). These stimuli produced characteristic SA-dominated, RA-dominated, or PC-dominated responses, respectively. In some penetrations, single vibrotactile modalities were predominant; however, in other penetrations, mixed responses were obtained.

◘ Figure 1-7

Stimulation of digit tips. (a) Vibrotactile stimuli applied by a 3-mm diameter probe driven by a force feedback controlled motor. (b) Digits are secured by pegs glued to fingernails and inserted into plasticine. Two digit tips are stimulated with two probes controlled by two separate motors. (c) Temporal timecourse of each of three vibrotactile stimuli that induce pressure (1 Hz), flutter (30 Hz), and vibration (200 Hz). Images acquisition (3 s) of cortical response during each vibrotactile stimulus

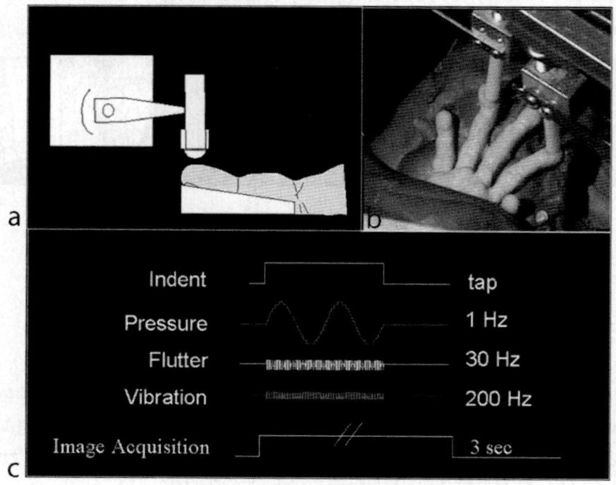

2.3.3 Vector Analysis

Intrinsic signal optical maps were obtained in response to each of these stimuli. A vector summation method was used to determine a pixel-by-pixel weighted response to the pressure, flutter, and vibration stimuli (similar to that used for visual cortical orientation maps, methodology details are described in Friedman et al., 2004). Clusters of pixels with saturated color would be evident only if one vector magnitude dominated the other two. Three examples of such pixel-wise SA/RA/PC vector summation are illustrated in

➋ Figure 1-8a–c. Pixel locations with a dominant SA response appear bright red, those with a dominant RA response appear bright green, and those with a dominant PC response appear bright blue. Patches of cortex that are coded white indicate areas exhibiting strong response to each of the pressure, flutter, and vibratory stimuli. In each map, we observed an irregular, interdigitating pattern of pressure (red), flutter (green), and vibration (blue) domains interspersed, in some maps, with domains of mixed preference (e.g., light blue). These domains were typically 200–300 µm in size in both Area 3b and Area 1 (➋ Figure 1-8d).

◘ Figure 1-8
Modality maps in SI. *Top*: Vibrotactile frequency-specific responses are plotted in different directions of color space (red: SA, 1 Hz; green: RA, 30 Hz; blue: PC, 200 Hz). a–c; Three cases: vector maps obtained through pixel-by-pixel vector summation. Domains dominated by single colors indicate regions preferentially responsive to a single vibrotactile frequency. Scale bar: 1 mm. (d) Size distribution of vibrotactile domains in Area 1 (*left*) and Area 3b (*right*)

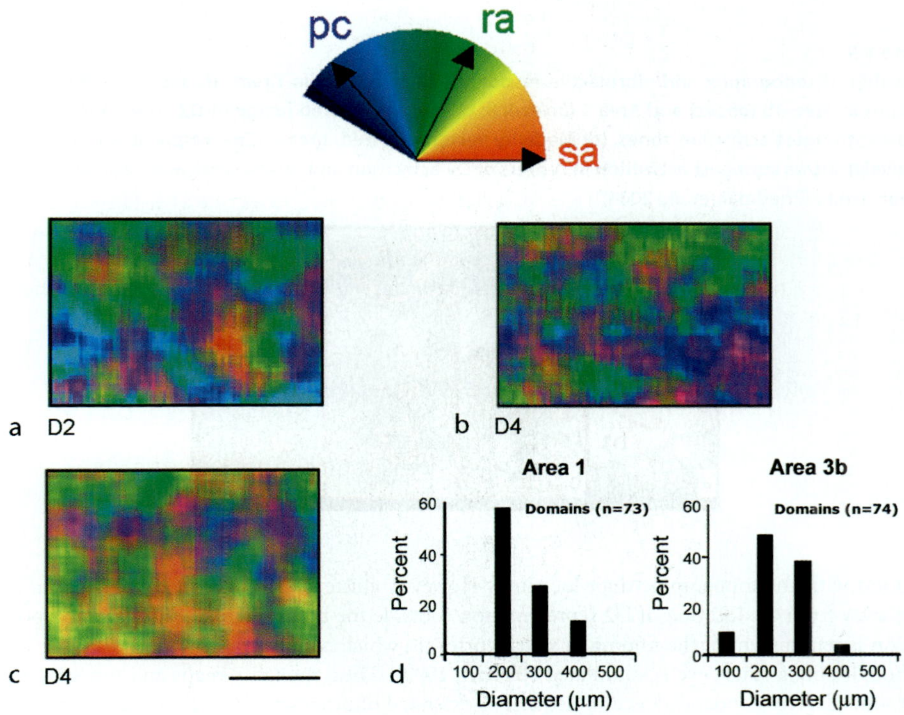

2.3.4 Neural Cortical Representation

Prior electrophysiology studies have not observed modular domains for vibrotactile stimuli in Area 1 (Costanzo and Gardner, 1980, Iwamura et al., 1993). A possible reason is that the small patch size of the modular domains in Area 1 revealed in our optical images would be difficult to discern solely with single and multiunit electrophysiology. In addition, previous studies focused on the adaptation (rapidly or slowly adapting) properties of neurons in Area 1 rather than on whether the neurons were integrating the information originating from SA, RA, or PC mechanoreceptors. When we examined these maps electrophysiologically, we found that, consistent with previous studies (Sur et al., 1984; Iwamura et al., 1993), single electrophysiological penetrations contained a mixture of neurons with SA, RA, and/or PC responses and single neurons that contained mixed responses. Often, we recorded neurons with solely rapidly adapting responses or mixed slowly and rapidly adapting responses. Thus, our data suggest that individual domains in Areas 3b and 1 contain neurons with a range of response properties, but coding of function within each domain is dominated by one modality.

2.4 Relationship of Vibrotactile Domains to Topography

It is evident that the modality-specific response extends beyond the classically defined topographic map as revealed by simple indentation stimuli. As shown in *Figure 1-9b*, an indentation stimulus (which activates all three receptors types) to digit D2 produces a fairly focal activation in both Area 3b and Area 1 (outline in red in *Figure 1-9c*). In response to pressure, flutter, and vibration stimulation, the vector summation map reveals that the strongest responses (most saturated red, green, and blue regions)

Figure 1-9

Relationship of topography and vibrotactile response. D2 activation in Areas 3b and 1. (a) Vessel map. D2 activations in Area 3b (*above*) and Area 1 (*below*). (b,c) Raw and filtered image of D2 indentation. Red outline delineates strongest activation zones. (d) Modality vector map (red, green, blue: response for SA, RA, and PC, respectively) shows strongest activation in centers of D2 activation and weaker activation outside of D2 zones. Scale bar: 1 mm. (Friedman et al., 2004)

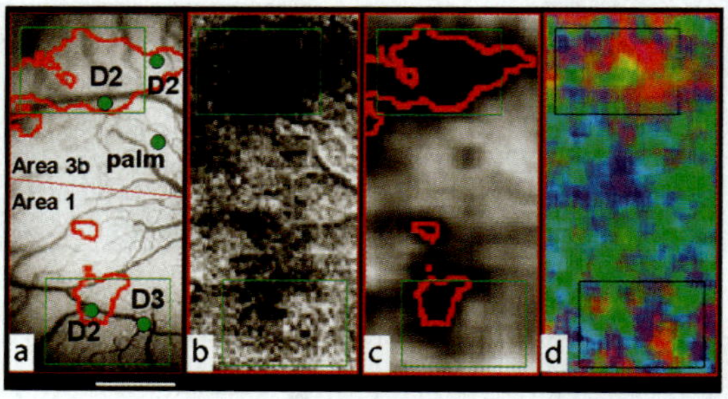

correspond with the topographic digit locations. However, clustered responses, though weaker, are also evident away from the location of D2 representation (outside the boxes). This additional nontopographic activation is reminiscent of the finding in visual cortex, in which complete orientation maps are obtained even though only a single eye is stimulated (Blasdel, 1992). Thus, modality maps and topographic maps exhibit some degree of independence. Whether this extended nontopographic region of activation is due to spiking and/or subthreshhold activity remains to be determined.

2.5 Are There Pinwheels?

In optical imaging studies of visual cortex, orientation pinwheels have received a great deal of attention. Pinwheels are locations (singularities) in orientation maps around which the orientation preference changes smoothly. Although the locations, density, stability, and function of pinwheels in optical maps has been a topic of much controversy (e.g., Bonhoeffer and Grinvald, 1991; Bartfeld and Grinvald, 1992; Obermayer et al., 1997; Swindale, 2000; Schummers et al., 2002; Polimeni et al., 2005), they have become a cornerstone of cortical functional architecture. The key feature of pinwheels is that they are points around which a parameter (such as contour orientation) varies continuously. Although further studies are needed, the possibility that vibrotactile maps in SI contain pinwheel-like features has been suggested (Friedman et al., 2004). In contrast to some visual cortical maps that demonstrate clear discontinuities, response preferences in SI vibrotactile maps appear to vary continuously, giving rise to an appearance more reminiscent of visual orientation maps. As shown in *Figure 1-10*, possible singularity sites are

Figure 1-10
Possible pinwheel centers in SI. (a) Same map shown in *Figure 1-9c*. (b) Expanded and thresholded portion of (a) Possible pinwheels marked by white dots, shown expanded in nearby panels. Note clockwise and counter-clockwise SA (red), RA (green), PC (blue) rotations

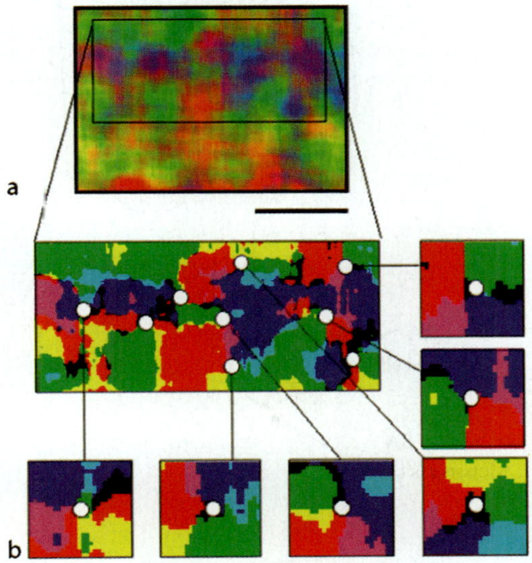

those around which a rotation of vibrotactile frequency can be observed. Whether these rotations are continuous changes in frequency or simply convergence points of discrete frequency clusters remains to be investigated.

3 Summary Model of SI Organization

The finding that somatosensory cortical domains are roughly 200–300 μm in size strengthens the view that modularity is a common organizational feature of cortical representation. Cortical domains of similar size have been described in multiple cortical areas [in V1 and V2 (see Roe, 2003 for review), and V4 (Felleman et al., 1997), IT (Tsunoda et al., 2001), Area 7 (Siegel et al., 2003), and in prefrontal areas (Kritzer and Goldman-Rakic, 1995)]. These findings suggest a revision of previous views of SI organization, which were based primarily on electrophysiological recordings. Previously, each cortical "hypercolumn" was thought to contain segregated Sa and Ra columns innervated predominantly by thalamocortical fibers of a single vibrotactile modality (*Figure 1-11*, Sur et al., 1980). Optical imaging evidence now suggests a modification of this view (*Figure 1-12*). As shown by reconstruction of single thalamocortical arbors (Garraghty et al., 1989), inputs to SI have multiple arbors (200–300 μm in size) that span several millimeters of cortex (see also Jones et al., 1982). These and other corticocortical arbors are likely to give rise to clustered activations that have been observed in 2-deoxyglucose studies (Juliano 1981, 1990; Juliano and Whistel, 1987; cf. Burton and Fabri, 1995). Such arbors could give rise either to an array of discrete clusters or, by

Figure 1-11
Traditional view of SI organization. (from Kandel & Schwartz)

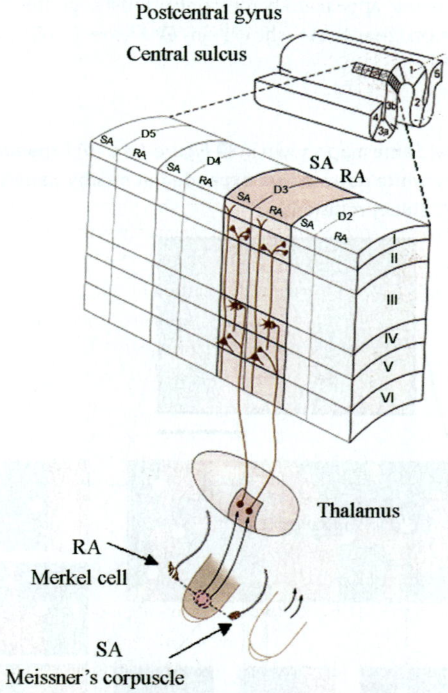

varying arbor overlap, continuous modality maps. Thus, regions dominated by single SA, RA, or PC inputs would give rise to SA (red), RA (green), or PC (blue) domains. Regions of some overlap would appear as magenta or yellow–green colors (not depicted). In addition, regions of high SA, RA, and PC overlap could

Figure 1-12
Proposed model of vibrotactile representation in SI. (a) Thalamocortical arbors in SI have multiple clusters extending across millimeters of cortex (Garraghty et al., 1989). (b) Single SA, RA, or PC fibers terminate in layer IV with three arbors (colored disks). These arbors project in turn to superficial layers II/III. Thus, some cortical columns are dominated by a single SA, RA, or PC input, whereas others have mixed input (overlapping disks)

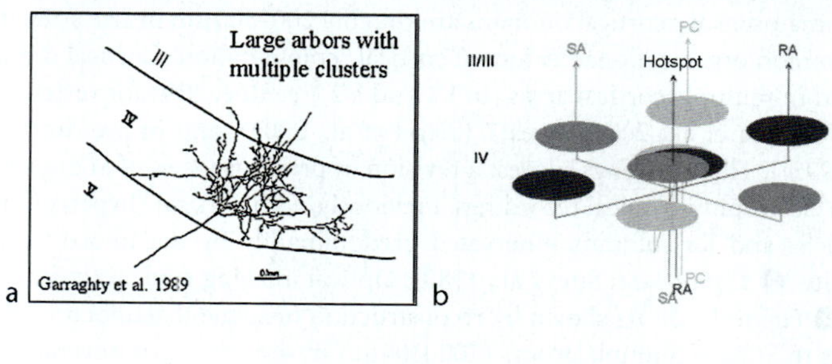

underlay the so-called hotspots commonly described in SI receptive fields (in optical images, these locations would appear as black, gray, or white domains); these would be well activated by broadband stimuli such as skin indentation. Other arbors extend to nontopographic locations away from the hotspot and locally establish some degree of modality-specific dominance (cf. ➋ *Figure 1-10*). Thus, not unlike the way horizontal iso-orientation networks in V1 give rise to resulting orientation map structure, the observed maps result from overlapping horizontal networks of patchy, modality-specific dominance. In sum, in the revised view, each digit representation is served by collections of interdigitating SA, RA, and PC columns. The effect of topographic stimulation is therefore no longer so discrete and can, under certain stimulation conditions, have nontopographic consequences.

References

Bakin JS, Kwon MC, Masino SA, Weinberger NM, Frostig RD. 1996. Suprathreshold auditory cortex activation visualized by intrinsic signal optical imaging. Cereb Cortex 6: 120-130.

Bartfeld E, Grinvald A. 1992. Relationships between orientation-preference pinwheels, cytochrome oxidase blobs, and ocular-dominance columns in primate striate cortex. Proc Natl Acad Sci USA 89(24): 11905-11909.

Blake DT, Hsiao SS, Johnson KO. 1997. Neural coding mechanisms in tactile pattern recognition: The relative contributions of slowly and rapidly adapting mechanoreceptors to perceived roughness. J Neurosci 17: 7480-7489.

Blasdel GG, Salama G. 1986. Voltage-sensitive dyes reveal a modular organization in monkey striate cortex. Nature 321: 579-585.

Blasdel GG. 1992. Differential imaging of ocular dominance and orientation selectivity in monkey striate cortex. J Neurosci 12: 3115-3138.

Bolanowski SJ Jr, Gescheider GA, Verrillo RT, Checkosky CM. 1988. Four channels mediate the mechanical aspects of touch. J Acoust Soc Am 84: 1680-1694.

Bonhoeffer T, Grinvald A. 1991. Iso-orientation domains in cat visual cortex are arranged in pinwheel-like patterns. Nature 353: 429-431.

Bonhoeffer T, Kim DS, Malonek D, Shoham D, Grinvald A. 1995. Optical imaging of the layout of functional domains in Area 17 and across the Area 17/18 border in cat visual cortex. Eur J Neurosci 7: 1973-1988.

Bonhoeffer T, Grinvald A. 1996. Brain mapping: The methods. Toga AW, Mazziotta JC, editors. New York: Academic Press; pp. 55-97.

Burton H, Fabri M. 1995. Ipsilateral intracortical connections of physiologically defined cutaneous representations in areas 3b and 1 of macaque monkeys: Projections in the vicinity of the central sulcus. J Comp Neurol 355: 508-538.

Burton H, Sinclair RJ, Hong SY, Pruett JR Jr, Whang KC. 1997. Tactile-spatial and cross-modal attention effects in the second somatosensory and 7b cortical areas of rhesus monkeys. Somatosens Mot Res 14(4): 237-267.

Cannestra AF, Black KL, Martin NA, Cloughesy T, Burton JS, et al. 1998. Topographical and temporal specificity of human intraoperative optical intrinsic signals. Neuroreport 9: 2557-2563.

Carlson M. 1981. Characteristics of sensory deficits following lesions of Brodmann's areas 1 and 2 in the postcentral gyrus of Macaca mulatta. Brain Res 204(2): 424-430.

Carlson M. 1984. Development of tactile discrimination capacity in Macaca mulatta. II. Effects of partial removal of primary somatic sensory cortex (SmI) in infants and juveniles. Brain Res 318: 83-101.

Chen LM, Friedman RM, Ramsden BM, La Motte RH, Roe AW. 2001. Fine-scale organization of primary somatosensory cortex (Area 3b) in the squirrel monkey revealed with intrinsic optical imaging. J Neurophysiol 86: 3011-3029.

Chen LM, Heider B, Healy FL, Ramsden BR, Williams GV, et al. 2002. A chamber and artificial dura method for long-term optical imaging in primates. J Neurosci 113: 41-49.

Chen LM, Friedman RM, Roe AW. 2005. Optical imaging of SI topography in anesthetized and awake squirrel monkeys. J Neurosci 25(33): 7648-7659.

Cohen RH, Vierck CJ. 1993. Population estimates for responses of cutaneous mechanoreceptors to a vertically indenting probe on the glabrous skin of monkeys. Exp Brain Res 94: 105-119.

Connor CE, Johnson KO. 1992. Neural coding of tactile texutre: Comparison of spatial and temporal mechanisms for roughness perception. J Neursoci 12: 3414-3426.

Costanzo RM, Gardner EP. 1980. A quantitative analysis of repsonses of direction sensitive neurons in somatosensory cortex of awake monkeys. J Neurophysiol 43: 139-1341.

Craig AD. 2003. Pain mechanisms: labeled lines versus convergence in central processing. Annu Rev Neurosci 26: 1-30.

Cusick CG, Steindler DA, Kaas JH. 1985. Corticocortical and collateral thalamocortical connections of postcentral somatosensory cortical areas in squirrel monkeys: A double-labeling study with radiolabeled wheatgerm agglutinin and wheatgerm agglutinin conjugated to horseradish peroxidase. Somatosens Res 3: 1-31.

Debowy DJ, Ghosh S, Ro JY, Gardner EP. 2001. Comparison of neuronal firing rates in somatosensory and posterior parietal cortex during prehension. Exp Brain Res 137(3–4): 269-291.

Dong WK, Chudler EH, Sugiyama K, Roberts VJ, Hayashi T. 1994. Somatosensory, multisensory, and task-related neurons in cortical area 7b (PF) of unanesthetized monkeys. J Neurophysiol 72(2): 542-564.

Duhamel JR, Colby CL, Goldberg ME. 1998. Ventral intraparietal area of the macaque: Congruent visual and somatic response properties. J Neurophysiol 79(1): 126-136.

Dykes RW, Sur M, Merzenich MM, Kaas JH, Nelson RJ. 1981. Regional segregation of neurons responding to quickly adapting, slowly adapting, deep and pacinian receptors within thalamic ventroposterior lateral and ventroposterior inferior nuclei in the squirrel monkey (Saimiri sciureus). Neurosci 6: 1687-1692.

Felleman DJ, Xiao Y, McClendon E. 1997. Modular organization of occipito-temporal pathways: Cortical connections between visual area 4 and visual area 2 and posterior inferotemporal ventral area in macaque monkeys. J Neurosci 17(9): 3185-3200.

Friedman RM, Chen LM, Roe AW. 2004. Modality maps within primate somatosensory cortex. Proc Natl Acad Sci USA 101: 12724-12729.

Garraghty PE, Pons TP, Sur M, Kaas JH. 1989. The arbors of axons terminating in middle cortical layers of somatosensory area 3b in owl monkeys. Somatosens Mot Res 6(4): 401-411.

Garraghty PE, Florence SL, Kaas J. 1990. Ablations of areas 3a and 3b of monkey somatosensory cortex abolish cutaneous responsivity in area 1. Brain Res 528: 165-169.

Gescheider GA, Frisina RD, Verrillo RT. 1979. Selective adaptation of vibrotactile thresholds. Sens Processes 3: 37-48.

Gescheider GA, Sklar BF, Van Doren CL, Verrillo RT. 1985. Vibrotactile forward masking: Psychophysical evidence for a triplex theory of cutaneous mechanoreception. J Acoust Soc Am 78(2): 534-543.

Goldreich D, Peterson BE, Merzenich MM. 1998. Optical imaging and electrophysiology of rat barrel cortex. II. Responses to paired-vibrissa deflections. Cereb Cortex 8: 184-192.

Grinvald A, Lieke E, Frostig RD, Gilbert CD, Wiesel TN. 1986. Functional architecture of cortex revealed by optical imaging of intrinsic signals. Nature 324: 361-364.

Harel N, Mori N, Sawada S, Mount RJ, Harrison RV. 2000. Three distinct auditory areas of cortex (AI, AII, and AAF) defined by optical imaging of intrinsic signals. Neuroimage 11: 302-312.

Harris JA, Harris IM, Diamond ME. 2001. The topography of tactile learning in humans. J Neurosci 21: 1056-1061.

Hyvarinen J, Poranen A. 1978. Movement-sensitive and direction and orientation-selective cutaneous receptive fields in the hand area of the post-central gyrus in monkeys. J Physiol 283: 523-537.

Issa NP, Trepel C, Stryker MP. 2000. Spatial frequency maps in cat visual cortex. J Neurosci 20: 8504-8514.

Iwamura Y, Tanaka M, Sakamoto M, Hikosaka O. 1983. Converging patterns of finger representation and complex response properties of neurons in Area 1 of the first somatosensory cortex of the conscious monkey. Exp Brain Res 51: 327-337.

Johansson RS, Landstrom U, Lundstrom R. 1982. Responses of mechanoreceptive afferent units in the glabrous skin of the human hand to sinusoidal skin displacements. Brain Res 244: 17-25.

Jones EG, Powell TPS. 1969. Connexions of the somatic sensory cortex of the rhesus monkey. I. Ipsilateral connections. Brain 92: 477-502.

Jones EG, Coulter JD, Hendry SHC. 1978. Intracortical connectivity of architectonic fields in the somatic sensory, motor and parietal cortex of monkeys. J Comp Neurol 181: 291-347.

Jones EG, Friedman DP, Hendry SHC. 1982. Thalamic basis of place- and modality-specific columns in monkey somatosensory cortex: A correlative anatomical and physiological study. J Neurophysiol 48: 545-568.

Juliano SL, Whitsel BL. 1987. A combined 2-deoxyglucose and neurophysiological study of primate somatosensory cortex. J Comp Neurol 263: 514-525.

Juliano SL, Hand PF, Whitsel BL. 1981. Patterns of increased metabolic activity in somatosensory cortex of monkeys Macaca fascicularis, subjected to controlled cutaneous stimulation: A 2-deoxyglucose study. J Neurophysiol 46: 1260-1284.

Juliano SL, Friedman DP, Eslin DE. 1990. Corticocortical connections predict patches of stimulus-evoked metabolic activity in monkey somatosensory cortex. J Comp Neurol 298: 23-39.

Kaas JH, Nelson RJ, Sur M, Lin CS, Merzenich MM. 1979. Multiple representations of the body within the primary somatosensory cortex of primates. Science 204: 521-523.

Kritzer MF, Goldman-Rakic PS. 1995. Intrinsic circuit organization of the major layers and sublayers of the dorsolateral prefrontal cortex in the rhesus monkey. J Comp Neurol 359: 131-143.

Krubitzer L, Clarey J, Tweedale R, Elston G, Calford M. 1995. A redefinition of somatosensory areas in the lateral sulcus of macaque monkeys. J Neurosci 15: 3821-3839.

La Motte RH, Mountcastle VB. 1975. Capacities of humans and monkeys to discriminate vibratory stimuli of different frequency and amplitude: A correlation between neural events and psychological measurements. J Neurophysiol 38: 539-559.

Malach R, Tootell RBH, Malonek D. 1994. Relationship between orientation domains, cytochrome oxidase stripes, and intrinsic horizontal connections in squirrel monkey area V2. Cereb Cortex 4: 151-165.

Masino SA, Kwon MC, Dory Y, Frostig RD. 1993. Characterization of functional organization within rat barrel cortex using intrinsic signal optical imaging through a thinned skull. Proc Natl Acad Sci USA 1993 Nov 1; 90(21): 9998-10002.

Masino SA, Frostig RD. 1996. Quantitative long-term imaging of the functional representation of a whicker in rat barrel cortex. Proc Natl Acad Sci USA 93: 4942-4947.

Merzenich MM, Nelson RJ, Kaas JH, Stryker MP, Jenkins WM, et al. 1987. Variability in hand surface representations in areas 3b and 1 in adult owl and squirrel monkeys. J Comp Neurol 258: 281-296.

Mountcastle VB, Powell TPS. 1959. Neural mechanisms subserving cutaneous sensibility, with special reference to the role of afferent inhibition in sensory perception and discrimination. Bull Johns Hopkins Hosp 105: 201-232.

Mountcastle VB, La Motte RH, Carli G. 1972. Detection thresholds for stimuli in humans and monkeys: Comparison with threshold events in mechanoreceptive afferent nerve fibers innervating the monkey hand. J Neurophysiol 35: 122-136.

Murray EA, Mishkin M. 1984. Relative contributions of SII and area 5 to tactile discrimination in monkeys. Behav Brain Res 11: 67-83.

Nelson RJ, Sur M, Felleman DJ, Kaas JH. 1980. Representations of the body surface in postcentral parietal cortex of Macaca fascicularis. J Comp Neurol 192(4): 611-643.

Nelson RJ, Smith BN, Douglas VD. 1991. Relationships between sensory responsiveness and premovement activity of quickly adapting neurons in areas 3b and 1 of monkey primary somatosensory cortex. Exp Brain Res 84: 75-90.

Obermayer K, Blasdel GG. 1997. Singularities in primate orientation maps. Neural Comput 9: 555-575.

Paul RL, Merzenich M, Goodman H. 1972. Representation of slowly and rapidly adapting cutaneous mechanoreceptors of the hand in Brodmann's areas 3 and 1 of Macaca mulatta. Brain Res 36: 229-249.

Polimeni JR, Granquist-Fraser D, Wood RJ, Schwartz EL. 2005. Physical limits to spatial resolution of optical recording: clarifying the spatial structure of cortical hypercolumns. Proc Natl Acad Sci USA 102(11): 4158-4163.

Pons TP, Garraghty PE, Cusick CG, Kaas JH. 1985. A sequential representation of the occiput, arm, forearm and hand across the rostrocaudal dimension of areas 1, 2 and 5 in macaque monkeys. Brain Res 335(2): 350-353.

Pons TP, Wall JT, Garraghty PE, Cusick CG, Kaas JH. 1987. Consistent features of the representation of the hand in area 3b of macaque monkeys. Somatosens Res 4: 309-331.

Powell TP, Mountacstle VB. 1959. The cytoarchitecture of the postcentral gyrus of the monkey Macaca mulatta. Bull Johns Hopkins Hosp 105: 108-131.

Rao SC, Toth LJ, Sur M. 1997. Optically imaged maps of orientation preference in primary visual cortex of cats and ferrets. J Comp Neurol 87: 358-370.

Robinson CJ, Burton H. 1980. Organization of somatosensory receptive fields in cortical areas 7b, retroinsula, postauditory and granular insula of M. fascicularis. J Comp Neurol 192(1): 69-92.

Robinson CJ, Burton H. 1980b. Somatic submodality distribution within the second somatosensory (SII), 7b, retroinsular, postauditory, and granular insular cortical areas of M. fascicularis. J Comp Neurol 192(1): 93-108.

Roe AW, Ts'o DY. 1995. Visual topography in primate V2: multiple representation across functional stripes. J Neurosci 15: 3689-3715.

Roe AW, Ts'o DY. 1999. Specificity of color connectivity between primate V1 and V2. J Neurophysiol 82(5): 2719-2730.

Roe AW. 2003. Modular complexity of area V2 in the Macaque monkey. The primate visual system. Collins C, Kaas J, editors. New York : CRC Press; pp. 109-138.

Roe AW, Fritsches K, Pettigrew JD. 2005. Optical imaging of functional organization of V1 and V2 in marmoset visual cortex. Anat Rec A Discov Mol Cell Evol Biol 287(2): 1213-25.

Romo R, Hernandez A, Zainos A, Salinas E. 1998. Somatosensory discrimination based on cortical microstimulation. Nature 392: 387-390.

Romo R, Hernandez A, Zainos A, Brody CD, Lemus L. 2000. Sensing without touching: Psychophysical performance based on cortical microstimulation. Neuron 26: 273-278.

Schneider RJ, Friedman DP, Mishkin M. 1993. A modality-specific somatosensory area within the insula of the rhesus monkey. Brain Res 621(1): 116-120.

Schummers J, Marino J, Sur M. 2002. Synaptic integration by V1 neurons depends on location within the orientation map. Neuron 36(5): 969-978.

Schwartz TH, Chen LM, Friedman RM, Spencer DD, Roe AW. 2004. Intraoperative optical imaging of face topography in human somatosensory cortex. Neuroreport 15: 1527-1532.

Sheth BR, Moore CI, Sur M. 1998. Temporal modulation of spatial borders in rat barrel cortex. J Neurophysiol 79: 464-470.

Shoham D, Grinvald A. 2001. The cortical representation of the hand in macaque and human Area S-I: High resolution optical imaging. J Neurosci 21: 6820-6835.

Siegel RM, Raffi M, Phinney RE, Turner JA, Jando G. 2003. Functional architecture of eye position gain fields in visual association cortex of behaving monkey. J Neurophysiol 90(2): 1279-1294.

Spitzer MW, Calford MB, Clarey JC, Pettigrew JD, Roe AW. 2001. Spontaneous and stimulus-evoked intrinsic optical signals in primary auditory cortex of the cat. J Neurophysiol 85: 1283-1298.

Sretavan D, Dykes RW. 1983. The organization of two cutaneous submodalities in the forearm region of area 3b of cat somatosensory cortex. J Comp Neurol 213: 381-398.

Sur M. 1980. Receptive fields of neurons in Areas 3b and 1 of somatosensory cortex in monkeys. Brain Res 198: 465-471.

Sur M, Merzenich MM, Kaas JH. 1980. Magnification, receptive-field area, and "hypercolumn" size in areas 3b and 1 of somatosensory cortex in owl monkeys. J Neurophysiol 44: 295-311.

Sur M, Nelson RJ, Kaas JH. 1982. Representations of the body surface in cortical areas 3b and 1 of squirrel monkeys: comparisons with other primates. J Comp Neurol 211: 177-192.

Sur M, Garraghty PE, Bruce C. 1985. Somatosensory cortex in macaque monkeys: laminar differences in receptive field size in areas 3b and 1. Brain Res 342: 391-395.

Sur M, Wall JT, Kaas JH. 1981. Modular segregation of functional cell classes within the postcentral somatosensory cortex of monkeys. Science 212: 1059-1061.

Sur M, Wall JT, Kaas JH. 1984. Modular distribution of neurons with slowly adapting and rapidly adapting responses in area 3b of somatosensory cortex in monkeys. J Neurophysiol 51: 724-744.

Swindale NV. 2000. How many maps are there in visual cortex? Cereb Cortex 10: 633-643.

Talbot WH, Darian-Smith I, Kornhuber HH, Mountcastle VB. 1968. The sense of flutter-vibration: comparison of the human capacity with response patterns of mechanoreceptive afferents from the monkey hand. J Neurophysiol 31: 301-334.

Tanji J, Wise SP. 1981. Submodality distribution in sensorimotor cortex of the unanesthetized monkey. J Neurophysiol 45: 467-481.

Tommerdahl M, Delemos KA, Vierck CJ, Favorov OV, Whitsel BL. 1996. Anterior parietal cortical response to tactile and skin-heating stimuli applied to the same skin site. J Neurophysiol 75: 2662-2670.

Tommerdahl M, Delemos KA, Favorov OV, Metz CB, Vierck CJ, et al. 1998. Response of anterior parietal cortex to different modes of same-site skin stimulation. J Neurophysiol 80: 3272-3283.

Tommerdahl M, Delemos KA, Whitsel BL, Favorov OV, Metz CB. 1999. Response of anterior parietal cortex to cutaneous flutter versus vibration. J Neurophysiol 82: 16-33.

Tommerdahl M, Favorov O, Whitsel BL. 2002. Optical imaging of intrinsic signals in somatosensory cortex. Behav Brain Res 135: 83-91.

Torebjork HE, Ochoa JL. 1980. Specific sensations evoked by activity in single identified sensory units in man. Acta Physiol Scand 110: 445-447.

Ts'o DY, Frostig RD, Lieke EE, Grinvald A. 1990. Functional organization of primate visual cortex revealed by high resolution optical imaging. Science 249: 417-420.

Tsunoda K, Yamane Y, Nishizaki M, Tanifuji M. 2001. Complex objects are represented in macaque inferotemporal cortex by the combination of feature columns. Nat Neurosci 4(8): 832-838.

Vallbo AB. 1981. Sensations evoked from the glabrous skin of the human hand by electrical stimulation of unitary mechanosensitive afferents. Brain Res 215(1-2): 359-363.

Vallbo AB, Johansson RS. 1984. Properties of cutaneous mechanoreceptors in the human hand related to touch sensation. Hum Neurobiol 3(1): 3-14.

Verrillo RT. 1966. Vibrotactile thresholds for hairy skin. J Exp Psychol 72(1): 47-50.

Warren S, Hamalainen HA, Gardner EP. 1986. Objective classification of motion- and direction-sensitive neurons in primary somatosensory cortex of awake monkeys. J Neurophysiol 56(3): 598-622.

Woolsey CN, Marshall WH, Bard P. 1942. Representation of cutaneous tactile sensibility in the cerebral cortex of the monkey as indicated by evoked potentials. Bull Johns Hopkins Hosp 70: 399-441.

Xu X, Bosking W, Sary G, Stefansic J, Shima D, et al. 2004. Functional organization of visual cortex in the owl monkey. J Neurosci 24(28): 6237-6247.

2 Determinants of Molecular Mechanisms in Neuroretinal Development

R. K. Sharma · D. A. Johnson

Abstract: Increasing understanding about retinal development is not an academic curiosity alone; the developmental pathways may hold the key to the success of regenerative and reconstructive therapies being investigated to combat retinal degeneration. Development of the retina is a highly organized process. Retinal pigment epithelium (RPE), as well as the neural retina, develops from neuroectoderm. The neural retina itself comprises a number of functionally and morphologically diverse neurons and the radial glia, the Müller cells. Retinal progenitors give rise to this diverse array of cells. The ontogenic development of retinal neurons is defined and preserved across species. The sequence of neuronal birth in retina follows the phylogenetic evolution of these neurons. This is achieved by cues from environment and perhaps increasingly restricted potential of the retinal progenitors. Retinal cells are generated in two waves. In the first wave, neurons of the cone pathways that are phylogenetically older are generated, whereas the second gives rise to the neurons of rod pathways, which developed later during evolution. Proliferation in the early stages of retinal development produces progenitors that are able to migrate laterally. In the later stages, when the cells of rod pathways are generated, the postmitotic cells migrate only radially on the scaffold provided by the Müller cell. Subsequently, specific connections and synapses are formed and the plexiform layers develop. The processes of the nonprojecting neurons of the retina are guided toward the partners on the railroads provided by the Müller cell processes. This happens due to the differential expression of cell adhesion molecules perhaps on the Müller cells. Cell adhesion molecules participate not only in guiding the neurites, but also in their growth. There is also evidence that neurotransmitters might be important for various aspects of retinal organization such as cell migration, synaptogenesis, and dendritic pruning.

The fully mature retina is arranged as thin sheet of cells divided into three concentric layers of cell bodies, each containing specific cell types. These layers are separated from each other by synaptic or plexiform layers containing specific retinal connections. Retinal neurons are arranged in repeating functional units spanning all retinal layers that form a mosaic across the retina. In this way, all points within the retinal field are able to sample visual signals appropriately. There is evidence that certain photoreceptors may act as a template for the development of these mosaics. In addition, migration of neurons and apoptosis also play important roles. Organization of specialized regions of the retina such as the fovea is more complex. Evidence suggests that it involves migration of neurons. How a rod-free fovea is created is a mystery. It is possible that the cytogenesis is truncated before the rods are generated. Alternatively, it is possible that some local factors induce the differentiation of the neurons generated there into the cones. Phylogenetic evolution of retinal neurons also determines the retinal circuitry. The neurons that evolved later found connections secondary to already existing ones. In this chapter, retinal development is discussed in the context of its phylogeny, ontogeny, and embryogenesis.

List of Abbreviations: AMD, Age-related macular degeneration; BDNF, brain-derived neurotrophic factor; CAM, cell adhesion molecule; Cdk, cyclin dependent kinases; CNS, central nervous system; FGF, fibroblastic growth factor; GABA, gamma-aminobutyric acid; GCL, Ganglion cell layer; INL, Inner nuclear layer; MAG, Myelin-associated glycoprotein; NCAM, neural cell adhesion molecule; NMDA, *N*-methyl-D-aspartate; ONL, Outer nuclear layer; OPL, Outer plexiform layer; PLCγ, phospholipase Cγ; PSA, polysialic acid (PSA); Rb, retinoblastoma protein; RPE, Retinal pigment epithelium; RPTP, receptor protein tyrosine phosphatase; TGF, Transforming growth factor

1 Introduction

The retina contributes to major sensory input in humans, engaging about one-third of the human brain in its processing. For clinicians, the retina is the most readily visible part of the central nervous system (CNS). CNS blood vessels at the arteriolar level can be directly visualized in situ only in retina. This allows a direct inspection of pathological changes in the CNS and the vascular system. For researchers, the retina is a well-organized part of the CNS with many remarkable features, making it an excellent tissue for scientific study.

Degenerative diseases of the retina are a major cause of blindness in the world. Age-related macular degeneration (AMD) alone affects more than 35 million people worldwide and has a huge socioeconomic impact on communities. Unless a treatment is found, these numbers are likely to grow because of increasing life expectancy and increasing world population. Hereditary retinal degenerations cause blindness in another 1.5 million people (Ammann et al., 1965; Boughman et al., 1980; Hu, 1982; Jay, 1982). In addition, the retina can also be involved in vascular diseases such as diabetes and hypertension. In glaucoma, although the primary risk factor is increased intraocular pressure, eventually it is death of retinal ganglion cells that causes visual impairment.

Finding a cure for retinal degeneration is an urgent goal for the scientific community. In order to achieve these goals, it is important to understand the molecular mechanisms of retinal degeneration. Interestingly, many of the pathways that cells utilize for development are also important in degeneration. Therefore, an understanding of retinal development is paramount for finding a treatment for degenerative retinal diseases.

2 Phylogeny of Retinal Neurons

Retinal phylogeny provides clues to some of the most basic mechanisms of retinal development, such as the sequence of cytogenesis, thus elucidating how the cells generated during different time periods interact with each other.

In the primitive nervous system, sensory cells evolved from general epithelial cells. Primitive nervous systems of modern echinoderms and lower deuterostomes are still composed of three cell types that include the primary sensory cells, the neurons that connect the sensory cells to distal targets, and a supporting cell that serves the special physiological needs of such a system (Lacalli, 2001). The basic structural plan of the retina is comparable to such a primitive nervous system. In the course of evolution, the photoreceptive system developed specialized photoreceptor cells (rods and cones), intra-retinal second-order neurons (bipolar cells), and tertiary output neurons (ganglion cells). This evolution perhaps took place in photopic conditions; therefore early photoreceptor cells were more like cones.

There is ample evidence that scotopic (rod) pathways evolved subsequent to the evolution of photopic (cone) pathways. Bipolar cells may have evolved first as a general type of light-sensing cell that projected directly to CNS targets. Once more specialized photoreceptors evolved, these cells lost their sensory capability but continued to serve the purpose of connectivity. Bipolar cells still express some of the proteins involved in phototransduction (e.g., recoverin; Sharma et al., 2003). The Müller cells have a number of morphological features that resemble the supporting cells of the primitive deuterostomian neuroepithelium. Müller cell densities more closely follow the densities of the cone photoreceptors rather than the rods, suggesting that Müller cells evolved before rods. Thus the most primitive retinas, which developed for functioning in photopic conditions, contained cone photoreceptors, connecting output neurons (bipolar and ganglion cells), and supporting cells (Müller cells). With the passage of time, retinas became equipped for ambient light (scotopic vision) conditions with the addition of rod pathways. Interneurons such as the amacrine cells and the horizontal cells evolved later as the purpose that they serve, namely contrast sensitivity and motion detection, became useful after a basic system for light perception was in place.

The sequence of differentiation of retinal neurons offers a good test for the biogenetic law (Haeckel, 1905). The law states that ontogeny recapitulates phylogeny. Indeed, the birth order of retinal neurons generally reflects their evolutionary sequence. A consequence of late evolution and hence late differentiation of rods is that they do not become mature until the cone pathway is complete. Thus rod connections are formed secondary to the already existing cone pathways and piggyback on them to convey scotopic information to the brain. For example, rods contact rod bipolar cells that do not directly synapse with ganglion cells. Rather, they synapse on amacrine cells that carry both rod and cone information to ganglion cells. In contrast, cones contact cone bipolar cells that have direct input to ganglion cells as well as indirect input through amacrine cells. Additionally, the cells belonging to the cone pathways and rod pathways are not only generated in different phases of proliferation, but they might even be different in the way they migrate and fit into the retinal architecture. Cells belonging to these two phases use different isotypes of structural proteins, unrelated to their function or morphology, suggesting that they might use distinct developmental programs inherited from their lineage (Sharma and Netland, 2005).

Even the most primitive vertebrate eyes can have elaborate specializations. Primarily, eyes have undergone adaptive specialization rather than further evolution. All vertebrate eyes have a comparable basic structure: they all have three nuclear layers, pigment epithelium, and identically derived dioptric and nutritional mechanisms. An excellent example of adaptive changes in the eye is the evolution of color vision. Vertebrates developed five major families of visual pigments (one rod and four spectrally distinct cone classes) early during evolution (350–400 million years ago) resulting in tetrachromatic color vision. Later, as the mammals evolved, perhaps due to their nocturnal ancestral environment, color vision became less important, and was reduced to dichromatism, with only two spectral classes of cones. Much later (35 million years ago) as the primates evolved, the need for distinguishing yellow-orange fruits from green foliage (Mollon, 1989; Nagle and Osorio, 1993) gave rise to the need of a third pigment. It was produced as a duplication and modification of one of the existing longer-wavelength cone pigments gene. Spectral tuning (Neitz et al., 1991; Williams et al., 1992) of these pigments added one more dimension to their color perception, thus creating trichromatic vision. In the most evolved primates, the humans, supremacy of vision lies not in the structure of the eye but in the ability of the mind to appreciate it.

3 Embryogenesis of the Retina

The neural plate is derived from ectoderm and represents the embryonic precursor of the nervous system, including the retina. One region of the ectoderm forms the skin and the other the neural plate. At the time of formation of the neural plate, different areas of ectoderm are committed to form specific parts of the CNS, including the neural elements of the eye. Soon after formation, the neural plate rolls up along the midline axis of the embryo to form the neural tube. In humans, the eyes can be distinguished as early as the third week of gestation (the 2.6-mm stage), when the brain is still at the three-vesicle stage. Two pits appear in the transverse neural folds. Within a few days (the 3.2-mm stage) these pits develop into optic vesicles. As the primary brain vesicles subdivide, the optic stalk becomes visible at the junction of the telencephalon and the diencephalon. If optic pits fail to develop, this results in primary anophthalmia. In secondary anophthalmia, the developmental arrest takes place after the optic vesicles have been formed, and therefore some rudimentary evidence of visual elements can usually be found. Ocular adnexa are fairly well developed in cases of anophthalmia, suggesting that they develop independent of the optic vesicles.

During the fourth to the sixth week (the 4- to 5-mm and 15- to 18-mm stages) the optic vesicles enlarge and begin to invaginate (➋ *Figure 2-1*). A developing lens is important for this step, probably exerting its effect through extracellular matrix components. The outer and the lower walls of the vesicle invaginate and come to rest on the upper and posterior walls, forming a two-layered cup that opens laterally with a fissure located in its lower nasal quadrant. Failure to invaginate results in congenital cystic eye. At later stages in normal development, the fissure closes. Failure to close results in a typical coloboma located in the inferior nasal quadrant of the eye. Different parts of the optic cup follow deviant paths of differentiation. The invaginating portion or the inner wall lining the cup gives rise to the neural retina (and its Müller glia) and the outer wall, the RPE. The optic stalk produces the glia of the optic nerve.

The process of induction plays an important role in retinal development. The development of highly specialized organs in the body is a stepwise procedure in which one previously formed part induces neighboring regions to differentiate into a related tissue. The process is repetitive, and developing anlagens must continually communicate with and guide each other in the process of development. A large number of polypeptides belonging to different families that mediate induction have now been identified. Interestingly, each family appears to have arisen from a single gene during evolution.

All nucleated cells possess the genetic information, i.e., the DNA sequences, to synthesize tens of thousands of different proteins, only a tiny fraction of which are expressed. Different cell types, and consequently tissue types, arise because of the different sets of genes expressed. Cells can change the pattern of gene expression in response to signals arising from other cells in its environment. Although the expression of genes can be regulated at any step (from transcription to the formation of proteins),

Figure 2-1
Schematic diagram of early events that result in the positioning of photoreceptors in the distal part of the neural retina in close proximity to the pigmented epithelium. RPE provides physiological and functional support to the neural retina. Photoreceptors receive incoming light after it passes through more proximal retinal layers. Invagination of the optic vesicle (a) forms the optic cup (b) The outer wall gives rise to RPE, whereas the neural retina (including Müller glia) develop from the inner wall (Adapted from Sharma and Ehinger, 2003)

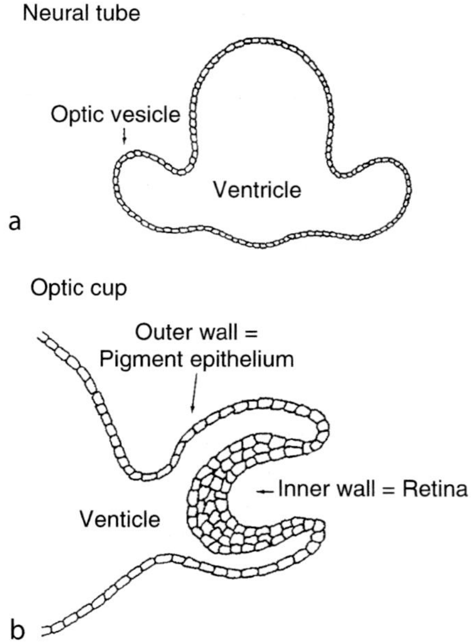

transcription is an extremely important control site. Gene regulatory proteins help control gene expression. These proteins recognize defined DNA sequences and determine which one of thousands of genes in the cell will be transcribed. During the course of development, induction brought about by polypeptide growth factors results in expression of certain regulatory genes. The expression of specific genes results in the formation of anlagens.

When the optic cup is formed (the 10-mm stage), retinal differentiation has already begun. The outer wall of the optic cup forms a single layer of pigmented cells identifiable as pigment epithelium at the eighth week. It appears that the developing neural retina influences the development of the RPE. For some time after the formation of the optic cup, it is possible to induce differentiating RPE cells to become retinal neurons after exposure to certain growth factors. On the distal surface, the sheet of RPE is firmly attached to Bruch's membrane, which is formed by extracellular secretions from both RPE and elements of the choroid. Bruch's membrane begins to develop by the 14- to 18-mm stage and is well formed by the sixth month. When they reach the 12-mm stage, two nuclear layers, called the inner and the outer neuroblastic layers, are established at the posterior pole of the retina. They are separated by a narrow acellular strip called the Chievitz's transient fiber layer. Development is initiated in the posterior pole and spreads peripherally as proliferation continues in the neuroblastic layer.

4 Ontogeny of Retinal Neurons

As proliferating neuroblasts leave the cell cycle, they migrate to their appropriate locations. With differentiation of these cells, a stereotypic pattern of axonal projections and connections needed for signal processing is established. This is achieved by guidance mechanisms that allow neuronal fibers to find and follow specified paths to their targets. Although the mechanisms and cellular cues involved are largely unknown, recent work has begun to reveal the molecular basis of these mechanisms (Messersmith et al., 1995; Wadsworth and Hedgecock, 1996; Sharma and Johnson, 2000; Sharma and Ehinger, 2003) and suggest that diffusible growth factors attract or repel growing axons by acting on cell surface receptors (Colamarino and Tessier-Lavigne, 1995; De Felipe et al., 1995; Marx, 1995; Tear et al., 1996).

4.1 Proliferation of Cells in the Neural Retina

The outer wall of the formed optic cup gives rise to the RPE, while proliferation in its inner wall gives rise to all of the retinal neurons and the Müller glial cells. Similar to the development of the cerebral cortex, proliferating retinal cells migrate radially through all layers of the developing retina, completing their migration from apical to basal surfaces and back again during each mitotic cycle. As a result, each stage of the cell cycle occurs while the cell traverses a specific retinal layer. In contrast to the cells of the cerebral cortex that undergo *cellular* migration along radial glia scaffolding, retinal cells undergo *nuclear* migration. The cell itself does not migrate but rather the nucleus moves within the cell. Cell division (M phase) occurs when nuclei are nearest the scleral or apical surface. A growing apical process of the daughter cells then projects to the vitreal or basal surface. As the nucleus follows and migrates basally, it enters G1 phase. When the nucleus reaches the basal surface, it begins DNA synthesis (S phase) and pauses (G2 phase) before migrating back to the apical surface to enter M phase or exit the cell cycle (Sharma and Ehinger, 1997b; ➋ *Figure 2-2* and ➋ *Figure 2-3*). The set of cells that make up a given subtype exit the mitotic cycle at roughly the same time, designated as their birthday. The birth order of retinal subtypes is generally the same across species with earlier evolved subtypes, such as cones, generally having earlier birth dates than later evolved cells, such as some types of amacrine cells (➋ *Figure 2-2* and ➋ *Figure 2-3*). Although there appear to be

Figure 2-2

Ganglion cells located in the proximal or the innermost layers of the retina are among the first cells to leave the cell cycle as seen in this embryonic day 15 rabbit retina. Other cells that also differentiate early, cone photoreceptors and the horizontal cells found in the outer retina, are much less numerous and thus are not readily recognized among the much larger cohort of undifferentiated cells in the outer retina. (a) A hematoxylin and eosin stained section showing a multilayered neuroblastic cell mass (NCM) and a thin anuclear layer (AL). (b) Cell proliferation in the same animal shown by immunostaining Ki-67 antigen. The proliferating cells are located in the apical (distal) two-third of the NCM and postmitotic cells in the basal (proximal) one-third suggesting that ganglion cells have already started differentiating. Certain large rounded nuclei in the distal most rows of the NCM (arrowheads) show cells in M phase. Immunoreactive cells are also seen in the pigment epithelium (arrows). Bar = 50 μm, IR = Immuno reactive, NR = non-reactive (Reproduced with permission from the publishers of Sharma and Ehinger, 1997b)

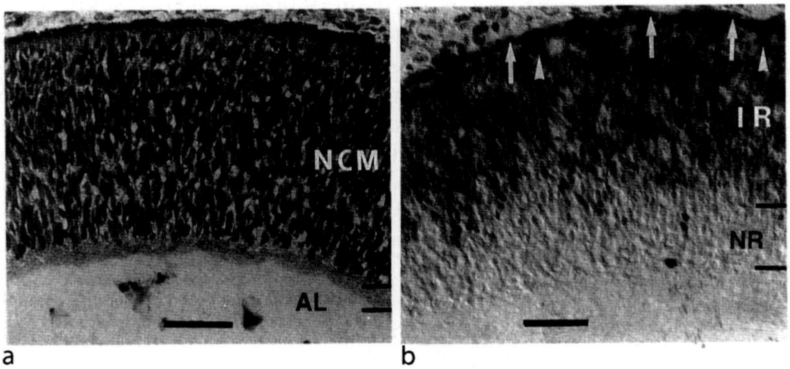

Figure 2-3
Proliferation of cells as demonstrated by immunostaining for Ki-67 antigen at embryonic day 25 rabbit retina. (a) An hematoxylin and eosin stained section showing the formation of inner plexiform layer (IPL) and the GCL. Some cells are observed in the IPL (arrow heads). (b) Ki-67 immunostaining shows proliferation in the same animal. Proliferation has ceased in the basal (proximal) part of the NCM consistent with the differentiation of ganglion and certain amacrine cells. Immunoreactive cells (IR) are accumulated close to the nonreactive layer, followed distally by the region of diffuse reactivity (DR). Deeply stained nuclei (M-phase) are always concentrated in the distal retina (arrow heads) suggesting that cells divide in this strata but then undergo interkinetic nuclear migration and DNA synthesis proximally. Some immunoreactive cells are also seen in the GCL or nerve fiber layer (arrows) perhaps in the glial cells (astrocytes). IPL, inner plexiform layer, GCL, Ganglion cell. Bar = 50 μm (Reproduced with permission from the publishers of Sharma and Ehinger, 1997b)

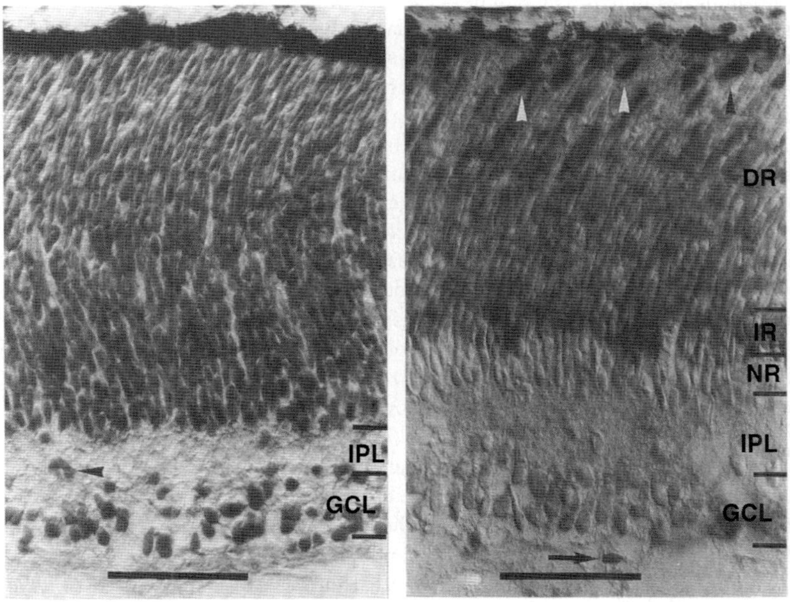

a b

some exceptions in the ontogenic recapitulation of phylogeny (perhaps due to our incomplete knowledge of cell type evolution), the basic birth order or retinal neurons does not vary across species as described subsequently. The generally held view is that the interplay between genetic readout from an individual cell and time-dependent changes in the extracellular environment comprise the necessary instructions for cell differentiation. Once the cell decides to leave the cell cycle, the nucleus migrates to the specified location and the cytoplasm detaches itself from both the surfaces (perikaryal translocation), leaving the cell at its destined location. The migration of cells is probably influenced by acetylcholine involving the muscarinic receptors (Yamashita et al., 1994).

Factors controlling proliferation are complex and not fully understood. There are both stimulatory and inhibitor stimuli. Transforming growth factor (TGF)-α, acidic and basic fibroblast growth factors, and epidermal growth factors stimulate the proliferation of retinal progenitor cells in cultures. These factors regulate cell proliferation via intracellular signaling cascades ultimately influencing the cell-cycle control system. There is also evidence that conventional fast-acting neurotransmitter systems may inhibit or at least regulate proliferation. Gamma-aminobutyric acid (GABA) and acetylcholine are good examples. Some of them are highly expressed during development and before any conventional nerve signaling occurs.

A sequence of cyclin–cyclin dependent kinases (Cdk) activities triggers the orderly progression of cell-cycle events. During G1 phase, Cdk activity is inhibited by Cdk inhibitors, cyclin proteolysis, and decreased transcription of cyclin genes. In late G1 this inhibition is overcome by increasing concentrations of G1- and G1/S-Cdks that trigger S-Cdk activation initiating DNA synthesis. After the completion of S phase, M-Cdk is activated, ushering in mitosis (M phase). In this phase, the cell assembles a mitotic spindle and prepares for segregation of the duplicated chromosomes that are held together. In anaphase, the proteins that hold the duplicate chromosomes together are lysed. M-Cdk is then inactivated by cyclin proteolysis, which leads to cytokinesis. This ends M phase. Progression of events through the cell cycle is tightly regulated by various inhibitory mechanisms. The cell cycle is arrested at specific checkpoints if DNA is damaged, events are unsuccessfully executed, or if the extracellular environment is not conducive.

The progression from G1 to S-phase is tightly controlled. Its disruption leads to unchecked cell-cycle entry and cell proliferation, resulting in cancers. The retinoblastoma protein (Rb) is required for entry into S phase and mutations in the Rb protein lead to formation of retinoblastoma tumors, the most common form of ocular tumor in children. The actions of Rb have been studied in detail. Mammalian cells suppress Cdk activity in G1 by Hct1 activation, p27 accumulation, and by transcription inhibition of cyclin gene. The effects of G1-Cdk activity are mediated by E2F, which is a gene regulatory protein. E2F binds to the promoter regions of several genes that encode proteins necessary for S-phase entry. The efficiency of E2F to activate transcription is inhibited by retinoblastoma protein (Rb) binding and perhaps related proteins such as p107 and p130. In G1, Rb bind to E2F and block the transcription of genes required for S phase. Signals for cell division cause active G1-Cdk accumulation causing phosphorylation of Rb, decreasing its affinity for E2F, which in turn activates S-phase gene expression. Rb protein remains phosphorylated throughout the S, G2, and M phases by Cdk2- and Cdk1-cyclin complexes. Following completion of mitosis when cells enter G1 or G0, Cdk-cyclin levels fall causing dephosphorylation of Rb protein by unopposed phosphatases. Resulting hypophosphorylated Rb protein is available to inhibit E2F activity in early G1 of the following cycle. Inherited or somatic mutations in Rb gene cause many types of cancers including retinoblastoma. The loss of function in both copies of the Rb gene causes excessive cell proliferation in the immature retina. In addition to its well-established role in regulating the cell cycle, Rb has recently been shown to be required for differentiation of rod photoreceptors (Donovan and Dyer, 2004; Zhang et al., 2004).

Proliferation ceases in the central retina first and then subsequently in more peripheral parts (➋ *Figure 2-4*). Consequently, the cellular differentiation and the development of the retinal nuclear and plexiform layers begins in the central retina and spreads peripherally (Sharma et al., 2003; ➋ *Figure 2-5*). In rabbits, proliferation ceases in the central retina around the time of birth, whereas it continues in the peripheral part for up to 15 days after birth (Sharma and Ehinger, 1997b; Sharma, 1999). The first cells to leave the cell cycle are the ganglion cells along with the cones (➋ *Figure 2-6*), and perhaps the horizontal cells in the central retina. Because of the larger number of ganglion cells generated compared with the number of cones and horizontal cells, postmitotic cells are more apparent earlier in the more proximal (vitreal) region of the retina than in the distal (scleral) region. Ganglion cells and early-generated forms of amacrines also form synapses in the inner plexiform layer (IPL) earlier than cones and horizontal cells do in the outer plexiform layer (OPL). Thus, the IPL is visible prior to the OPL. In rabbit, cells in the ganglion cell layer (GCL) can already be seen at embryonic day 15. By embryonic days 22–25, the IPL develops and postmitotic cells can be seen on both sides of this plexiform layer, the cells of the GCL being on the proximal side and the amacrine cells on the distal. By embryonic day 29, the OPL develops and the horizontal cells are localized on its proximal side. Even after the formation of the OPL, cells in different phases of cell cycle, including those in the mitotic phase, can be seen both in the inner and in the outer nuclear layer (ONL). In humans, the GCL becomes visible at 12 weeks, and it is well established by the fifth month. The OPL forms at the fourth month, and cells located between the outer and the IPLs then consolidate to form the inner nuclear layer (INL), erasing most of the Chievitz's transient fiber layer, except a remnant at the macula. This also disappears as the macula matures. Chievitz's layer leaves no identifiable mark in the adult retina.

Retinal cells in marsupials are generated in two phases. The first cells to generate are the ganglion cells, which themselves are generated in two sub-phases (Allodi et al., 1992). This is followed by genesis of certain amacrine, horizontal, and cone cells. In the second phase bipolar cells, more amacrine cells, Müller cells, and rods are generated (Harman et al., 1992). In other mammals also, such as mice (Sidman, 1961; Carter-Dawson and LaVail, 1979), cats (Zimmerman et al., 1988), monkeys (LaVail et al., 1991), and rabbits

Figure 2-4
Proliferation of cells in the retina ceases initially in the central retina and later in peripheral as evident in this postnatal day 0 rabbit retina showing Ki-67 antigen in proliferating cells at different eccentricities, central (a) mid periphery (b) periphery (c) and retinal edge at the ora serrata (d). Most of the reactive cells in the outer nuclear layer (ONL) are concentrated close to the outer plexiform layer (OPL) (IR2) perhaps giving rise to rod photoreceptors. Large deeply stained cells in M phase are situated within the distal regions (c and d, arrowheads). Immunoreactive cells in the inner nuclear layer (INL) (IR1) are confined mostly to the middle parts of this layer, where Müller-cell nuclei are located. Some large deeply stained cells (M-phase) can be seen (c, arrow) in this layer suggesting a separate zone of proliferation within INL. Cells in the proximal and the distal parts of this layer are nonreactive (c). In Figure d there are many larger and deeply stained cells (arrowheads) than in the more central retina. Bar = 50 μm (Reproduced with permission from the publishers of Sharma and Ehinger, 1997b)

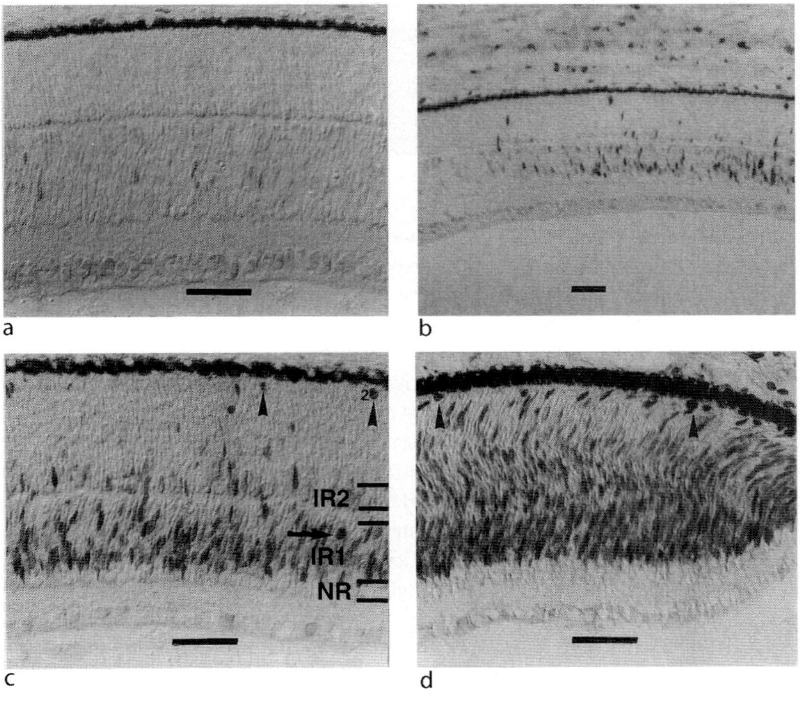

(Reichenbach et al., 1994b; Sharma and Ehinger, 1997b) the same sequence of cytogenesis is found. There are no estimates as to how many amacrine cells are produced in the early phase and how many in the late phase of proliferation. In species in which the retinal development takes place over a short period of time (Xenopus), there is overlapping in the birth dates of cells, whereas species in which retinal development is extended (monkeys), birthdays are more sharply demarcated. As discussed earlier, the birth order of retinal neurons is conserved during the evolution and reflects the evolutionary sequence of retinal neurons.

4.2 Differentiation of Retinal Neurons

In both the waves of proliferation, the eventual fate of the neurons is ultimately determined by microenvironmental factors. The retinal progenitor cells appear to be multipotent up to the last division. When the progenies of these cells are traced, they include various retinal neurons and the Müller cells. It is not clear if the cells produced in the early and the late phases of proliferation arise from the different progenitors or

Figure 2-5
Following the cessation of proliferation first in the central retina and then in peripheral, the differentiation of cells also follows the same pattern. In this micrograph, recoverin immunoreactivity demonstrates differentiation of photoreceptors in PN 0 retina with a distinct center to peripheral gradient. (a) The extreme periphery of the retina shows only a few immunoreactive cells. (b) Immunoreactivity in the mid periphery of the same retina shows a larger number of immunoreactive cells. (c) Immunoreactivity in the central retina shows numerous immunoreactive photoreceptor cells, some with short processes. Bar = 100 μm (Reproduced with permission from the publishers of Sharma et al., 2003)

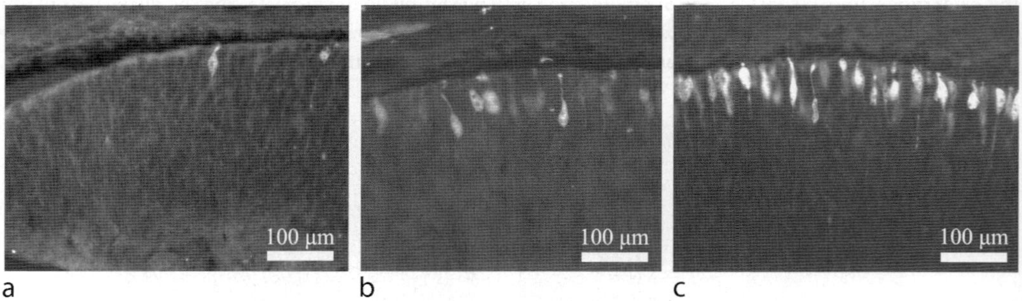

if the progenitor cells change their properties giving rise to a narrowing range of neurons later in development. Morphological similarities, protein expression, and the capability of Müller cells to re-enter the cell cycle have prompted speculations that the progenitors of the late phase could in fact be the immature Müller cells, which were produced in the earlier phase but whose maturation is retarded, and these cells re-enter the cell cycle to give rise to cells of the late phase.

Tracing of retinal clones in postnatal rodent retinas showed the genesis of cells of the late phase including neurons as well as the Müller cells (Turner and Cepko, 1987). Interestingly, the frequencies of the individual cell types in these clones do not exceed the numbers found in association with a single Müller cell in a developed retina (Turner and Cepko, 1987; Fields-Berry et al., 1992), suggesting that the progenitor cells produce the cells of the radial column associated with a single Müller cell. Several differences have been noticed in the progenitor cells of the early and late phases of proliferation. These progenitor cells are responsive to different growth factors (Lillien and Cepko, 1992). Progenitor cells of the ganglion cells and those of cones show neuron-specific filament proteins (Guillemot and Cepko, 1992) and cone-specific proteins (Liou et al., 1994) before their final mitosis, whereas those of the late phase show vimentin immunoreactivity and not that for neurofilaments or rod-specific proteins (Liou et al., 1994). Even the mature cells of these two phases of proliferation show different proteins (Barnstable, 1987; Reichenbach et al., 1994a; Sharma and Netland, 2005). For example, class III β tubulin, a component of microtubule, is found in adult retinal cells associated with retinal neurons produced in the first phase (Sharma and Netland, 2005), suggesting that these cells have some common elements in their developmental programs. Further evidence that the offspring of early and late phases of proliferation have different fates comes from the fact that retinoic acid promotes the formation of cones in the cells born in the early phase and the formation of rods in the cells produced in the late phase (Kelley et al., 1994).

Tracing the progeny of a single progenitor cell in nonmammalian retina also shows that while the progeny of the earlier phase of proliferation migrate tangentially, that of later proliferation results in separate radial columns (Williams and Goldowitz, 1992; Reese et al., 1995). This suggests that early- and late-born neurons may differ in their migration characteristics. The neurons born in the early phase reach their final destination by perikaryal translocation (Snow and Robson, 1995), and once the Müller cells have matured, lateral migration might become less prominent and the cells produced in the late phase migrate radially using the Müller cells for path guidance. This might explain why cells generated in the earlier phase of proliferation are not associated with radial columns. Another possibility is that these cells are produced early when the retina has not undergone expansion. Thus, these cells develop more central-to-peripheral gradient than the cells that are produced later and therefore are less likely to be affected by the expansion of

Figure 2-6
Differentiation of photoreceptors tracked by recoverin immunoreactivity in developing mouse retina. Shown here are changes in the numbers and types of immunoreactive cells during development. (a) At E 18.5 evidence of immunoreactivity was present in a subset of cells located in the distal most part of the neuroblastic mass (arrow). (b) At PN 0 retina, immunoreactivity was present in more distal rows of cells with some immunoreactive fibers projecting proximally (arrowhead; marked box has been enlarged in the inset). (c) At PN 3 the pattern of immunoreactivity was similar to that seen at PN 0 (d) At PN 7 immunoreactive cells could be seen throughout the ONL, but most of the cell bodies were located in the distal most cells. Stained axons projected toward the newly formed OPL. The timing of differentiated photoreceptor terminals (large arrowheads) reaching the OPL coincides with the development of a plexus by the differentiating horizontal cells (Sharma et al., 2003). (e) At PN 14, virtually all cells in the ONL showed immunoreactivity. Photoreceptor outer segments and the OPL were distinctly immunoreactive. In addition certain cells in the INL (bipolar cells) were also immunoreactive. (f) Immunoreactivity in the adult retina is comparable with that at PN 14. Immunoreactivity is seen in ONL and certain bipolar cells in the INL (see inset). OS, outer segments, ONL, outer nuclear layer, OPL, outer plexiform layer, INL, inner nuclear layer, IPL, inner plexiform layer, GCL, ganglion cell layer, NFL, nerve fiber layer, bar = 20 μm (Reproduced with permission from the publishers of Sharma et al., 2003)

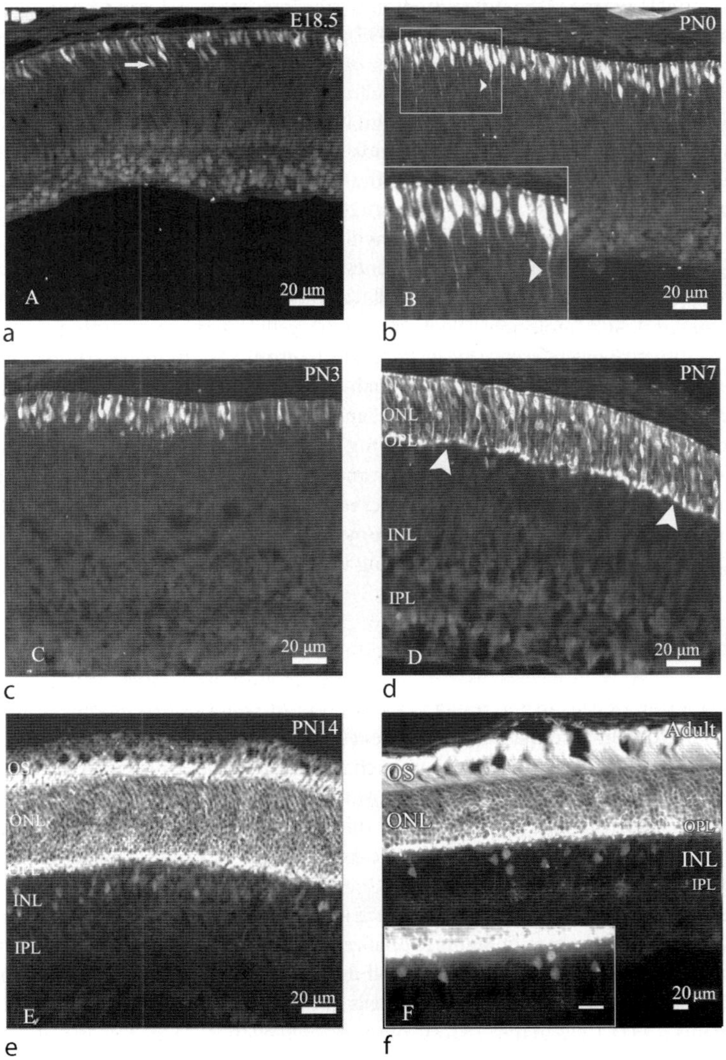

the retina (Williams and Goldowitz, 1992; Reese et al., 1995). It is noteworthy that ganglion cells, which are the first cells to be born, show the maximal central-to-peripheral gradient.

Even though the progenitor cells are multipotent, once they become postmitotic they become receptive to signals from other cells. The time of birth, local environment, specific growth factors, neurotransmitters, extracellular matrix molecules, and perhaps as yet undefined differences in progenitor subtypes all contribute in instructing the postmitotic cells to follow the path to their final destiny. It has been suggested that the first postmitotic cells may differentiate into ganglion cells or cones (Adler and Hatlee, 1989) by default (Reh, 1992). These cells then consequently determine the fate of other cells in a hierarchical fashion. As cells differentiate, subsequent postmitotic cells encounter an altered environment and cellular interactions (Turner and Cepko, 1987; Turner et al., 1990). The new interactions induce the proliferating cells to differentiate on different paths. Thus, divergent retinal cell types are produced in response to the changing inducing signal. The nature of the inducing signal is not known, but some factors seem to play a role. *Sonic hedgehog* protein is a putative mitogen for the precursor cells (Jensen and Raff, 1997). Laminin (Albini et al., 1992; Hunter et al., 1992) and acidic FGF (fibroblastic growth factor; Guillemot and Cepko, 1992) perhaps induce the differentiation of ganglion cells via notch-pathway (Austin et al., 1995; Bao and Cepko, 1997). Similarly, factors such as S-laminin (Hunter et al., 1992), basic FGF (Mack and Fernald, 1993), and retinoic acid (Kelley et al., 1994) promote the differentiation of photoreceptors. Thyroid hormone has been shown to promote cone differentiation (Kelley et al., 1995). Factors that promote the differentiation of rods decrease the proportion of bipolar cells. Promotion of bipolar cell (by TGF-α) and Müller cell (by bFGF) differentiation has an inverse effect on the other cell type, respectively (Lillien and Cepko, 1992). Differentiated cells themselves perhaps provide an inhibitory feedback to prevent overproduction of one particular cell type. Retinal cell transplantation studies have shown that once the cells are committed to their neurotransmitters, receptors and other structural and functional proteins develop according to their intrinsic timetable (Sharma et al., 1997; Sharma, 2000a, b). In summary, both the genetic and the environmental factors contribute to the final destiny of the cell.

Considering the complexities involved in cytogenesis in retina, it is obvious that there has to be a system for the distribution of cues needed for it. Müller cells seem to be a good candidate because they possess the required morphological and molecular elements. These cells are radially arranged in columns, thus providing a three-dimensional reference system for the neurons. Additionally, Müller cells are equipped with receptors for a number of growth factors, such as nerve growth factor (Chakrabarti et al., 1990), epidermal growth factor, TGF-α (Sagar et al., 1991), and insulin growth factor binding proteins (Lee et al., 1992). Therefore, these cells are capable of responding to a number of stimuli. Bcl-2, a survival protein in apoptotic pathways, is expressed in Müller cells early and remains expressed in adult life (Sharma, 2001a, c). RPE is another important source of factors that affect retinal development (Campochiaro, 1993; Rothermel et al., 1997), and responsiveness of Müller cells to some RPE cell derived factors (Jaynes et al., 1995) raises the possibility that even these factors might be acting via Müller cells.

4.3 Distribution of Cells

The relative distribution of cells in the retina is a reflection of relationships among various neurons. The density of cells in the vertebrate retina varies with the retinal eccentricity. Most prominently, this variation is caused by the development of specialized regions such as area centralis (or fovea in primates). Much of the "general-purpose" retina (serving the purposes of low visual acuity and motion detection), typically in the mid-periphery, shows that the packing densities of the cells are related to each other. Their ratios can be correlated with the number of cell divisions that the progenitor cells would have to go through to achieve the observed cell number.

The densities of Müller cells are strikingly similar in all mammalian retinas, with the central-to-peripheral ratio being in the range of 2:1 except in highly specialized regions. Cone photoreceptor packing densities are also fairly consistent (mean 4000–8000/mm sq.) in rod-dominated retinas. However, the rod-packing densities are inconsistent across the species. The densities of cells in the INL are correlated with the cones in the ONL (Jeffery and Williams, 1994). Cells in the ONL are outnumbered by those in the INL in rod-dominated retinas.

It is estimated that for every cone there are 2–3 bipolars, 1 amacrine, and less than 1 horizontal cell. In primate fovea, there may be more than 4 bipolars for every cone (Martin and Grunert, 1992).

The densities of cells in the GCL are closely correlated to cone density rather than rod density. For example, ganglion cell density peaks in the foveal region as does cone density, but no such correlation occurs near the retinal rim, where rod density is high. In most vertebrate retinas approximately one GCL cell may originally be produced for every cone; however, developmentally programed cell death reduces this number to 0.3–0.7. (Jeffery and Williams, 1994; Reichenbach et al., 1994b). The consistency of cone packing densities across species as compared with rods and their close relationship with the cell densities of the INL suggest that cones are phylogenetically older than rods, which lack these characteristics.

Analysis of the neuron to Müller cell ratio shows that it is not randomly distributed, but peaks at 16 or 32 neurons per Müller cell. Moreover, the number of Müller cells correlates with the number of photoreceptors (since most of the photoreceptors are rods, it correlates with the number of rods) rather than the number of ganglion cells (Reichenbach et al., 1994b). Comparison of the ratios of various retinal cells in different species suggests that the phylogeny and the ontogeny of these retinas can be explained by accounting for the number of divisions of progenitor cell (Reichenbach and Robinson, 1995).

Retinal cells are not generated independent of each other, but in a coordinated way so that a numerical relationship is maintained for creating functional and morphological units within the retina. In addition, the fact that the numerical relationship can be explained by the number of divisions in the progenitor cell suggests that the basis of this relationship resides in the phylogeny and thus the ontogeny of these neurons.

5 Tissue Organization of Retina

5.1 Arrangement of Cells: Formation of Nuclear Layers

5.1.1 Columnar Development

As already pointed out, cells in the late phases of proliferation are born in radial columns of which the Müller cells form the core. The columnar organization is also found in other areas in the brain such as cortex (Rakic, 1972; Luskin et al., 1988), spinal cord (Leber et al., 1990), and optic tectum (Gray and Sanes, 1991) where the radial glia act as a scaffold for the migration of neurons (Hatten, 1990, 1993; Rakic et al., 1994; Rakic, 1995). Müller cells probably play an important role in guiding the clonal cells in columns.

5.1.1.1 Role of Müller Cells in Tissue Architecture of the Retina A number of studies, including those involving retinal culture and in oculo embryonic retinal transplantation, suggest that Müller cells are involved in the organization of the retina. Chimeric stratospheroids in culture prepared by mixing the retinal cells from two species in order to follow their fate show that Müller cells inhibit lateral migration of neurons and encourage cells to align in chimeric columns. Interestingly, the fibers of the neurons incorporated into the columns are able to cross the interspecies border, perhaps using the branches of the Müller cell processes. This suggests that Müller cell processes may provide scaffold for the formation of fine architecture and sublamination of the IPL that is occupied by processes of specific neurons (Layer et al., 1997). If the Müller cells in these experiments were damaged by specific toxins, the neurons lost their columnar arrangement and started to migrate (Willbold et al., 1995). Thus, it seems Müller cells are important in stabilizing retinal columns. Studies in which fragmented pieces of embryonic neural retina were transplanted into the subretinal space also suggest that Müller cells might be important for the genesis of retinal architecture (Ehinger et al., 1996; Sharma, 1996, 1999; Sharma et al., 1995). Similar to the results obtained in retinal cultures, fragmented pieces of embryonic neural retina transplanted *in oculo*, soon organize themselves into small spherical structures called rosettes. Such rosetted grafts contain all retinal layers found in a normal retina including two distinct layers of cell bodies. The first nuclear layer, apparently of developing photoreceptor cells, is situated toward the luminal side of the rosettes. The second layer, consisting of cells from the INL, is located more peripherally (Sharma et al., 1997; Sharma, 2000a, b). A plexiform layer equivalent to the OPL also develops between the two nuclear layers. At other places, another plexiform layer, most likely equivalent to the IPL, also develops. Thus, the rosettes in the transplant develop the histotypic structure of retina consisting of two cell-rich layers and two layers resembling the outer and inner nuclear and plexiform layers. Müller cells, when labeled with antibody for vimentin, are radially arranged similar to spokes of a wheel. Similar organization is achieved in retinal cultures. Interestingly, cultures from other regions of the brain, even though they develop the spheroids, fail to develop the histotypic layers, and their radial glial cells, unlike the Müller cells in rosettes, are randomly arranged (Willbold et al., 1997a). The ability of Müller cells to organize radially in the rosettes is perhaps responsible for the histotypic layering seen in the rosettes formed by the retinal cells in culture and after transplantation.

An insight into the early events governing the formation of rosettes (or retinal layering) can be gained by studying the pattern of cell proliferation in retinal cells transplanted to the subretinal space. The first sign of rosette formation is seen when the mitotic cells cluster together. Subsequently, proliferating cells adopt the general proliferation pattern seen in normal retinogenesis around these clusters. As mentioned previously, the mitotic figures in the normal developing retina are observed at the outer limiting membrane (the distal retina; Sharma and Ehinger, 1997b). This is because proliferating retinal cells normally undergo interkinetic migration, which means that they synthesize their DNA away from the ventricular surface (Sharma and Ehinger, 1997b) and then migrate to the ventricular surface, where they undergo mitosis. In retinal cell transplants, the same pattern is maintained, albeit around the rosettes. Cells in metaphase are found on the luminal surface of the rosettes, which corresponds to the ventricular (apical) surface of the normal retina. This shows that the proliferating cells undergo interkinetic migration within the rosettes, similar to that which is seen in the normal retina. (Sharma and Ehinger, 1997a). As the graft matures, vimentin-stained Müller cells are found radially arranged like the spokes of a wheel in the rosettes along the path of interkinetic migration, and the hystotypic retinal layers are also arranged in the same plane.

5.1.1.2 Müller Cells and Cell Recognition Müller cells express a number of cell recognition molecules, consistent with their proposed role in the migration and organization of the radial columns. These molecules include L1/NgCAM (Drazba and Lemmon, 1990), 5A11 (a cell recognition molecule which is involved in neuronal glial interactions; Fadool and Linser, 1993), and F11 (Willbold et al., 1997b). F11 is a member of immunoglobulin super family primarily expressed in brain and is involved in cell contact dependent modulation of neuronal differentiation such as neurite outgrowth and fasciculation (Brummendorf and Rathjen, 1995). In brain, it is expressed in association with neural processes, whereas in retina it is found in plexiform layers. In addition, it has also been found in radially organized Müller cells, implying interaction between the neurons and the glia (Willbold et al., 1997b). L1/NgCAM is a ligand for F11, and it is involved in the radial migration of the neuronal precursors in forebrain and the cerebellum (Chuong et al., 1987; Barami et al., 1994). In retina, L1/NgCAM has been shown to be involved in the growth of ganglion cell processes (Drazba and Lemmon, 1990). The presence of F11 on the Müller cell radial processes may reflect its role in radial migration of late-phase retinal neurons. In addition to these molecules, which provide positive cues for the guidance, there are certain barrier proteins that provide the negative cues and form a repulsive barrier. Certain barrier proteins, such as embryonic avian polypeptide a 300-kDa protein or clustrin, have also been found on the Müller cells (McCabe and Cole, 1992).

5.1.1.3 Mosaic Formation Interactions among events occurring during development form the basis of the retinal architecture. Specifically, factors contributing importantly to retinal histology include proliferation of cells and their migration to pre-specified locations in the retina, axonal growth and their guidance to form specific connections between various retinal neurons, and patterning of cell bodies to form regular retinal mosaics. The resulting neuronal network permits extensive parallel processing of the visual stimuli. This makes it possible for the local neuronal circuits to simultaneously analyze different static and dynamic aspects of information such as color, brightness, and contrast from a complex and ever-changing visual scene. Regular repetition of these circuits forms mosaics in the retina. Mosaics are not unique to the retina. There are indications that the architecture of retina is principally similar to that of the diencephalic region of the brain, from which it arises. Similar mosaics of neurons have also been identified in avian midbrain tectum.

5.1.1.4 Mechanisms of Mosaic Formation Dendritic competition plays an important role in determining cell distribution in retina; however, additional mechanisms must exist. For example, cell bodies of α and β ganglion cells have similar distributions, yet the degree of dendritic overlap varies. There is significant dendritic overlap of ON and OFF subtypes of β ganglion cells in the cat retina, but limited overlap produced by the α cells. (Wassle et al., 1981a, b). It seems that photoreceptors may serve as a template for the mosaic formation, a mechanism that is readily apparent in insects. In the fly eye, the fate of the cells is determined by a series of inductive interactions, where in an initial featureless developing retina, the first photoreceptors, are formed in a spatially regular pattern by a process of lateral inhibition (Honda et al., 1990). These photoreceptors then act as a template for subsequent construction of ommatidia. In lower vertebrates (fishes) also, it is found that the first opsin-expressing cells develop and spread around the retina in an orderly fashion (Stenkamp et al., 1996). These cells act as a spatial template for the adjoining cells. A similar mechanism may also be at work in higher mammals such as macaque monkeys (Wikler and Rakic 1991; Wikler et al., 1997). Cones are among the first cells to differentiate even when the outer retina is still a neuroblastic cell mass (Sharma et al., 2003; ➋ *Figure 2-6*). It has been noted that a subpopulation constituting about 10% of immature cones synthesize cone opsins before they undergo synaptogenesis. These cones are distributed in a regular pattern across the retina and could act as modules for mosaics. Mosaics of cones expressing other nonopsin antibodies have been observed even at earlier ages (Wikler et al., 1997).

Tangential displacement of the other neurons during development may also contribute to mosaic formation. Cells generated in the late phase, namely the rods, bipolar cells, and Müller cells (which make up 85% of retinal neurons in mouse), maintain a radial alignment consistent with the argument that Müller cells are the developmental cores of the radial units (Bilous et al., 1991). However, retinal neurons that are produced early during development such as cones, horizontal cells, amacrine cells, and ganglion cells (which make up 15% of retinal cells in mouse) are unrelated to the radial units. Half of these early-born cells are found to be in clonally unrelated columns, suggesting that these cells become tangentially displaced during development (Reese et al., 1995; Reese et al., 1999). The estimated magnitude of displacement that these cells undergo to form mosaics is small (around 100 microns), but it is sufficient to place these cells in regular mosaics. Displacement of early-born cells has also been reported in chick retina (Straznicky and Chehade, 1987). The clonal displacement of the cells is found to be greater in the peripheral retina; this can be explained by the fact that the peripheral retina undergoes more expansion during development than does the central region.

Amacrine cells can be observed in the process of mosaic formation by ISLET1 antibody labeling. This antibody labels amacrine cells while they are still migrating to their final destination. These labeled cells are randomly arranged during their migration, but once they reach their adult location they appear regularly arranged. These cells rearrange to give space to new arriving cells (Galli-Resta et al., 1997). In later development, the retinas expand nonuniformly resulting in variation in the spatial density of their mosaics, but not in the regularity.

Apoptotic cell death of retinal neurons also contributes to mosaic formation. Rather than obliterating the regularity that the neurons have attained during development, programed cell death actively targets neurons so that a more regular pattern of cell distribution is reached (Jeyarasasingam et al., 1998).

Development of culture systems in which distribution of various neurons could be studied under controlled conditions contributes to our understanding of these mechanisms. However, such experiments need reliable methods of labeling live neurons. Moreover, information regarding how the retinal mosaics respond to various pathological conditions, such as retinal degeneration (Sharma, 2001a, b; Sharma et al., 2001), enhances our understanding of mosaic formation and compensatory mechanisms that may reside in the retina.

5.1.1.5 Asymmetries in the Retina Cell distribution across the retina is not symmetrical. In part, this is due to the asymmetrical growth of retina after birth. In humans, retina grows by a factor of 1.8 after birth until 6 years of age. This is largely due to enlargement of the peripheral retina and results in decreased cell density peripherally. However, there are other asymmetries, such as the accumulation of cones in fovea and that of rods in the rod ring that surrounds the fovea. The cone and ganglion cell density is also higher in the nasal retina than the temporal, even before birth. Further, ganglion cells from different regions of the retina send their processes either to the ipsilateral or the contralateral side of the brain to form topographic maps. This suggests the presence of important positional information in the developing retina, but the details remain to be established.

Central retina in many animals, especially the primates, is highly specialized because it is responsible for the acuity of vision. In humans, central retina develops over a prolonged period of time. In the early stages, the macular region is actually elevated (thicker) due to accumulation of the ganglion cells there. By 6 months of gestation, there are up to nine rows of foveal ganglion cells. After this stage, the macular area thins. This happens because the ganglion cells and other cells of the inner retina are displaced centrifugally (Sharma and Ehinger, 2003). The processes of these cells in the OPL are accordingly elongated to allow this displacement, which results in a depression in the center of the macula. Thus, the foveal pit is formed by the migration of ganglion cells, cells of the INL, and Müller cells toward the periphery while still maintaining the synaptic contacts made at an earlier stage (Hendrickson and Yuodelis 1984). At the time of birth, the cone density of the human fovea is only 20% of the adult value; it reaches the adult value only 4–5 years after birth. This is probably achieved by centripetal migration of cones toward the center of the fovea. In addition, the foveal area also decreases during development, contributing to increased cone density. As the fovea matures, the cones change their shape and become thinner so that they are more compactly arranged. Foveal cones continue to change their shape and develop outer segments well into adulthood. The human fovea is rod free. How a rod-free fovea is formed is not clear. The absence of rods in the foveal region is already evident by the time cytogenesis ceases. There is no evidence of selective rod apoptosis in this region. It is therefore likely that the progenitor cells produce only cones in the foveal region. This could be possible if cytogenesis ceases in this region before the beginning of the second phase of cytogenesis, when rods are normally produced. In primates, cell proliferation in the central region ceases relatively early compared to other mammals. A second mechanism that could produce a rod-free fovea would be for cells produced in the second phase to be induced to differentiate into cones instead of rods by certain factors. In this regard, it is significant to note that in Albinism, the fovea is missing. This might imply that changes in RPE caused by Albinism alter secretion of factors that could be involved in fovea formation.

5.2 Retinal Circuitry: Formation of Plexiform Layers

One of the intriguing events in the development of the nervous system, including the retina, is the formation of specific cell-to-cell contacts through mechanisms that guide neuronal fibers to their appropriate target areas. Naturally, cell recognition and cell-to-cell interactions are important. Identification of the specific cell recognition molecules involved has helped clarify how these molecules help mediate communications between cells and their environment. Such communications are needed to sort out the cells and guide them during migration and synapses formation.

5.2.1 Cell Recognition Molecules and Neurite Guidance

The expression pattern of axon associated CAMs in retinal cells, such as G4, neurofascin, and F11, suggest that these molecules play a role in the growth of axons. Certain CAMs provide the positive cues for the growth and guidance of the axons, whereas families such as semaphorins and certain tyrosine kinase receptors probably provide the negative cues (Brambilla and Klein, 1995; Kolodkin, 1996). Axonal growth involves interaction of numerous CAMs of which NCAM, L1, and N-cadherin are important and better understood. The expression pattern of NCAM suggests that these molecules provide the substrate for the growth of the ganglion cell axons along the endfeet of the radial glia (Silver, 1994). Such molecules are also responsible for the affinity of the growth cones toward each other and the formation of axon fascicles. The specificity of cell–cell adhesion probably results from different combinations of adhesion molecules creating a vast variety of surface properties of growth cones allowing them to select complex pathways depending upon the combination of molecules on the surface of the cells along the way.

In addition, there are inhibitory molecules, such as sulfated proteoglycans, that could create a barricade to the advancing neurites (Snow et al., 1990). Removal of chondrotin sulfate profoundly affects the direction of ganglion cell projections. Many trophic molecules are capable of binding proteoglycans, suggesting proteoglycans may have inhibitory effects by shielding the action of trophic molecules. It has also been reported that the administration of an antibody called IN-1 in vitro can encourage long distance growth of certain neurons after injury (Schnell and Schwab, 1990). Myelin also acts as an inhibitory substrate for the axon growth. Myelin-associated glycoprotein (MAG), a known component of myelin, inhibits axonal growth in CNS neurons (McKerracher et al., 1994; Mukhopadhyay et al., 1994), including retinal cells (DeBellard et al., 1996). Removal of sialic acid from the neuron or addition of oligosaccharides in the media can neutralize the inhibitory property of MAG. It is very likely that MAG is only one of the many molecules that have inhibitory effect on the axonal regeneration (Keynes and Cook, 1995). The presence of similar molecules may be responsible for the inability of the CNS neurons to regenerate after the injury.

Most investigations of axon guidance mechanisms have concentrated on retinal ganglion cells, which project their axons for long distances along the endfeet of Müller glia that line the vitreal surface, and along astroglial cells of the optic nerve to reach their targets in the brain (Stuermer and Bastmeyer, 2000). Certain growth factors, such as brain-derived neurotrophic factor (BDNF), are important for the guidance of the growing axons of ganglion cells to the target areas in the brain (Cohen-Cory and Fraser, 1994). However, most neurons in the retina project for shorter distances to reach targets within the retina, and these neurons never come in contact with the Müller endfeet. However, there is close contact between intra-retinal processes and highly branched Müller cell processes present in both plexiform layers. There is a paucity of information regarding the guidance of neurites of these nonprojecting neurons that include bipolar, amacrine, and photoreceptor cells. Evidence suggests that various types of nonprojecting neurons are guided by different factors, such as specific cell adhesion molecules, and that they have specific substrate requirements for their growth (Adler et al., 1985; Kljavin and Reh, 1991; Kljavin et al., 1994; Politi et al., 1998). For selective targeting, receptors on their axons interact with a complex array of molecules either released by target cells or expressed by neighboring cells. For the interaction of nonprojecting neurons, the CAMs are also important. One example of selective responsiveness is laminin, which is important for the amacrine cells, but not for the photoreceptors (Carbonetto et al., 1983; Kljavin et al., 1994). Laminin might be exerting its effect via protein kinase C (PKC; Politi et al., 1998).

Studies also show that Müller cells are important for morphogenesis and neurite guidance of rods in the retina. In these studies, rods prefer Müller cells for their neurite outgrowth suggesting that Müller cells selectively support neurite outgrowth from different classes of retinal neurons (Kljavin and Reh, 1991). Since rods are generated in association with Müller cells in retinal columns, there is ample opportunity for interactions between them. Such preferences have also been noted in other parts of the CNS where neurons from a particular region of the brain prefer glia from the same region for neurite growth (Denis-Donini and Estenoz, 1988). The ability of any cell adhesion molecule to act depends upon not only the cell type, but also the developmental period. One study showed that postnatal retinal ganglion cells respond to L1 and N-cadherin, but not to NCAM. In the same study, amacrine cells from earlier and later postnatal ages were shown to be responsive to these molecules, but the percentage of cells responding differed with developmental period. It should be kept in mind that each major cell type in retina is composed of multiple subtypes and that developmental programs must exist not only for cell fate determination, but also for spatial and temporal control of synaptogenesis in each subtype. The most heterogeneous group is the amacrine cells, which comprise approximately 50 functionally and morphologically distinct subtypes. Each has a distinct shape and dendritic field size and projection pattern in the IPL and thus may respond to different cues. Since L1 is usually present on the axonal surface projecting into long fiber tracts and not on the cell bodies or the dendrites (Lemmon and McLoon, 1986), it is possible that amacrine cells that have somewhat long projections might contain and/or respond to it. L1 has been shown to localize in two or three substrata of the IPL (Lemmon and McLoon, 1986; Kljavin et al., 1994), raising the possibility that this and similar molecules may provide cues for the amacrine cell fibers to project to the appropriate lamina of the IPL. Likewise, widespread distribution of NCAM and N-cadherin in the IPL suggests that amacrine cells with small and highly branched neurites may contain and respond to these molecules. NCAM is more diffusely distributed in the IPL and therefore could be responsible for cell sorting rather than guidance of the fibers (Grunwald et al., 1982; Hoffman et al., 1986). However, other possible functions of NCAM should not be ruled out.

There are distinct sublaminae in both plexiform layers, each containing defined sets of synaptic connections. This sublamination could result from selective guidance of growth cones in response to specific cell adhesion molecules. There are a number of candidates that have been implicated, including L1 (Persohn and Schachner, 1987; Martini and Schachner, 1988), neurofascin (Rathjen et al., 1987), F3 (Durbec et al., 1992), DM-DRASP/SCI/IC7 (Burns et al., 1991; Tanaka et al., 1991), Tag-1/Axinin-1 (Dodd and Jessell, 1988; Stoeckli et al., 1991), and P84 (Chuang and Lagenaur, 1990).

5.2.2 Cell Recognition Molecules and Neurite Growth

In addition to playing a pivotal role in neurite guidance, CAMs also stimulate neurite growth. Certain CAMs, such as NCAM, function mainly via homophilic binding interactions in the process of axonal growth, whereas others such as axonin-1 and F11 bind to NgCAM (Kuhn et al., 1991; Brummendorf et al., 1993), meaning that these act by heterophilic interactions.

A unique property of NCAM is that it is posttranslationally modified by addition of polysialic acid (PSA) at its fifth Ig domain by an enzyme polysialyltransferase (PST1). The amount of PSA in NCAM, but not the total amount of NCAM itself, changes with the development and loss of synaptic plasticity (Hoffman et al., 1982). The amount of PSA on NCAM probably is related to the level of neuronal activity and is itself regulated at the level of transcription (Bruses et al., 1995). During the development of retinal ganglion cell axons, removal of PSA from the ganglion cell axons and regions between the optic chiasma to the tectum results in a decrease in the size of axonal fascicles (Doherty et al., 1990; Yin et al., 1995). It is possible that removal of PSA causes an increase in homophilic NCAM (i.e., NCAM–NCAM) interaction resulting in its decreased interaction with other molecules responsible for axonal growth.

CAMs exert their effect on the axonal growth by activation of second messenger pathways (Doherty and Walsh, 1994). These pathways are not well understood, but a variety of calcium channel antagonists are potent inhibitors of NCAM, N-cadherin, and L1-mediated neurite growth. It has been suggested that tyrosine kinases are associated with these pathways (Beggs et al., 1994). Activity of kinases itself is regulated by phosphatases. The role of receptor phosphatases in cell adhesion and axonal growth is under investigation, especially that of receptor protein tyrosine phosphatase (RPTP) μ and κ, both of which can bind in a homophilic manner and contain Ig domains, FNIII motifs, and an MAM (or meprin) domain. MAM domains may be responsible for their binding specificity (Zondag et al., 1995). In addition to homophilic binding, phosphatases may interact at the cell surface through their hetrophilic partner contactin (Peles et al., 1995). Contactin is a glycosyl-phosphatidylinositol anchored CAM related to F3/F11. Receptor phosphatases also have intracellular binding partners. RPTPμ can bind to E-cadherin and possibly regulate phosphorylation to control the adhesion state of cadherin–catenin complex. Cadherin-mediated interactions are important during migration and cell–cell contacts. There is also evidence that CAMs (NCAM, N-cadherin, and L1) may stimulate neurite outgrowth via activation of the FGF receptor through phospholipase Cγ (PLCγ; Doherty and Walsh, 1994). Mutational analyses of CAMs suggest that the cytoplasmic domains of CAMs are important for their effect on axon growth (Saffell et al., 1995).

Certain neurotransmitters such as acetylcholine have been shown to have an effect on neurite outgrowth from ganglion cells. This effect may be mediated by nicotinic receptors (Lipton et al., 1988).

5.2.3 Synaptogenesis

Once neurites have grown and found their way to the target cells, they should form synaptic contacts. Although not well understood, it is noted that certain neurotransmitters and neuromodulators are expressed in the retina well before synaptic activity is present (Madtes and Redburn, 1983; Sharma et al., 1997), suggesting that neurotransmitters might have a developmental role. Neonatal horizontal cells contain all the components of a GABAergic cell, but soon after birth these markers are downregulated

Figure 2-7
Certain neurotransmitters such as GABA are transiently expressed in the retina during development suggesting a role in retinal maturation. A contiguous layer of horizontal cells (arrowheads) are stained with biotinylated anti rat GABA antisera at birth (a) and postnatal day 3 (b). In contrast few horizontal cells are stained at PN 5 (c) and 7 (d) or in the adults (e). A bilaminar immunoreactive band was observed at PN 3 through adult in the IPL (asterisks). In the inner retina amacrine cells are labeled at all ages. Bar = 20 μm (Reproduced with permission from the publishers of Mitchell and Redburn, 1996)

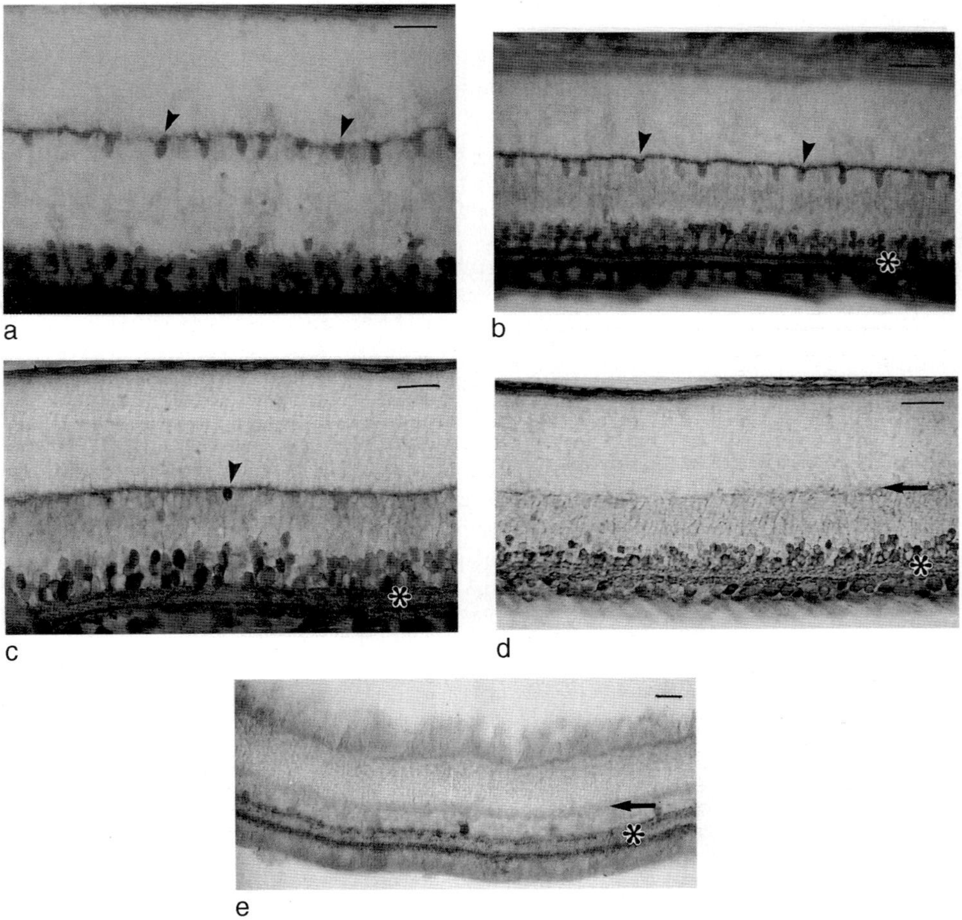

(Redburn and Madtes, 1986; Mitchell and Redburn, 1996; *Figure 2-7*). If horizontal cells are destroyed or certain GABA receptors (GABA$_A$ receptors) are blocked, cone synaptogenesis with the horizontal cells is disrupted (Messersmith and Redburn, 1990). Interestingly, GABA$_A$ receptors (β 2 and 3 subunits) are expressed early during development (Mitchell and Redburn, 1996; Hu et al., 1998; Sharma, 2000b) particularly in growth cones of cone photoreceptors at the time of cone-horizontal cell synaptogenesis (Greferath et al., 1994; *Figure 2-8*). The pharmacological properties of the GABA receptors can change over time as various subunits of the receptors are differentially expressed. Indeed, GABA$_A$ receptor activation causes increased intracellular Ca in neonatal neurons, whereas the opposite is true for adult neurons (Huang and Redburn, 1996). It is possible that the influx of calcium in the growth cone tips, mediated by GABA$_A$ receptors, when they come in contact with the horizontal cell in the OPL, may cause their collapse and trigger synaptogenesis (Sharma and Johnson, 2000; *Figure 2-9*).

Figure 2-8

Not only GABA but its receptors are expressed transiently in outer retinal regions suggesting a ligan–receptor interaction important for development. Here $GABA_A$ receptor in developing rabbit retina are immunolabeled. Staining was present in the OPL and the IPL. Immunoreactivity in the OPL increased after birth (a) at PN 3 (b) through day 5 (c) remained high at day 7 (d) and decreased in adult (e). A trilaminar band in the IPL and the certain amacrine cell bodies were labeled in the adult retina. Bar (a,b) 40 μm, (c–e) 20 μm (Reproduced with permission from the publishers of Mitchell and Redburn, 1996)

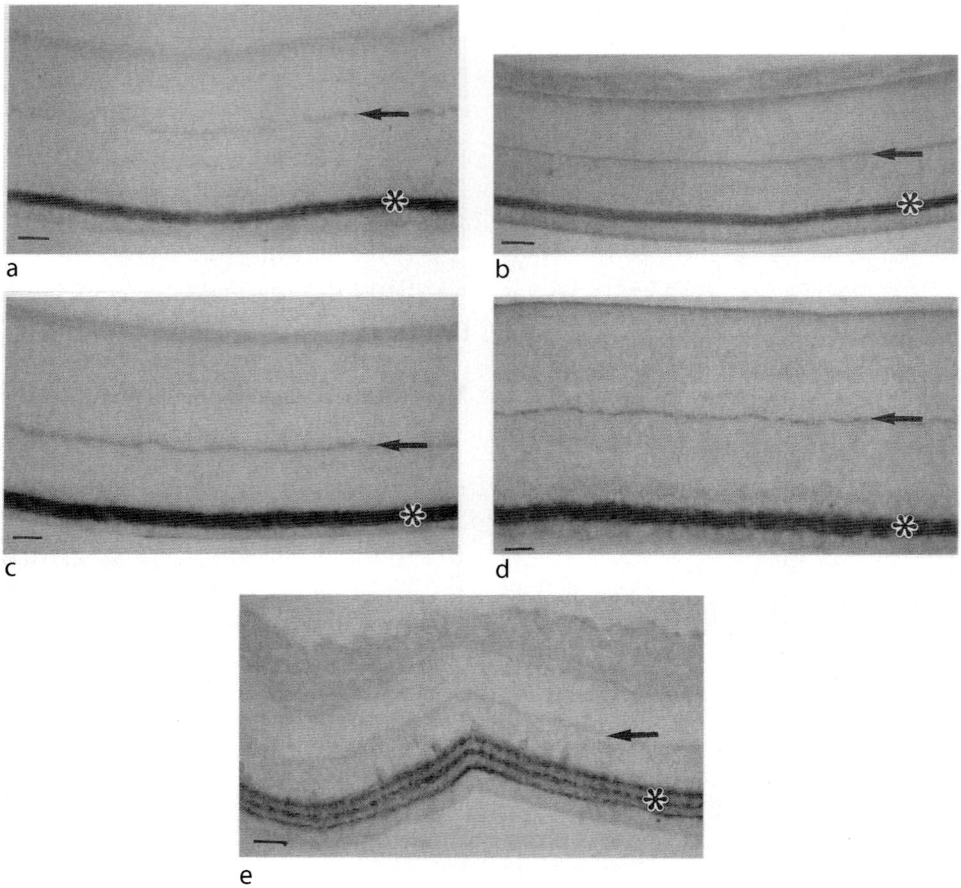

There is also evidence that the neuronal activity plays an important part in synaptogenesis (as well as selective programed cell death) in the nervous system. Even at early developmental stages when the retina is not yet light-sensitive, rhythmic spontaneous, rhythmic neuronal firing may guide the connectivity of retinal neurons. Possibly neurotransmitters such as GABA, and their receptors, that are expressed before photic responses are developed, might be involved in such spontaneous activity (*Figure 2-7*).

Retinal circuitry is further refined by eliminating inappropriate synapses. Glutamate, through NMDA receptors, is known to be involved in synaptic pruning in the CNS, and there is evidence that it also does so in the retina.

High levels of extracellular pools of glutamate, which could be toxic to adult neurons, are found in the developing retina, apparently doing no harm to these immature ones (Redburn et al., 1992; Haberecht et al., 1997). Additionally, non-NMDA metabotrophic receptor agonists were found to disrupt the lamellar distribution of ganglion cells to the ON and OFF sublamina of the IPL (Constantine-Paton et al., 1990).

Figure 2-9
(a) Calcium signaling may play a significant role during development specially in synaptogenesis. Evidence presented in this figure suggests that the role of GABA during retinal development may be mediated through calcium signaling. Here wholemount retinal explants from PN 5 rabbit were double labeled with Peanut aglutinin-rhodamine (to stain cone photoreceptors) shown in red and fluo-3 (to label calcium) shown in green. (b) An optical section of the double-labeled retinal explant at the photoreceptor layer taken approximately 1 s after 100 μM GABA application. The GABA-induced increase in free calcium occurs in cones (arrows) seen as overlapping PNA-rhodamine and Fluo-3 (green) images (yellow). Bar = 2 μm (Reproduced with permission from the publishers of Huang and Redburn, 1996)

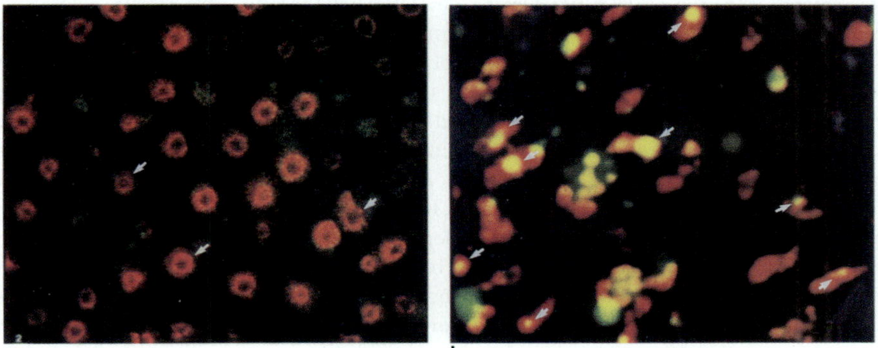

a b

6 Programed Cell Death in Retinal Development

In order to fine-tune the process of development, selected cells are removed by programed cell death or apoptosis. During the early development of the eye, apoptosis sculpts the optic primodium and forms the optic fissure. More importantly, it may be responsible for creating passages where optic fibers grow (Silver and Hughes, 1973). During maturation of cell types in the neural retina, especially during the establishment of their synaptic contacts, cell death fine-prunes the number of cells of each subtype. Among all the cells in the neural retina, ganglion cells have the highest rate of apoptosis. Approximately 70% of ganglion cells in developing human retina die (Provis et al., 1985a, b), but different subclasses of ganglion cells die at different rates. Some ganglion cells die because they project to the wrong target in the brain but this number is small. More cell loss is associated with establishment of the retinotopic map of the visual system. (O'Leary et al., 1986; Catsicas et al., 1987). Cell death may also be a contributing factor in establishing the ganglion cell density gradient in the central versus peripheral retina. Lack of efferent trophic support from target tissue is an important trigger for apoptosis in ganglion cells, as is neuronal activity. The effect of afferent trophic support appears to be less profound, but it is also less investigated. Evidence suggests that ganglion cell dendrites also compete for the synaptic contacts (or their growth factors) in the IPL. These dendrite-to-dendrite interactions may fine-tune the regular spacing and dendritic coverage of the ganglion cells over the retina (Dhingra et al., 1997). It is possible that other retinal cell types may also be regulated by dendrite-to-dendrite interactions in order to form regular mosaics in the retina. Such regular mosaics are obviously needed so that all the places in the retina possess the neural elements needed to extract all attributes of visual information. The presence of pyknotic cells in other locations indicates that other cell types such as amacrine cells also undergo cell death, but in smaller magnitudes (Robinson, 1987).

7 Development of Retinal Vasculature

How retina derives its nourishment varies among species. Certain mammalian species, including primates, have vascularized retina. In others, for example rabbit, the retina is nonvascularized and derives most of its nourishment via diffusion from choroidal vasculature. In humans, the outer retina draws nutrients from the choroid plexus, a network of small capillaries under the retina. The inner layers have their own blood supply from blood vessels coming from the optic nerve head. The inner retinal blood vessels form circular arcades around the macula. The center-most portion of the macula, the fovea, has no blood vessels and derives its oxygen and nutrients from the choroid plexus.

The human retinal vasculature develops in the third trimester as endothelial cells emerge from the central retinal artery and form retinal blood vessels. There is evidence that the retinal astrocytes are associated with vascularization (Prat et al., 2001). Astrocytes are found only in vascularized retinas, and also they are found in close proximity to retinal blood vessels (Gariano et al., 1996). Retinal vascularization develops as astrocytes emerge from the optic nerve. During development, the astrocytes migrate just ahead of the developing vascular network. The exact molecular events and interactions between the developing retinal vessels and other cells, including astrocytes, are not known. However, as already discussed in previous sections (➋ *Section 5.1.1.2*), glial cells can provide a substrate for the migration of cells. Several cytokines have been implicated as the mediators for endothelial cell migration during the formation of blood vessels.

In adults, formation of new vessels, or angiogenesis, is highly regulated and occurs only in limited circumstances, including wound healing, pregnancy, and menstrual cycles. In angiogenesis the new blood vessels grow from preexisting capillaries. However, in certain diseases, such as malignancies, new vessel formation plays a significant role in disease pathology. In many eye diseases, abnormal angiogenesis or neovascularization is the major cause of vision loss. In conditions such as diabetic retinopathy, age-related macular degeneration, retinopathy of prematurity, and rubeotic glaucoma, unregulated growth of retinal or uveal vessels in the eye leads to macular edema, hemorrhage (and eventually, tractional retinal detachment) and increased intraocular pressure, resulting in severe loss of vision.

8 Development of Histological Organization in the Retina

Once the retina has fully developed, it consists of six neuron types: photoreceptor cells, horizontal cells, bipolar cells, amacrine cells, interplexiform cells (possibly a subtype of amacrine cell), and ganglion cells. Retinal neurons are in intimate physical and physiological contact with Müller glial cells (➋ *Figure 2-10*). In addition, the photoreceptor cells' distal parts specialize to capturing the photons—the outer segments. The RPE cells functionally support the neural retina, particularly photoreceptors whose modified outer segments are responsible for phototransduction. RPE cells provide metabolic intermediates (11-*cis* retinal) that are necessary for synthesis of the rod photopigment, rhodopsin. Processing of visual information in birds and cold-blooded vertebrates occurs primarily in the retina, whereas in mammals, more of the visual processing takes place in the cortex. This has resulted in a decreased diversity in the amacrine and horizontal cell types in mammals compared to lower vertebrates. These cells originally evolved to process information at the retinal level.

Histologically, a developed retina consists of ten layers:

1. The RPE comprising a single layer of pigmented cells.
2. The receptor layer comprising the outer and the inner segments of the photoreceptor cells.
3. The outer limiting membrane, which is not a true membrane but a narrow zone containing numerous zonulae adherentes between Müller cells, and between Müller and the photoreceptor cells.
4. The ONL containing the photoreceptor cell bodies and their fibers.
5. The OPL formed by fibers of and synapses between photoreceptor cells, bipolar cells, and horizontal cells.
6. The INL, which contains the cell bodies (and fibers) of horizontal, bipolar, amacrine, interplexiform, and Müller cells.
7. The IPL formed by fibers, and synapses between bipolar, amacrine, and ganglion cells.
8. The ganglion cell layer, which along with the retinal ganglion cells also contains amacrine cells; so-called displaced amacrine cells.
9. The nerve fiber layer containing axons from ganglion cells on their way to the optic nerve.
10. The inner limiting membrane formed by the endfeet of Müller cells.

Figure 2-10

A fully developed retina consists of six neuron types and the Müller glial cells as shown in this schematic representation. A, Amacrine cells; B, bipolar cells; C, cone photoreceptor cells; G, ganglion cells; H, horizontal cells; I, interplexiform cells; M, Müller cells; PE, pigment epithelium; R, rod photoreceptor cell; INL, inner nuclear layer; IPL, inner plexiform layer; ONL, outer nuclear layer; Ph, photoreceptors; RPE, RPE (Reproduced with permission from the publishers of Sharma and Ehinger, 2003)

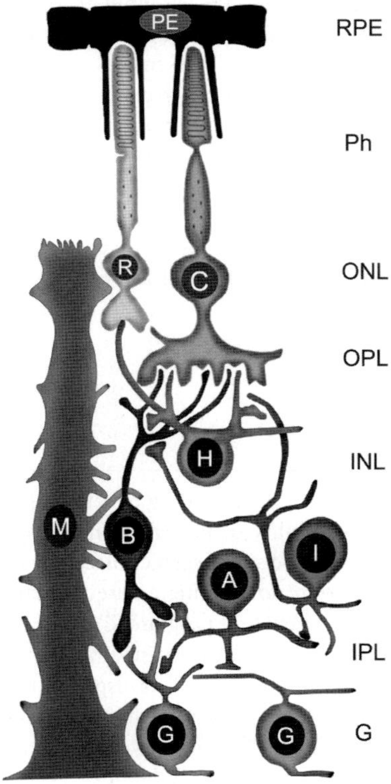

9 Development of Functional Organization in the Retina

In addition to its distinctive molecular, cellular, and histological organization, the retina also follows the basic functional organization of other special sensory systems, including the use of receptor cells to send sensory information encoded in graded potentials and action potentials along a series of neuronal relays to the appropriate cortical receiving areas. In retina, photoreceptors convert the energy in photons into a cellular event (a membrane resistance change). By convention, the photoreceptor is considered to be a true neuron and thus the primary sensory neuron in the visual system pathway. Bipolar cells and ganglion cells, respectively, represent secondary and tertiary neurons in the pathway. A fuller appreciation of the demanding requirements of developmental programs in retina can be gained by briefly considering the complexity of the functional circuits for transmitting and processing visual information.

1. The modular arrangement of neuronal columns in retina assures that each point in visual space can be continually monitored via the capture of photons at corresponding points in the retinal field. This is the basis of the retinotopic organization of the visual pathway.
2. Information is passed to the brain in parallel circuits that are further divided at each synaptic relay. In the OPL, information from cones is divided into two pathways. One depolarizes in response to increases in photon capture (the ON channel, composed of bipolar cells, ganglion cells, and many cells of the lateral geniculate and visual cortex). The other depolarizes in response to decreases in

photon capture (the OFF channel, composed of a corresponding set of bipolars, ganglion cells, etc.) These channels transmit information about the shape of an object and its position and movement in space. Spectral information from the three types of cones is processed through lateral inhibitory circuits of horizontal cells before being sent to color-sensing regions of the visual cortex. Contrast sensitivity is also established partially through actions of horizontal cell circuits that create center versus surround receptive fields for bipolar cells and hence for much of the remaining visual pathway.

3. In the IPL, inhibitory and excitatory amacrine cell inputs sort and enhance the temporal aspects of visual input, thus establishing movement-specific (directionally sensitive) responses.
4. The retinal output neuron, the ganglion cell, assembles processed information from bipolar and amacrine cells and further sorts it into two main components. The M pathway, originating from large, magnocellular ganglion cells, transmits information regarding movement and position in space. The P pathway, originating from small, parvocellular ganglion cells, transmits information about fine detail and color.

Many of these processing attributes emerge only after years of visual experience.

For example, color vision is not fully mature until the age 12–14 in humans. Thus, activity-driven neuronal plasticity rather than pre-programed developmental mechanisms are thought to be the driving force for final establishment of the mature visual pathway.

10 Concluding Comments

There has been a vast improvement in our understanding of retinal development in the recent past. It has been made possible because of the availability of newer technologies such as improved labeling agents, immunohistochemistry, retinal culture models, transgenic mice, and RNA silencing technologies. Significantly, this knowledge is equally important for finding a cure for retinal degeneration, a major cause of blindness. Mechanisms of retinal development and programed cell death provide a platform to understand the molecular events that might be used to design treatment strategies for regeneration and reconstruction of the degenerated retina. Retinal progenitor/stem cell transplantation provides hope for replacing lost neurons. In order to succeed, these transplanted cells must first differentiate into appropriate neuron types. This still remains a challenge, but knowledge of retinal development is proving invaluable in overcoming this hurdle. Differentiated neurons must integrate with the host retina. Once again, what we have learned about neurite growth and synaptogenesis is proving very useful.

Acknowledgments

This work was supported in part by generous grant from Gail and Richard Siegal, the Gerwin Scholar Award, Research to Prevent Blindness, the Hyde Foundation, the Lion's Club, and NEI Vision Core Grant. The authors thank Peter A. Netland, M. D., Ph. D. for his advice and support, Lee Thompson for help with graphics and Marina Kedrov for help with preparation of the manuscript. Figures were reproduced with kind permission from Elsevier and Cambridge University Press.

References

Adler R, Hatlee M. 1989. Plasticity and differentiation of embryonic retinal cells after terminal mitosis. Science 243: 391-393.

Adler R, Jerdan J, Hewitt AT. 1985. Responses of cultured neural retinal cells to substratum-bound laminin and other extracellular matrix molecules. Dev Biol 112: 100-114.

Albini A, Noonan DM, Melchiori A, Fassina GF, Percario M, et al. 1992. Laminin-induced retinoblastoma cell differentiation: Possible involvement of a 100-kDa cell-surface laminin-binding protein. Proc Natl Acad Sci USA 89: 2257-2261.

Allodi S, Cavalcante LA, Hokoc JN, Bernardes RF. 1992. Genesis of neurons of the retinal ganglion cell layer in the opossum. Anat Embryol (Berl) 185: 489-499.

Ammann F, Klein D, Franceschetti A. 1965. Genetic and epidemiological investigations on pigmentary degeneration of the retina and allied disorders in Switzerland. J Neurol Sci 2: 183-196.

Austin CP, Feldman DE, Ida JA Jr, Cepko CL. 1995. Vertebrate retinal ganglion cells are selected from competent progenitors by the action of Notch. Development 121: 3637-3650.

Bao ZZ, Cepko CL. 1997. The expression and function of Notch pathway genes in the developing rat eye. J Neurosci 17: 1425-1434.

Barami K, Kirschenbaum B, Lemmon V, Goldman SA. 1994. N-cadherin and Ng-CAM/8D9 are involved serially in the migration of newly generated neurons into the adult songbird brain. Neuron 13: 567-582.

Barnstable CJ. 1987. A molecular view of vertebrate retinal development. Mol Neurobiol 1: 9-46.

Beggs HE, Soriano P, Maness PF. 1994. NCAM-dependent neurite outgrowth is inhibited in neurons from Fyn-minus mice. J Cell Biol 127: 825-833.

Bilous AM, McKay M, Milliken J. 1991. A comparison between Ki-67 antibody reactivity and other pathological variables in breast carcinoma. Pathology 23: 282-285.

Boughman JA, Conneally PM, Nance WE. 1980. Population genetic studies of retinitis pigmentosa. Am J Hum Genet 32: 223-235.

Brambilla R, Klein R. 1995. Telling axons where to grow: A role for Eph receptor tyrosine kinases in guidance. Mol Cell Neurosci 6: 487-495.

Brummendorf T, Hubert M, Treubert U, Leuschner R, Tarnok A, et al. 1993. The axonal recognition molecule F11 is a multifunctional protein: Specific domains mediate interactions with Ng-CAM and restrictin. Neuron 10: 711-727.

Brummendorf T, Rathjen FG. 1995. Cell adhesion molecules 1: Immunoglobulin superfamily. Protein Profile 2: 963-1108.

Bruses JL, Oka S, Rutishauser U. 1995. NCAM-associated polysialic acid on ciliary ganglion neurons is regulated by polysialytransferase levels and interaction with muscle. J Neurosci 15: 8310-8319.

Burns FR, von Kannen S, Guy L, Raper JA, Kamholz J, et al. 1991. DM-GRASP, a novel immunoglobulin superfamily axonal surface protein that supports neurite extension. Neuron 7: 209-220.

Campochiaro PA. 1993. Cytokine production by retinal pigmented epithelial cells. Int Rev Cytol 146: 75-82.

Carbonetto S, Gruver MM, Turner DC. 1983. Nerve fiber growth in culture on fibronectin, collagen, and glycosaminoglycan substrates. J Neurosci 3: 2324-2335.

Carter-Dawson LD, LaVail MM. 1979. Rods and cones in the mouse retina. II. Autoradiographic analysis of cell generation using tritiated thymidine. J Comp Neurol 188: 263-272.

Catsicas S, Thanos S, Clarke PG. 1987. Major role for neuronal death during brain development: Refinement of topographical connections. Proc Natl Acad Sci USA 84: 8165-8168.

Chakrabarti S, Sima AA, Lee J, Brachet P, Dicou E. 1990. Nerve growth factor (NGF), proNGF and NGF receptor-like immunoreactivity in BB rat retina. Brain Res 523: 11-15.

Chuang W, Lagenaur CF. 1990. Central nervous system antigen P84 can serve as a substrate for neurite outgrowth. Dev Biol 137: 219-232.

Chuong CM, Crossin KL, Edelman GM. 1987. Sequential expression and differential function of multiple adhesion molecules during the formation of cerebellar cortical layers. J Cell Biol 104: 331-342.

Cohen-Cory S, Fraser SE. 1994. BDNF in the development of the visual system of Xenopus. Neuron 12: 747-761.

Colamarino SA, Tessier-Lavigne M. 1995. The axonal chemoattractant netrin-1 is also a chemorepellent for trochlear motor axons. Cell 81: 621-629.

Constantine-Paton M, Cline HT, Debski E. 1990. Patterned activity, synaptic convergence, and the NMDA receptor in developing visual pathways. Annu Rev Neurosci 13: 129-154.

De Felipe C, Pinnock RD, Hunt SP. 1995. Modulation of chemotropism in the developing spinal cord by substance P. Science 267: 899-902.

DeBellard ME, Tang S, Mukhopadhyay G, Shen YJ, Filbin MT. 1996. Myelin-associated glycoprotein inhibits axonal regeneration from a variety of neurons via interaction with a sialoglycoprotein. Mol Cell Neurosci 7: 89-101.

Denis-Donini S, Estenoz M. 1988. Interneurons versus efferent neurons: Heterogeneity in their neurite outgrowth response to glia from several brain regions. Dev Biol 130: 237-249.

Dhingra NK, Ramamohan Y, Raju TR. 1997. Developmental expression of synaptophysin, synapsin I and syntaxin in the rat retina. Brain Res Dev Brain Res 102: 267-273.

Dodd J, Jessell TM. 1988. Axon guidance and the patterning of neuronal projections in vertebrates. Science 242: 692-699.

Doherty P, Walsh FS. 1994. Signal transduction events underlying neurite outgrowth stimulated by cell adhesion molecules. Curr Opin Neurobiol 4: 49-55.

Doherty P, Cohen J, Walsh FS. 1990. Neurite outgrowth in response to transfected N-CAM changes during development and is modulated by polysialic acid. Neuron 5: 209-219.

Donovan SL, Dyer MA. 2004. Developmental defects in Rb-deficient retinae. Vision Res 44: 3323-3333.

Drazba J, Lemmon V. 1990. The role of cell adhesion molecules in neurite outgrowth on Müller cells. Dev Biol 138: 82-93.

Durbec P, Gennarini G, Goridis C, Rougon G. 1992. A soluble form of the F3 neuronal cell adhesion molecule promotes neurite outgrowth. J Cell Biol 117: 877-887.

Ehinger B, Juliusson B, Bergström A, Sharma RK. 1996. Photoreceptor development in retinal cell transplants. Kato S, Osborne NN, Tamai M, editors. Retinal degenerations and regeneration. Kugler Publications; Amsterdam/New York: pp. 155-164.

Fadool JM, Linser PJ. 1993. 5A11 antigen is a cell recognition molecule which is involved in neuronal-glial interactions in avian neural retina. Dev Dyn 196: 252-262.

Fields-Berry SC, Halliday AL, Cepko CL. 1992. A recombinant retrovirus encoding alkaline phosphatase confirms clonal boundary assignment in lineage analysis of murine retina. Proc Natl Acad Sci USA 89: 693-697.

Galli-Resta L, Resta G, Tan SS, Reese BE. 1997. Mosaics of islet-1-expressing amacrine cells assembled by short-range cellular interactions. J Neurosci 17: 7831-7838.

Gariano RF, Sage EH, Kaplan HJ, Hendrickson AE. 1996. Development of astrocytes and their relation to blood vessels in fetal monkey retina. Invest Ophthalmol Vis Sci 37: 2367-2375.

Gray GE, Sanes JR. 1991. Migratory paths and phenotypic choices of clonally related cells in the avian optic tectum. Neuron 6: 211-225.

Greferath U, Grunert U, Müller F, Wassle H. 1994. Localization of GABAA receptors in the rabbit retina. Cell Tissue Res 276: 295-307.

Grunwald GB, Pratt RS, Lilien J. 1982. Enzymic dissection of embryonic cell adhesive mechanisms. III. Immunological identification of a component of the calcium-dependent adhesive system of embryonic chick neural retina cells. J Cell Sci 55: 69-83.

Guillemot F, Cepko CL. 1992. Retinal fate and ganglion cell differentiation are potentiated by acidic FGF in an in vitro assay of early retinal development. Development 114: 743-754.

Haberecht MF, Mitchell CK, Lo GJ, Redburn DA. 1997. N-methyl-D-aspartate-mediated glutamate toxicity in the developing rabbit retina. J Neurosci Res 47: 416-426.

Haeckel E. 1905. The Evolution of Man. New York: Putnam.

Harman AM, Sanderson KJ, Beazley LD. 1992. Biphasic retinal neurogenesis in the brush-tailed possum, Trichosurus vulpecula: Further evidence for the mechanisms involved in formation of ganglion cell density gradients. J Comp Neurol 325: 595-606.

Hatten ME. 1990. Riding the glial monorail: A common mechanism for glial-guided neuronal migration in different regions of the developing mammalian brain. Trends Neurosci 13: 179-184.

Hatten ME. 1993. The role of migration in central nervous system neuronal development. Curr Opin Neurobiol 3: 38-44.

Hendrickson A, Yuodelis C. 1984. The morphological development of the human fovea. Ophthalmology 91: 603-612.

Hoffman S, Friedlander DR, Chuong CM, Grumet M, Edelman GM. 1986. Differential contributions of Ng-CAM and N-CAM to cell adhesion in different neural regions. J Cell Biol 103: 145-158.

Hoffman S, Sorkin BC, White PC, Brackenbury R, Mailhammer R, et al. 1982. Chemical characterization of a neural cell adhesion molecule purified from embryonic brain membranes. J Biol Chem 257: 7720-7729.

Honda H, Tanemura M, Yoshida A. 1990. Estimation of neuroblast numbers in insect neurogenesis using the lateral inhibition hypothesis of cell differentiation. Development 110: 1349-1352.

Hu DN. 1982. Genetic aspects of retinitis pigmentosa in China. Am J Med Genet 12: 51-56.

Hu M, Bruun A, Ehinger B. 1998. The expression of GABA(A) receptors during the development of the rabbit retina. Acta Ophthalmol Scand 76: 515-520.

Huang BO, Redburn DA. 1996. GABA-induced increases in [Ca2+]i in retinal neurons of postnatal rabbits. Vis Neurosci 13: 441-447.

Hunter DD, Murphy MD, Olsson CV, Brunken WJ. 1992. S-laminin expression in adult and developing retinae: A potential cue for photoreceptor morphogenesis. Neuron 8: 399-413.

Jay M. 1982. On the heredity of retinitis pigmentosa. Br J Ophthalmol 66: 405-416.

Jaynes CD, Sheedlo HJ, Agarwal N, O'Rourke K, Turner JE. 1995. Müller and retinal pigment epithelial (RPE) cell expression of NGFI-A and c-fos mRNA in response to medium conditioned by the RPE. Brain Res Mol Brain Res 32: 329-337.

Jeffery G, Williams A. 1994. Is abnormal retinal development in albinism only a mammalian problem? Normality of a hypopigmented avian retina. Exp Brain Res 100: 47-57.

Jensen AM, Raff MC. 1997. Continuous observation of multipotential retinal progenitor cells in clonal density culture. Dev Biol 188: 267-279.

Jeyarasasingam G, Snider CJ, Ratto GM, Chalupa LM. 1998. Activity-regulated cell death contributes to the formation of ON and OFF alpha ganglion cell mosaics. J Comp Neurol 394: 335-343.

Kelley MW, Turner JK, Reh TA. 1994. Retinoic acid promotes differentiation of photoreceptors in vitro. Development 120: 2091-2102.

Kelley MW, Turner JK, Reh TA. 1995. Ligands of steroid/thyroid receptors induce cone photoreceptors in vertebrate retina. Development 121: 3777-3785.

Keynes RJ, Cook GM. 1995. Repulsive and inhibitory signals. Curr Opin Neurobiol 5: 75-82.

Kljavin IJ, Reh TA. 1991. Müller cells are a preferred substrate for in vitro neurite extension by rod photoreceptor cells. J Neurosci 11: 2985-2994.

Kljavin IJ, Lagenaur C, Bixby JL, Reh TA. 1994. Cell adhesion molecules regulating neurite growth from amacrine and rod photoreceptor cells [published erratum appears in J Neurosci 1994 Oct; 14(10): Following table of contents]. J Neurosci 14: 5035-5049.

Kolodkin AL. 1996. Growth cones and the cues that repel them. Trends Neurosci 19: 507-513.

Kuhn TB, Stoeckli ET, Condrau MA, Rathjen FG, Sonderegger P. 1991. Neurite outgrowth on immobilized axonin-1 is mediated by a heterophilic interaction with L1(G4). J Cell Biol 115: 1113-1126.

Lacalli TC. 2001. New perspectives on the evolution of protochordate sensory and locomotory systems, and the origin of brains and heads. Philos Trans R Soc Lond B Biol Sci 356: 1565-1572.

LaVail MM, Rapaport DH, Rakic P. 1991. Cytogenesis in the monkey retina. J Comp Neurol 309: 86-114.

Layer PG, Berger J, Kinkl N. 1997. Cholinesterases precede "ON-OFF" channel dichotomy in the embryonic chick retina before onset of synaptogenesis. Cell Tissue Res 288: 407-416.

Leber SM, Breedlove SM, Sanes JR. 1990. Lineage, arrangement, and death of clonally related motoneurons in chick spinal cord. J Neurosci 10: 2451-2462.

Lee WH, Javedan S, Bondy CA. 1992. Coordinate expression of insulin-like growth factor system components by neurons and neuroglia during retinal and cerebellar development. J Neurosci 12: 4737-4744.

Lemmon V, McLoon SC. 1986. The appearance of an L1-like molecule in the chick primary visual pathway. J Neurosci 6: 2987-2994.

Lillien L, Cepko C. 1992. Control of proliferation in the retina: Temporal changes in responsiveness to FGF and TGF alpha. Development 115: 253-266.

Liou GI, Wang M, Matragoon S. 1994. Timing of interphotoreceptor retinoid-binding protein (IRBP) gene expression and hypomethylation in developing mouse retina. Dev Biol 161: 345-356.

Lipton SA, Frosch MP, Phillips MD, Tauck DL, Aizenman E. 1988. Nicotinic antagonists enhance process outgrowth by rat retinal ganglion cells in culture. Science 239: 1293-1296.

Luskin MB, Pearlman AL, Sanes JR. 1988. Cell lineage in the cerebral cortex of the mouse studied in vivo and in vitro with a recombinant retrovirus. Neuron 1: 635-647.

Mack AF, Fernald RD. 1993. Regulation of cell division and rod differentiation in the teleost retina. Brain Res Dev Brain Res 76: 183-187.

Madtes P Jr, Redburn DA. 1983. GABA as a trophic factor during development. Life Sci 33: 979-984.

Martin PR, Grunert U. 1992. Spatial density and immunoreactivity of bipolar cells in the macaque monkey retina. J Comp Neurol 323: 269-287.

Martini R, Schachner M. 1988. Immunoelectron microscopic localization of neural cell adhesion molecules (L1, N-CAM, and myelin-associated glycoprotein) in regenerating adult mouse sciatic nerve. J Cell Biol 106: 1735-1746.

Marx J. 1995. Helping neurons find their way [news]. Science 268: 971-973.

McCabe CF, Cole GJ. 1992. Expression of the barrier-associated proteins EAP-300 and claustrin in the developing central nervous system. Brain Res Dev Brain Res 70: 9-24.

McKerracher L, David S, Jackson DL, Kottis V, Dunn RJ, et al. 1994. Identification of myelin-associated glycoprotein as a major myelin-derived inhibitor of neurite growth. Neuron 13: 805-811.

Messersmith EK, Leonardo ED, Shatz CJ, Tessier-Lavigne M, Goodman CS, et al. 1995. Semaphorin III can function as a selective chemorepellent to pattern sensory projections in the spinal cord. Neuron 14: 949-959.

Messersmith EK, Redburn DA. 1990. Kainic acid lesioning alters development of the outer plexiform layer in neonatal rabbit retina. Int J Dev Neurosci 8: 447-461.

Mitchell CK, Redburn DA. 1996. GABA and GABA-A receptors are maximally expressed in association with cone synaptogenesis in neonatal rabbit retina. Brain Res Dev Brain Res 95: 63-71.

Mollon JD. 1989. "Tho' she kneel'd in that place where they grew..." The uses and origins of primate colour vision. J Exp Biol 146: 21-38.

Mukhopadhyay G, Doherty P, Walsh FS, Crocker PR, Filbin MT. 1994. A novel role for myelin-associated glycoprotein as an inhibitor of axonal regeneration. Neuron 13: 757-767.

Nagle MG, Osorio D. 1993. The tuning of human photopigments may minimize red-green chromatic signals in natural conditions. Proc Biol Sci 252: 209-213.

Neitz M, Neitz J, Jacobs GH. 1991. Spectral tuning of pigments underlying red-green color vision. Science 252: 971-974.

O'Leary DD, Fawcett JW, Cowan WM. 1986. Topographic targeting errors in the retinocollicular projection and their elimination by selective ganglion cell death. J Neurosci 6: 3692-3705.

Peles E, Nativ M, Campbell PL, Sakurai T, Martinez R, et al. 1995. The carbonic anhydrase domain of receptor tyrosine phosphatase beta is a functional ligand for the axonal cell recognition molecule contactin. Cell 82: 251-260.

Persohn E, Schachner M. 1987. Immunoelectron microscopic localization of the neural cell adhesion molecules L1 and N-CAM during postnatal development of the mouse cerebellum. J Cell Biol 105: 569-576.

Politi LE, Insua F, Buzzi E. 1998. Selective outgrowth and differential tropism of amacrine and photoreceptor axons to cell targets during early development in vitro. J Neurosci Res 52: 105-117.

Prat A, Biernacki K, Wosik K, Antel JP. 2001. Glial cell influence on the human blood-brain barrier. Glia 36: 145-155.

Provis JM, Van Driel D, Billson FA, Russell P. 1985a. Development of the human retina: Patterns of cell distribution and redistribution in the ganglion cell layer. J Comp Neurol 233: 429-451.

Provis JM, Van Driel D, Billson FA, Russell P. 1985b. Human fetal optic nerve: Overproduction and elimination of retinal axons during development. J Comp Neurol 238: 92-100.

Rakic P. 1972. Extrinsic cytological determinants of basket and stellate cell dendritic pattern in the cerebellar molecular layer. J Comp Neurol 146: 335-354.

Rakic P. 1995. A small step for the cell, a giant leap for mankind: A hypothesis of neocortical expansion during evolution. Trends Neurosci 18: 383-388.

Rakic P, Cameron RS, Komuro H. 1994. Recognition, adhesion, transmembrane signaling and cell motility in guided neuronal migration. Curr Opin Neurobiol 4: 63-69.

Rathjen FG, Wolff JM, Chang S, Bonhoeffer F, Raper JA. 1987. Neurofascin: A novel chick cell-surface glycoprotein involved in neurite-neurite interactions. Cell 51: 841-849.

Redburn DA, Madtes P Jr. 1986. Postnatal development of 3H-GABA-accumulating cells in rabbit retina. J Comp Neurol 243: 41-57.

Redburn DA, Agarwal SH, Messersmith EK, Mitchell CK. 1992. Development of the glutamate system in rabbit retina. Neurochem Res 17: 61-66.

Reese BE, Harvey AR, Tan SS. 1995. Radial and tangential dispersion patterns in the mouse retina are cell- class specific. Proc Natl Acad Sci USA 92: 2494-2498.

Reese BE, Necessary BD, Tam PP, Faulkner- Jones B, Tan SS. 1999. Clonal expansion and cell dispersion in the developing mouse retina. Eur J Neurosci 11: 2965-2978.

Reh TA. 1992. Cellular interactions determine neuronal phenotypes in rodent retinal cultures. J Neurobiol 23: 1067-1083.

Reichenbach A, Robinson SR. 1995. Phylogenetic Constraints on Retinal Organisation and Development. Prog Retin Eye Res 15: 139-171.

Reichenbach A, Kasper M, Schnitzer J, Osborne NN, Pritz Hohmeier S. 1994a. A set of early-born neurons is distinctly labeled by several defined antibodies in the adult rabbit retina. J Hirnforsch/International Journal of Brain Research and Neurobiology 35: 391-395.

Reichenbach A, Ziegert M, Schnitzer J, Pritz-Hohmeier S, Schaaf P, et al. 1994b. Development of the rabbit retina. V. The question of 'columnar units'. Brain Res Dev Brain Res 79: 72-84.

Robinson SR. 1987. Ontogeny of the area centralis in the cat. J Comp Neurol 255: 50-67.

Rothermel A, Willbold E, deGrip WJ, Layer PG. 1997. Pigmented epithelium induces complete retinal reconstitution from dispersed embryonic chick retinae in reaggregation culture. Proc R Soc Lond B Biol Sci 264: 1293-1302.

Saffell JL, Doherty P, Tiveron MC, Morris RJ, Walsh FS. 1995. NCAM requires a cytoplasmic domain to function as a neurite outgrowth-promoting neuronal receptor. Mol Cell Neurosci 6: 521-531.

Sagar SM, Edwards RH, Sharp FR. 1991. Epidermal growth factor and Transforming growth factor alpha induce c-fos gene expression in retinal Müller cells in vivo. J Neurosci Res 29: 549-559.

Schnell L, Schwab ME. 1990. Axonal regeneration in the rat spinal cord produced by an antibody against myelin-associated neurite growth inhibitors. Nature 343: 269-272.

Sharma RK. 1996. Development of retinal transplants. Lic Med Sc Thesis, University of Lund Sweden (ISBN 91-628-2082-6). http://ask.lub.lu.se/archive/00018221/01/SharmaLMSc.pdf

Sharma RK. 1999. Retinal Transplants: Growth, Differentiation, Integration, Organization and Survival, Ph.D. Thesis, University of Lund. (ISBN 91-628-3430-4). http://www.lub.lu.se/luft/diss/med302_transit.html

Sharma RK. 2000a. Development of glycine-accumulating neurons in retinal transplants. Ophthalmologica 214: 264-270.

Sharma RK. 2000b. Developmental Expression of GABA(A) Receptors in Retinal Transplants. Ophthalmic Res 32: 45-51.

Sharma RK. 2001a. Bcl-2 expression during the development and degeneration of RCS rat retinae. Brain Res Dev Brain Res 132: 81-86.

Sharma RK. 2001b. Development and survival of tyrosine hydroxylase containing neurons in RCS rat retinae. Curr Eye Res 23: 256-262.

Sharma RK. 2001c. Expression of Bcl-2 during the development of rabbit retina. Curr Eye Res 22: 208-214.

Sharma RK, Ehinger B. 1997a. Cell proliferation in retinal transplants. Cell Transplant 6: 141-148.

Sharma RK, Ehinger B. 1997b. Mitosis in developing rabbit retina: An immunohistochemical study. Exp Eye Res 64: 97-106.

Sharma RK, Ehinger B. 2003. Development and Structure of the Retina. Adler's Physiology of the Eye: Clinical Application. Kaufman PL, Alm A, editors. St. Louis: Mosby; pp. 319-347.

Sharma RK, Johnson DA. 2000. Molecular signals for development of neuronal circuitry in the retina. Neurochem Res 25: 1257-1263.

Sharma RK, Netland PA. 2005. Class IIIbeta-tubulin isotype expression in developing retina and identification of cell types expressing it. Invest Ophthalmol Vis Sci (ARVO abstract)

Sharma RK, Bergström A, Ehinger B. 1995. Retinal cell transplants. Prog Retin Eye Res 15: 197-230.

Sharma RK, Perez MT, Ehinger B. 1997. Immunocytochemical localisation of neuronal nitric oxide synthase in developing and transplanted rabbit retinas. Histochem Cell Biol 107: 449-458.

Sharma RK, Warfvinge K, Ehinger B. 2001. Expression of nitric oxide synthase during the development of RCS rat retinas. Ophthalmologica 215: 222-228.

Sharma RK, O'Leary TE, Fields CM, Johnson DA. 2003. Development of the outer retina in the mouse. Brain Res Dev Brain Res 145: 93-105.

Sidman RL. 1961. Histiogenesis of mouse retina studied with thymidine-H^3. The Structure of the Eye. Smelser GK, editor. London: Academic Press; pp. 487-506.

Silver J. 1994. Inhibitory molecules in development and regeneration. J Neurol 242: S22-S24.

Silver J, Hughes AF. 1973. The role of cell death during morphogenesis of the mammalian eye. J Morphol 140: 159-170.

Snow DM, Lemmon V, Carrino DA, Caplan AI, Silver J. 1990. Sulfated proteoglycans in astroglial barriers inhibit neurite outgrowth in vitro. Exp Neurol 109: 111-130.

Snow RL, Robson JA. 1995. Migration and differentiation of neurons in the retina and optic tectum of the chick. Exp Neurol 134: 13-24.

Stenkamp DL, Hisatomi O, Barthel LK, Tokunaga F, Raymond PA. 1996. Temporal expression of rod and cone opsins in embryonic goldfish retina predicts the spatial organization of the cone mosaic. Invest Ophthalmol Vis Sci 37: 363-376.

Stoeckli ET, Kuhn TB, Duc CO, Ruegg MA, Sonderegger P. 1991. The axonally secreted protein axonin-1 is a potent substratum for neurite growth. J Cell Biol 112: 449-455.

Straznicky C, Chehade M. 1987. The formation of the area centralis of the retinal ganglion cell layer in the chick. Development 100: 411-420.

Stuermer CA, Bastmeyer M. 2000. The retinal axon's pathfinding to the optic disk. Prog Neurobiol 62: 197-214.

Tanaka H, Matsui T, Agata A, Tomura M, Kubota I, et al. 1991. Molecular cloning and expression of a novel adhesion molecule, SC1. Neuron 7: 535-545.

Tear G, Harris R, Sutaria S, Kilomanski K, Goodman CS, et al. 1996. Commissureless controls growth cone guidance across the CNS midline in Drosophila and encodes a novel membrane protein. Neuron 16: 501-514.

Turner DL, Cepko CL. 1987. A common progenitor for neurons and glia persists in rat retina late in development. Nature 328: 131-136.

Turner DL, Snyder EY, Cepko CL. 1990. Lineage-independent determination of cell type in the embryonic mouse retina. Neuron 4: 833-845.

Wadsworth WG, Hedgecock EM. 1996. Hierarchical guidance cues in the developing nervous system of C. elegans. Bioessays 18: 355-362.

Wassle H, Boycott BB, Illing RB. 1981a. Morphology and mosaic of on- and off-beta cells in the cat retina and some functional considerations. Proc R Soc Lond B Biol Sci 212: 177-195.

Wassle H, Peichl L, Boycott BB. 1981b. Morphology and topography of on- and off-alpha cells in the cat retina. Proc R Soc Lond B Biol Sci 212: 157-175.

Wikler KC, Rakic P. 1991. Relation of an array of early-differentiating cones to the photoreceptor mosaic in the primate retina. Nature 351: 397-400.

Wikler KC, Rakic P, Bhattacharyya N, Macleish PR. 1997. Early emergence of photoreceptor mosaicism in the primate retina revealed by a novel cone-specific monoclonal antibody. J Comp Neurol 377: 500-508.

Willbold E, Berger J, Reinicke M, Wolburg H. 1997a. On the role of Müller glia cells in histogenesis: Only retinal spheroids, but not tectal, telencephalic and cerebellar spheroids develop histotypical patterns. J Hirnforsch /Int J Brain Res Neurobiol 38: 383-396.

Willbold E, Brummendorf T, Rathjen FG, Schwarz H, Weiss B. 1997b. The neural cell recognition molecule F11 is expressed on Müller cells and Schwann cells in vitro. J Hirnforsch /Int J Brain Res Neurobiol 38: 71-80.

Willbold E, Reinicke M, Lance-Jones C, Lagenaur C, Lemmon V, et al. 1995. Müller glia stabilizes cell columns during retinal development: Lateral cell migration but not neuropil growth is inhibited in mixed chick-quail retinospheroids. Eur J Neurosci 7: 2277-2284.

Williams AJ, Hunt DM, Bowmaker JK, Mollon JD. 1992. The polymorphic photopigments of the marmoset: Spectral tuning and genetic basis. EMBO J 11: 2039-2045.

Williams RW, Goldowitz D. 1992. Structure of clonal and polyclonal cell arrays in chimeric mouse retina. Proc Natl Acad Sci USA 89: 1184-1188.

Yamashita M, Yoshimoto Y, Fukuda Y. 1994. Muscarinic acetylcholine responses in the early embryonic chick retina. J Neurobiol 25: 1144-1153.

Yin X, Watanabe M, Rutishauser U. 1995. Effect of polysialic acid on the behavior of retinal ganglion cell axons during growth into the optic tract and tectum. Development 121: 3439-3446.

Zhang J, Gray J, Wu L, Leone G, Rowan S, et al. 2004. Rb regulates proliferation and rod photoreceptor development in the mouse retina. Nat Genet 36: 351-360.

Zimmerman RP, Polley EH, Fortney RL. 1988. Cell birthdays and rate of differentiation of ganglion and horizontal cells of the developing cat's retina. J Comp Neurol 274: 77-90.

Zondag GC, Koningstein GM, Jiang YP, Sap J, Moolenaar WH, et al. 1995. Homophilic interactions mediated by receptor tyrosine phosphatases mu and kappa. A critical role for the novel extracellular MAM domain [published erratum appears in J Biol Chem 1995 Oct 13; 270(41): 24621]. J Biol Chem 270: 14247-14250.

3 Molecular Neurobiology of Retinal Degeneration

R. K. Sharma

Abstract: Retinal degeneration is a major cause of blindness in the elderly worldwide. The problem is likely to increase as people live longer lives, imposing a unique challenge to the scientific community to find a cure. In recent years, our understanding of molecular mechanisms behind retinal degenerations has vastly increased. Several key mechanisms have been identified. These include ischemia, oxidative stress, excitotoxicity, autoimmunity, and inflammation. These mechanisms often act in concert with each other, and the final converging point for most of them appears to be apoptosis pathways. Diseases causing retinal degeneration, through their unique pathophysiology, evoke the mechanisms mentioned earlier. This chapter first describes the general mechanisms of retinal degeneration, and then discusses how the disease-specific pathology relates to these mechanisms.

List of Abbreviations: AMPA, amino-3-hydroxy-5-methyl-4-isoxazole propionic acid; AMD, age-related macular degeneration; AA, arachidonic acid; adRP, autosomal dominant retinitis pigmentosa; bFGF, basic fibroblast growth factor; CNV, choroidal neovascularization; CatD, Cathepsin D; CFH, compliment factor H; nNOS, Neuronal NOS; eNOS, endothelial NOS; FKHR, Forkhead-related transcription factor; GCL, ganglion cell layer; GS, glutamine synthetase; GC, guanylate cyclase; HIF-1, Hypoxia inducible factor; HNE, 4-hydroxynonenal; INL, inner nuclear layer; IPL, inner plexiform layer; IGF-I, insulin-like growth factor I; iNOS, inducible nitric oxide synthase; IAP, Inhibitor of apoptosis proteins; ICG, indocyanine green; LCA-1, Leber congenital amaurosis-1; MAP kinase, mitogen-activated protein kinase; MD, malondialdehyde; mGluRs, Metabotropic glutamate receptors; mtDNA, mitochondrial DNA; MMPs, matrix metalloproteinases; NMDA, *N*-methyl-D-aspartate; NO, Nitric oxide; NOS, nitric oxide synthase; nicotinamide; NMNAT, mononucleotide adenylyl transferase; OPL, outer plexiform layer; OH, hydroxyl; OAT, ornithine-delta-aminotransferase; PDGF, platelet derived growth factor; PCNA, proliferating cell nuclear antigen; PKC, protein kinase C; PEDF, pigmented epithelium derived factor; RPE, retinal pigment epithelium; ROS, Reactive oxygen species; RP3, X-linked retinitis pigmentosa; REP-1, Rab escort protein-1; TNF-[α], tumor necrosis factor-[α]; TGF-β, transforming growth factor-β; TIMPs, tissue inhibitors of metalloproteinases; TNF-α, tumor necrosis factor-a; TIMP-3, tissue inhibitor of metalloproteinases-3; Tyr402His, histidine at amino acid 402; UbE4b, ubiquitination factor E4b; VEGF, vascular endothelial growth factor; XLRD, X-linked retinal degenerations

1 Introduction

With a relative control over infectious and inflammatory diseases having been gained, degenerative and inherited disorders are increasingly becoming important causes of morbidity. The number of patients suffering from degenerative disorders is on the rise because of increasing population size and life expectancy, putting an enormous economic and social burden on societies. Finding a cure for degenerative disorders, including retinal degeneration, is one of the most important challenges facing the scientific community (Sharma and Ehinger, 1999).

The retina is an extension of the central nervous system, and it converts light into nerve impulses. It comprises different types of neurons, glial cells, and components of the vascular system. Photoreceptors, the cells that capture light, are of two types: rods (for scotopic vision) and cones (for photopic vision). Photoreceptors have highly specialized outer segments where photons are captured. Photoreceptors form contacts with horizontal and bipolar cells in the outer plexiform layer (OPL). Coupling between neighboring rods and cones in OPL allows the first stage of visual processing. The inner nuclear layer (INL) contains cell bodies of the bipolar, the horizontal, amacrine, and Muller glial cells (Sharma and Ehinger, 2003). Extensive lateral processing between neighboring bipolar cells takes place in the inner plexiform layer (IPL), through synaptic contacts between retinal ganglion cells and amacrine cells. The ganglion cell layer (GCL) contains the cell bodies of retinal ganglion cells and certain displaced amacrine cells (➋ *Figure 3-1*). Inside the eye, ganglion cell fibers run along the retinal surface toward the optic nerve head (Sharma and Ehinger, 2003). Once the ganglion cell axons enter the optic nerve, a myelin sheath formed by astrocytes encapsulates the fiber. This decreases the membrane conductance, allowing increased conduction velocity and the distance over which impulses can be conducted. Most of these fibers synapse on neurons in the lateral

Figure 3-1
Structure and organization of retinal neurones

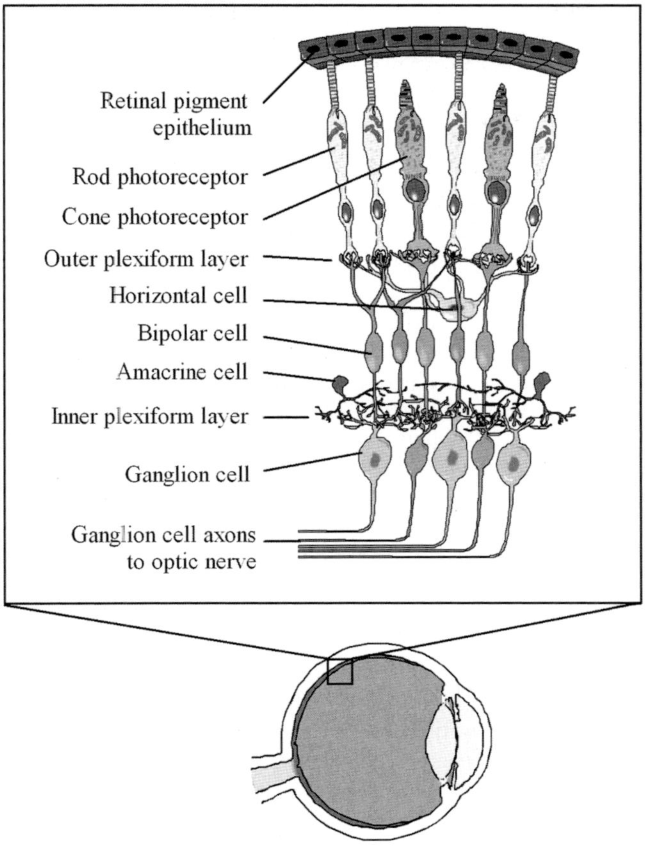

geniculate nucleus of the thalamus. The axons of these neurons run to the primary visual cortex. The neural retina also contains several types of supporting cells. Muller cells balance the electrolytes in the extracellular space in response to the activity of the retinal neurons and can play an important role in the pathogenesis of retinal diseases. The retinal pigment epithelium (RPE) cells provide the supportive role necessary to sustain the high metabolic demands of the photoreceptors (Sharma et al., 1997; Mandel et al., 1999). RPE cells supply nutrients and oxygen, regenerate phototransduction products, and digest debris shed by the photoreceptors. Dysfunction of RPE has serious consequences on the health of photoreceptors.

Photoreceptors receive nutrients via a network of small capillaries under the retina, the choroid plexus. The inner layers have their own blood supply coming from the blood vessels in the optic nerve head. The blood vessels in the inner retina form circular arcades around the central retina, the macula. The center-most portion of the macula (the fovea) contains no blood vessels, and derives oxygen and nutrients from the choroid plexus.

For its protection, the retina is physiologically and immunologically segregated from the rest of the body by the blood retinal barrier formed by tight junctions between vascular endothelial cells and RPE cells. Only small molecules can cross this barrier, making it difficult for many drugs to reach ocular tissue. In addition, intraocular tissue is an immune privileged site. This protects the ocular tissue from the innocent bystander effect of inflammation.

A variety of diseases originating from retinal neurons, blood vessels, or the support system (RPE cells) can affect the retina and the optic nerve. These diseases could be infectious, inflammatory, traumatic, neoplastic, vascular, or degenerative. These diseases can selectively damage certain cell types in the retina.

There can be several reasons for this, including the differential vulnerability of retinal cells to various stresses, metabolic adaptations, and the nature of the blood supply. The diseases affecting the outer retina include age-related macular degeneration (AMD), pathologic myopia, and hereditary retinal degenerations. Inner retinal diseases include retinal vascular diseases such as diabetic retinopathy, retinal venous occlusive disease, and retinopathy of prematurity. Retinal vascular diseases are a leading cause of blindness. Ocular infections (retinitis, endophthalmitis) cause a more widespread damage affecting all retinal layers. Optic nerve diseases include glaucoma, optic neuritis, and ischemic optic neuropathy.

2 General Mechanisms of Retinal Degeneration

There has been great progress in understanding the molecular pathways connecting disease processes to neuronal dysfunction or death in degenerative diseases. Several key mechanisms have been identified. These include ischemia, oxidative stress, excitotoxicity, the action of cytokines, and autoimmunity. Although for the sake of clarity these mechanisms are described separately, it is important to understand that they act in tandem with each other and that involved pathways overlap. For example, an ischemia insult could cause metabolic dysfunction in the cells, but it also elicits excitotoxicity, oxidative stress, and cytokine-mediated cellular damage. On the other hand, excitotoxicity can cause oxidative stress. Interestingly, most of these molecular pathways initiated by various degenerative disease processes converge on certain cellular events such as apoptosis, finally leading to neat disintegration of the cells. Important mechanisms at work in retinal degeneration are described in the following sections.

2.1 Retinal Ischemia

Ischemia has been implicated in a large number of retinal diseases such as diabetic retinopathy, anterior ischemic optic atrophy, retinal and choroidal vessel occlusions, retinopathy of prematurity, glaucoma, and traumatic optic neuropathy. All of these diseases result in ganglion cell death and thinning of the nerve fiber layer (Osborne et al., 1999b). Retinal neurons are relatively more resistant to ischemic injury than those of the brain (Hayreh and Weingeist, 1980). One possible reason is that brain is enclosed in bony cranium; therefore, oedematous brain is incapable of expanding and would compress the vessels in the brain, causing further ischemia. Edematous retina could expend into the vitreous cavity without obstructing the blood flow (Ames III et al., 1968). There is also evidence that the resistance of retinal neurons to ischemia could have metabolic bases (Casson et al., 2004). There are two clinically important pathological consequences of ischemia in the retina: (1) neuronal death and (2) neovascularization.

2.1.1 Neuronal Death in Retinal Ischemia

Although retinal ganglion cells are more sensitive to ischemia than other retinal neurons, other classes of retinal neurons are also susceptible (Osborne et al., 1999a, b, c). Experimental models involving retinal ischemia demonstrate injury to the retinal ganglion cells (Akiyama et al., 2002; Zhang et al., 2003). The ischemic paradigm has been used to study retinal ganglion cell diseases such as glaucoma. Within various classes of ganglion cells, parvocellular ganglion cells (foveal) appear to be more resilient to ischemia as compared to magnocellular ganglion cells. Molecular bases of this difference could lie in the density of *N*-methyl-D-aspartate (NMDA) receptors on these neurons. The magnocellular neurons are more susceptible to NMDA toxicity and have a higher density of NMDA receptors (Sucher et al., 1997). Interestingly, the central vision is often spared in macular infarction.

Photoreceptors are less sensitive to ischemia than the ganglion cells. Reasons for this are not clear. High densities of inner segment mitochondria that are normally maintained at a low pO_2 could suggest that photoreceptors may have evolved a compensatory mechanism to survive in a low-oxygen environment. Significantly, neuroglobin, a neuron-specific respiratory protein, is present in high amounts in the retina, especially in association with the photoreceptors (Schmidt et al., 2003). Photoreceptor sensitivity to ischemia is also influenced by light adaptation. Light-adapted photoreceptors are less sensitive due to diminished metabolic demands. Rods are more sensitive to ischemia than cones, possibly because the latter have a more efficient Ca^{2+} buffering ability.

Amacrine cells, an important class of interneurons of the inner retina, are also sensitive to ischemia and glutamate excitotoxicity. Ischemia upregulates expression of proinflammatory mediators such as COX-2 and nNOS in amacrine cells (Ju et al., 2003). These molecules also are involved in ischemic injury to the ganglion cells, suggesting that common mechanisms of ischemic injury exist in different cells. Both of these cell types die by apoptosis (Singh et al., 2001). Amacrine cells are morphologically, functionally, and molecularly diverse, thus not all amacrine cells are equally susceptible to ischemia or excitotoxicity (Osborne and Larsen, 1996).

2.1.2 Neovascularization

In addition to neuronal death, another important clinical consequence of retinal ischemia is neovascularization. Retinal and choroidal neovascularization, resulting from ischemic conditions of the retina, cause blindness in millions due to the wet type of AMD, diabetic retinopathies, and other conditions. It has been known for a long time that retinal ischemia leads to neovascularization through a mysterious diffusable factor (called Wise's "factor x"). This mysterious factor has now been identified as vascular endothelial growth factor (VEGF) (Connolly et al., 1989). It is important to realize, however, that it may not act alone, and there is considerable evidence implicating several other factors. Other soluble factors that have been demonstrated to stimulate angiogenesis include the fibroblast growth factor (FGF) family (Abraham et al., 1986), tumor necrosis factor-[α] (TNF-[α]) (Leibovich et al., 1987), insulin-like growth factor I (IGF-I) (Grant et al., 1993), hepatocyte growth factor (HGF) (Laterra et al., 1997), and others. However, VEGF plays a key role. It is upregulated by hypoxia (Plate et al., 1992), and its levels increase in the retina and vitreous of patients (Adamis et al., 1994) and laboratory animals (Miller et al., 1994) with ischemic retinopathies. VEGF is present in fibroblast-like and trans-differentiated RPE cells in choroidal neovascular membranes (Kvanta et al., 1996). In laser-induced choroidal neovascularization, VEGF mRNA levels increase in RPE, choroidal vascular endothelial cells, and fibroblast-like cells in the lesions (Ogata et al., 1996). Interestingly, increased expression of VEGF in photoreceptors does not result in CNV (Okamoto et al., 1997). FGF2 seems to be less important in stimulating retinal neovascularization. Factors that have antiangiogenetic effects have also been identified. These include members of the transforming growth factor-β (TGF-β) family. TGF-β may also inhibit endothelial cell migration and repair after injury (Madri et al., 1989).

Angiogenesis involves vasodilatation of existing vessels, increased vascular permeability, and degradation of the surrounding matrix. This allows activated endothelial cells to proliferate, migrate, and form new tubes (Witmer et al., 2003). The molecular basis of neovascularization is complex and involves the interplay of controlling factors. For example, endothelial cell activation involves urokinase-type plasminogen activator, matrix metalloproteinase, and cysteine proteinase pathways. These pathways themselves are under the control of other factors. Proteinase gene expression is controlled by cytokines and proangiogenic factors such as basic fibroblast growth factor (bFGF) and VEGF. In addition to the activation of proproteinases, proteolysis is controlled by downregulation of protease inhibitors such as plasminogen activator inhibitors, tissue inhibitors of metalloproteinases (TIMPs), and cystatins. Recruitment of pericytes and smooth muscle cells, important for the developing vessel, is controlled by platelet-derived growth factor (PDGF) (Mignatti and Rifkin, 1996).

Hypoxia is an integral part of ischemia and plays a vital role in its pathophysiology. Hypoxia inducible factor (HIF-1) mediates transcription of several genes. Activation of HIF-1 also stimulates the production of VEGF, erythropoietin, bFGF, and other factors associated with neovascularization (Semenza, 2000; Maxwell and Ratcliffe, 2002; Vincent et al., 2002). Another potential factor for neovascularization is adenosine, a neuromodulator that can act on specific receptors (putatively A2A and A2B receptors) on the endothelial cells, and through nitric oxide mediation it can stimulate cell migration and tube formation (Lutty and Mcleod, 2003).

2.1.3 Modes of Ischemic Injury to Neurons

Ischemic injuries have been extensively studied in the brain in the context of stroke; however, much remains to be learnt about the pathophysiological mechanisms. Excitotoxicity is an important contributing mechanism, especially in the areas where the blood supply is not critically reduced. In addition, ischemia affects cell bodies and the axons differently (Petty and Wettstein, 1999). These differences are important in understanding the pathophysiology of glaucoma, where the key events may be occurring in the ganglion cell axons rather than in the cell bodies. Ischemia could injure the neurons by several different mechanisms or their combinations.

2.1.3.1 Effects of Ischemia on Neuronal Metabolism In ischemia, the oxygen and glucose supply to the retina is blocked, and this disrupts cellular energy metabolism, resulting in falling intracellular ATP levels. This jeopardizes membrane function and ion homeostasis, triggering the cascades that ultimately kill the cells (Zeevalk and Nicklas, 1990). One of the effects of energy deprivation is the inhibition of Na^+/K^+ ATPase that eventually disrupts membrane potential regulation, and cell volume (Lipton, 1999). Pharmacological Na^+/K^+ ATPase inhibition produces neurotoxicity (Lees, 1991). Reduced Na^+/K^+ ATPase transport inhibits repolarization of axon and synaptic membranes, decreasing the voltage-dependent Mg^{2+} block of NMDA receptors (Zeevalk and Nicklas, 1992). This causes NMDA receptors to be activated by raised external glutamate levels, increasing depolarization and Ca^{2+} influx. Ionotropic amino-3-hydroxy-5-methyl-4-isoxazole propionic acid (AMPA)/kainate-type glutamate receptors are also activated, leading to the influx of Ca^{2+}, Na^+, Cl^-, and osmotically obligated water. The influx of water leads to tissue edema and cell lysis (Lipton, 1999).

2.1.3.2 Effect of Ischemia on Neurotransmitter Systems Neurons expend most of their energy maintaining the membrane potential. Released neurotransmitters are quickly removed by the neurons and the glia. Therefore, in the normal physiological state, levels of neurotransmitters in the extracellular space are very low. However, during retinal ischemia this balance is disturbed, causing extracellular glutamate accumulation. Release of other neurotransmitters, such as the major inhibitory neurotransmitter GABA (Neal et al., 1994), glycine, dopamine, acetylcholine (Neal et al., 1994), and the neuromodulator adenosine (Roth et al., 1996), has also been reported.

The increased levels of neurotransmitters in the extracellular space can trigger death or survival pathways depending on the distribution of neurotransmitter receptors on the neuron. Accumulation of glutamate in the extracellular space, which is even more profound during reperfusion, is especially injurious to the neurons (Louzada-Junior et al., 1992). Neurons of the inner retina, in particular the ganglion cells, express high levels of ionotropic glutamate receptors (Brandstatter et al., 1994). Activation of these receptors causes excessive depolarization and cell death. Inhibitory transmitters, such as GABA, on the other hand, cause hyperpolarization and could therefore counteract the depolarization caused by glutamate, (Schwartz-Bloom and Sah, 2001) and thus could have neuroprotective effects (Green et al., 2000). However, the protective effects of inhibitory neurotransmitters are not consistently observed (Chen et al., 1999), suggesting complexities in interactions.

Although there is some controversy as to whether the glutamate levels in the nervous system actually increase after ischemia (Zeevalk and Nicklas, 1990), most studies confirm that cerebral ischemia (Drejer et al., 1985) and hypoglycemia (Burke and Nadler, 1989) cause endogenous release and extracellular accumulation of glutamate.

The mechanism for the release of glutamate is only partially understood. In focal cerebral ischemia, extracellular glutamate increases in a biphasic manner (Wahl et al., 1994). In earlier phases, the ischemia-induced membrane depolarization causes glutamate release by activation of voltage-dependent Ca^{2+} channels; however, once the intracellular ATP levels are too low to support exocytotic glutamate release, the release becomes Ca^{2+} independent, suggesting alternative means of glutamate release (Nishizawa, 2001). Reversed transport of glutamate by glutamate transporter proteins is a possible mechanism for the Ca^{2+}-independent glutamate increase. Glutamate from the synaptic cleft and the extracellular spaces is removed by uptake into the cells (Barbour et al., 1988). This uptake is ATP independent and driven by the

Na^+ gradient across the cell membrane together with the concentration of K^+ and pH-changing anions (Barbour et al., 1988). With reduced ATP supply to the Na^+/K^+-ATPase pump during ischemia, K^+ accumulates in the extracellular space. This depolarizes the cells, decreasing the Na^+ gradient and reducing the glutamate uptake, allowing glutamate to accumulate in the extracellular space (Rossi et al., 2000).

Normally, glutamate is taken up by glia, converted by glutamine synthetase (GS), and then recycled back to the neuron (Thoreson and Witkovsky, 1999). Retinal ischemia impairs glutamate uptake by Müller cells (Barnett et al., 2001). Ischemia also lowers glial intracellular ATP levels, reducing the GS activity. Astrocytes, abundant in retina (Hernandez, 2000), can also contribute to extracellular glutamate. Ca^{2+} influx into astrocytes during ischemia (Fern, 1998) can release glutamate by a prostaglandin-mediated mechanism (Bezzi et al., 1998).

2.1.3.3 Intracellular Influx of Ca^{2+} in Retinal Ischemia Ischemic or energy-deprived cells cannot maintain homeostasis, causing rapid and massive fluxes of Ca^{2+}, K^+, Na^+, and Cl^- (Hansen, 1985). Under these conditions, the extracellular Ca^{2+} concentration is rapidly reduced (Nicholson et al., 1977) while intracellular Ca^{2+} increases (Silver and Erecinska, 1990). Not much is known about how extracellular Ca^{2+} enters the cells in ischemic conditions. Possible mechanisms include voltage-dependent and receptor-operated Ca^{2+} channels, release of Ca^{2+} from internal stores, and Ca^{2+} buffering mechanisms (Wang et al., 2002).

The Ca^{2+} channel antagonists, e.g., nifedipine (Crosson et al., 1990), lomerizine (Toriu et al., 2000), flunarizine (Takahashi et al., 1992), and levemopamil (Block and Schwarz, 1998), have been shown to be neuroprotective in different animal models. In addition to the influx of extracellular Ca^{2+}, mobilization of Ca^{2+} from intracellular stores also contributes to the ischemia-induced Ca^{2+} overload (Kristian and Siesjo, 1998). This mobilization is triggered by IP3 following activation of phospholipase C-coupled surface receptors such as metabotropic glutamate receptors.

It is commonly believed that Ca^{2+} is a major mediator of neuronal cell death in ischemia, especially in areas where the release of glutamate causes secondary excitotoxic damage. There are also data that show that a slight rise in intracellular Ca^{2+} can protect a neuron from adverse events of ischemia (Wang et al., 2002). Since ischemia triggers a plethora of other adverse events, some of which do not directly involve Ca^{2+}, the specific role of intracellular Ca^{2+} in triggering neurotoxic cascades has mostly been studied in cell culture models. Direct extrapolation of these results for in vivo conditions is problematic.

2.1.3.4 Free Radical Generation in Retinal Ischemia Several mechanisms are possible for the generation of free radicals during ischemia-reperfusion. Large quantities of superoxide radicals ($\cdot O_2^-$) are produced at the beginning of reperfusion. The concentrations of hypoxanthine and xanthine increase following ischemia–reperfusion due to degradation of ATP (Roth et al., 1997; Banin et al., 2000). In addition, increased intracellular calcium activates the Ca^{2+}-dependent protease calpain, which converts xanthine dehydrogenase into xanthine oxidase. Xanthine oxidase activity has been shown to increase soon after reperfusion of retina (Marak et al., 1990). Xanthine oxidase, in turn, oxidizes the accumulated hypoxanthine to uric acid, resulting in the release of $\cdot O_2^-$, which in turn is converted to hydroxyl radical ($\cdot OH$) by the Haber Weiss reaction. This reaction is catalyzed by iron. The needed iron is released from its protein-bound stores at the low pH conditions prevailing during ischemia. In addition, ischemia produces nitric oxide, which interacts with $\cdot O_2^-$ forming peroxynitrite, nitrosyl radical, and eventually $\cdot OH$ (Gilgun-Sherki et al., 2002).

In addition to being generated in the neuronal mitochondria, free radicals are also generated in the activation of glial cells. There is infiltration of leukocytes during the early phases of ischemic injury. They release inflammatory mediators such as arachidonic acid, nitric oxide, and cytokines; these mediators also play an important role in the formation of free radicals.

In ischemia, the endogenous free radical quenching capability of the neurons diminishes further adding to cell stress. Consequently, free radical scavengers such as extract of Ginkgo biloba (Szabo et al., 1991), α-lipoic acid (Block and Schwarz 1997), vitamin E (Celebi et al., 2002), thioredoxin (Shibuki et al., 2000), an ascorbic acid derivative (Kuriyama et al., 2001), mannitol (Gupta and Marmor, 1993), and the iron chelator desferrioxamine (Ophir et al., 1994) have all been shown to be neuroprotective in ischemia-reperfusion injury.

2.1.3.5 The Role of Inflammatory Cytokines in Retinal Ischemia It has become increasingly evident in recent years that acute inflammatory responses contribute to ischemic brain injury, especially following reperfusion through a variety of mechanisms (Danton and Dietrich, 2003). These mechanisms are discussed in detail in ➋ *Section 2.4.*

2.1.3.6 Acidosis in Retinal Ischemia Ischemia causes a rapid drop in intracellular and extracellular pH. This happens due to the accumulation of lactic acid, formed because of the shift from oxidative phosphorylation to anaerobic glycolysis. The amount of lactate produced during ischemia is related to the available glucose and glycogen; therefore, preexisting hyperglycemia should increase lactate production, while hypoglycemia should decrease it. If so, elevated glucose concentrations should worsen the ischemic injury, and restricted glucose should have a beneficial effect. Some studies have shown this to be the case (Lipton, 1999). However, unlike its effect on neurons in the brain, elevated glucose in the retina protects neurons against ischemic damage (Romano et al., 1993). There are fundamental differences between retinal and cerebral energy metabolism, especially in terms of reliance on mitochondrial metabolism (retina relies less on mitochondrial metabolism than the brain). Some studies also suggest that mild acidosis is protective (Tombaugh and Sapolsky, 1993).

Acidosis can affect cell viability in other ways as well. Increased intracellular H^+ can accelerate lipid peroxidation and free radical generation. Also, acidosis can exaggerate the rise in intracellular calcium by inhibiting intracellular calcium sequestration. Severe acidosis can also cause necrosis and damage to glial cells and the microvasculature. Acidosis inhibits the mitochondrial respiratory chain, lactate oxidation, and H^+ extrusion. Acidosis also generates cell edema (Staub et al., 1993).

2.1.4 Cell Death in Ischemia

Ischemic cell death was once considered as classical necrotic cell death because it met the morphological criterion of cell swelling. Now, evidence suggests that apoptosis, or a closely related mechanism, is a prominent mode of ischemic cell death. There are numerous reports demonstrating oligonucleosomal DNA fragmentation either by gel electrophoresis ("DNA laddering") or by the TUNEL technique (Linnik et al., 1993). Apoptosis has also been demonstrated in retinal ischemia caused by elevated intraocular pressure by DNA fragmentation (Rosenbaum et al., 1997). In addition, the expression of molecules involved in apoptosis pathways strongly suggests that ischemic neuronal cell death is apoptotic. These molecules include c-Jun or cyclin 1 (Kuroiwa et al., 1998), p-JNK (Ettaiche et al., 1999), caspase-1, caspase-3 (Katai and Yoshimura, 1999), caspase-2 (Kurokawa et al., 1999), p53 (Joo et al., 1999), and proliferating cell nuclear antigen (PCNA) (Ju et al., 2001) and IL-1. Furthermore, caspase inhibitors prevent ischemia-induced retinal damage (Katai and Yoshimura, 1999; Kurokawa et al., 1999).

In focal ischemia, apoptotic cell death is mainly seen at the ischemic border-zone (Linnik et al., 1993). When the ischemic insult is more severe, apoptotic cells appear earlier than after mild insults (Endres et al., 1998). It is possible that some of the apoptosis seen during ischemia could be a result of growth factor deprivation rather than the direct effect of ischemia. Ischemia can damage certain cells, neuronal or glial, that normally provide growth factor support to other neurons. Once this support is gone, the deprived neurons die by apoptosis.

2.2 Oxidative Stress in Retina

Reactive oxygen species (ROS) are by-products of aerobic metabolism (➋ *Figure 3-2*). Under physiological conditions, it is estimated that up to 1% of the mitochondrial electron flow leads to the formation of superoxide (O_2^-.), the primary oxygen free-radical produced by mitochondria. Interference with electron transport can dramatically increase O_2^-. production. ROS are toxic to cells.

Prooxidants are endogenous or exogenous substances that can cause oxidation of target molecules either directly by abstraction of electrons or indirectly by production of highly reactive intermediate

Figure 3-2

Mitochondria are a major source of ROS. Mitochondrial membrane and DNA are vulnerable to damage due to their proximity to the site of ROS generation

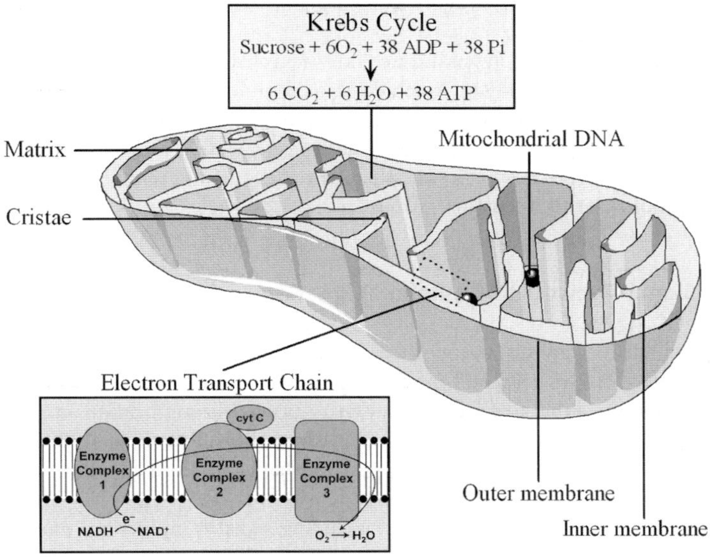

chemical entities. Free radicals represent a class of such entities whose reactivity is derived from the presence of unpaired electrons in their atomic structure, but which are capable of independent existence only for a brief time (Halliwell, 1997). ROS collectively include all reactive forms of oxygen, including both the radical and nonradical species that participate in the initiation and propagation of radical chain reactions.

Several forms of ROS contain nitrogen or chlorine, in addition to oxygen. Nitrogen dioxide and nitric oxide (NO) are free radicals with odd numbers of electrons. Nitrogen dioxide is a strong oxidant, whereas NO acts as a weak reducing agent. It can react with the superoxide radical and give rise to a powerful oxidant, the nonradical peroxynitrite. NO is a vasodilator and hyperpolarizing agent in endothelial cells (Moncada and Higgs, 1991; Vanhoutte, 2004). Hypochlorous acid is a powerful oxidant. It is produced by activated neutrophils from hydrogen peroxide by the heme-containing enzyme, myeloperoxidase (Gutteridge, 1986). Other free radicals such as sulfhydryl (thiyl) do not contain the oxygen atom, but undergo electron transfer and hydrogen transfer reactions with a variety of biological molecules.

2.2.1 Sources of ROS

ROS can be generated from multiple sources. These include mitochondrial electron transport, NOS activity, arachidonic acid metabolism, and xanthine oxidase (produced by hydrolysis of xanthine dehydrogenase). Moreover, altered binding of transition metals such as iron due to acidic conditions in ischemia may increase their participation in the Haber-Weiss reaction (Halliwell, 1992). In addition, P450 enzymes and NAD (P)H oxidase are possible sources of ROS.

Mitochondria as a site of ROS production has been known for a long time. ROS production from mitochondria can be increased by stimulating the electron-transport activity or by disturbing electron transfer down the transport chain. Increased intracellular Ca^{2+} levels and exposure to fatty acids that alter the properties of the mitochondrial membrane can increase the leak of ROS from mitochondria.

Another important source of ROS is the enzyme NOS, especially in pathological conditions. NOS converts arginine and molecular oxygen to citrulline and NO, a free radical gas. This process generates free radicals (Beckman and Koppenol, 1996). NO itself has weak radical reactivity; however, it combines readily with O_2 and possibly H_2O_2 to produce a highly oxidizing, nonradical compound, peroxynitrite (Beckman and Koppenol, 1996).

Production of eicosanoids from polyunsaturated fatty acids such as arachidonic acid may also generate ROS (Schreiber et al., 1986; Kontos, 1987). Arachidonic acid undergoes a series of reactions to convert to biologically active lipids, the eicosanoids. These reactions can generate free radicals. In particular, cyclooxygenase isoforms and 5-lipoxygenase containing heme iron generate a low concentration of superoxide anion constitutively. 12-Lipoxygenases possess nonheme iron. It does not release superoxide anion, but can induce lipid peroxidation after translocation to membranes.

2.2.2 ROS and Signal Transduction—Redox Signaling

It was long thought that free radicals are involved only in damaging reactions, but the discovery of the endogenous generation of NO in cells, and the finding that this freely diffusing gas participates in signal transduction pathways implied that free radicals play a role in cell communication and signaling processes. This novel role has now been suspected for other ROS such as H_2O_2. Reactions of ROS at high concentrations, which are cytotoxic, are different from those at low concentrations prevalent under physiological conditions (Droge, 2002; Forman et al., 2002).

ROS act as modulators of biological processes, including signal transduction, transcription, and programmed cell death. Signal transduction is a process by which cellular components receive a message from within or outside the cells, and transmit that information intracellularly to elicit a response. Typically, hormones, cytokines, neurotransmitters, and other molecules that interact with receptors in or on the cell carry this information. However, NO can act as a signal transducer and affects metabolic processes in several cell types including vascular smooth muscles and neurons (Bondy, 1995; Dawson and Dawson, 1995; Sharma et al., 1997; Ali and Mann, 2004). Signaling systems present in cells rely on complex kinase cascades, protease cascades, and second messengers (such as calcium ions, etc.) to regulate cytoplasmic and nuclear proteins, which activate transcription factors such as AP-1 and NFκ–B and affect various cell functions such as cell growth and apoptosis (Palmer and Paulson, 1997; Ali and Mann, 2004).

There is only limited evidence for ROS other than NO acting as second messengers. H_2O_2 can activate EGF receptors. TNF has been shown to increase intracellular H_2O_2 and to induce ROS production by activating enzymes such as NADPH oxidase. Membrane receptors are often coupled to cytosolic signal transduction via the small protein, *ras*. Activated *ras* stimulates the small G protein, *rac*, which in turn binds to and activates the membrane-bound NADPH oxidase complex to produce ROS. ROS have been shown to influence mitogen-activated protein kinase (MAP kinase) cascades and the transcription factors that are controlled by these kinase cascades such as AP-1 and NF-κB (➋ *Figure 3-3*). ROS are also implicated second messengers in the activation of NF-κB/Rel transcription factors, implicated in cell viability (Powis et al., 1997). In RPE cells, low doses of H_2O_2 can alter multiple cellular functions such as cell viability and cell attachment (Sharma et al., 2004). Cytotoxic reactions of ROS are utilized by the body for defense against intrusion of foreign pathogens (Babior et al., 2003).

2.2.3 Cellular Defense Mechanisms Against Oxidative Stress

The cell has evolved effective defenses against oxidative stress and repair mechanisms to deal with associated oxidative damage. Generally, cells defend themselves against oxidative stress at three different levels: (1) They prevent the damage by preventing ROS-generating electron leakage; (2) they intercept potentially dangerous ROS, with the help of scavenging antioxidant molecules; and (3) they repair the damage, by removing the damaged molecules. These defense mechanisms utilize enzymes (superoxide dismutase, catalase, peroxidase, and peroxiredoxin), low molecular weight antioxidant species (Vitamin E, ascorbate,

Figure 3-3
Redox signaling might play an important role in RPE cell functions. Lower levels of oxidative stress may evoke endogenous cytoprotection. In addition, redox signaling may also affect other important cellular functions. However, if the cellular defense mechanism is unable to effectively deal with ROS, the cells undergo apoptosis

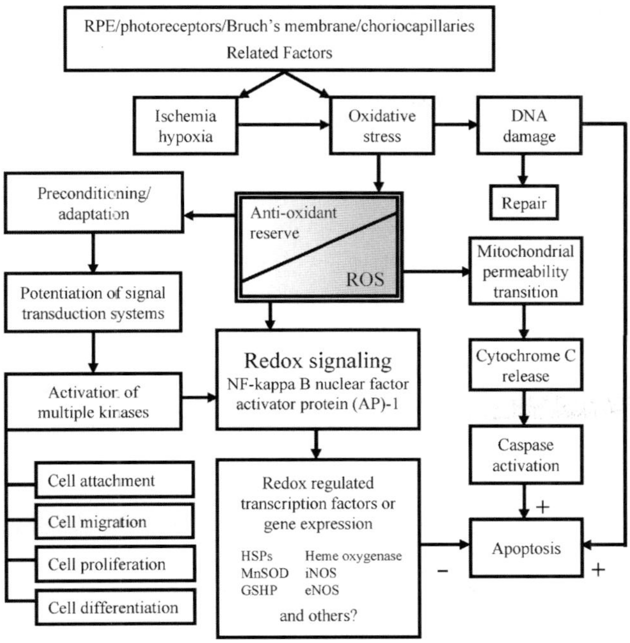

glutathione), and more complex forms of protection such as systems for metal transport and buffering and induction of transcription factors. Biological antioxidants, including glutathione, α-tocopherol (vitamin E), carotenoids, and ascorbic acid, react with most oxidants. In addition, the antioxidant enzymes catalase and glutathione peroxidase detoxify H_2O_2 by converting it to O_2 and H_2O.

2.2.4 Redox Homeostasis and Oxidative Stress

Homeostasis is a key element to all healthy physiologic functions throughout the body; loss of homeostasis often leads to disease. Redox homeostasis describes the normal physiologic process of reduction and oxidation to repair unstable, damaging, reduced ROS. In the physiological state, ROS generation in normal cells is under tight homeostatic control (*Figure 3-4*). However, when ROS levels exceed the antioxidant capacity of a cell, oxidative stress occurs. Oxidative stress arises because of imbalance between the production of potentially toxic ROS and the scavenging activities. Because of their high chemical reactivity, excess levels of ROS might damage the cell's structural and functional integrity. Although cells possess a variety of defense mechanisms and repair systems against ROS, if the ROS production overwhelms the antioxidant defenses, oxidative stress occurs. Therefore, oxidative stress can be regarded as an imbalance between prooxidant/free radical production and opposing antioxidant defenses. Acute and chronic oxidative stress have been implicated in a wide variety of degenerative diseases such as atherosclerosis, diabetes, ischemia/reperfusion injury, inflammatory diseases, cancer, neurological diseases, hypertension, pulmonary diseases, and hematological diseases (Maxwell, 1995; Opara, 2004). Age-related loss of physiological functions has also been attributed to the chronic effects of ROS on various biological macromolecules (Droge, 2003). In the eye also, oxidative stress is implicated in a large number of diseases (*Figure 3-5*). These include cataract, senile muscular degeneration, diabetic retinopathy, retrolental fibroplasia, uveitis,

Figure 3-4
Oxidative metabolism is an integral part of cell biology, and cells have developed effective defense mechanisms to deal with it, including antioxidants. Redox homeostasis is the normal physiological process of oxidation and reduction. Loss of redox homeostasis occurs when there is excess of ROS or impaired antioxidant defense capacity. This can lead to cell death and disease

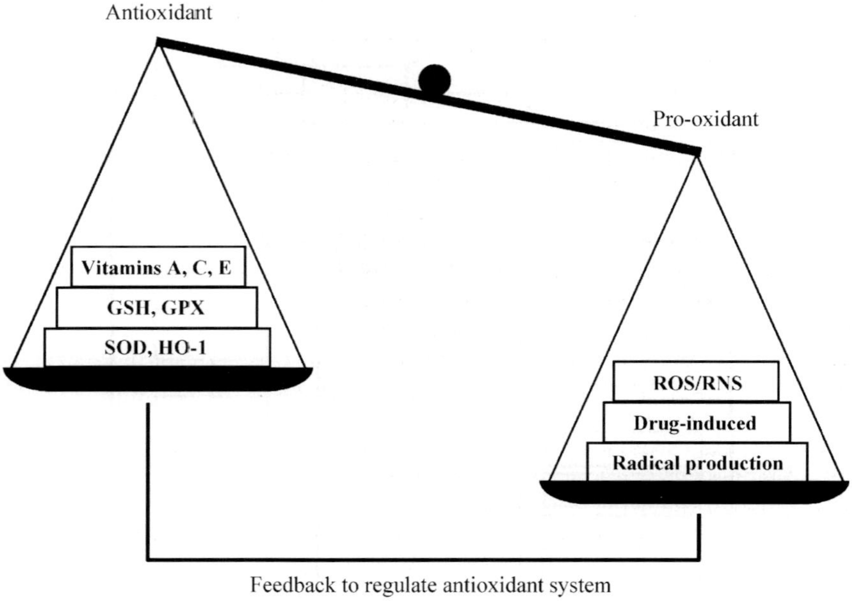

Figure 3-5
Schematic interactions of oxidative stress leading to disease

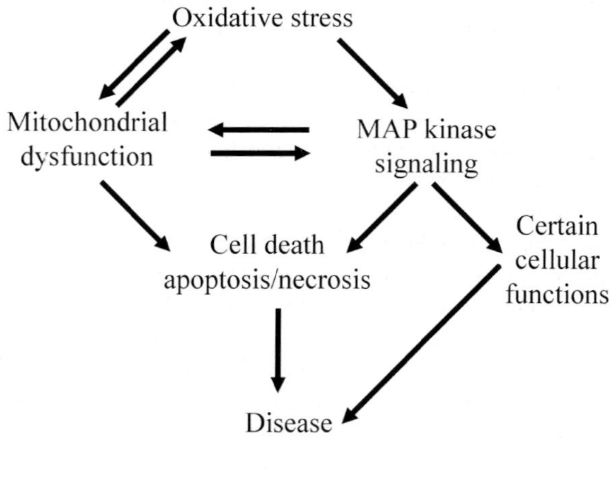

and glaucoma. Unchecked, excessive ROS can cause damage to cellular components, including lipids, protein, and DNA; ultimately the cells die by apoptosis or necrosis (Kannan and Jain, 2000). With increasing age, the antioxidant defenses of the cells decrease (Cassarino and Bennett Jr, 1999), leading to oxidative stress and damage in the cells.

2.2.5 Vulnerability of the Retina

Neurons in general are particularly vulnerable to oxidative stress (Reiter, 1995; Halliwell, 2001). Factors predisposing retina to oxidative stress are as follows:

1. Retina has one of the highest rates of oxygen consumption of any tissue in the body.
2. The retina is exposed to sunlight, which can cause photooxidative stress.
3. Photoreceptors are rich in PUFAs, which are vulnerable to ROS damage.

Glutamate is an important neurotransmitter in the retina. Glutamate and calcium are involved in the transmission of signals from photoreceptor cells to bipolar cells and ganglion cells. Light activated photoreceptors modulate the release of glutamate, which in turn activates bipolar cells. Bipolar cells also release glutamate onto retinal ganglion cells and amacrine cells. In certain pathologies, the extracellular glutamate levels increase, which induces a cascade of reactions resulting in the formation of ROS. Activation of the NMDA receptors, for example, stimulates the release and metabolism of arachidonic acid (AA) via lipoxygenase pathways. AA hydroperoxides, such as 12-HETE and 12-HPETE, produced during the AA cascade–lipoxygenase pathways not only act as second messengers, but are also a recognized source of ROS. Glutamate also promotes the production of ROS through Ca signaling.

1. NOS is found in certain amacrine cells in the retina (Sharma et al., 1997; Sharma et al., 2001). NOS produces NO, and the interaction of NO with the superoxide radical is implicated in normal neuronal metabolism as well as neurodegeneration.
2. Retinal neurons are nonreplicating cells and any damage by the ROS tends to be cumulative over time.
3. At least a few amacrine cells in the retina are dopaminergic (Sharma, 2001b; Sharma and Ehinger, 2003). ROS generation during the oxidation of dopamine by monoamine oxidase in the nerve terminals produces increased oxidative stress (e.g., in brain regions, such as the substantia nigra).

2.2.6 Modes of Cell Damage in Oxidative Stress

2.2.6.1 Damage to the Mitochondria Mitochondria are very susceptible to oxidative damage (Wallace, 2001; Beal, 2003; Inoue et al., 2003; Sadek et al., 2003). The inner mitochondrial membrane is a major source of ROS. In addition, iron released during the degradation of heme-containing cytochrome complexes of the respiratory chain accelerates hydroxyl radical formation. Because hydroxyl radicals are highly reactive and have a short half-life, they cause damage close to the site of production, i.e., mitochondria themselves. In addition, mitochondrial membranes are rich in phospholipids and PUFAs (Parker et al., 1990), which are preferred substrates for free radical peroxidation. Lipid peroxides can be degraded to cytotoxic aldehydes and hydrocarbons in the presence of readily available iron originating at the time of degradation of mitochondrial cytochromes. These interactions cause membrane disruption affecting permeability to cations.

The mtDNA is also especially vulnerable to structural damage by the ROS produced in the mitochondria, especially because mitochondria lack an adequate DNA repair mechanism. The steady-state level of oxidized bases in mtDNA is approximately 16 times higher than that in nuclear DNA. To complicate the matter further, the genetic information in mtDNA is closely packed; both the heavy and light strands are transcribed, and there are very few noncoding sequences. Therefore, any unrepaired structural damage to mtDNA can lead to severe mitochondrial dysfunction that can increase mitochondrial ROS release and generate an energy crisis due to failure of oxidative phosphorylation. Mitochondrial dysfunction is especially harmful to postmitotic cells, such as neurons, that draw most of their energy from mitochondrial electron transport (Jellinger, 2003). Defective mitochondrial energy metabolism has other effects also. These include decrease in high-energy phosphate stores, weakened antioxidant defense, abnormal membrane potentials, and disruption of calcium homeostasis (causing excitotoxicity) (Annunziato et al., 2003).

Age-related decline in mitochondrial function due to cumulative oxidative stress plays an important role in the functional impairments in the aging brain and in age-related neurodegenerative disorders

(Wallace, 1992; Sastre et al., 2003). Furthermore, mitochondria produce H_2O_2 and O_2^- at an increased rate with age (Sohal et al., 1994). Consequently, the levels of mtDNA deletions also increase with age in the human brain (Soong et al., 1992). The most common 4.9-kb human mtDNA deletion that accumulates with aging includes genes coding for cytochrome *c* oxidase [four genes coding for complex I subunits and one gene for subunit 3 (complex IV)]. Complex I is vulnerable to the hydroxyl radical, and complex IV to H_2O_2 (Zhang et al., 1990). An age-dependent functional impairment of complexes I and IV, but not complex II, exclusively encoded by the nuclear DNA, has been reported in primate brain (Bowling et al. 1993).

2.2.6.2 Damage to Other Cellular Membranes The photoreceptor outer segments are unusually rich in PUFAs, and they are constantly phagocytized by the RPE cells. By reacting with PUFAs in the cellular membranes, ROS such as hydroxyl (OH) and peroxynitrite (ONOO) produce a variety of lipid peroxidation products (LPP), including 4-hydroxynonenal (HNE) and malondialdehyde (MD). Once formed, these peroxidation metabolites stay active in the cells for long periods of time (minutes to hours) and interact with multiple cellular components. Evidence suggests that lipid peroxidation and lipid peroxidation products contribute significantly to the neuronal changes and neurotoxicity in neurodegenerative conditions. Additionally, lipid peroxidation products can modulate a wide array of activities within the neural tissue, including long-term potentiation, neurite outgrowth, and proliferation.

2.2.6.3 Oxidative Stress and Excitotoxicity Excitotoxicity is an important contributor to oxidative stress (➔ *Figure 3-6*). However, the converse is also true. In addition to the involvement of free radicals at several points in the apoptotic cascade, including serving as initiators, early signals, and possibly late effectors, ROS can also modulate excitotoxic pathways. It has been demonstrated that ROS generated in neurons can disrupt the glutamate transport function of neighboring astrocytes (Rao et al., 2003). An ROS-induced decrease in glutamate transport would further increase extracellular glutamate and increase excitotoxicity. Similarly, this cycle could be initiated by ROS arising in microglial cells.

Figure 3-6
Excitatory amino acids can cause oxidative stress. Excitation of glutamate receptors, such as NMDA receptors, causes intracellular calcium influx. Calcium can activate various pathways to produce free radicals

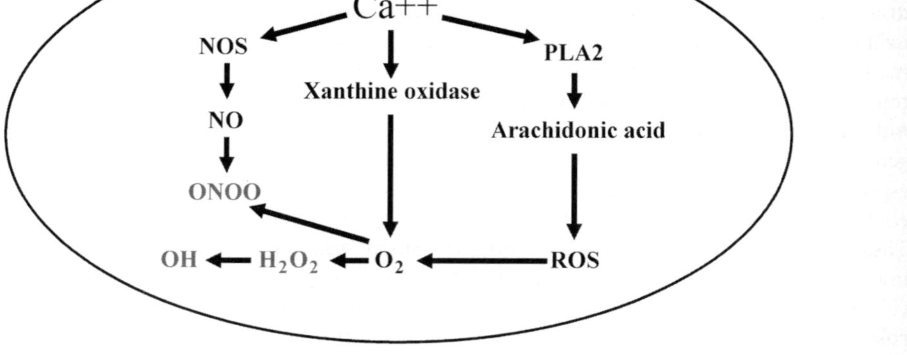

2.2.6.4 Protein Damage The attack by free radicals on sulfhydryl protein bonds leads to the destruction of amino acids and polypeptide chains, cross-linking, or vulcanization of sulfhydryl rich proteins. These changes are responsible for the stiffening of the proteins during aging, specifically collagen of the extracellular matrix.

2.2.6.5 DNA Damage The damaging effect of ROS on mitochondrial DNA has already been discussed. Free radicals can also fragment nuclear DNA molecules. This fragmentation leads to the activation of poly (ADP-ribose) polymerase, cytokines, and NF-κB, which are likely to be instrumental in upregulations of iNOS and COX-2, subsequent release of glutamate, and effects on Ca^{2+} homeostasis (Lipton, 1999).

2.3 Excitotoxicity

Glutamate is the major excitatory retinal neurotransmitter in retina. It is released by photoreceptors, bipolar cells, and ganglion cells (Sharma and Ehinger, 2003). Normally, the released glutamate remains in the synaptic cleft only for a short time (a few milliseconds). If glutamate levels remain elevated for a prolonged period of time, this can excite neurons to death. This mechanism of cell death is referred to as excitotoxicity.

Excitotoxicity has shown to be involved in several acute forms of neuronal death, including hypoglycemia, trauma, and seizures. However, the strongest evidence for the involvement of excitotoxicity in brain injury comes from ischemic injury (stroke). During ischemia, the extracellular levels of glutamate and aspirate increase. Glutamate receptor antagonists can prevent ischemic cell death in neurons. These findings suggest that excessive stimulation of glutamate receptors somehow triggers a chain of events leading to neuronal death.

Excitotoxicity, especially the pharmacology of glutamate receptors, has become a topic of intense research because of its possible implications in finding ways to treat or prevent neuronal death. Unfortunately, clinical trials of glutamate receptor antagonists have not been very successful.

2.3.1 Excitotoxicity in Retina

Excitotoxicity has been implicated in a number of retinal disorders, especially the ones in which ischemia plays a key role. These conditions include glaucoma and diabetic retinopathy. The role of excitotoxicity in glaucoma has been controversial, but in animal models excitotoxicity has been shown to cause ganglion cell damage.

Glutamate or its receptor agonist causes histological damage to retina in vitro (Heidinger et al., 1998). This excitotoxic damage is similar to that observed in experimental ischemia (Zeevalk and Nicklas, 1990). In addition, ischemic damage to the retina can be reversed with the addition of NMDA and AMPA or kainate antagonists (Osborne et al., 1999; Osborne et al., 1999a). Similar to observations in vitro, glutamate administration destroys the inner retinal layers in neonatal and adult animals (Vorwerk et al., 1996). Retinal damage following NMDA injection into the adult rat eye produces a dose-dependent loss of ganglion and cholinergic amacrine cells (Siliprandi et al., 1992; Osborne et al., 1999a) similar to what is seen in retinal ischemia. There are, however, limitations in adopting excitotoxicity as a model for ischemia, especially in vivo. Ischemia, in vivo, involves reduction/obliteration of blood flow. This generates a complex pathological response that cannot be matched in culture conditions. This provides a possible explanation as to why glutamate antagonists never provide the level of protection in vivo as in vitro. Ischemia, in vivo, is not simply an excitotoxicity, but a more complex phenomenon.

2.3.2 The Role of Different Glutamate Receptors in Excitotoxicity

Glutamate exerts its action via a number of receptors that include NMDA, AMPA/kainate-type ionotropic glutamate receptors, and metabotropic receptors (Hollmann and Heinemann, 1994). Glutamate receptors mediate excitotoxicity (Choi, 1992). Retinal ganglion cells express both kainate-type (Lin et al., 2002) and NMDA-type ionotropic receptors (Brandstatter et al., 1994). Consequently, both kainate and NMDA cause loss of these neurons. However, the NMDA subtype is more important for retinal ganglion cell death (Siliprandi et al., 1992). NMDA receptors are ligand-gated ion channels composed of NR1 and NR2 subunits and have binding sites for a number of positive and negative modulators (Dingledine et al., 1999).

Studies suggest that glutamate causes neurotoxicity by different mechanisms. The first involves an increase in intracellular calcium. Activation of AMPA/kainate and to a lesser extent NMDA receptors, causes Na^+ influx into the neuron, depolarizing the plasma membrane. This allows the influx of Ca^{2+} ion through voltage-gated Ca^{2+} channels. Voltage-gated Ca^{2+} channel blockers, such as nifedipine or betaxolol (Melena et al., 1999), attenuate the kinate neurotoxicity. AMPA/kainate receptors lacking the GluR2 subunit are, themselves, Ca^{2+} permeable, and their activation can directly raise the levels of intracellular Ca (Anzai et al., 2003).

There are suggestions that glutamate toxicity is not due to increased intracellular Ca^{2+} alone and that additional mechanisms exist. For example, raising extracellular K^+ also depolarizes neurons through voltage-gated Ca^{2+} channels, but this does not cause as severe a damage as glutamate (Li et al., 2001). When either Na^+ or Cl^- ions are removed, there is no toxicity (Rothman and Olney, 1995), including that in the chick embryonic retina (Olney et al., 1986). Excess glutamate promotes a marked and rapid Na^+ influx via its ionotropic receptors, and this is followed by an obligatory Cl^- influx. An outward anion efflux cannot compensate for this Cl^- influx since the cell membrane is not permeable to the majority of intracellular ions. This increases intracellular osmolarity, drawing water into the cell and causing death by lysis. Indeed, the first sign of glutamate toxicity is neuronal swelling and edema (Lipton, 1999).

Glutamate can also have a direct effect on glia. Elevated extracellular glutamate can induce intracellular Na^+ influx through glutamate transporters, or can activate receptors on glial cells causing Ca^{2+} influx. Glutamate receptors have also been identified on cultured RPE cells (Fragoso and Lopez-Colome, 1999), but it is not known how such cells are affected by excessive levels of glutamate.

Metabotropic glutamate receptors (mGluRs) are another group of important receptors in the retina. Eight distinct subtypes of mGluRs belonging to three functional groups have been identified: group I (mGlu1 and mGlu5), group II (mGlu2 and mGlu3), and group III (mGlu4 and mGlu6-8). Retina expresses all of the subtypes. These receptors are located in the plexiform layers in association with the majority of retinal cell types (Thoreson and Witkovsky, 1999). Most importantly, multiple mGluR subtypes are found on amacrine and ganglion cells (Hartveit et al., 1995). The group I mGluRs are often expressed postsynaptically and are responsible for enhancing cell excitability. Through their coupling via Gq/G11 to polyphosphoinositide hydrolysis, they mobilize intracellular Ca^{2+} and activate Protein Kinase C. Some group I mGluRs are negatively coupled to K^+ channels. In contrast, group II and III mGluRs are found presynaptically. Their transduction pathway involves a negative coupling to adenylate cyclase, and their activation suppresses glutamate release and thus synaptic excitation (Nicoletti et al., 1996). Therefore, whether stimulation of mGluRs will have damaging or protective effects on neurons will depend on the receptor subtype activated. As expected, some studies in the brain show that antagonism of group I receptors or activation of group II/III receptors is neuroprotective in ischemia (Bruno et al., 2001).

2.3.3 Role of Calcium in Neuronal Death

There is overwhelming evidence that Ca^{2+} overload is injurious to cell viability (Kristian and Siesjo, 1998; Lipton, 1999; Sattler and Tymianski, 2001). Glutamate excitotoxic cell death to some extent depends on the influx of extracellular Ca^{2+} (Sucher et al., 1997; Kuriyama et al., 2001). Elevated intracellular Ca^{2+}, if sustained for a sufficient time, can initiate multiple detrimental cascades. Ca^{2+} can set into motion the pathways that generate free radicals, produce NO, activate phospholipase A2, damage DNA, activate proteases, and damage the cytoskeleton (Kristian and Siesjo, 1998; Lipton, 1999; Sattler and Tymianski, 2001). Raised intracellular Ca^{2+} in ischemia may also alter gene expression either directly or via the endoplasmic reticulum (Schonthal et al., 1991) to trigger oncogene expression (c-fos and c-jun) in ischemic neurons (Gerlach et al., 2002).

Although the significance of calcium in neuronal cell death is well established, only recently has it emerged that other metals such as zinc may also be important (Weiss et al., 2000). Studies have shown that the inter-neuronal movement of Zn^{2+} is important in ischemic neuronal death and glutamate toxicity (Frederickson and Bush, 2001). The role of Zn^{2+} in the pathogenesis of retinal ischemia is complex (Nakamichi et al., 2003). Chelatable Zn^{2+} is primarily found in association with photoreceptors and the RPE. Low levels of Zn^{2+} could accumulate in the extracellular space around neurons, such as ganglion cells, during ischemia and possibly influence cell viability through downregulation of NMDA receptors (Christine and Choi, 1990). Zn^{2+} is essential for the viability of RPE cells, although high doses are toxic (Wood and Osborne, 2001).

2.4 Cytokines and Mediators of Inflammation in Retinal Injury

Cytokines are small proteins synthesized de novo. They mediate and regulate functions such as immunity and inflammation. Cytokines bind to specific membrane receptors, signaling the cell via second messengers (often tyrosine kinases) to alter gene expression. Typical effects of cytokines include altered expression of membrane proteins (including cytokine receptors), proliferation, and secretion of effector molecules.

Injury to the CNS leads to neuroinflammation. This involves activation of resident glial cells such as astrocytes and microglia in addition to influx of blood-derived monocytes/macrophages. Inflammation is mediated by the release of proinflammatory cytokines, many of which activate microglia/macrophages and astrocytes. These include interleukin-1 (Kong et al., 1997) and tumor necrosis factor-α (TNF-α (Pusztai et al., 1994; Taupin et al., 1997). Once reactive, microglia and astrocytes acquire cytotoxic capabilities and play direct or indirect roles in the pathogenesis of neuronal degeneration (Hellendall and Ting 1997). As a result of activation, microglia and astrocytes express inducible nitric oxide synthase (iNOS), synthesizing high levels of NO. Neuroinflammation can cause neuronal damage, but it also has neuroprotective and neurotrophic effects.

2.4.1 Nitric Oxide

NO is an important neuromodulator in the retina, and is implicated in many physiological processes (Goldstein et al., 1996). NO is synthesized from arginine via the action of nitric oxide synthase (NOS). Three distinct isoforms of NOS have been identified. Neuronal NOS (nNOS) and endothelial NOS (eNOS) are Ca^{2+}-dependent. nNOS is constitutively expressed by certain types of amacrine cells in the retina. These cells often have long projections in the innerplexiform layer (Sharma et al., 1997; Sharma et al., 2001). eNOS is expressed by the endothelial cells of blood vessels (Cheon et al., 2003). iNOS is Ca^{2+}-independent and expressed in Muller and RPE cells in response to certain stimuli (Lopez-Costa et al., 1997). Activation of the NMDA receptor leads to an increase in intracellular calcium levels, which can induce expression of, and activate NOS isoforms, either directly (nNOS) or via the activation of calcium-dependent protein kinase C (PKC) (Lipton, 1999) (➋ *Figure 3-6*).

During hereditary retinal degeneration, nNOS immunoreactive cells survive even when the retina is extensively degenerated (Sharma et al., 2001). In ischemic retina, expressions of all NOS isoforms have been reported to increase. Activated macrophages and other inflammatory cells that infiltrate the retina are a major source of NO produced by iNOS. In addition, retinal astrocytes and Müller cells also contribute (Neufeld et al., 2002).

It is not yet clear whether the increase in NO production is beneficial or detrimental to the retina. There are contradicting reports. In general, the results indicate that nNOS and iNOS are detrimental presumably due to the formation of peroxynitrite (Dawson, 1995), but eNOS is beneficial to neurons (Iadecola, 1997). The nonspecific inhibitors of NOS have been shown to be neuroprotective against ischemic injury (Ju et al., 2000; Neufeld et al., 2002). Microglia and Müller cells have been shown to express iNOS following ganglion cell axotomy, and its inhibition promotes survival, suggesting that glial-derived NO is cytotoxic to injured ganglion cells (Koeberle and Ball, 1999).

2.4.2 Arachidonic Acid

The phospholipase A2 group of enzymes specifically hydrolyze fatty acids under physiological and pathophysiological conditions. This yields lysophospholipids and free fatty acids, such as arachidonic acid. Lysophospholipids are the precursors of potent bioactive mediators (e.g., platelet-activating factor). Arachidonic acid is metabolized by a number of enzymes, especially by cyclooxygenases to bioactive prostanoids. Arachidonic acid also has other functions. It regulates enzyme activities (protein kinases) and modulates ion channels and gene expression. These pathways are required for maintaining normal cellular functions. In pathological conditions such as ischemia, phospholipase A2 is activated and promotes free radical formation (Farooqui et al., 2001; Osborne et al., 2004). Phospholipase A2 can be induced by a variety of stimuli to facilitate arachidonic acid production. These stimuli include elevated intracellular Ca^{2+}, IL-1, TNF, and glutamate (Kudo and Murakami, 2002).

Oxidative metabolism of arachidonic acid produces several potent vasoconstrictive eicosanoids such as prostaglandin F2, thromboxane A2, and leukotriene D4. The resulting vasoconstriction can further compromise the blood flow in an ischemic tissue. These products also increase neutrophil adhesion to the endothelium and vascular permeability (Sapirstein and Bonventre, 2000). Nevertheless, not all the effects of arachidonic acid in the ischemic retina are likely to be detrimental. Arachidonic acid can protect neurons by depressing the non-NMDA receptor currents. Recent work has shown that low concentrations of arachidonic acid can reduce glutamate neurotoxicity in ganglion cells (Kawasaki et al., 2002). Involvement of arachidonic acid in retinal ischemia is implied by increased cyclooxygenase-2 signals in retinal ischemia, and blocking these signals elicits neuroprotection to ganglion cells (Ju et al., 2003).

2.4.3 Cytokines

IL-1 and tumor necrosis factor (TNF) are proinflammatory cytokines. They are produced in response to adverse stimuli. Each cytokine exists as two well-characterized isoforms: IL-1α and IL-1β, and TNF-α and TNF-β, respectively. There is strong evidence for the involvement of IL-1β and TNF α in the pathogenesis of experimental brain ischemia (Hallenbeck, 2002; Patel et al., 2003). In the retina, transient ischemia causes upregulation of TNF-α (Fontaine et al., 2002). In the early phase of reperfusion, TNF-α is primarily upregulated in ganglion cells, amacrine cells, and Müller cells. There is no consensus about the overall effect of TNF-α on retinal cell viability. There is an indication that activation of TNF receptor 2 is neuroprotective, whereas activation of TNF receptor 1 augments neuronal death.

The mechanism of action of IL-1β is presently uncharacterized. It is possible that it acts through excitotoxic and iNOS pathways. Inhibition of IL-1 is protective against glutamate neurotoxicity in retinae in vitro (Yoneda et al., 2001). IL-1 inhibition also reduces iNOS expression, free radical formation, and neuronal damage in ischemic brain models (Ohtaki et al., 2003; Osborne et al., 2004) (➋ *Figure 3-7*).

2.5 Autoimmunity in Retinal Degeneration

One of the emerging concepts in neural degeneration is the role of immunity in modulating the degenerative process. There is evidence that activated T cells enter the normal central nervous system as part of immune surveillance (Hickey et al., 1991; Raivich et al., 1998). The recruitment of T cells could have a positive effect on the disease progression because this immune response helps the resident immune cells (microglia) clear the site of injury of cell debris and other deleterious matter, such as breakdown products of degenerating nerves (Kipnis and Schwartz, 2002). The concept of protective immunity is based on the premise that the autoreactive T cells could protect the neurons from degeneration. It is suspected that one of the physiological functions of autoimmunity is to protect the body against harmful self-components (Schwartz, 2000; Schori et al., 2001) including those produced during CNS injury (Yoles and Schwartz, 1998), and an autoimmune disease might result from imbalance of this controlled response (Kipnis et al., 2002). It is suggested that immunological intervention can boost the body's ability to counteract the toxicity of

Figure 3-7
Activation of microglia by cytokines can produce NO (by iNOS) and superoxide that can diffuse into the neighboring neurones and cause damage. Microglia play an important role in retinal degeneration. In glaucoma, astrocytes in the optic nerve exhibit functional and biochemical characteristics of activation

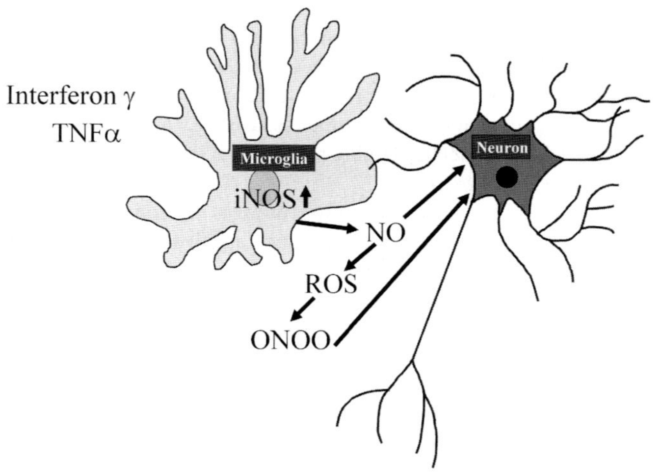

physiologic compounds acting as stress signals (Schwartz et al., 2003; Schwartz, 2003b). There is evidence that the beneficial effects of T-cell-mediated protection may arise from increased production of cytokines, including interferon-[γ] and neurotrophins (Aharoni et al., 2003).

Vaccination with copolymer-1 (Cop-1; Copaxone; Teva Pharmaceutical Industries, Petah Tikva, Israel) protects retinal ganglion cells from death due to mechanical injury to their axons, or injury to the perikarya by glutamate or increased intraocular pressure (Kipnis et al., 2002). Cop-1 appears to act as an antigen that cross-reacts with, and activates, a wide range of self-reacting T cells (Hafler, 2002; Kipnis and Schwartz, 2002). This can boost the T-cell effect and achieve autoimmunity comparable to that evoked by T cells that are specific to hidden epitopes from antigens at the site of stress (Kipnis and Schwartz, 2002).

2.6 Cell Disintegration

In recent years, our understanding of cellular death pathways has increased enormously. Apoptosis has been increasingly shown to be one of the common mechanisms employed by cells in the majority of circumstances. In specific cases, the mode of cell death may not exhibit all the characteristics of apoptosis. It has also been speculated that neuronal death in excitotoxicity is conducted by apoptosis. However, there is much disagreement regarding how excitotoxic and apoptotic cell death processes relate to one another. There is also support for the notion that excitotoxic and apoptotic neurodegeneration are two separate and distinct cell death processes. It is also suggested that ischemia directly triggers excitotoxic neurodegeneration, and apoptotic neurodegeneration follows subsequently because the remaining neurons are deprived of synaptic connections needed for their viability (Young et al., 2004).

3.6.1 Apoptosis

Apoptotic cell death comprises a well-defined sequence of morphological and molecular events leading to the death of the cell. Apoptosis is derived from a Greek term meaning "dropping off" as leaves fall from a tree (Kerr et al., 1974). Cells undergoing apoptosis shrink, condense, and fragment, releasing no intracellular contents that could be deleterious to the neighboring cells. Other cells phagocytose small membrane-bound apoptotic bodies. This contrasts with necrotic cell death (so-called accidental cell death), in which cells swell, burst, and release their contents, which can cause inflammation and damage the surrounding cells (Majno and Joris, 1995).

A family of protease enzymes called caspases consists of the effector molecules in apoptotic pathways. The specific intracellular targets of caspases include proteins of the nuclear lamina and cytoskeleton. Once these targets are cleaved, the cell dies. Activation of caspases is a common feature of most cell-death programs. In the cell, caspases are synthesized as inactive procaspases. Once activated, caspases cleave and thereby activate other procaspases, resulting in a proteolytic cascade. Finally, some of the activated caspases cleave proteins vital for cell integrity. For example, some caspases cleave the nuclear laminas, whereas others cleave a protein that keeps DNAse inactive, freeing the DNAse that cuts the DNA in the nucleus. The protease cascade, once activated, is irreversible.

Activation of procaspase can be triggered in a number of ways. (1) Activation of death receptors, Fas, on the cell surface by Fas ligand produced by another cell (such as killer lymphocytes). This can activate the caspase cascade by activating caspase-8. Interestingly, some stressed or damaged cells can produce both the Fas ligand and the Fas protein. (2) Stressed or damaged cells can activate the caspase cascade by releasing cytochrome *c* from the mitochondria. This pathway is commonly used in most pathological conditions. Normally, cytochrome *c* is localized between the inner and outer mitochondrial membrane, but in a dysfunctional cell cytochrome *c* is released into the cytosol where it binds with and activates an adaptor protein called Apaf-1. (3) DNA damage can trigger apoptosis through the involvement of p53. This activates the transcription of genes encoding for Bcl-2 family proteins. These proteins promote the release of cytochrome c from mitochondria.

Intracellular proteins belonging to the Bcl-2 family regulate the activation of caspases. Some members of this family such as Bcl-2 and Bcl-xL block cytochrome release and thus are prosurvival. Others promote procaspase activation and therefore promote cell death. They can have different modes of action; for example Bad, an apoptosis-promoting protein, acts by binding to and inactivating the death-inhibiting members of the family, whereas Bax and Bak stimulate cytochrome *c* release. Inhibitor of apoptosis proteins (IAP) can inhibit apoptosis by binding to some pro-caspases, thereby preventing their activation, and by binding to caspases, inhibiting their activity. When mitochondria release cytochrome *c*, they also release a protein that blocks IAPs to enhance proapoptotic activity.

A number of extracellular factors can influence apoptotic pathways by modulating the activities of Bcl-2 and IAP family members. In addition, other trophic factors, including nerve growth factor, can activate PI-3 kinase pathways, which through a downstream kinase called Akt can phosphorylate Bad. In other cell types, different trophic factors may promote cell survival by posttranslational modification of other components of the cell-death machinery (see pathways described in ➋ *Section 2.7*).

3.6.2 Axonal Degeneration

Neurons also appear to have programs molecularly distinct from apoptosis for the destruction of their axons. Wallerian degeneration, both in the PNS as well as the CNS, occurs in part of the axon that is severed from the cell body, thereby allowing a neuron to eliminate a part of itself while keeping the remainder intact. This part of the axon disassembles in a characteristic and orderly fashion. Events include breakdown of endoplasmic reticulum, mitochondrial swelling, and neurofilament degradation. The axon breaks up into fragments that are phagocytosed (Griffin et al., 1996). In neurodegenerative diseases, a somewhat similar process called dying back occurs. In this case, an unhealthy axon slowly (over weeks or months) and progressively degenerates from the distal end toward the cell body (Schaumburg et al., 1974; Azzouz et al., 1997; Iseki et al., 2001). Axonal degeneration also occurs during normal development. The axonal branches reaching inappropriate regions of the CNS are lost by branch elimination, perhaps because of lack of appropriate trophic signals (O'Leary and Koester, 1993; Sharma and Johnson, 2000; Sharma and Ehinger, 2003).

Although the mechanisms of Wallerian degeneration are not fully understood, it is at least known that the breakdown of neurofilaments depends on Ca^{2+} influx and the activation of the Ca^{2+}-dependent protease calpain (George et al., 1995). The discovery of a spontaneous mutation in Wallerian degeneration slow (Wlds) mice suggests that the degeneration is not passive but rather an active process (Lunn et al., 1989). In these seemingly normal mice, Wallerian degeneration in the PNS as well as the CNS is greatly slowed. The distal portion of a transected Wlds axon can conduct action potentials for weeks (Perry et al., 1990). The Wlds mutation is dominant and has been mapped to the distal end of chromosome 4 (Lyon et al., 1993), where there is an 85-kb tandem triplication that results in the production of an abnormal fusion protein (Conforti et al., 2000). This fusion protein contains the intact enzyme nicotinamide mononucleotide adenylyl transferase (NMNAT), which functions in the synthetic pathway for nicotinamide adenine dinucleotide (NAD+), as well as the ubiquitination factor E4b (UbE4b).

There is evidence that both Wallerian degeneration and axonal degeneration occur by mechanisms distinct from apoptosis. The expression of bcl-2 transgene in mouse neurons blocks the axotomy-induced apoptosis of developing retinal ganglion cells, but not the Wallerian degeneration of their axons (Dubois-Dauphin et al., 1994). In addition, Wallerian degeneration and axonal degeneration apparently do not depend on caspases (Finn et al., 2000). Wlds neurons have normal apoptosis capabilities (Deckwerth and Johnson Jr, 1994).

Dying-back axonal degeneration could well be a response to a wide variety of neuronal insults. Possibly, the magnitude of an insult determines whether a neuron undergoes caspase-dependent apoptosis or caspase-independent axonal destruction to conserve energy. The importance of understanding this program is that once the cell bodies, ganglion cells in glaucoma for example, start to undergo apoptosis it is too late to be of clinical importance. Degeneration of the neuron's axons often precedes the death of the cell body, and intervention made at this stage could be clinically important.

2.7 Preconditioning

Tolerance or preconditioning is an adaptive process by which noncytotoxic exposure renders the tissue resistant to lethal doses of a neurotoxic agent. This phenomenon, first described in the heart, has attracted attention in the brain and, more recently, in the retina. There is evidence that ischemic preconditioning in the retina occurs by mechanisms similar to those that occur in other parts of the CNS. This involves adenosine and ATP-sensitive K^+ channels (Li and Roth, 1999; Sakamoto et al., 2001). Similar protection has been described for low-level exposure to oxidative stress. When RPE cells are preconditioned with sublethal doses of hydrogen peroxide, they show remarkable resistance to subsequent lethal exposures. Protective pathways remain active for some time even after the preconditioning stimulus is withdrawn. Interestingly, nonlethal exposure also affects other cellular functions, implying a role of redox signaling (Sharma and Ehinger, 2003). Redox signaling may play a significant role in the pathogenesis of AMD (➋ *Figure 3-3*).

Although very little is known about the endogenous protective mechanisms in RPE cells at present, information from other systems suggests that phosphoinositide 3-kinase (PI-3-K) pathways and the NFκB transcriptional factor might be important. PI-3-Ks are lipid kinases that generate PtdIns3P, 4P2, and PtdIns3, 4, 5P3 (Wymann and Pirola, 1998). These lipids act as second messengers that bind to Akt (protein kinase B; PKB) and protein kinase 1. PI-3-K activation plays a role in survival signaling in a number of cell types (Cantley, 2002), but its role in RPE is not well characterized. There are several known downstream effectors of PI-3-K. These include Rac, p70s6k, certain isoforms of PKC, and, most relevant to cell survival, Akt/PKB (Wymann and Pirola, 1998; Cantley, 2002). Akt has been shown to be responsible for PI-3-K-dependent cell survival (Cantley, 2002; Vivanco and Sawyers, 2002).

However, not all survival-promoting signals act via Akt. For example, cAMP-mediated neuronal survival occurs in an Akt-independent manner (Li et al., 2000). The mechanisms of Akt-induced survival are only now falling in place. Akt phosphorylates the proapoptotic Bcl-2 family member, Bad. Phosphorylation of Bad results in decreased formation of the Bad-Bcl-XL heterodimer. This allows free Bcl-xL to protect the cell from apoptosis (Cantley, 2002; Vivanco and Sawyers, 2002). Although this appears to be an important mechanism for PI-3-K-induced cell survival, alternate mechanisms might exist. Bad is expressed in RPE cells (Mukherjee et al., 2004). In certain cell types where Bad is not expressed, PI-3-K/Akt still prevents apoptosis. Possible alternate mechanisms by which PI-3-K may promote cell survival are as follows: (1) Akt also phosphorylates the Forkhead-related transcription factor (FKHR) in the cytoplasm.

Retention of FKHR in the cytoplasm inhibits its transcriptional gene targets, which include proapoptotic proteins such as BIM and Fas ligand. (2) Akt may indirectly increase the function of the NFκB transcription factor complex, which has prosurvival functions. RPE cells are resistant to TNF-α-induced cell death, even after NFκB activation is specifically blocked (Yang et al., 2004). This suggests that responses of RPE cell to survival/death signal differ considerably from that typically seen in other cell types. (3) Akt may increase p53 degradation through phosphorylation of MDM2 (Cantley, 2002; Vivanco and Sawyers, 2002).

2.8 Pathophysiology of Retinal Degeneration

2.8.1 Pathophysiology of Hereditary Retinal Degeneration

Retinitis pigmentosa is a set of genetic diseases leading to progressive photoreceptor degeneration and thus severe visual impairment. One out of three to four thousand people is affected by the disease (Ammann et al., 1965; Boughman et al., 1980; Hu, 1982; Jay 1982; Bunker et al., 1984; Haim et al., 1992; Rosner et al., 1993). It is estimated that every human carries at least one recessive gene predisposing the offspring to hereditary disease, and one in 50 people carries a gene for recessive retinitis pigmentosa (Boughman et al., 1980). Some experimental treatments are being tried, but no effective treatment is available (Sharma et al., 1995; Sharma and Ehinger, 1997a, b; Sharma, 2000a).

A large number of mutations on a large number of genes for retinal degeneration have been identified. This list is continuously growing, and an updated catalog is maintained at RetNet (http://www.sph.uth.tmc.edu/retnet/). Despite the success in identifying the responsible genes, the mechanisms by which these mutations lead to photoreceptor death are not well understood (Pierce, 2001; Pacione et al., 2003). Several major mechanisms can be identified. These include (1) disruption of photoreceptor outer segment morphogenesis, (2) metabolic overload, (3) dysfunction of retinal pigment epithelial cells, and (4) chronic activation of phototransduction (Pierce, 2001).

Hereditary disorders are caused by defects on the genetic coding, which in turn give rise to defective proteins. If the protein concerned is active only in one type of cell, the primary effect will be confined to that cell type, although other cells may be affected secondarily. On the other hand, if the gene product is responsible for some general metabolic function, the defect will cause a more generalized disorder. In the heterozygous state of dominant disorders, the effect may be due to the abnormal gene product and/or a deficient product from the normal gene. In X-linked retinal degenerations (XLRD), the random inactivation of the X chromosome in the second or third week of embryonic development (Lyon, 1961, 1962) can influence the course of the disease. The inactivation in general can involve the affected or the normal chromosome. The severity of the manifestations in the carriers can depend upon the number of normal and abnormal X-chromosomes that are inactivated.

2.8.1.1 Heterogeneity in Retinitis Pigmentosa It is becoming increasingly clear that different mutations may present themselves with the same phenotype. This can be viewed as the limited response of the retina to a variety of genetic insults. In another pattern, different mutations in the same gene may have different clinical manifestations. In yet other cases, the same mutation may be present as clinically distinct diseases in different individuals. There are many possible genetic mechanisms that can explain this. (1) The expression of one gene can be influenced by another gene in what is called a compound heterozygous state, where a phenotypic presentation of a mutation on one gene is affected by a different mutation on the allelic gene. For example, common genetic variation can provide protection in sickle cell disease in which there is a hereditary persistence of high fetal hemoglobin concentrations (Conley et al., 1963; Herman Jr and Conley, 1960). (2) Two different mutations may produce an apparently dominant disease with incomplete penetrance or a recessive disease (Kajiwara et al., 1994). (3) The inheritance of the same mutation from father or mother may produce different diseases (Nicholls et al., 1989; Magenis et al., 1990) or phenotypic characteristics by the mechanism of genomic imprinting (Moore and Haig, 1991; Willison, 1991). (4) In some inherited disorders (such as myotonic dystrophy), the disease becomes more severe over generations because a repeating nucleotide sequence becomes longer by repeated insertion (Hunter et al., 1992).

2.8.1.2 Possible Mechanisms in Known Genetic Defects in Retinal Degeneration Interestingly, mutations leading to human photoreceptor degeneration are not unique to humans. Many of them cause retinal degeneration in a variety of animals ranging from fruit fly (*Drosophila melanogaster*) to mammals. Much of what we know about the pathobiology of retinal degenerations has been learned from animal models.

Rhodopsin. It is estimated that about 25–30% of patients with autosomal dominant retinitis pigmentosa (adRP) carry a mutation in the rhodopsin gene (Dryja et al., 1990a; Sung et al., 1991a). A large number of mutations in the rhodopsin gene result in adRP (Dryja et al., 1991; Farrar et al., 1991; Gal et al., 1991; Inglehearn et al., 1991; Sung et al., 1991b; Chelva et al., 1992; Inglehearn et al., 1992). Pro23His was originally reported as the most common of all the rhodopsin mutations, accounting for about 60% of all rhodopsin mutations and 12–15% of all adRP families in the United States (Dryja et al., 1990b; Sung et al., 1991b). This mutation has not been found in Europe, and it is believed that all patients carrying this mutation might have emerged from one source (Dryja et al., 1990a).

Rhodopsin is a glycoprotein composed of 348 amino acid residues in a single polypeptide chain. It is a membrane protein with seven transmembrane regions, present in rod cells of the retina. The protein opsin carries 11-*cis*-retinal, covalently attached to residue Lys296, which is responsible for absorbing the photons as the first step of the visual transduction cascade. Mutations causing retinal degenerations have been found in the transmembrane, cytoplasmic, and intradiscal regions of the protein. Many of them affect amino acids that are located at important functional or structural sites, such as those for glycosylation (Dryja et al., 1991; Bunge et al., 1993), phosphorylation, attachment of 11-*cis*-retinal, transducin binding, or a disulfide bridge. It seems that there are differences in the mechanisms by which different mutations result in disease.

Generally, the way cells handle abnormal protein determines the effect of a mutation. Depending on the effect of a mutation on protein folding (Wright et al., 1987), the abnormal protein may or may not pass from the rough endoplasmic reticulum to the Golgi apparatus. The proteins that do not pass may accumulate in the endoplasmic reticulum or may be destroyed (Carlson et al., 1988; Bonifacino et al., 1990). The accumulation of the protein in one part of the cell and its failure to reach an appropriate target may then be detrimental to the cell, perhaps because of the resulting physical instability of cell membranes (Karnik et al., 1988; Karnik and Khorana, 1990). There is evidence, at least in some cases, of defective protein transport in the connecting cilium (Liu et al., 1997). Another possible explanation for rhodopsin malfunction is that mutations at the binding site for 11-*cis*-retinal result in failure to form complexes (Robinson et al., 1992). This prevents proper dark adaption. A third possibility is that mutations in the C-terminal region affect transduction efficiency, resulting in reduced signals and thus reduced sensitivity.

There has been an effort to classify proteins produced by rhodopsin mutations. They have been classified into two categories by in vitro expression studies (Sung et al., 1991a). Class I mutant proteins resemble wild-type rhodopsin in expression yield, capacity for regeneration with 11-*cis*-retinal, and plasma membrane localization. Class II mutants, which regenerate with 11-*cis* retinal variably, or not at all, are inefficiently transported to the plasma membrane and accumulate rhodopsin at the cell surface to a lesser extent. The Pro347Leu substitution, causing diffuse retinitis pigmentosa, has been designated as a type I mutation, and the mutations Pro23His, Thr58Arg, and Gly106Arg, causing regional retinitis pigmentosa and slow recovery from light adaptation, have been designated type II mutations.

Peripherin/rds. Peripherin/rds is a photoreceptor cell-specific transmembrane glycoprotein localized to the outer segment disc membrane of both rods and cones. The name *peripherin* was also given to a protein originating in peripheral nerves but without any relation to the photoreceptor protein, which is therefore, now called peripherin/rds. A defect in this gene causes photoreceptor degeneration in rds mice (Travis et al., 1989; Travis et al., 1991).

Because the protein is expressed in both rods and cones, it is not surprising to see that the phenotype of a peripherin/rds gene mutation can fluctuate between rod and cone dystrophies. A functionally important noncovalent link between peripherin/rds and the ROM1 protein (which is found exclusively in rods; Bascom et al., 1992) suggests a different structural association of this protein in rods and in cones. A mutation at the binding site of ROM1 in rods and an equivalent protein in cones may be what determines whether the damage appears in rods or cones. This defect is very common in human photoreceptor degeneration and causes over 40 types of human retinal degeneration (Keen and Inglehearn, 1996) ranging from retinitis pigmentosa to progressive macular degenerations (Wells et al., 1993).

cGMP phosphodiesterase. The gene encoding the β-subunit of rod cGMP phosphodiesterase, like the rhodopsin gene, encodes a protein involved in the phototransduction cascade. Its homolog causes photoreceptor degeneration in rd mice and Irish setter dogs with rod/cone dysplasia (Bowes et al., 1990; Pittler and Baehr, 1991; Suber et al., 1993).

ABCA4 gene. ABCA4 gene mutation(s) cause Stargardt disease (Azarian and Travis, 1997) (➋ *Section 3.8.7.5*). The ABCA4 knockout mouse (Weng et al., 1999) exhibits electrophysiological and morphological characteristics of retinal degeneration. Deposition of lipofuscin fluorophore A2E in RPE cells perhaps makes them dysfunctional, causing secondary photoreceptor degeneration. Studies have also shown that abcr–/– mutant mice raised in darkness have diminished accumulation of A2E, suggesting the role of light in certain forms of retinal degenerations (Mata et al., 2000). Evidence suggests that the ABCA4 protein is a specific target for photoxidative damage and that compromised ABCR activity could be involved in light-induced damage (Sun and Nathans, 2001).

RPGR mutation. A murine model for X-linked retinitis pigmentosa (RP3) has been created using gene knockout techniques (Hong et al., 2000). These mice are RPGR-deficient and exhibit rod and cone degeneration. The RPGR mutation causes disruption of the unidirectional movement of opsin in the photoreceptor cell. This destabilizes the cells and causes photoreceptor death.

MERTK gene. Defects in the gene MERTK abolish the ingestion phase of photoreceptor outer segment phagocytosis by RPE and eventually cause progressive photoreceptor degeneration in the RCS rat (D'Cruz et al., 2000; Sharma et al., 2001; 2001a, b). Mutations in the human ortholog to this gene cause hereditary retinal degeneration in humans (Gal et al., 2000).

Guanylate cyclase gene. The ability of photoreceptors to transduce light into nerve signals depends on the regulation of cGMP and calcium levels within the outer segments. CGMP levels in the outer segments change after the sequential activation of cGMP phosphodiesterase and guanylate cyclase (GC) during a light signal event (Ridge et al., 2003). Mutations in regions of the GC1 gene that encode the extracellular, dimerization, and catalytic domains of the enzyme cause Leber congenital amaurosis (LCA)-1 (Rozet et al., 2001). Patients with LCA have severely diminished sight or blindness at birth. These mutations reduce or abolish guanylate cyclase activity. GC1 gene mutations in the dimerization domain (Duda et al., 1999; Tucker et al., 1999) and mutations affecting GCAP1 ability to bind to calcium and regulate GC activity (Payne et al., 1998; Wilkie et al., 2000) have been linked to cone-rod dystrophies. It appears that GC1-GCAP1 interactions play a central role in phototransduction and cone survival.

2.8.2 Choroideremia

The gene responsible for choroideremia is expressed in various tissues including retina, choroid, and/or RPE. The gene product, Rab escort protein-1 (REP-1), is a component of the enzyme complex, Rab-geranylgeranyl transferase (Rab-GGtase), which is responsible for catalyzing the attachment of geranylgeranyl groups to Rab proteins. These are believed to control various membrane vesicle transport mechanisms by regulating the fusion events that underlie endocytosis and exocytosis (Pfeffer, 1992).

2.8.3 Hereditary Retinal Degenerations with Known Metabolic Causes

Many hereditary metabolic diseases are associated with retinal degeneration. In these syndromes, the metabolic defect has been identified and in many instances useful therapies have been implemented (Sharma and Ehinger, 1999).

2.8.3.1 Refsum's Disease In a specific form of autosomal recessive retinitis pigmentosa, the inborn error of metabolism results in the accumulation of exogenous phytanic acid in the body, most significantly from dairy products and fish fat (Refsum, 1946). These patients develop peripheral neuropathy, ataxia, and retinitis pigmentosa. The patients have raised serum phytanic acid levels. The biochemical defect has been identified as the inability to convert phytanic acid to α-hydroxy phytanic acid (Eldjarn et al., 1966) because of the absence or deficiency of a mitochondrial enzyme, phytanic acid-α-hydroxylase. Phytanic acid accumulates in the RPE (Toussaint and Danis, 1971) and in a variety of other tissues. Since phytanic acid comes exclusively from exogenous sources, reduction in intake of food rich in phytanic acid (animal fat and milk products) and phytol (green leafy vegetables) helps in controlling serum phytanic acid levels (Stokke et al., 1986).

2.8.3.2 Gyrate Atrophy In gyrate atrophy of the choroid and retina, mutations are found in the ornithine-delta-aminotransferase (OAT) gene situated on chromosome 10 (O'Donnell et al., 1988). Many point mutations (Mitchell et al., 1988; Ramesh et al., 1988; Inana et al., 1989; Mitchell et al., 1989; Brody et al., 1992) and nine base pair deletions (McClatchey et al., 1990) have been described in this gene. As a result of the mutations, OAT is unable to convert ornithine and α-ketoglutarate properly to pyrroline-5-carboxylate and glutamate (Weleber et al., 1981), resulting in a 10- to 20-fold elevation of plasma ornithine levels (Takki and Simell, 1974; Sipilä et al., 1981). The absence of OAT products, primarily lysine or creatine, may also play a role in pathogenesis. Supplementing creatine results in marked improvement of muscular abnormalities (Sipilä et al., 1981; Vannas-Sulonen et al., 1985). Plasma ornithine levels could be controlled by facilitating its excretion by giving α-amino-isobutyric acid (Behrens Baumann et al., 1982).

2.8.3.3 Kearns-Sayre Syndrome In the Kearns-Sayre syndrome, retinitis pigmentosa is associated with ophthalmoplegia presenting as ptosis or chronic progressive ophthalmoplegia. There are also respiratory distress and heart conduction blocks (Eagle Jr et al., 1982; Shoffner et al., 1989). The syndrome is associated with a mitochondrial DNA mutation. The respiratory distress seen in this syndrome may benefit from treatment with coenzyme Q10, but its efficacy in treating the retinal manifestations remains to be established (Shoffner et al., 1989).

2.8.3.4 Sorsby's Fundus Dystrophy This is a rare autosomal dominant disorder primarily affecting the macula. It is caused by mutations in the gene for the tissue inhibitor of metalloproteinases-3 (TIMP-3). Vitamin A has been shown to have beneficial effects in Sorsby's fundus dystrophy (Jacobson et al., 1995).

2.8.3.5 Abetalipoproteinemia Abetalipoproteinemia (Bassen-Kornzweig disease) is a rare autosomal recessive disorder in which retinitis pigmentosa develops in the third decade of life. The disease is caused by a failure in the synthesis of β-apoprotein or in the intracellular assembly of β-apoprotein with lipid (β-apolipoprotein). This makes the transmembrane transport of fat-soluble vitamins ineffective and hence they are not absorbed in the gut properly. Consequently, patients with this disorder have hypovitaminosis A and E, resulting in photoreceptor degeneration. High doses of vitamins A and E and essential fatty acids prevent or stabilize the ocular and neurological complications of the disease (Gouras et al., 1971; Bieri et al., 1984).

2.8.4 Light Damage to Photoreceptor Cells

Light damage to the photoreceptor cell is well known (Noell et al., 1966). Light-induced photoreceptor apoptosis requires a functional rhodopsin, suggesting that phototransduction cascade is involved (Naash et al., 1996). Induction HO-1, a specific heat shock protein, suggests that light injury causes oxidative damage to the retina (Kutty et al., 1995). The transcriptional factor AP-1 also plays an important role in light-induced cell death (Suter et al., 2000).

2.8.5 Apoptosis in Hereditary Retinal Degeneration

In retinitis pigmentosa, mutations in many different genes ultimately lead to apoptotic death of photoreceptors. It seems photoreceptor cell death is a stereotypical response of these cells in response to a wide variety of insults. However, there are many indications that the photoreceptor death in retinal dystrophies is not always a direct result of a mutation (Huang et al., 1993). The death of cones resulting from mutations in rhodopsin as well as the death of photoreceptors in RCS rats, where the primary defect lies in pigment epithelium, (Mullen and LaVail, 1976) are some examples.

2.8.6 Pathophysiology of Aging Retina

In CNS, it is estimated that up to half of the neurons may be lost during a person's life span (Weale, 1975; Henderson et al., 1980). Age-related loss of neurons occurs in retina also. The sensitivity of the peripheral field of vision declines more rapidly with aging than that of the central field. Rods are more vulnerable than cones, affecting scotopic vision more than photopic vision (Curcio and Allen, 1990; Gao and Hollyfield, 1992). There are age-related functional changes in photoreceptors, especially the rods. There is a delayed dark adaptation in the elderly and in AMD patients (Jackson and Owsley, 2000). An estimated half of all rods is lost between the second and fourth decades (Gao and Hollyfield, 1992). The density of rods in the central retina continues to decline throughout life. Numbers of cones in the macular area are relatively stable (Curcio and Allen, 1990), but some studies suggest that the foveal cones decrease in number with age as well (Farber et al., 1985; Curcio et al., 1987). Lipofuscin accumulates in the inner segments of cones (Tucker, 1986). The photoreceptor outer segments become convoluted with age, perhaps due to dysfunctional aging RPE cells that are unable to phagocytize them properly. Photoreceptors may be vulnerable because they are subject to oxidative damage and they are exposed to light, including ultraviolet light (Winkler et al., 1999). This can damage mitochondrial DNA (mtDNA). Progressive accumulation of mtDNA mutations with age, particularly in the foveal region, can induce apoptosis.

The number of ganglion cells also decreases during aging throughout the retina (Gao and Hollyfield, 1992). The nerve fibers in the optic nerve are replaced by connective tissue (Dolman et al., 1980). Hyalin bodies (corpora amylacea) appear in the peripapillary nerve fiber layer, optic nerve head, and the optic nerve (Avendano et al., 1980). Ganglion cells and bipolar cells accumulate lipids.

There are changes indicative of aging in the retinal glial cells also. The Müller cells and the internal limiting membrane become thickened (Gartner, 1970). Astrocytes display higher levels of glial fibrillary acidic protein and more cytoplasmic organelles (Ramirez et al., 2001).

In addition to their normally high metabolic rate, RPE cells also have to remove the outer segments of photoreceptor cells. When unable to cope, aging RPE cells start to accumulate incomplete digestion products of photoreceptor outer segments, which are rich in polyunsaturated fatty acids and vitamin A. The phagocytosed material is not entirely degraded within the RPE lysosome. This results in the accumulation of lipofuscin. The name *lipofuscin* is used for a heterogeneous group of complex, autofluorescent lipid/protein aggregates present in a variety of tissues, both neuronal and nonneuronal (Boulton, 1991). Lipofuscin is toxic and a source of ROS (Marshall, 1987; Eldred and Lasky, 1993; Wassell et al., 1998). Lipofuscin is light sensitive, and when exposed, it generates superoxide ions, singlet oxygen, hydrogen peroxide, and lipid peroxides (Wassell et al., 1999). The lipofuscin fluorophore, A2E, mediates the blue light-induced apoptosis of RPE cells by inhibiting the degradative capacity of lysosomes, by disrupting membrane integrity and by virtue of its phototoxicity (Schutt et al., 2000; Sparrow et al., 2000). There are also suggestions that lipofuscin is an inert substance and that the damage it causes is by virtue of cytoplasmic congesting, as it can occupy a considerable volume of cytoplasm in certain tissues in elderly people (Feeney-Burns et al., 1984).

The waste products are partially deposited on the Bruch's membrane (Young, 1987) in the form of drusen. The accumulation of lipofuscin in RPE cells appears detrimental to its function (Flood et al., 1984) and causes photoreceptor death (Dorey et al., 1989). With age, the number of RPE cells decreases in the central retina, and they become pleomorphic (Dorey et al., 1989). Other changes are also frequent. They include atrophy, depigmentation, hyperplasia, hypertrophy, and cell migration. The melanin concentration in the RPE cells decreases with age, especially in Caucasians, but also in blacks (Feeney-Burns et al., 1984). The melanin granules are slowly (over decades) digested by lysosomes (Burns and Feeney-Burns 1980).

Bruch's membrane deposits start to appear in as early as childhood (Burns and Feeney-Burns, 1980). At this age, only occasional deposits are seen in the inner collagenous zone. Deposits become more frequent with increasing age (Feeney-Burns and Ellersieck, 1985), causing thickening, hyalinization, and basophilia in the Bruch's membrane (Killingsworth, 1987). The lipid contents of the Bruch's membrane increase throughout life (Pauleikhoff et al., 1990). Alterations in the Bruch's membrane are comparable to aging-related changes in arterial intima. The resulting decline in the hydraulic conductivity of the Bruch's membrane obstructs diffusion between the RPE and choroidal vessels (Curcio et al., 2000). This change is implicated in AMD. Dysfunctional RPE cells shed their cytoplasm on the Bruch's membrane; from there it is probably removed by the macrophages and choriocapillary pericytes. If the removal is not efficient, the deposits accumulate. These changes are most marked in the macular and peripapillar region, and account for the angioid streaks seen in aged eyes. Later, perhaps due to altered filtration across the thickened Bruch's membrane, deposits called basal laminar deposits also appear between the plasma membrane and the basal membrane, especially in the vicinity of drusen. These deposits and macular drusen are uncommon in Asian eyes (Hoshino et al., 1984).

Proteasomal pathways degrade the oxidized and misfolded proteins. The partial loss of proteasomal expression and function has been described in the retinae of elderly rats (Louie et al., 2002). Oxidative and metabolic stress can have a cumulative effect. Accumulation of oxidatively damaged molecules can lead to the dysfunction of various metabolic and signaling pathways.

Newly available microarray techniques have begun to unveil the global changes in gene expression that accompany aging in retinal cells. Early results show interesting trends. For example, several genes involved in cell growth and protein processing are preferentially expressed in the retinae of younger people when compared with genes involved in stress responses and energy metabolism that are preferentially expressed in aging retinae (Yoshida et al., 2002). The data also indicate the possible involvement of IL-1, suggesting a role for inflammatory processes (Jenssen et al., 2001; Yoshida et al., 2002).

2.8.7 Pathophysiology of Age-Related Macular Degeneration

AMD is a major cause of blindness in the elderly population, affecting an estimated 30 to 35 million people worldwide. In the beginning, the vision loss is minimal, but as the disease progresses, vision loss can be catastrophic. Soft drusen and RPE abnormalities signal increased risk of vision loss (Bressler, 1993). Clinically, AMD is of two types: dry and wet. Some people go blind due to the dry form of AMD, but choroidal neovascularization (CNV), a major blinding complication in wet AMD that develops in 1 out of every 150–200 people above the age of 65 years, is a major cause of blindness.

Several lines of evidence suggest that oxidative stress, ischemia, and senescence play an important role in the pathogenesis of AMD. The association of cardiovascular risk factors with AMD suggests the role of oxidative stress and ischemia. These risk factors include smoking (Hyman et al., 1983; Christen et al., 1996), which is known to compromise the antioxidants and the choroidal blood flow, therefore possibly affecting the metabolism of RPE cells. The literature on the association of hypertension with AMD is inconsistent. Some studies show the association, but population-based studies do not (Klein et al., 1997). Hypertension could damage the choroidal vasculature and thus contribute to AMD. Environmental risk factors for AMD include exposure to ultraviolet radiation and antioxidant status. The racial and ethnic determinants of AMD are associated with pigmentation, suggesting a role of photooxidation. Prevalence is lower in some non-Caucasian racial groups. The wet form of AMD is rare in Barbadians of African descent (Schachat et al., 1995) and African Americans (Sommer et al., 1991).

Pathology in AMD involves the RPE, Bruch's membrane, the photoreceptors, and perhaps the choriocapillaris (Sarks, 1976; Green and Enger, 1993). AMD is a complex, multifactorial disease. The factors associated with the pathogenesis of the disease include smoking, family history, antioxidant status, cardiovascular disease, and genotype. As yet, the exact relationships between these factors is not clear, nor is it clear as to what is the primary site of injury.

2.8.7.1 Choriocapillaris The suggestion of vascular involvement in AMD comes from the association of cardiovascular risk factors in AMD. Changes in choriocapillaris are already apparent in eyes older than 50 years. Filling of the choriocapillaris with indocyanine green (ICG) dye is delayed in these eyes. Areas of hypofluorescence are noted in macula of AMD patients, especially when associated with CNV (Ross et al.,

1998). Measurement of choroidal blood flow by laser doppler in the macula also shows that the blood flow and blood volumes are reduced in old eyes and even more so in AMD patients. When RPE cells are removed from Bruch's membrane, the choriocapillaris degenerates, raising the question of whether the changes in choriocapillaris are primary or secondary to the changes in RPE (Burke and Soref, 1988). Finally, RPE cells produce substances that stimulate both the formation and the regression of CNV in animal models.

2.8.7.2 RPE/Bruch's Membrane Compromised RPE cell functions could seriously jeopardize photoreceptor health. RPE senescence has most widely been studied in the model of replicative senescence. It results from repeated division of the RPE cells in culture (Flood et al., 1980; Burke and Soref, 1988; Sheedlo et al., 1997). These studies have established a relationship between donor age and the location of the RPE cells (central versus peripheral) and replicative lifespan. There are also studies describing changes in gene expression or the alteration of enzymatic activities during the replicative senescence of RPE cells in vitro (Tombran-Tink et al., 1995).

Telomere length shortens in replicative senescence (Matsunaga et al., 1999). Senescent cells also lose their proliferative response to mitogens. The cell arrest in replicative senescence is distinct from the quiescent or G0 state observed in early passage cells achieved by serum deprivation or contact inhibition. The expression of the EPC-1 gene is detected in early passage, serum-deprived cells (Cristofalo and Pignolo, 1996). EPC-1 is G0-specific and inhibitory for cellular proliferation. The product of this gene, a 50 kD secreted protein, is similar to the pigmented epithelium derived factor (PEDF) produced by RPE cells. It also exhibits significant sequence similarity to serpins (a family of secreted mammalian serine protease inhibitors). Interestingly, the expression of EPC-1 is lost in senescent fibroblasts and perhaps in RPE cells as well. While the mechanisms of EPC-1 in senescence are still being worked out, an interesting question remains as to whether the loss of expression of EPC-1 prevents the senescent cells from entering G0. In addition, EPC-1 might have other biological roles such as tumor suppression and neuronal differentiation.

Another important RPE function that might be altered by senescence, or by other stresses, is lysosomal enzyme activity. RPE cells digest the continuously growing outer segments (Sharma and Ehinger, 2003). A decrease in lysosomal enzyme activity has been reported to accompany the aging process. Impairment of RPE lysosomal enzymatic activity could play an important role in the development of AMD (Berson, 1973; Yamada et al. 1990; Boulton et al., 1994). Cathepsin D (CatD) is an aspartic protease lysosomal enzyme involved in opsin proteolysis. CatD is present in various retinal cell types, especially the RPE cells (Yamada et al., 1990). Its impairment could play a role in AMD.

As already pointed out, deposits of waste products (presumably derived from RPE) on the Bruch's membrane were an early finding in cases of AMD. There is evidence that the transport across Bruch's membrane is impaired due to these deposits, resulting in more deposits (Feeney-Burns and Ellersieck, 1985). It is not clear if choriocapillary dysfunction and dropout is the cause, or the result, of these deposits (Bird, 1992). Dropout of choriocapillaris could cause ischemia in RPE, changing their gene expression and perhaps causing them to produce growth factors responsible for choroidal neovascularization.

2.8.7.3 Photoreceptors Photoreceptors are the final victims in the pathophysiology of AMD. The extent to which they are the perpetuators, however, is not clear. Photooxidative stress perhaps plays an important role. A high rate of respiration of photoreceptor inner segments, high content of PUFAs in the photoreceptor outer segments (which are phagocytized by RPE cells), and exposure to light make the environment of photoreceptors and the RPE cells rich in free radicals (Young, 1976; Ahmed et al., 1993). Several lines of evidence suggest the role of oxidative stress, including accumulation of lipofuscin, in retinal degeneration.

In addition to photoreceptors, the choriocapillaris is another possible target for oxidative damage that can contribute to the pathogenesis of AMD (Gottsch et al., 1990). It has also been proposed that photoactivation of hemoglobin precursors in red blood cells passing through the choriocapillaris may generate ROS, which may damage the RPE and Bruch's membrane.

Naturally occurring pigments in the macula limit retinal oxidative damage by absorbing incoming blue light and quenching ROS. Several risk factors for AMD are linked to the macular pigment. These include female gender and light iris color. Studies have also found association between high plasma levels of lutein and zeaxanthin with reduced risk of neovascular AMD.

2.8.7.4 Inflammatory Processes Analysis of drusen and deposits on the Bruch's membrane shed light on the metabolic changes in RPE in AMD. A number of drusen-associated molecules, including complement proteins, are important components in immune and/or inflammatory processes, suggesting the involvement of inflammation in the pathogenesis of AMD (Hageman and Mullins, 1999). Observations also suggest the involvement of monocytic cells on the choroidal side in the Bruch's membrane. Reactivity to specific CD antigens suggests that these are dendritic cells belonging to DC1 lineage. DC1 cells are antigen-presenting cells that are thought to participate in the induction of immunity.

Recently, Edward et al. (2005), Haines et al. (2005), and Klein et al. (2005) have identified a common variant in complement factor H (CFH) gene (on chromosome 1q31) that suggests a direct causal relationship between polymorphism in this gene and lifetime risk for AMD. Klein et al. (2005) used the haplotype structure of human chromosomes (HapMap data) in the CFH region to identify the specific variant, whereas Edwards et al. (2005) and Haines et al. (2005) developed haplotype data on the basis of their samples. In all cases, the data pointed to a specific CFH single-nucleotide polymorphism that results in the replacement of the amino-acid tyrosine with histidine at amino acid 402 (Tyr402His). Individuals that carry a single copy of the histidine allele in Tyr402His polymorphism have —two to fourfold, where as those who carry two copies have —five to sevenfold increased risk of AMD. A role for complement in AMD has long been suspected. The complement system is an essential component of immune response and this polymorphism is likely to affect complement activity. The complement system could be a potential target for modulatory drugs in the treatment of AMD.

2.8.7.5 Genetic Processes Evidence suggests familial association in AMD (Ferris III, 1983; Seddon et al., 1997). Several retinal diseases clinically resemble AMD. In the recent past, causative genes for some of these diseases have been identified, although none was linked with AMD. A better understanding of the pathophysiology of hereditary macular degenerations would throw additional light on the pathophysiology of AMD and perhaps lead to the discovery of its cure. Macular diseases for which genes have been identified are as follows:

RDS-associated pattern dystrophy. The RDS gene encodes a protein that interacts with the gene-product of ROM1 to stabilize the rhodopsin-bearing membranous discs in the outer segments. In rare cases, sequence variations in the RDS and ROM1 genes could be additive and result in retinitis pigmentosa through a true digenic mechanism (Dryja et al., 1996). It is also a good example of how multiple phenotypes can result from mutations in the same gene. Mutations in RDS gene are capable of causing a wide variety of retinal degenerations including maculopathies.

Best disease. The mutation lies in the bestrophin gene. The function of bestrophin, a VMD2 gene product, is not yet well understood. It is localized in the basolateral plasma membrane of the RPE (Kajiwara et al., 1991).

Stargardt disease. Stargardt disease is a common Mendelian maculopathy, occurring in approximately 1 in every 10,000 people. It is caused by mutation in a gene called ABCA4, which encodes for a photoreceptor protein (Allikmets, 1997). The function of this protein appears to be the translocation of *N*-retinylidene phosphatidylethanolamine, an intermediate of the visual cycle, from the intradiscal space to the cytoplasm (Sun et al., 1999; Weng et al., 1999). Failure of this translocation results in the formation of A2E, which can severely damage cell membranes (Eldred and Lasky, 1993; Schutt et al., 2000) and induce apoptosis (Schutt et al., 2000; Sparrow et al., 2000; Suter et al., 2000). This results in the loss of RPE and photoreceptor cells similar to what happens in AMD. It has been shown that for A2E to accumulate, light exposure is needed in mice lacking the ABCA4 gene; this suggests that control of light exposure might change the outcome of the disease (Mata et al., 2000). Mutations in the ABCA4 gene can produce a variety of phenotypes varying from Stargardt disease to retinitis pigmentosa. It is a highly polymorphic gene (Webster et al., 2001), and it is suggested that variations in the ABCA4 gene may also be responsible for some cases of AMD (Allikmets et al., 1997), but this hypothesis is not proven (Stone et al., 1998).

Sorsby fundus dystrophy. Sorsby fundus dystrophy progresses peripherally in contrast to central involvement in AMD. The associated gene encodes a tissue inhibitor of metalloproteinases (TIMP-3) (Weber et al., 1994). Most of the mutations that cause this disease alter the tertiary structure of the protein. The altered TIMP-3 then causes an accumulation of collagenous material beneath the RPE (Capon et al., 1989; Chong et al., 2000).

Malattia leventinese. The gene (EFEMP1) was identified in 1999 (Stone et al., 1999). A single mutation, Arg345Trp, is responsible for all cases of this phenotype in the world, probably originating from a common ancestor. This is likely to disrupt one of the six calcium-binding EGF-like domains of the molecule. The gene product is an extracellular matrix protein that is expressed in the brain, spleen, heart, kidney, and abundantly in the eye and lung.

Stargardt-like dominant macular dystrophy. The disease gene ELOVL4 was identified in 2001 (Zhang et al., 2001). This gene encodes a protein that is expressed in photoreceptors. ELOVL4 has significant structural and topological similarity to members of the ELO gene family that are involved in the elongation of fatty acids in yeast and mice.

2.8.7.6 Pathogenesis of Choroidal Neovascularization Choroidal neovascularization is the most sight-threatening sequela in AMD. Hypoxia and ischemia play an important role in its development (Shimizu et al., 1981). As discussed before, the blood flow to the macular region is compromised in aging and AMD (Grunwald et al., 1998; Ross et al., 1998). In addition, thickened Bruch's membrane can also obstruct nutrient flow, causing ischemia. Furthermore, smoking, a risk factor in AMD, could exacerbate vascular disease and ischemia. Smoking can also contribute to choroidal neovascularization by causing oxidative stress that can result in deposition of ECM along Bruch's membrane (Gottsch et al., 1993) and induce production of VEGF and FGF2 in RPE cells (Hackett et al., 1997). Exposure of cultured RPE cells to ECM molecules such as thrombospondin-1 increases VEGF and to a lesser extent FGF2 in cell supernatants (Mousa et al., 1999).

Generally, angiogenesis is a result of imbalance between pro and antiangiogenic factors (see ➋ *Section 2.1.2*). Proangiogenic factors affect several processes in endothelial cells, including cell proliferation, migration, and proteolytic activity (Sato and Rifkin, 1988). In addition to soluble factors, extracellular matrix molecules can have an impact on angiogenesis. Several mechanisms are possible. ECM molecules can sequester the soluble angiogenic factors and thus prevent them from exerting their action unless they are released by proteolysis (Park et al., 1993). ECM molecules can also bind to integrins and other cell surface receptors, thus affecting intracellular signaling (Dike and Ingber, 1996). The proangiogenic factors may also act by primarily altering the expression of integrins on retinal vascular endothelial cells (Friedlander et al., 1995). Proangiogenic factors affect the proteolytic activity of endothelial cells. Two distinct proteolytic systems have been implicated in the breakdown of ECM during angiogenesis, one involving the urokinase type of plasminogen activator (Pepper et al., 1987) and the other matrix metalloproteinases (MMPs) (Cornelius et al., 1995). The exact role of these two systems in CNV is not clear. In Sorsby's Fundus Dystrophy, a disease caused by mutation in TIMP-3 gene, some patients have deposits along Bruch's membrane and high incidences of CNV. The TIMP-3 gene product is involved in regulation of ECM turnover (Weber et al., 1994). All of these data strongly suggest that alteration of the ECM of RPE cells may play an important role in the development of CNV in patients with AMD. The molecular signals for angiogenesis could reside in the abnormal extracellular matrix or the RPE. Possibly, the abnormalities of the extracellular matrix promote a proangiogenic RPE phenotype.

2.8.8 Pathophysiology of Glaucoma

The retinal ganglion cells (their soma, dendrites, or axons) are the primary cell types involved in glaucomatous injury to the retina. Functional (psychophysical) testing (Sun and Nathans, 2001) and histological studies (Yucel et al., 2003) suggest that all the subsets of ganglion cells are affected. However, neuronal death in glaucoma is not limited to ganglion cells alone (Weber et al., 2000). Neurons in the lateral geniculate nucleus and the visual cortex are also lost (Yucel et al., 2003).

The pathobiology of neurodegeneration in glaucoma is only partially understood. Increased intraocular pressure is certainly an important risk factor; however, even pressures considered normal in some patients may cause glaucomatous damage in others (Quigley, 1993). Pathophysiological processes in glaucoma take place in two different sites. Firstly, there are processes responsible for intraocular pressure regulation. This involves changes at the level of aqueous outflow, especially the increased resistance in trabecular meshwork.

Secondly, there are processes that are involved in the retinal ganglion cell, dendrite, or axonal death. Several mechanisms have been identified. (1) With intraocular pressure above physiological levels, the pressure gradient across the lamina cribrosa increases. The lamina cribrosa remodels in response to increased intraocular pressure (Pena et al., 2001). However, if the remodeling is insufficient it causes deformation of, and mechanical stress to, the ganglion cell axons (Bellezza et al., 2003). Compression of optic nerve axons at the lamina cribrosa results in an impaired axonal protein transport (Fechtner and Weinreb, 1994; Quigley et al., 2000). Impairment of the flow of trophic factor along the axonal transport can cause ganglion cell death by trophic insufficiency (2) The retina has high metabolic needs and thus its survival depends upon the blood supplied to meet them. The outer retina draws its nourishment from choriocapillaris, but the inner retina depends on retinal vessels. Retinal ischemia-hypoxia, perhaps due to dysfunction of blood-flow autoregulation, can have deleterious effects on inner retinal neurons, especially ganglion cells (Siliprandi et al., 1992; Dreyer et al., 1996; Chew and Ritch, 1997; Lipton, 2001). Due to technical limitations, it has been difficult to unequivocally establish the role of ischemia in glaucoma. (3) Excitotoxicity due to excessive stimulation of the glutamatergic system, specifically the NMDA subtypes of glutamate receptors, has also been implicated in ganglion cell death in glaucoma (Dreyer et al., 1996; Lipton, 2003). Increased glutamate levels have been reported in the vitreous of glaucoma patients and experimental monkeys (Dreyer et al., 1996). Excessive accumulation of glutamate in the extracellular space leads to an influx of calcium, resulting in excitotoxic neuronal death (Choi, 1992; Lipton, 2001). Retinal ischemia, one of the proposed mechanisms, also releases glutamate (Louzada-Junior et al., 1992). However, the issue of excitotoxicity in glaucoma pathophysiology is far from settled. There is still debate on whether excess glutamate has protective or damaging effects on retinal ganglion cells. Also, different classes of cells could respond differently to glutamate depending upon their repertoire of glutamate receptors. Dysfunctional cellular pumps and glutamate transporters have also been implicated. (4) Oxidative stress and formation of free radicals have been implicated in retinal ganglion cell death in glaucoma. (5) Inflammatory cytokines (tumor necrosis factor and nitric oxide) have been implicated in the pathogenesis of glaucoma (Liu and Neufeld, 2001). It is possible that astrocytes, activated by raised intraocular pressure or by other mechanisms, alter the environment and produce a milieu that causes axonal degeneration or death of the retinal ganglion cells (Pena et al., 2001). (6) It has been suggested that aberrant autoimmunity can play a significant role in glaucoma (Schwartz, 2003a).

2.8.9 Pathophysiology of Diabetic Retinopathy

Diabetic retinopathy, a microangiopathy, is a leading cause of blindness in the West (Brinchmann-Hansen et al., 1992). It is characterized by retinal capillary microaneurysms, retinal exudate, macular edema, and, in advanced stages, neovascularization. The molecular pathophysiology of diabetic retinopathy is not fully elucidated. Hyperglycemia plays a pivotal role in the development and progression of the disease, but the exact mechanisms are not known. Several mechanisms have been proposed. (1) Progressive diabetic retinopathy is marked by thrombosis of the microaneurysms that leads to ischemia. Retinal ischemia plays a very significant role in the pathogenesis of diabetic retinopathy. Increasing ischemia is accompanied by the release of VEGF. In diabetic retinopathy, expression of other angiogenic factors also increases. These include insulin-like growth factor, placenta growth factor, platelet-derived growth factor, and fibroblast growth factor-2 (Robbins et al., 1994; Spirin et al., 1999; Burgos et al., 2000). Release of these growth factors leads to abnormal vasculature and increased permeability. The role of ischemia in neovascularization has already been described in ➋ *Section 2.1.2.* (2) Oxidative stress levels increase in diabetic retinae due to increased generation of free radicals and/or a compromised antioxidant defense system. Antioxidants can inhibit retinal abnormalities in animal models, suggesting the role of oxidative stress. (3) Ischemia can also release glutamate in the extracellular space and cause excitotoxicity (Louzada-Junior et al., 1992). In diabetic retinopathy, glutamate levels in vitreous and retina are elevated (Ambati et al., 1997; Kowluru et al., 2001). Glutamate levels in the retina are controlled by the enzyme glutamine synthetase in retinal Müller cells, but in experimental and human diabetes the expression of glutamine synthetase is decreased, further suggesting the role of glutamate in diabetic retinopathy

(Mizutani et al., 1998). The toxic effects of glutamate in the retina are well known (Lucas and Newhouse, 1957; Olney, 1969).

In addition to neovascularization, there is increased vascular permeability in diabetic retinopathy. Clinically, increased vascular permeability manifests as retinal exudates and diabetic maculopathy. In diabetic macular edema, there is accumulation of extracellular fluid in Henle's layer and the inner nuclear layer. The boundaries of macular edema are partly determined by the barrier properties of the inner and outer plexiform layers. The extracellular fluid in diabetic retinopathy comes from retinal blood vessels primarily due to breakdown of blood retinal barriers. Changes to retinal blood flow may also contribute. Polarized cells, such as vascular endothelial and RPE cells, develop effective barriers by forming tight junctions that are specialized regions of the junctional complex that comprises transmembrane proteins, peripheral membrane structural proteins, and a variety of associated regulatory proteins. Tight junctions contribute to the blood retinal barriers. These barriers fail in a variety of diseases, destabilizing the normal cellular environments and leading to organ failure. Breakdown of the blood retinal barriers is likely to involve the tight junction proteins such as occludin and ZO-1 (Harhaj et al., 2002). Increased levels of growth factors are often found in diseases exhibiting altered tissue permeability. This suggests that growth factors, particularly VEGF, play a role in increasing permeability. The vascular endothelial growth factor is produced in response to chronic hyperglycemia, but the mechanism by which growth factors increase permeability remains unclear.

3 Concluding Remarks

Our increased understanding of the molecular mechanisms behind degenerative conditions of retina has provided the essential basis for comprehending the pathophysiology of these important causes of blindness. However, much remains to be learned. Newer concepts are emerging. There are a number of experimental treatment strategies that have generated hope. Among them, attempts to reconstruct the degenerated retina with the help of retinal transplants (Sharma et al., 1997; Sharma and Ehinger, 1997b; Sharma, 2000b), and possibly with stem cells, are noteworthy. There are numerous other strategies that have shown promising results in the laboratory (Sharma and Ehinger, 1999); these include use of growth factors, gene transfer, and gene expression control. Neuroprotection is a potentially useful concept that could provide hope for millions with retinal degeneration, but manipulation of glutamate receptors to achieve this effect has not been successful so far. Antioxidants have had some limited success. Endogenous cytoprotection and autoimmunity are emerging fields that have potential for clinical application.

Acknowledgments

This work is dedicated to Vani. This work was supported by a generous grant by Gail and Richard Siegal, Gerwin Scholar Award, and Research to Prevent Blindness. The author thanks Dr. Peter A. Netland for his advice and support, Dr. Dianna Johnson and Ozlen Konu for valuable comments on the manuscript and Lee Thompson for help with graphics.

References

Abraham JA, Mergia A, Whang JL, Tumolo A, Friedman J, et al. 1986. Nucleotide sequence of a bovine clone encoding the angiogenic protein, basic fibroblast growth factor. Science 233: 545-548.

Adamis AP, Miller JW, Bernal MT, D'Amico DJ, Folkman J, et al. 1994. Increased vascular endothelial growth factor levels in the vitreous of eyes with proliferative diabetic retinopathy. Am J Ophthalmol 118: 445-450.

Aharoni R, Kayhan B, Eilam R, Sela M, Arnon R. 2003. Glatiramer acetate-specific T cells in the brain express T helper 2/3 cytokines and brain-derived neurotrophic factor in situ. Proc Natl Acad Sci USA 100: 14157-14162.

Ahmed J, Braun RD, Dunn R Jr, Linsenmeier RA. 1993. Oxygen distribution in the macaque retina. Invest Ophthalmol Vis Sci 34: 516-521.

Akiyama H, Nakazawa T, Shimura M, Tomita H, Tamai M. 2002. Presence of mitogen-activated protein kinase in retinal Muller cells and its neuroprotective effect ischemia-reperfusion injury. Neuroreport 13: 2103-2107.

Ali S, Mann DA. 2004. Signal transduction via the NF-kappaB pathway: A targeted treatment modality for infection, inflammation and repair. Cell Biochem Funct 22: 67-79.

Allikmets R. 1997. A photoreceptor cell-specific ATP-binding transporter gene (ABCR) is mutated in recessive Stargardt macular dystrophy. Nat Genet 17: 122.

Allikmets R, Shroyer NF, Singh N, Seddon JM, Lewis RA. 1997. Mutation of the Stargardt disease gene (ABCR) in age-related macular degeneration. Science 277: 1805-1807.

Ambati J, Chalam KV, Chawla DK, D'Angio CT, Guillet EG, et al. 1997. Elevated gamma-aminobutyric acid, glutamate, and vascular endothelial growth factor levels in the vitreous of patients with proliferative diabetic retinopathy. Arch Ophthalmol 115: 1161-1166.

Ames A III, Wright RL, Kowada M, Thurston JM, Majno G. 1968. Cerebral ischemia. II. The no-reflow phenomenon. Am J Pathol 52: 437-453.

Ammann F, Klein D, Franceschetti A. 1965. Genetic and epidemiological investigations on pigmentary degeneration of the retina and allied disorders in Switzerland. J Neurol Sci 2: 183-196.

Annunziato L, Amoroso S, Pannaccione A, Cataldi M, Pignataro G, et al. 2003. Apoptosis induced in neuronal cells by oxidative stress: Role played by caspases and intracellular calcium ions. Toxicol Lett 139: 125-133.

Anzai T, Tsuzuki K, Yamada N, Hayashi T, Iwakuma M, et al. 2003. Overexpression of Ca^{2+}-permeable AMPA receptor promotes delayed cell death of hippocampal CA1 neurons following transient forebrain ischemia. Neurosci Res 46: 41-51.

Avendano J, Rodrigues MM, Hackett JJ, Gaskins R. 1980. Corpora amylacea of the optic nerve and retina: A form of neuronal degeneration. Invest Ophthalmol Vis Sci 19: 550-555.

Azarian SM, Travis GH. 1997. The photoreceptor rim protein is an ABC transporter encoded by the gene for recessive Stargardt's disease (ABCR). FEBS Lett 409: 247-252.

Azzouz M, Leclerc N, Gurney M, Warter JM, Poindron P, et al. 1997. Progressive motor neuron impairment in an animal model of familial amyotrophic lateral sclerosis. Muscle Nerve 20: 45-51.

Babior BM, Takeuchi C, Ruedi J, Gutierrez A, Wentworth P Jr. 2003. Investigating antibody-catalyzed ozone generation by human neutrophils. Proc Natl Acad Sci USA 100: 3031-3034.

Banin E, Berenshtein E, Kitrossky N, Pe'er J, Chevion M. 2000. Gallium-desferrioxamine protects the cat retina against injury after ischemia and reperfusion. Free Radic Biol Med 28: 315-323.

Barbour B, Brew H, Attwell D. 1988. Electrogenic glutamate uptake in glial cells is activated by intracellular potassium. Nature 335: 433-435.

Barnett NL, Pow DV, Bull ND. 2001. Differential perturbation of neuronal and glial glutamate transport systems in retinal ischaemia. Neurochem Int 39: 291-299.

Bascom RA, Garcia Heras J, Hsieh CL, Gerhard DS, Jones C, et al. 1992. Localization of the photoreceptor gene ROM1 to human chromosome 11 and mouse chromosome 19: Sublocalization to human 11q13 between PGA and PYGM. Am J Hum Genet 51: 1028-1035.

Beal MF. 2003. Mitochondria, oxidative damage, and inflammation in Parkinson's disease. Ann NY Acad Sci 991: 120-131.

Beckman JS, Koppenol WH. 1996. Nitric oxide, superoxide, and peroxynitrite: The good, the bad, and ugly. Am J Physiol 271: C1424-C1437.

Behrens BW, Konig U, Schroder K, Hausmann I, Langenbeck U. 1982. Biochemical and therapeutic studies in a case of gyrate atrophy. Von Graefe's Arch Klin Exp Ophthalmol 218: 21-24.

Bellezza AJ, Rintalan CJ, Thompson HW, Downs JC, Hart RT, et al. 2003. Deformation of the lamina cribrosa and anterior scleral canal wall in early experimental glaucoma. Invest Ophthalmol Vis Sci 44: 623-637.

Berson EL. 1973. Retinitis pigmentosa, the electroretinogram, and Mendel's laws. Trans Pa Acad Ophthalmol Otolaryngol 26: 109-113.

Bezzi P, Carmignoto G, Pasti L, Vesce S, Rossi D, et al. 1998. Prostaglandins stimulate calcium-dependent glutamate release in astrocytes. Nature 391: 281-285.

Bieri JG, Hoeg JM, Schaefer EJ, Zech LA, Brewer HB Jr. 1984. Vitamin A and vitamin E replacement in abetalipoproteinemia. Ann Int Med 100: 238-239.

Bird AC. 1992. Bruch's membrane change with age. Br J Ophthalmol 76: 166-168.

Block F, Schwarz M. 1997. Effects of antioxidants on ischemic retinal dysfunction. Exp Eye Res 64: 559-564.

Block F, Schwarz M. 1998. The b-wave of the electroretinogram as an index of retinal ischemia. Gen Pharmacol 30: 281-287.

Bondy SC. 1995. The relation of oxidative stress and hyperexcitation to neurological disease. Proc Soc Exp Biol Med 208: 337-345.

Bonifacino JS, Suzuki CK, Klausner RD. 1990. A peptide sequence confers retention and rapid degradation in the endoplasmic reticulum. Science 247: 79-82.

Boughman JA, Conneally PM, Nance WE. 1980. Population genetic studies of retinitis pigmentosa. Am J Hum Genet 32: 223-235.

Boulton ME. 1991. Aging of the retinal pigment epithelium. Prog Retin Eye Res 11: 125-151.

Boulton M, Moriarty P, Jarvis-Evans J, Marcyniuk B. 1994. Regional variation and age-related changes of lysosomal enzymes in the human retinal pigment epithelium. Br J Ophthalmol 78: 125-129.

Bowes C, Li T, Danciger M, Baxter LC, Applebury ML, et al. 1990. Retinal degeneration in the rd mouse is caused by a defect in the beta subunit of rod cGMP-phosphodiesterase: See comments. Nature 347: 677-680.

Bowling AC, Mutisya EM, Walker LC, Price DL, Cork LC, et al. 1993. Age-dependent impairment of mitochondrial function in primate brain. J Neurochem 60: 1964-1967.

Brandstatter JH, Hartveit E, Sassoe-Pognetto M, Wassle H. 1994. Expression of NMDA and high-affinity kainate receptor subunit mRNAs in the adult rat retina. Eur J Neurosci 6: 1100-1112.

Bressler SB. 1993. Does wavelength matter when photocoagulating eyes with macular degeneration or diabetic retinopathy? Arch Ophthalmol 111: 177-180.

Brinchmann-Hansen O, Dahl-Jorgensen K, Sandvik L, Hanssen KF. 1992. Blood glucose concentrations and progression of diabetic retinopathy: The seven year results of the Oslo study. Br Med J 304: 19-22.

Brody LC, Mitchell GA, Obie C, Michaud J, Steel G, et al. 1992. Ornithine delta-aminotransferase mutations in gyrate atrophy. Allelic heterogeneity and functional consequences. J Biol Chem 267: 3302-3307.

Bruno V, Battaglia G, Copani A, D'Onofrio M, Di Iorio P, et al. 2001. Metabotropic glutamate receptor subtypes as targets for neuroprotective drugs. J Cereb Blood Flow Metab 21: 1013-1033.

Bunge S, Wedemann H, David D, Terwilliger DJ, van den Born LI, et al. 1993. Molecular analysis and genetic mapping of the rhodopsin gene in families with autosomal dominant retinitis pigmentosa. Genomics 17: 230-233.

Bunker CH, Berson EL, Bromley WC, Hayes RP, Roderick TH. 1984. Prevalence of retinitis pigmentosa in Maine. Am J Ophthalmol 97: 357-365.

Burgos R, Mateo C, Canton A, Hernandez C, Mesa J, et al. 2000. Vitreous levels of IGF-I, IGF binding protein 1, and IGF binding protein 3 in proliferative diabetic retinopathy: A case-control study. Diabetes Care 23: 80-83.

Burke SP, Nadler JV. 1989. Effects of glucose deficiency on glutamate/aspartate release and excitatory synaptic responses in the hippocampal CA1 area in vitro. Brain Res 500: 333-342.

Burke JM, Soref C. 1988. Topographical variation in growth in cultured bovine retinal pigment epithelium. Invest Ophthalmol Vis Sci 29: 1784-1788.

Burns RP, Feeney-Burns L. 1980. Clinico-morphologic correlations of drusen of Bruch's membrane. Trans Am Ophthalmol Soc 78: 206-225.

Cantley LC. 2002. The phosphoinositide 3-kinase pathway. Science 296: 1655-1657.

Capon MR, Marshall J, Krafft JI, Alexander RA, Hiscott PS, et al. 1989. Sorsby's fundus dystrophy. A light and electron microscopic study. Ophthalmology 96: 1769-1777.

Carlson JA, Rogers BB, Sifers RN, Hawkins HK, Finegold MJ, et al. 1988. Multiple tissues express alpha 1-antitrypsin in transgenic mice and man. J Clin Invest 82: 26-36.

Cassarino DS, Bennett JP Jr. 1999. An evaluation of the role of mitochondria in neurodegenerative diseases: Mitochondrial mutations and oxidative pathology, protective nuclear responses, and cell death in neurodegeneration. Brain Res Brain Res Rev 29: 1-25.

Casson RJ, Chidlow G, Wood JP, Osborne NN. 2004. The effect of hyperglycemia on experimental retinal ischemia. Arch Ophthalmol 122: 361-366.

Celebi S, Dilsiz N, Yilmaz T, Kukner AS. 2002. Effects of melatonin, vitamin E and octreotide on lipid peroxidation during ischemia-reperfusion in the guinea pig retina. Eur J Ophthalmol 12: 77-83.

Chelva E, McLaren TL, Kay SM, Collins DW, Black JL, et al. 1992. A retinitis pigmentosa register for western Australia. Aust NZ J Ophthalmol 20: 311-317.

Chen Q, Moulder K, Tenkova T, Hardy K, Olney JW, et al. 1999. Excitotoxic cell death dependent on inhibitory receptor activation. Exp Neurol 160: 215-225.

Cheon EW, Park CH, Kang SS, Cho GJ, Yoo JM, et al. 2002. Nitric oxide synthase expression in the transient ischemic rat retina: Neuroprotection of betaxolol. Neurosci Lett 330: 265-269.

Chew SJ, Ritch R. 1997. Neuroprotection: The next breakthrough in glaucoma? Proceedings of the third annual optic nerve rescue and restoration think tank. J Glaucoma 6: 263-266.

Choi DW. 1992. Excitotoxic cell death. J Neurobiol 23: 1261-1276.

Chong NH, Alexander RA, Gin T, Bird AC, Luthert PJ. 2000. TIMP-3, collagen, and elastin immunohistochemistry and histopathology of Sorsby's fundus dystrophy. Invest Ophthalmol Vis Sci 41: 898-902.

Christen WG, Glynn RJ, Manson JE, Ajani UA, Buring JE. 1996. A prospective study of cigarette smoking and risk of age-related macular degeneration in men. JAMA 276: 1147-1151.

Christine CW, Choi DW. 1990. Effect of zinc on NMDA receptor-mediated channel currents in cortical neurons. J Neurosci 10: 108-116.

Conforti L, Tarlton A, Mack TG, Mi W, Buckmaster EA, et al. 2000. A Ufd2/D4Cole1e chimeric protein and overexpression of Rbp7 in the slow Wallerian degeneration (WldS) mouse. Proc Natl Acad Sci USA 97: 11377-11382.

Conley CL, Weatherall DJ, Richardson SN, Shepard MK, Charache S. 1963. Hereditary persistence of fetal hemoglobin: A study of 79 affected persons in 15 Negro families in Baltimore. Blood 21: 261-281.

Connolly DT, Heuvelman DM, Nelson R, Olander JV, Eppley BL, et al. 1989. Tumor vascular permeability factor stimulates endothelial cell growth and angiogenesis. J Clin Invest 84: 1470-1478.

Cornelius LA, Nehring LC, Roby JD, Parks WC, Welgus HG. 1995. Human dermal microvascular endothelial cells produce matrix metalloproteinases in response to angiogenic factors and migration. J Invest Dermatol 105: 170-176.

Cristofalo VJ, Pignolo RJ. 1996. Molecular markers of senescence in fibroblast-like cultures. Exp Gerontol 31: 111-123.

Crosson CE, Willis JA, Potter DE. 1990. Effect of the calcium antagonist, nifedipine, on ischemic retinal dysfunction. J Ocul Pharmacol 6: 293-299.

Curcio CA, Allen KA. 1990. Topography of ganglion cells in human retina. J Comp Neurol 300: 5-25.

Curcio CA, Owsley C, Jackson GR. 2000. Spare the rods, save the cones in aging and age-related maculopathy. Invest Ophthalmol Vis Sci 41: 2015-2018.

Curcio CA, Sloan KR, Packer O Jr, Hendrickson AE, Kalina RE. 1987. Distribution of cones in human and monkey retina: Individual variability and radial asymmetry. Science 236: 579-582.

D'Cruz PM, Yasumura D, Weir J, Matthes MT, Abderrahim H, et al. 2000. Mutation of the receptor tyrosine kinase gene Mertk in the retinal dystrophic RCS rat. Hum Mol Genet 9: 645-651.

Danton GH, Dietrich WD. 2003. Inflammatory mechanisms after ischemia and stroke. J Neuropathol Exp Neurol 62: 127-136.

Dawson VL. 1995. Nitric oxide: Role in neurotoxicity. Clin Exp Pharmacol Physiol 22: 305-308.

Dawson VL, Dawson TM. 1995. Physiological and toxicological actions of nitric oxide in the central nervous system. Adv Pharmacol 34: 323-342.

Deckwerth TL, Johnson EM Jr. 1994. Neurites can remain viable after destruction of the neuronal soma by programmed cell death (apoptosis). Dev Biol 165: 63-72.

Dike LE, Ingber DE. 1996. Integrin-dependent induction of early growth response genes in capillary endothelial cells. J Cell Sci 109 (Pt 12): 2855-2863.

Dingledine R, Borges K, Bowie D, Traynelis SF. 1999. The glutamate receptor ion channels. Pharmacol Rev 51: 7-61.

Dolman CL, McCormick AQ, Drance SM. 1980. Aging of the optic nerve. Arch Ophthalmol 98: 2053-2058.

Dorey CK, Wu G, Ebenstein D, Garsd A, Weiter JJ. 1989. Cell loss in the aging retina. Relationship to lipofuscin accumulation and macular degeneration. Invest Ophthalmol Vis Sci 30: 1691-1699.

Drejer J, Benveniste H, Diemer NH, Schousboe A. 1985. Cellular origin of ischemia-induced glutamate release from brain tissue in vivo and in vitro. J Neurochem 45: 145-151.

Dreyer EB, Zurakowski D, Schumer RA, Podos SM, Lipton SA. 1996. Elevated glutamate levels in the vitreous body of humans and monkeys with glaucoma. Arch Ophthalmol 114: 299-305.

Droge W. 2002. Free radicals in the physiological control of cell function. Physiol Rev 82: 47-95.

Droge W. 2003. Oxidative stress and aging. Adv Exp Med Biol 543: 191-200.

Dryja TP, Hahn LB, Cowley GS, Mcgee TL, Berson EL. 1991. Mutation spectrum of the rhodopsin gene among patients with autosomal dominant retinitis pigmentosa. Proc Natl Acad Sci USA 88: 9370-9374.

Dryja TP, Hahn LB, Reboul T, Arnaud B. 1996. Missense mutation in the gene encoding the alpha subunit of rod transducin in the Nougaret form of congenital stationary night blindness. Nat Genet 13: 358-360.

Dryja TP, Mcgee TL, Hahn LB, Cowley GS, Olsson JE, et al. 1990a. Mutations within the rhodopsin gene in patients with autosomal dominant retinitis pigmentosa. N Engl J Med 323: 1302-1307.

Dryja TP, Mcgee TL, Reichel E, Hahn LB, Cowley GS, et al. 1990b. A point mutation of the rhodopsin gene in one form of retinitis pigmentosa. Nature 343: 364-366.

Dubois-Dauphin M, Frankowski H, Tsujimoto Y, Huarte J, Martinou JC. 1994. Neonatal motoneurons overexpressing the bcl-2 protooncogene in transgenic mice are protected from axotomy-induced cell death. Proc Natl Acad Sci USA 91: 3309-3313.

Duda T, Venkataraman V, Goraczniak R, Lange C, Koch KW. 1999. Functional consequences of a rod outer segment membrane guanylate cyclase (ROS-GC1) gene mutation linked with Leber's congenital amaurosis. Biochemistry 38: 509-515.

Eagle RC Jr, Hedges TR, Yanoff M. 1982. The atypical pigmentary retinopathy of Kearns–Sayre syndrome. A light and electron microscopic study. Ophthalmology 89: 1433-1440.

Edwards AO, Ritter R III, Abel KJ, Manning A, Panhuysen C, et al. 2005. Complement factor H polymorphism and age-related macular degeneration. Science 308: 421-424.

Eldjarn L, Stokke O, Try K. 1966. Alpha-oxidation of branched chain fatty acids in man and its failure in patients with Refsum's disease showing phytanic acid accumulation. Scand J Clin Lab Invest 18: 694-695.

Eldjarn L, Try K, Stokke O. 1966. The ability of patients with herdopathia atactica polyneuritiformis to alpha-oxidize and degrade several isoprenoid branch-chained fatty structures. Scand J Clin Lab Invest 18: 141-150.

Eldred GE, Lasky MR. 1993. Retinal age pigments generated by self-assembling lysosomotropic detergents. Nature 361: 724-726.

Endres M, Namura S, Shimizu-Sasamata M, Waeber C, Zhang L, et al. 1998. Attenuation of delayed neuronal death after mild focal ischemia in mice by inhibition of the caspase family. J Cereb Blood Flow Metab 18: 238-247.

Ettaiche M, Fillacier K, Widmann C, Heurteaux C, Lazdunski M. 1999. Riluzole improves functional recovery after ischemia in the rat retina. Invest Ophthalmol Vis Sci 40: 729-736.

Farber DB, Flannery JG, Lolley RN, Bok D. 1985. Distribution patterns of photoreceptors, protein, and cyclic nucleotides in the human retina. Invest Ophthalmol Vis Sci 26: 1558-1568.

Farooqui AA, Yi OW, Lu XR, Halliwell B, Horrocks LA. 2001. Neurochemical consequences of kainate-induced toxicity in brain: Involvement of arachidonic acid release and prevention of toxicity by phospholipase A(2) inhibitors. Brain Res Brain Res Rev 38: 61-78.

Farrar GJ, Kenna P, Redmond R, Shiels D, McWilliam P, et al. 1991. Autosomal dominant retinitis pigmentosa: A mutation in codon 178 of the rhodopsin gene in two families of Celtic origin. Genomics 11: 1170-1171.

Fechtner RD, Weinreb RN. 1994. Mechanisms of optic nerve damage in primary open angle glaucoma. Surv Ophthalmol 39: 23-42.

Feeney-Burns L, Ellersieck MR. 1985. Age-related changes in the ultrastructure of Bruch's membrane. Am J Ophthalmol 100: 686-697.

Feeney-Burns L, Hilderbrand ES, Eldridge S. 1984. Aging human RPE: Morphometric analysis of macular, equatorial, and peripheral cells. Invest Ophthalmol Vis Sci 25: 195-200.

Fern R. 1998. Intracellular calcium and cell death during ischemia in neonatal rat white matter astrocytes in situ. J Neurosci 18: 7232-7243.

Ferris FL III. 1983. Senile macular degeneration: Review of epidemiologic features. Am J Epidemiol 118: 132-151.

Finn JT, Weil M, Archer F, Siman R, Srinivasan A, et al. 2000. Evidence that Wallerian degeneration and localized axon degeneration induced by local neurotrophin deprivation do not involve caspases. J Neurosci 20: 1333-1341.

Flood MT, Gouras P, Kjeldbye H. 1980. Growth characteristics and ultrastructure of human retinal pigment epithelium in vitro. Invest Ophthalmol Vis Sci 19: 1309-1320.

Flood MT, Haley JE, Gouras P. 1984. Cellular aging of human retinal epithelium in vivo and in vitro. Monogr Dev Biol 17: 80-93.

Fontaine V, Mohand-Said S, Hanoteau N, Fuchs C, Pfizenmaier K, et al. 2002. Neurodegenerative and neuroprotective effects of tumor necrosis factor (TNF) in retinal ischemia: Opposite roles of TNF receptor 1 and TNF receptor 2. J Neurosci22, RC216.

Forman HJ, Torres M, Fukuto J. 2002. Redox signaling. Mol Cell Biochem 234-235: 49-62.

Fragoso G, Lopez-Colome AM. 1999. Excitatory amino acid-induced inositol phosphate formation in cultured retinal pigment epithelium. Vis Neurosci 16: 263-269.

Frank RN, Amin RH, Eliott D, Puklin JE, Abrams GW. 1996. Basic fibroblast growth factor and vascular endothelial growth factor are present in epiretinal and choroidal neovascular membranes. Am J Ophthalmol 122: 393-403.

Frederickson CJ, Bush AI. 2001. Synaptically released zinc: Physiological functions and pathological effects. Biometals 14: 353-366.

Friedlander M, Brooks PC, Shaffer RW, Kincaid CM, Varner JA, et al. 1995. Definition of two angiogenic pathways by distinct alpha V integrins. Science 270: 1500-1502.

Gal A, Artlich A, Ludwig M, Niemeyer G, Olek K, et al. 1991. Pro-347-Arg mutation of the rhodopsin gene in autosomal dominant retinitis pigmentosa. Genomics 11: 468-470.

Gal A, Li Y, Thompson DA, Weir J, Orth U, et al. 2000. Mutations in MERTK, the human orthologue of the RCS rat retinal dystrophy gene, cause retinitis pigmentosa (in Process Citation). Nat Genet 26: 270-271.

Gao H, Hollyfield JG. 1992. Aging of the human retina. Differential loss of neurons and retinal pigment epithelial cells. Invest Ophthalmol Vis Sci 33: 1-17.

Gartner J. 1970. Electron microscopic observations on the cilio-zonular border area of the human eye with particular reference to the aging changes. Z Anat Entwicklungsgesch 131: 263-273.

George EB, Glass JD, Griffin JW. 1995. Axotomy-induced axonal degeneration is mediated by calcium influx through ion-specific channels. J Neurosci 15: 6445-6452.

Gerlach R, Beck M, Zeitschel U, Seifert V. 2002. MK 801 attenuates c-fos and c-jun expression after in vitro ischemia in rat neuronal cell cultures but not in PC 12 cells. Neurol Res 24: 725-729.

Gilgun-Sherki Y, Rosenbaum Z, Melamed E, Offen D. 2002. Antioxidant therapy in acute central nervous system injury: Current state. Pharmacol Rev 54: 271-284.

Goldstein IM, Ostwald P, Roth S. 1996. Nitric oxide: A review of its role in retinal function and disease. Vis Res 36: 2979-2994.

Gottsch JD, Bynoe LA, Harlan JB, Rencs EV, Green WR. 1993. Light-induced deposits in Bruch's membrane of protoporphyric mice. Arch Ophthalmol 111: 126-129.

Gottsch JD, Pou S, Bynoe LA, Rosen GM. 1990. Hematogenous photosensitization. A mechanism for the development of age-related macular degeneration. Invest Ophthalmol Vis Sci 31: 1674-1682.

Gouras P, Carr RE, Gunkel RD. 1971. Retinitis pigmentosa in abetalipoproteinemia: Effects of vitamin A. Invest Ophthalmol Vis Sci 10: 784-793.

Grant MB, Mames RN, Fitzgerald C, Ellis EA, Caballero S, et al. 1993. Insulin-like growth factor I as an angiogenic agent. In vivo and in vitro studies. Ann NY Acad Sci 692: 230-242.

Green WR, Enger C. 1993. Age-related macular degeneration histopathologic studies. The 1992 Lorenz E. Zimmerman Lecture. Ophthalmology 100: 1519-1535.

Green AR, Hainsworth AH, Jackson DM. 2000. GABA potentiation: A logical pharmacological approach for the treatment of acute ischaemic stroke. Neuropharmacology 39: 1483-1494.

Griffin JW, George EB, Chaudhry V. 1996. Wallerian degeneration in peripheral nerve disease. Baillieres Clin Neurol 5: 65-75.

Grunwald JE, Hariprasad SM, DuPont J, Maguire MG, Fine SL, et al. 1998. Foveolar choroidal blood flow in age-related macular degeneration. Invest Ophthalmol Vis Sci 39: 385-390.

Gupta LY, Marmor MF. 1993. Mannitol, dextromethorphan, and catalase minimize ischemic damage to retinal pigment epithelium and retina. Arch Ophthalmol 111: 384-388.

Gutteridge JM. 1986. Iron promoters of the Fenton reaction and lipid peroxidation can be released from haemoglobin by peroxides. FEBS Lett 201: 291-295.

Hackett SF, Schoenfeld CL, Freund J, Gottsch JD, Bhargave S, et al. 1997. Neurotrophic factors, cytokines and stress increase expression of basic fibroblast growth factor in retinal pigmented epithelial cells. Exp Eye Res 64: 865-873.

Hafler DA. 2002. Degeneracy, as opposed to specificity, in immunotherapy. J Clin Invest 109: 581-584.

Hageman GS, Mullins RF. 1999. Molecular composition of drusen as related to substructural phenotype. Mol Vis 5: 28.

Haim M, Holm NV, Rosenberg T. 1992. Prevalence of retinitis pigmentosa and allied disorders in Denmark. I Main results. Acta Ophthalmol (Copenh) 70: 178-186.

Haines JL, Hauser MA, Schmidt S, Scott WK, Olson LM, et al. 2005. Complement factor H variant increases the risk of age-related macular degeneration. Science 308: 419-421.

Hallenbeck JM. 2002. The many faces of tumor necrosis factor in stroke. Nat Med 8: 1363-1368.

Halliwell B. 1992. Reactive oxygen species and the central nervous system. J Neurochem 59: 1609-1623.

Halliwell B. 1997. Antioxidants: The basics–what they are and how to evaluate them. Adv Pharmacol 38: 3-20.

Halliwell B. 2001. Role of free radicals in the neurodegenerative diseases: Therapeutic implications for antioxidant treatment. Drugs Aging 18: 685-716.

Hansen AJ. 1985. Effect of anoxia on ion distribution in the brain. Physiol Rev 65: 101-148.

Harhaj NS, Barber AJ, Antonetti DA. 2002. Platelet-derived growth factor mediates tight junction redistribution and increases permeability in MDCK cells. J Cell Physiol 193: 349-364.

Hartveit E, Brandstatter JH, Enz R, Wassle H. 1995. Expression of the mRNA of seven metabotropic glutamate receptors (mGluR1 to 7) in the rat retina. An in situ hybridization study on tissue sections and isolated cells. Eur J Neurosci 7: 1472-1483.

Hayreh SS, Weingeist TA. 1980. Experimental occlusion of the central artery of the retina. I. Ophthalmoscopic and fluorescein fundus angiographic studies. Br J Ophthalmol 64: 896-912.

Heidinger V, Dreyfus H, Sahel J, Christen Y, Hicks D. 1998. Excitotoxic damage of retinal glial cells depends upon normal neuron–glial interactions. Glia 23: 146-155.

Hellendall RP, Ting JP. 1997. Differential regulation of cytokine-induced major histocompatibility complex class II expression and nitric oxide release in rat microglia and astrocytes by effectors of tyrosine kinase, protein kinase C, and cAMP. J Neuroimmunol 74: 19-29.

Henderson G, Tomlinson BE, Gibson PH. 1980. Cell counts in human cerebral cortex in normal adults throughout life using an image analysing computer. J Neurol Sci 46: 113-136.

Herman EC Jr, Conley CL. 1960. Hereditary persistence of fetal hemoglobin. A family study. Am J Med 29: 9-17.

Hernandez MR. 2000. The optic nerve head in glaucoma: Role of astrocytes in tissue remodeling. Prog Retin Eye Res 19: 297-321.

Hickey WF, Hsu BL, Kimura H. 1991. T-lymphocyte entry into the central nervous system. J Neurosci Res 28: 254-260.

Hollmann M, Heinemann S. 1994. Cloned glutamate receptors. Annu Rev Neurosci 17: 31-108.

Hong DH, Pawlyk BS, Shang J, Sandberg MA, Berson EL, et al. 2000. A retinitis pigmentosa GTPase regulator (RPGR)-deficient mouse model for X-linked retinitis pigmentosa (RP3). Proc Natl Acad Sci USA 97: 3649-3654.

Hoshino M, Mizuno K, Ichikawa H. 1984. Aging alterations of retina and choroid of Japanese: Light microscopic study of macular region of 176 eyes. Jpn J Ophthalmol 28: 89-102.

Hu DN. 1982. Genetic aspects of retinitis pigmentosa in China. Am J Med Genet 12: 51-56.

Huang PC, Gaitan AE, Hao Y, Petters RM, Wong F. 1993. Cellular interactions implicated in the mechanism of photoreceptor degeneration in transgenic mice expressing a mutant rhodopsin gene. Proc Natl Acad Sci USA 90: 8484-8488.

Hunter A, Tsilfidis C, Mettler G, Jacob P, Mahadevan M, et al. 1992. The correlation of age of onset with CTG trinucleotide repeat amplification in myotonic dystrophy. J Med Genet 29: 774-779.

Hyman LG, Lilienfeld AM, Ferris FL III, Fine SL. 1983. Senile macular degeneration: A case-control study. Am J Epidemiol 118: 213-227.

Iadecola C. 1997. Bright and dark sides of nitric oxide in ischemic brain injury. Trends Neurosci 20: 132-139.

Inana G, Chambers C, Hotta Y, Inouye L, Filpula D, et al. 1989. Point mutation affecting processing of the ornithine aminotransferase precursor protein in gyrate atrophy. J Biol Chem 264: 17432-17436.

Inglehearn CF, Bashir R, Lester DH, Jay M, Bird AC, et al. 1991. A 3-bp deletion in the rhodopsin gene in a family with autosomal dominant retinitis pigmentosa. Am J Hum Genet 48: 26-30.

Inglehearn CF, Keen TJ, Bashir R, Jay M, Fitzke F, et al. 1992. A completed screen for mutations of the rhodopsin gene in a panel of patients with autosomal dominant retinitis pigmentosa. Hum Mol Genet 1: 41-45.

Inoue M, Sato EF, Nishikawa M, Park AM, Kira Y, et al. 2003. Mitochondrial generation of reactive oxygen species and its role in aerobic life. Curr Med Chem 10: 2495-2505.

Iseki E, Kato M, Marui W, Ueda K, Kosaka K. 2001. A neuropathological study of the disturbance of the nigro-amygdaloid connections in brains from patients with dementia with Lewy bodies. J Neurol Sci 185: 129-134.

Jackson GR, Owsley C. 2000. Scotopic sensitivity during adulthood. Vis Res 40: 2467-2473.

Jacobson SG, Cideciyan AV, Regunath G, Rodriguez FJ, Vandenburgh K, et al. 1995. Night blindness in Sorsby's fundus dystrophy reversed by vitamin. Nat Genet 11: 27-32.

Jay M. 1982. On the heredity of retinitis pigmentosa. Br J Ophthalmol 66: 405-416.

Jellinger KA. 2003. General aspects of neurodegeneration. J Neural Transm Suppl 101-144.

Jenssen TK, Laegreid A, Komorowski J, Hovig E. 2001. A literature network of human genes for high-throughput analysis of gene expression. Nat Genet 28: 21-28.

Joo CK, Choi JS, Ko HW, Park KY, Sohn S, et al. 1999. Necrosis and apoptosis after retinal ischemia: Involvement of NMDA-mediated excitotoxicity and p53. Invest Ophthalmol Vis Sci 40: 713-720.

Ju WK, Kim KY, Neufeld AH. 2003. Increased activity of cyclooxygenase-2 signals early neurodegenerative events in the rat retina following transient ischemia. Exp Eye Res 77: 137-145.

Ju WK, Kim KY, Cha JH, Kim IB, Lee MY, et al. 2000. Ganglion cells of the rat retina show osteopontin-like immunoreactivity. Brain Res 852: 217-220.

Ju WK, Gwon JS, Kim KY, Oh SJ, Kim SY, et al. 2001. Up-regulated eNOS protects blood-retinal barrier in the L-arginine treated ischemic rat retina. Neuroreport 12: 2405-2409.

Kajiwara K, Berson EL, Dryja TP. 1994. Digenic retinitis pigmentosa due to mutations at the unlinked peripherin/RDS and ROM1 loci. Science 264: 1604-1608.

Kajiwara K, Hahn LB, Mukai S, Travis GH, Berson EL, et al. 1991. Mutations in the human retinal degeneration slow gene in autosomal dominant retinitis pigmentosa. Nature 354: 480-483.

Kannan K, Jain SK. 2000. Oxidative stress and apoptosis. Pathophysiology 7: 153-163.

Karnik SS, Khorana HG. 1990. Assembly of functional rhodopsin requires a disulfide bond between cysteine residues 110 and 187. J Biol Chem 265: 17520-17524.

Karnik SS, Sakmar TP, Chen HB, Khorana HG. 1988. Cysteine residues 110 and 187 are essential for the formation of correct structure in bovine rhodopsin. Proc Natl Acad Sci USA 85: 8459-8463.

Katai N, Yoshimura N. 1999. Apoptotic retinal neuronal death by ischemia-reperfusion is executed by two distinct caspase family proteases. Invest Ophthalmol Vis Sci 40: 2697-2705.

Kawasaki A, Han MH, Wei JY, Hirata K, Otori Y, et al. 2002. Protective effect of arachidonic acid on glutamate neurotoxicity in rat retinal ganglion cells. Invest Ophthalmol Vis Sci 43: 1835-1842.

Keen TJ, Inglehearn CF. 1996. Mutations and polymorphisms in the human peripherin-RDS gene and their involvement in inherited retinal degeneration. Hum Mutat 8: 297-303.

Kerr JF, Harmon B, Searle J. 1974. An electron-microscope study of cell deletion in the anuran tadpole tail during spontaneous metamorphosis with special reference to apoptosis of striated muscle fibers. J Cell Sci 14: 571-585.

Killingsworth MC. 1987. Age-related components of Bruch's membrane in the human eye. Graefes Arch Clin Exp Ophthalmol 225: 406-412.

Kipnis J, Schwartz M. 2002. Dual action of glatiramer acetate (Cop-1) in the treatment of CNS autoimmune and neurodegenerative disorders. Trends Mol Med 8: 319-323.

Kipnis J, Mizrahi T, Hauben E, Shaked I, Shevach E, et al. 2002. Neuroprotective autoimmunity: Naturally occurring CD4+CD25+ regulatory T cells suppress the ability to withstand injury to the central nervous system. Proc Natl Acad Sci USA 99: 15620-15625.

Klein R, Klein BE, Jensen SC, Meuer SM. 1997. The five-year incidence and progression of age-related maculopathy: The Beaver Dam eye study. Ophthalmology 104: 7-21.

Klein RJ, Zeiss C, Chew EY, Tsai JY, Sackler RS, et al. 2005. Complement factor H polymorphism in age-related macular degeneration. Science 308: 385-389.

Koeberle PD, Ball AK. 1999. Nitric oxide synthase inhibition delays axonal degeneration and promotes the survival of axotomized retinal ganglion cells. Exp Neurol 158: 366-381.

Kong LY, Lai C, Wilson BC, Simpson JN, Hong JS. 1997. Protein tyrosine kinase inhibitors decrease lipopolysaccharide-induced proinflammatory cytokine production in mixed glia, microglia-enriched or astrocyte-enriched cultures. Neurochem Int 30: 491-497.

Kontos HA. 1987. Oxygen radicals from arachidonate metabolism in abnormal vascular responses. Am Rev Respir Dis 136: 474-477.

Kowluru RA, Engerman RL, Case GL, Kern TS. 2001. Retinal glutamate in diabetes and effect of antioxidants. Neurochem Int 38: 385-390.

Kristian T, Siesjo BK. 1998. Calcium in ischemic cell death. Stroke 29: 705-718.

Kudo I, Murakami M. 2002. Phospholipase A2 enzymes. Prostaglandins Other Lipid Mediat 68-69: 3-58.

Kuriyama H, Nakagawa M, Tsuda M. 2001. Intracellular Ca(2+) changes induced by in vitro ischemia in rat retinal slices. Exp Eye Res 73: 365-374.

Kuriyama H, Waki M, Nakagawa M, Tsuda M. 2001. Involvement of oxygen free radicals in experimental retinal ischemia and the selective vulnerability of retinal damage. Ophthalmic Res 33: 196-202.

Kuroiwa S, Katai N, Shibuki H, Kurokawa T, Umihira J, et al. 1998. Expression of cell cycle-related genes in dying cells in retinal ischemic injury. Invest Ophthalmol Vis Sci 39: 610-617.

Kurokawa T, Katai N, Shibuki H, Kuroiwa S, Kurimoto Y, et al. 1999. BDNF diminishes caspase-2 but not c-Jun immunoreactivity of neurons in retinal ganglion cell layer after transient ischemia. Invest Ophthalmol Vis Sci 40: 3006-3011.

Kutty RK, Kutty G, Wiggert B, Chader GJ, Darrow RM, et al. 1995. Induction of heme oxygenase 1 in the retina by intense visible light: Suppression by the antioxidant dimethylthiourea. Proc Natl Acad Sci USA 92: 1177-1181.

Kvanta A, Algvere PV, Berglin L, Seregard S. 1996. Subfoveal fibrovascular membranes in age-related macular degeneration express vascular endothelial growth factor. Invest Ophthalmol Vis Sci 37: 1929-1934.

Laterra J, Nam M, Rosen E, Rao JS, Lamszus K. 1997. Scatter factor/hepatocyte growth factor gene transfer enhances glioma growth and angiogenesis in vivo. Lab Invest 76: 565-577.

Lees GJ. 1991. Inhibition of sodium-potassium-ATPase: A potentially ubiquitous mechanism contributing to central nervous system neuropathology. Brain Res Brain Res Rev 16: 283-300.

Leibovich SJ, Polverini PJ, Shepard HM, Wiseman DM, Shively V, et al. 1987. Macrophage-induced angiogenesis is mediated by tumour necrosis factor-alpha. Nature 329: 630-632.

Li B, Roth S. 1999. Retinal ischemic preconditioning in the rat: Requirement for adenosine and repetitive induction. Invest Ophthalmol Vis Sci 40: 1200-1216.

Li J, Kato K, Ikeda J, Morita I, Murota S. 2001. A narrow window for rescuing cells by the inhibition of calcium influx and the importance of influx route in rat cortical neuronal cell death induced by glutamate. Neurosci Lett 304: 29-32.

Li M, Wang X, Meintzer MK, Laessig T, Birnbaum MJ, et al. 2000. Cyclic AMP promotes neuronal survival by phosphorylation of glycogen synthase kinase 3beta. Mol Cell Biol 20: 9356-9363.

Lin B, Martin PR, Grunert U. 2002. Expression and distribution of ionotropic glutamate receptor subunits on parasol ganglion cells in the primate retina. Vis Neurosci 19: 453-465.

Linnik MD, Zobrist RH, Hatfield MD. 1993. Evidence supporting a role for programmed cell death in focal cerebral ischemia in rats. Stroke 24: 2002-2008.

Lipton P. 1999. Ischemic cell death in brain neurons. Physiol Rev 79: 1431-1568.

Lipton SA. 2001. Retinal ganglion cells, glaucoma and neuroprotection. Prog Brain Res 131: 712-718.

Lipton SA. 2003. Possible role for memantine in protecting retinal ganglion cells from glaucomatous damage. Surv Ophthalmol 48 (Suppl 1): S38-S46.

Liu B, Neufeld AH. 2001. Nitric oxide synthase-2 in human optic nerve head astrocytes induced by elevated pressure in vitro. Arch Ophthalmol 119: 240-245.

Liu X, Vansant G, Udovichenko IP, Wolfrum U, Williams DS. 1997. Myosin VIIa, the product of the Usher 1B syndrome gene, is concentrated in the connecting cilia of photoreceptor cells. Cell Motil Cytoskeleton 37: 240-252.

Lopez-Costa JJ, Goldstein J, Saavedra JP. 1997. Neuronal and macrophagic nitric oxide synthase isoforms distribution in normal rat retina. Neurosci Lett 232: 155-158.

Louie JL, Kapphahn RJ, Ferrington DA. 2002. Proteasome function and protein oxidation in the aged retina. Exp Eye Res 75: 271-284.

Louzada-Junior P, Dias JJ, Santos WF, Lachat JJ, Bradford HF, et al. 1992. Glutamate release in experimental ischaemiá of the retina: An approach using microdialysis. J Neurochem 59: 358-363.

Lucas DR, Newhouse JP. 1957. The toxic effect of sodium L-glutamate on the inner layers of the retina. Ama Arch Opthalmol 58: 193-201.

Lunn ER, Perry VH, Brown MC, Rosen H, Gordon S. 1989. Absence of Wallerian degeneration does not hinder regeneration in peripheral nerve. Eur J Neurosci 1: 27-33.

Lutty GA, Mcleod DS. 2003. Retinal vascular development and oxygen-induced retinopathy: A role for adenosine. Prog Retin Eye Res 22: 95-111.

Lyon MF. 1961. Gene action in the X-chromosome of the mouse (*MusmusculusL.*). Naturwissenschaften 190: 372-373.

Lyon MF. 1962. Sex chromatin and gene action in the mammalian X-chromosome. Am J Hum Genet 14: 135-148.

Lyon MF, Ogunkolade BW, Brown MC, Atherton DJ, Perry VH. 1993. A gene affecting Wallerian nerve degeneration maps distally on mouse chromosome 4. Proc Natl Acad Sci USA 90: 9717-9720.

Madri JA, Reidy MA, Kocher O, Bell L. 1989. Endothelial cell behavior after denudation injury is modulated by transforming growth factor-beta1 and fibronectin. Lab Invest 60: 755-765.

Magenis RE, Toth FS, Allen LJ, Black M, Brown MG, et al. 1990. Comparison of the 15q deletions in Prader-Willi and Angelman syndromes: Specific regions, extent of deletions, parental origin, and clinical consequences. Am J Med Genet 35: 333-349.

Majno G, Joris I. 1995. Apoptosis, oncosis, and necrosis. An overview of cell death. Am J Pathol 146: 3-15.

Mandel RJ, Gage FH, Clevenger DG, Spratt SK, Snyder RO, et al. 1999. Nerve growth factor expressed in the medial septum following in vivo gene delivery using a recombinant adeno-associated viral vector protects cholinergic neurons from fimbria-fornix lesion-induced degeneration. Exp Neurol 155: 59-64.

Marak GE, de Kozak Y, Faure JP. 1990. Free radicals and antioxidants in the pathogenesis of eye diseases. Adv Exp Med Biol 264; 513–527.

Marshall J. 1987. The ageing retina: Physiology or pathology. Eye 1 (Pt 2): 282-295.

Mata NL, Weng J, Travis GH. 2000. Biosynthesis of a major lipofuscin fluorophore in mice and humans with ABCR-mediated retinal and macular degeneration. Proc Natl Acad Sci USA 97: 7154-7159.

Matsunaga H, Handa JT, Aotaki-Keen A, Sherwood SW, West MD, et al. 1999. Beta-galactosidase histochemistry and telomere loss in senescent retinal pigment epithelial cells. Invest Ophthalmol Vis Sci 40: 197-202.

Maxwell SR. 1995. Prospects for the use of antioxidant therapies. Drugs 49: 345-361.

Maxwell PH, Ratcliffe PJ. 2002. Oxygen sensors and angiogenesis. Semin Cell Dev Biol 13: 29-37.

McClatchey AI, Kaufman DL, Berson EL, Tobin AJ, Shih VE, et al. 1990. Splicing defect at the ornithine aminotransferase (OAT) locus in gyrate atrophy. Am J Hum Genet 47: 790-794.

Melena J, Wood JP, Osborne NN. 1999. Betaxolol, a beta1-adrenoceptor antagonist, has an affinity for L-type Ca^{2+} channels. Eur J Pharmacol 378: 317-322.

Mignatti P, Rifkin DB. 1996. Plasminogen activators and matrix metalloproteinases in angiogenesis. Enzyme Protein 49: 117-137.

Miller JW, Adamis AP, Shima DT, D'Amore PA, Moulton RS, et al. 1994. Vascular endothelial growth factor/vascular permeability factor is temporally and spatially correlated with ocular angiogenesis in a primate model. Am J Pathol 145: 574-584.

Mitchell GA, Brody LC, Looney J, Steel G, Suchanek M, et al. 1988. An initiator codon mutation in ornithine-delta-aminotransferase causing gyrate atrophy of the choroid and retina. J Clin Invest 81: 630-633.

Mitchell GA, Brody LC, Sipila I, Looney JE, Wong C, et al. 1989. At least two mutant alleles of ornithine delta-aminotransferase cause gyrate atrophy of the choroid and retina in Finns. Proc Natl Acad Sci USA 86: 197-201.

Mizutani M, Gerhardinger C, Lorenzi M. 1998. Muller cell changes in human diabetic retinopathy. Diabetes 47: 445-449.

Moncada S, Higgs EA. 1991. Endogenous nitric oxide: Physiology, pathology and clinical relevance. Eur J Clin Invest 21: 361-374.

Moore T, Haig D. 1991. Genomic imprinting in mammalian development: A parental tug-of-war. Trends Genet 7: 45-49.

Mousa SA, Lorelli W, Campochiaro PA. 1999. Role of hypoxia and extracellular matrix–integrin binding in the modulation of angiogenic growth factors secretion by retinal pigmented epithelial cells. J Cell Biochem 74: 135-143.

Mukherjee PK, Marcheselli VL, Serhan CN, Bazan NG. 2004. Neuroprotectin D1: A docosahexaenoic acid-derived docosatriene protects human retinal pigment epithelial cells from oxidative stress. Proc Natl Acad Sci USA 101: 8491-8496.

Mullen RJ, La Vail MM. 1976. Inherited retinal dystrophy: Primary defect in pigment epithelium determined with experimental rat chimeras. Science 192: 799-801.

Naash ML, Peachey NS, Li ZY, Gryczan CC, Goto Y, et al. 1996. Light-induced acceleration of photoreceptor degeneration in transgenic mice expressing mutant rhodopsin. Invest Ophthalmol Vis Sci 37: 775-782.

Nakamichi N, Chidlow G, Osborne NN. 2003. Effects of intraocular injection of a low concentration of zinc on the rat retina. Neuropharmacology 45: 637-648.

Neal MJ, Cunningham JR, Hutson PH, Hogg J. 1994. Effects of ischaemia on neurotransmitter release from the isolated retina. J Neurochem 62: 1025-1033.

Neufeld AH, Kawai S, Das S, Vora S, Gachie E, et al. 2002. Loss of retinal ganglion cells following retinal ischemia: The role of inducible nitric oxide synthase. Exp Eye Res 75: 521-528.

Nicholls RD, Knoll JH, Butler MG, Karam S, Lalande M. 1989. Genetic imprinting suggested by maternal heterodisomy in nondeletion Prader–Willi syndrome. Nature 342: 281-285.

Nicholson C, Bruggencate GT, Steinberg R, Stockle H. 1977. Calcium modulation in brain extracellular microenvironment demonstrated with ion-selective micropipette. Proc Natl Acad Sci USA 74: 1287-1290.

Nicoletti F, Bruno V, Copani A, Casabona G, Knopfel T. 1996. Metabotropic glutamate receptors: A new target for the therapy of neurodegenerative disorders? Trends Neurosci 19: 267-271.

Nishizawa Y. 2001. Glutamate release and neuronal damage in ischemia. Life Sci 69: 369-381.

Noell WK, Walker VS, Soon Kang B, Berman S. 1966. Retinal damage by light in rats. Invest Ophthalmol 5: 450-473.

O'Donnell JJ, Vannas Sulonen K, Shows TB, Cox DR. 1988. Gyrate atrophy of the choroid and retina: Assignment of the ornithine aminotransferase structural gene to human chromosome 10 and mouse chromosome 7. Am J Hum Genet 43: 922-928.

O'Leary DD, Koester SE. 1993. Development of projection neuron types, axon pathways, and patterned connections of the mammalian cortex. Neuron 10: 991-1006.

Ogata N, Matsushima M, Takada Y, Tobe T, Takahashi K, et al. 1996. Expression of basic fibroblast growth factor mRNA in developing choroidal neovascularization. Curr Eye Res 15: 1008-1018.

Ohtaki H, Funahashi H, Dohi K, Oguro T, Horai R, et al. 2003. Suppression of oxidative neuronal damage after transient middle cerebral artery occlusion in mice lacking interleukin-1. Neurosci Res 45: 313-324.

Okamoto N, Tobe T, Hackett SF, Ozaki H, Vinores MA, et al. 1997. Transgenic mice with increased expression of vascular endothelial growth factor in the retina: A new model of intraretinal and subretinal neovascularization. Am J Pathol 151: 281-291.

Olney JW. 1969. Glutaate-induced retinal degeneration in neonatal mice. Electron microscopy of the acutely evolving lesion. J Neuropathol Exp Neurol 28: 455-474.

Olney JW, Price MT, Fuller TA, Labruyere J, Samson L, et al. 1986. The anti-excitotoxic effects of certain anesthetics, analgesics and sedative-hypnotics. Neurosci Lett 68: 29-34.

Opara EC. 2004. Role of oxidative stress in the etiology of type 2 diabetes and the effect of antioxidant supplementation on glycemic control. J Investig Med 52: 19-23.

Ophir A, Berenshtein E, Kitrossky N, Averbukh E. 1994. Protection of the transiently ischemic cat retina by zinc-desferrioxamine. Invest Ophthalmol Vis Sci 35: 1212-1222.

Osborne NN, Larsen AK. 1996. Antigens associated with specific retinal cells are affected by ischaemia caused by raised intraocular pressure: Effect of glutamate antagonists. Neurochem Int 29: 263-270.

Osborne NN, Safa R, Nash MS. 1999. Photoreceptors are preferentially affected in the rat retina following permanent occlusion of the carotid arteries. Vis Res 39: 3995-4002.

Osborne NN, DeSantis L, Bae JH, Ugarte M, Wood JP, et al. 1999a. Topically applied betaxolol attenuates NMDA-induced toxicity to ganglion cells and the effects of ischaemia to the retina. Exp Eye Res 69: 331-342.

Osborne NN, Wood JP, Chidlow G, Bae JH, Melena J, et al. 1999c. Ganglion cell death in glaucoma: What do we really know? Br J Ophthalmol 83: 980-986.

Osborne NN, Casson RJ, Wood JP, Chidlow G, Graham M, et al. 2004. Retinal ischemia: Mechanisms of damage and potential therapeutic strategies. Prog Retin Eye Res 23: 91-147.

Pacione LR, Szego MJ, Ikeda S, Nishina PM, McInnes RR. 2003. Progress toward understanding the genetic and biochemical mechanisms of inherited photoreceptor degenerations. Annu Rev Neurosci 26: 657-700.

Palmer HJ, Paulson KE. 1997. Reactive oxygen species and antioxidants in signal transduction and gene expression. Nutr Rev 55: 353-361.

Park JE, Keller GA, Ferrara N. 1993. The vascular endothelial growth factor (VEGF) isoforms: Differential deposition into the subepithelial extracellular matrix and bioactivity of extracellular matrix-bound VEGF. Mol Biol Cell 4: 1317-1326.

Parker WD Jr, Filley CM, Parks JK. 1990. Cytochrome oxidase deficiency in Alzheimer's disease. Neurology 40: 1302-1303.

Patel HC, Boutin H, Allan SM. 2003. Interleukin-1 in the brain: Mechanisms of action in acute neurodegeneration. Ann NY Acad Sci 992: 39-47.

Pauleikhoff D, Harper CA, Marshall J, Bird AC. 1990. Aging changes in Bruch's membrane. A histochemical and morphologic study. Ophthalmology 97: 171-178.

Payne AM, Downes SM, Bessant DA, Taylor R, Holder GE, et al. 1998. A mutation in guanylate cyclase activator 1A (GUCA1A) in an autosomal dominant cone dystrophy pedigree mapping to a new locus on chromosome 6p21.1. Hum Mol Genet 7: 273-277.

Pena JD, Agapova O, Gabelt BT, Levin LA, Lucarelli MJ, et al. 2001. Increased elastin expression in astrocytes of the lamina cribrosa in response to elevated intraocular pressure. Invest Ophthalmol Vis Sci 42: 2303-2314.

Pepper MS, Vassalli JD, Montesano R, Orci L. 1987. Urokinase-type plasminogen activator is induced in migrating capillary endothelial cells. J Cell Biol 105: 2535-2541.

Perry VH, Lunn ER, Brown MC, Cahusac S, Gordon S. 1990. Evidence that the rate of Wallerian degeneration is controlled by a single autosomal dominant gene. Eur J Neurosci 2: 408-413.

Petty MA, Wettstein JG. 1999. White matter ischaemia. Brain Res Brain Res Rev 31: 58-64.

Pfeffer SR. 1992. GTP-binding proteins in intracellular transport. Trends Cell Biol 2: 41-46.

Pierce EA. 2001. Pathways to photoreceptor cell death in inherited retinal degenerations. Bioessays 23: 605-618.

Pittler SJ, Baehr W. 1991. Identification of a nonsense mutation in the rod photoreceptor cGMP phosphodiesterase beta-subunit gene of the rd mouse. Proc Natl Acad Sci USA 88: 8322-8326.

Plate KH, Breier G, Weich HA, Risau W. 1992. Vascular endothelial growth factor is a potential tumour angiogenesis factor in human gliomas in vivo. Nature 359: 845-848.

Quigley HA. 1993. Open-angle glaucoma. N Engl J Med 328: 1097-1106.

Quigley HA, McKinnon SJ, Zack DJ, Pease ME, Kerrigan-Baumrind LA, et al. 2000. Retrograde axonal transport of BDNF in retinal ganglion cells is blocked by acute IOP elevation in rats. Invest Ophthalmol Vis Sci 41: 3460-3466.

Raivich G, Jones LL, Kloss CU, Werner A, Neumann H, et al. 1998. Immune surveillance in the injured nervous system: T-lymphocytes invade the axotomized mouse facial motor nucleus and aggregate around sites of neuronal degeneration. J Neurosci 18: 5804-5816.

Ramesh V, McClatchey AI, Ramesh N, Benoit LA, Berson EL, et al. 1988. Molecular basis of ornithine aminotransferase deficiency in B-6-responsive and-nonresponsive forms of gyrate atrophy. Proc Natl Acad Sci USA 85: 3777-3780.

Ramirez JM, Ramirez AI, Salazar JJ, de Hoz R, Trivino A. 2001. Changes of astrocytes in retinal ageing and age-related macular degeneration. Exp Eye Res 73: 601-615.

Rao SD, Yin HZ, Weiss JH. 2003. Disruption of glial glutamate transport by reactive oxygen species produced in motor neurons. J Neurosci 23: 2627-2633.

Refsum S. 1946. Heredopathia atactica polyneuritiformis: A familial syndrome not hitherto described. A contribution to the clinical study of the hereditary diseases of the nervous system. Acta Psychiatr Scand (Suppl 38): 1-303.

Reiter RJ. 1995. Oxidative processes and antioxidative defense mechanisms in the aging brain. FASEB J 9: 526-533.

Ridge KD, Abdulaev NG, Sousa M, Palczewski K. 2003. Phototransduction: Crystal clear. Trends Biochem Sci 28: 479-487.

Robbins SG, Mixon RN, Wilson DJ, Hart CE, Robertson JE, et al. 1994. Platelet-derived growth factor ligands and receptors immunolocalized in proliferative retinal diseases. Invest Ophthalmol Vis Sci 35: 3649-3663.

Robinson PR, Cohen GB, Zhukovsky EA, Oprian DD. 1992. Constitutively active mutants of rhodopsin. Neuron 9: 719-725.

Romano C, Price M, Bai HY, Olney JW. 1993. Neuroprotectants in Honghua: Glucose attenuates retinal ischemic damage. Invest Ophthalmol Vis Sci 34: 72-80.

Rosenbaum DM, Rosenbaum PS, Gupta A, Michaelson MD, Hall DH, et al. 1997. Retinal ischemia leads to apoptosis which is ameliorated by aurintricarboxylic acid. Vis Res 37: 3445-3451.

Rosner M, Hefetz L, Abraham FA. 1993. The prevalence of retinitis pigmentosa and congenital stationary night blindness in Israel (letter). Am J Ophthalmol 116: 373-374.

Ross RD, Barofsky JM, Cohen G, Baber WB, Palao SW, et al. 1998. Presumed macular choroidal watershed vascular filling, choroidal neovascularization, and systemic vascular disease in patients with age-related macular degeneration. Am J Ophthalmol 125: 71-80.

Rossi DJ, Oshima T, Attwell D. 2000. Glutamate release in severe brain ischaemia is mainly by reversed uptake. Nature 403: 316-321.

Roth S, Osinski JV, Park SS, Ostwald P, Moshfeghi AA. 1996. Measurement of purine nucleoside concentration in the intact rat retina. J Neurosci Methods 68: 87-90.

Roth S, Park SS, Sikorski CW, Osinski J, Chan R, et al. 1997. Concentrations of adenosine and its metabolites in the rat retina/choroid during reperfusion after ischemia. Curr Eye Res 16: 875-885.

Rothman SM, Olney JW. 1995. Excitotoxicity and the NMDA receptor—still lethal after eight years. Trends Neurosci 18: 57-58.

Rozet JM, Perrault I, Gerber S, Hanein S, Barbet F, et al. 2001. Complete abolition of the retinal-specific guanylyl cyclase (retGC-1) catalytic ability consistently leads to leber congenital amaurosis (LCA). Invest Ophthalmol Vis Sci 42: 1190-1192.

Sadek HA, Nulton-Persson AC, Szweda PA, Szweda LI. 2003. Cardiac ischemia/reperfusion, aging, and redox-dependent alterations in mitochondrial function. Arch Biochem Biophys 420: 201-208.

Sakamoto K, Kuwagata M, Nakahara T, Ishii K. 2001. Late preconditioning in rat retina: Involvement of adenosine and ATP-sensitive K(+) channel. Eur J Pharmacol 418: 89-93.

Sapirstein A, Bonventre JV. 2000. Specific physiological roles of cytosolic phospholipase A(2) as defined by gene knockouts. Biochim Biophys Acta 1488: 139-148.

Sarks SH. 1976. Ageing and degeneration in the macular region: A clinico-pathological study. Br J Ophthalmol 60: 324-341.

Sastre J, Pallardo FV, Vina J. 2003. The role of mitochondrial oxidative stress in aging. Free Radic Biol Med 35: 1-8.

Sato Y, Rifkin DB. 1988. Autocrine activities of basic fibroblast growth factor: Regulation of endothelial cell movement, plasminogen activator synthesis, and DNA synthesis. J Cell Biol 107: 1199-1205.

Sattler R, Tymianski M. 2001. Molecular mechanisms of glutamate receptor-mediated excitotoxic neuronal cell death. Mol Neurobiol 24: 107-129.

Schachat AP, Hyman L, Leske MC, Connell AM, Wu SY. 1995. Features of age-related macular degeneration in a black population. The Barbados eye study group. Arch Ophthalmol 113: 728-735.

Schaumburg HH, Wisniewski HM, Spencer PS. 1974. Ultrastructural studies of the dying-back process. I. Peripheral nerve terminal and axon degeneration in systemic acrylamide intoxication. J Neuropathol Exp Neurol 33: 260-284.

Schmidt M, Giessl A, Laufs T, Hankeln T, Wolfrum U, et al. 2003. How does the eye breathe? Evidence for neuroglobin-mediated oxygen supply in the mammalian retina. J Biol Chem 278: 1932-1935.

Schonthal A, Sugarman J, Brown JH, Hanley MR, Feramisco JR. 1991. Regulation of c-fos and c-jun protooncogene expression by the Ca(2+)-ATPase inhibitor thapsigargin. Proc Natl Acad Sci USA 88: 7096-7100.

Schori H, Kipnis J, Yoles E, Wolde Mussie E, Ruiz G, et al. 2001. Vaccination for protection of retinal ganglion cells against death from glutamate cytotoxicity and ocular hypertension: Implications for glaucoma. Proc Natl Acad Sci USA 98: 3398-3403.

Schreiber J, Eling TE, Mason RP. 1986. The oxidation of arachidonic acid by the cyclooxygenase activity of purified prostaglandin H synthase: Spin trapping of a carbon-centered free radical intermediate. Arch Biochem Biophys 249: 126-136.

Schutt F, Davies S, Kopitz J, Holz FG, Boulton ME. 2000. Photodamage to human RPE cells by A2-E, a retinoid component of lipofuscin. Invest Ophthalmol Vis Sci 41: 2303-2308.

Schwartz M. 2000. Autoimmune involvement in CNS trauma is beneficial if well controlled. Prog Brain Res 128: 259-263.

Schwartz M. 2003a. Neurodegeneration and neuroprotection in glaucoma: Development of a therapeutic neuroprotective vaccine: The Friedenwald lecture. Invest Ophthalmol Vis Sci 44: 1407-1411.

Schwartz M. 2003b. Neuroprotection as a treatment for glaucoma: Pharmacological and immunological approaches. Eur J Ophthalmol 13 (Suppl 3): S27-S31.

Schwartz M, Shaked I, Fisher J, Mizrahi T, Schori H. 2003. Protective autoimmunity against the enemy within: Fighting glutamate toxicity. Trends Neurosci 26: 297-302.

Schwartz-Bloom RD, Sah R. 2001. Gamma-aminobutyric acid(A) neurotransmission and cerebral ischemia. J Neurochem 77: 353-371.

Seddon JM, Ajani UA, Mitchell BD. 1997. Familial aggregation of age-related maculopathy. Am J Ophthalmol 123: 199-206.

Semenza GL. 2000. Expression of hypoxia-inducible factor 1: mechanisms and consequences. Biochem Pharmacol 59: 47-53.

Sharma RK. 2000a. Developmental expression of GABA(A) receptors in retinal transplants. Ophthalmic Res 32: 45-51.

Sharma RK. 2000b. Dissection and cotransplantation of large pieces of RPE and neural retina; effect of protease K on the development. Acta Ophthalmol Scand 78: 3-8.

Sharma RK. 2001a. Bcl-2 expression during the development and degeneration of RCS rat retinae. Brain Res Dev Brain Res 132: 81-86.

Sharma RK. 2001b. Development and survival of tyrosine hydroxylase containing neurons in RCS rat retinae. Curr Eye Res. 23(4): 256-262.

Sharma RK, Ehinger B. 1997a. Cell proliferation in retinal transplants. Cell Transplant 6: 141-148.

Sharma RK, Ehinger B. 1997b. Retinal cell transplants: How close to clinical application? Acta Ophthalmol Scand 75: 355-363.

Sharma RK, Ehinger B. 1999. Management of hereditary retinal degenerations: Present status and future directions. Surv Ophthalmol 43: 427-444.

Sharma RK, Ehinger B. 2003. Development and Structure of the Retina. Adler's physiology of the eye: Clinical application. Kaufman PL, Alm A, editors. St Louis: Mosby; pp. 319-347.

Sharma RK, Johnson DA. 2000. Molecular signals for development of neuronal circuitry in the retina. Neurochem Res 25: 1257-1263.

Sharma RK, Bergström A, Ehinger B. 1995. Retinal cell transplants. Prog Retin Eye Res 15: 197-230.

Sharma RK, Bergstrom A, Ehinger B. 1997. Influence of technique and transplantation site on rosette formation in rabbit retinal transplants. Acta Ophthalmol Scand 75: 3-10.

Sharma RK, Perez MT, Ehinger B. 1997. Immunocytochemical localisation of neuronal nitric oxide synthase in developing and transplanted rabbit retinas. Histochem Cell Biol 107: 449-458.

Sharma RK, Warfvinge K, Ehinger B. 2001. Expression of nitric oxide synthase during the development of RCS rat retinas. Ophthalmologica 215: 222-228.

Sharma RK, Netland P, Johnson DA, Kedrov M. 2004 Preconditioning Protect the ARPE-19 Cells from Oxidative Stress Induced Cell Death. Invest Ophthalmol Vis Sci. E-Abstract 759.

Sheedlo HJ, Wordinger RJ, Fan W, Turner JE. 1997. A transformed neonatal rat retinal pigment epithelial cell line: Secreted protein analysis and fibroblast growth factor and receptor expression. Curr Eye Res 16: 116-126.

Shibuki H, Katai N, Yodoi J, Uchida K, Yoshimura N. 2000. Lipid peroxidation and peroxynitrite in retinal ischemia-reperfusion injury. Invest Ophthalmol Vis Sci 41: 3607-3614.

Shimizu K, Kobayashi Y, Muraoka K. 1981. Midperipheral fundus involvement in diabetic retinopathy. Ophthalmology 88: 601-612.

Shoffner JM, Lott MT, Voljavec AS, Soueidan SA, Costigan DA, et al. 1989. Spontaneous Kearns-Sayre/chronic external ophthalmoplegia plus syndrome associated with a mitochondrial DNA deletion: A slip-replication model and metabolic therapy. Proc Natl Acad Sci USA 86: 7952-7956.

Siliprandi R, Canella R, Carmignoto G, Schiavo N, Zanellato A, et al. 1992. *N*-methyl-D-aspartate-induced neurotoxicity in the adult rat retina. Vis Neurosci 8: 567-573.

Silver IA, Erecinska M. 1990. Intracellular and extracellular changes of [Ca2+] in hypoxia and ischemia in rat brain in vivo. J Gen Physiol 95: 837-866.

Singh M, Savitz SI, Hoque R, Gupta G, Roth S, et al. 2001. Cell-specific caspase expression by different neuronal phenotypes in transient retinal ischemia. J Neurochem 77: 466-475.

Sipilä I, Rapola J, Simell O, Vannas A. 1981. Supplementary creatine as a treatment for gyrate atrophy of the choroid and retina. N Engl J Med 304: 867-870.

Sohal RS, Ku HH, Agarwal S, Forster MJ, Lal H. 1994. Oxidative damage, mitochondrial oxidant generation and antioxidant defenses during aging and in response to food restriction in the mouse. Mech Ageing Dev 74: 121-133.

Sommer A, Tielsch JM, Katz J, Quigley HA, Gottsch JD, et al. 1991. Racial differences in the cause-specific prevalence of blindness in east Baltimore. N Engl J Med 325: 1412-1417.

Soong NW, Hinton DR, Cortopassi G, Arnheim N. 1992. Mosaicism for a specific somatic mitochondrial DNA mutation in adult human brain. Nat Genet 2: 318-323.

Sparrow JR, Nakanishi K, Parish CA. 2000. The lipofuscin fluorophore A2E mediates blue light-induced damage to retinal pigmented epithelial cells. Invest Ophthalmol Vis Sci 41: 1981-1989.

Spirin KS, Saghizadeh M, Lewin SL, Zardi L, Kenney MC, et al. 1999. Basement membrane and growth factor gene expression in normal and diabetic human retinas. Curr Eye Res 18: 490-499.

Staub F, Mackert B, Kempski O, Peters J, Baethmann A. 1993. Swelling and death of neuronal cells by lactic acid. J Neurol Sci 119: 79-84.

Stokke O, Skjeldal OH, Hoie K. 1986. Disorders related to the metabolism of phytanic acid. Scand J Clin Lab Invest (Suppl 184): 3-10.

Stone EM, Webster AR, Vandenburgh K, Streb LM, Hockey RR, et al. 1998. Allelic variation in ABCR associated with Stargardt disease but not age-related macular degeneration. Nat Genet 20: 328-329.

Stone EM, Lotery AJ, Munier FL, Heon E, Piguet B, et al. 1999. A single EFEMP1 mutation associated with both Malattia Leventinese and Doyne honeycomb retinal dystrophy. Nat Genet 22: 199-202.

Suber ML, Pittler SJ, Qin N, Wright GC, Holcombe V, et al. 1993. Irish setter dogs affected with rod/cone dysplasia contain a nonsense mutation in the rod cGMP phosphodiesterase beta-subunit gene. Proc Natl Acad Sci USA 90: 3968-3972.

Sucher NJ, Lipton SA, Dreyer EB. 1997. Molecular basis of glutamate toxicity in retinal ganglion cells. Vis Res 37: 3483-3493.

Sun H, Nathans J. 2001. ABCR, the ATP-binding cassette transporter responsible for Stargardt macular dystrophy, is an efficient target of all-trans-retinal-mediated photooxidative damage in vitro. Implications for retinal disease. J Biol Chem 276: 11766-11774.

Sun H, Molday RS, Nathans J. 1999. Retinal stimulates ATP hydrolysis by purified and reconstituted ABCR, the photoreceptor-specific ATP-binding cassette transporter responsible for Stargardt disease. J Biol Chem 274: 8269-8281.

Sung CH, Schneider BG, Agarwal N, Papermaster DS, Nathans J. 1991b. Functional heterogeneity of mutant rhodopsins responsible for autosomal dominant retinitis pigmentosa. Proc Natl Acad Sci USA 88: 8840-8844.

Sung CH, Davenport CM, Hennessey JC, Maumenee IH, Jacobson SG, et al. 1991a. Rhodopsin mutations in autosomal dominant retinitis pigmentosa. Proc Natl Acad Sci USA 88: 6481-6485.

Suter M, Reme C, Grimm C, Wenzel A, Jaattela M, et al. 2000. Age-related macular degeneration. The lipofusion component *N*-retinyl-*N*-retinylidene ethanolamine detaches proapoptotic proteins from mitochondria and induces apoptosis in mammalian retinal pigment epithelial cells. J Biol Chem 275: 39625-39630.

Szabo ME, Droy-Lefaix MT, Doly M, Carre C, Braquet P. 1991. Ischemia and reperfusion-induced histologic changes in the rat retina. Demonstration of a free radical-mediated mechanism. Invest Ophthalmol Vis Sci 32: 1471-1478.

Takahashi K, Lam TT, Edward DP, Buchi ER, Tso MO. 1992. Protective effects of flunarizine on ischemic injury in the rat retina. Arch Ophthalmol 110: 862-870.

Takki K, Simell O. 1974. Genetic aspects in gyrate atrophy of the choroid and retina with hyperornithinaemia. Br J Ophthalmol 58: 907-916.

Thoreson WB, Witkovsky P. 1999. Glutamate receptors and circuits in the vertebrate retina. Prog Retin Eye Res 18: 765-810.

Tombaugh GC, Sapolsky RM. 1993. Evolving concepts about the role of acidosis in ischemic neuropathology. J Neurochem 61: 793-803.

Tombran-Tink J, Shivaram SM, Chader GJ, Johnson LV, Bok D. 1995. Expression, secretion, and age-related downregulation of pigment epithelium-derived factor, a serpin with neurotrophic activity. J Neurosci 15: 4992-5003.

Toriu N, Akaike A, Yasuyoshi H, Zhang S, Kashii S, et al. 2000. Lomerizine, a Ca^{2+} channel blocker, reduces glutamate-induced neurotoxicity and ischemia/reperfusion damage in rat retina. Exp Eye Res 70: 475-484.

Toussaint D, Danis P. 1971. An ocular pathologic study of Refsum's syndrome. Am J Ophthalmol 72: 342-347.

Travis GH, Sutcliffe JG, Bok D. 1991. The retinal degeneration slow (RDS) gene product is a photoreceptor disc membrane-associated glycoprotein. Neuron 6: 61-70.

Travis GH, Brennan MB, Danielson PE, Kozak CA, Sutcliffe JG. 1989. Identification of a photoreceptor-specific mRNA encoded by the gene responsible for retinal degeneration slow (RDS). Nature 338: 70-73.

Tucker GS. 1986. Refractile bodies in the inner segments of cones in the aging human retina. Invest Ophthalmol Vis Sci 27: 708-715.

Tucker CL, Woodcock SC, Kelsell RE, Ramamurthy V, Hunt DM, et al. 1999. Biochemical analysis of a dimerization domain mutation in RetGC-1 associated with dominant cone-rod dystrophy. Proc Natl Acad Sci USA 96: 9039-9044.

Vanhoutte PM. 2004. Endothelium-dependent hyperpolarizations: The history. Pharmacol Res 49: 503-508.

Vannas-Sulonen K, Sipilä I, Vannas A, Simell O, Rapola J. 1985. Gyrate atrophy of the choroid and retina. A five-year follow-up of creatinine supplementation. Ophthalmology 92: 1719-1727.

Vincent KA, Feron O, Kelly RA. 2002. Harnessing the response to tissue hypoxia: HIF-1 alpha and therapeutic angiogenesis. Trends Cardiovasc Med 12: 362-367.

Vivanco I, Sawyers CL. 2002. The phosphatidylinositol 3-kinase AKT pathway in human cancer. Nat Rev Cancer 2: 489-501.

Vorwerk CK, Kreutz MR, Dreyer EB, Sabel BA. 1996. Systemic L-kynurenine administration partially protects against NMDA, but not kainate-induced degeneration of retinal ganglion cells, and reduces visual discrimination deficits in adults rats. Invest Ophthalmol Vis Sci 37: 2382-2392.

Wahl F, Obrenovitch TP, Hardy AM, Plotkine M, Boulu R, et al. 1994. Extracellular glutamate during focal cerebral ischaemia in rats: Time course and calcium dependency. J Neurochem 63: 1003-1011.

Wallace DC. 1992. Mitochondrial genetics: A paradigm for aging and degenerative diseases? Science 256: 628-632.

Wallace DC. 2001. Mouse models for mitochondrial disease. Am J Med Genet 106: 71-93.

Wang C, Nguyen HN, Maguire JL, Perry DC. 2002. Role of intracellular calcium stores in cell death from oxygen-glucose deprivation in a neuronal cell line. J Cereb Blood Flow Metab 22: 206-214.

Wassell J, Davies S, Bardsley W, Boulton M. 1999. The photoreactivity of the retinal age pigment lipofuscin. J Biol Chem 274: 23828-23832.

Wassell J, Ellis S, Burke J, Boulton M. 1998. Fluorescence properties of autofluorescent granules generated by cultured human RPE cells. Invest Ophthalmol Vis Sci 39: 1487-1492.

Weale RA. 1975. Senile changes in visual acuity. Trans Ophthalmol Soc UK 95: 36-38.

Weber AJ, Chen H, Hubbard WC, Kaufman PL. 2000. Experimental glaucoma and cell size, density, and number in the primate lateral geniculate nucleus. Invest Ophthalmol Vis Sci 41: 1370-1379.

Weber BH, Vogt G, Wolz W, Ives EJ, Ewing CC. 1994. Sorsby's fundus dystrophy is genetically linked to chromosome 22q13-qter. Nat Genet 7: 158-161.

Webster AR, Heon E, Lotery AJ, Vandenburgh K, Casavant TL, et al. 2001. An analysis of allelic variation in the ABCA4 gene. Invest Ophthalmol Vis Sci 42: 1179-1189.

Weiss JH, Sensi SL, Koh JY. 2000. Zn(2+): A novel ionic mediator of neural injury in brain disease. Trends Pharmacol Sci 21: 395-401.

Weleber RG, Kennaway NG, Buist NR. 1981. Gyrate atrophy of the choroid and retina. Approaches to therapy. Int Ophthalmol 4: 23-32.

Wells J, Wroblewski J, Keen J, Inglehearn C, Jubb C, et al. 1993. Mutations in the human retinal degeneration slow (RDS) gene can cause either retinitis pigmentosa or macular dystrophy (see comments). Nat Genet 3: 213-218.

Weng J, Mata NL, Azarian SM, Tzekov RT, Birch DG, et al. 1999. Insights into the function of Rim protein in photoreceptors and etiology of Stargardt's disease from the phenotype in abcr knockout mice. Cell 98: 13-23.

Wilkie SE, Newbold RJ, Deery E, Walker CE, Stinton I, et al. 2000. Functional characterization of missense mutations at codon 838 in retinal guanylate cyclase correlates with disease severity in patients with autosomal dominant cone-rod dystrophy. Hum Mol Genet 9: 3065-3073.

Willison K. 1991. Opposite imprinting of the mouse Igf2 and Igf2r genes. Trends Genet 7: 107-109.

Winkler BS, Boulton ME, Gottsch JD, Sternberg P. 1999. Oxidative damage and age-related macular degeneration. Mol Vis 5: 32.

Witmer AN, Vrensen GF, Van Noorden CJ, Schlingemann RO. 2003. Vascular endothelial growth factors and angiogenesis in eye disease. Prog Retin Eye Res 22: 1-29.

Wood JP, Osborne NN. 2001. The influence of zinc on caspase-3 and DNA breakdown in cultured human retinal pigment epithelial cells. Arch Ophthalmol 119: 81-88.

Wright CF, McKenney K, Hamer DH, Byrd J, Winge DR. 1987. Structural and functional studies of the amino terminus of yeast metallothionein. J Biol Chem 262: 12912-12919.

Wymann MP, Pirola L. 1998. Structure and function of phosphoinositide 3-kinases. Biochim Biophys Acta 1436: 127-150.

Yamada T, Hara S, Tamai M. 1990. Immunohistochemical localization of cathepsin D in ocular tissues. Invest Ophthalmol Vis Sci 31: 1217-1223.

Yang P, McKay BS, Allen JB, Jaffe GJ. 2004. Effect of NF-kappa B inhibition on TNF-alpha-induced apoptosis in human RPE cells. Invest Ophthalmol Vis Sci 45: 2438-2446.

Yoles E, Schwartz M. 1998. Degeneration of spared axons following partial white matter lesion: Implications for optic nerve neuropathies. Exp Neurol 153: 1-7.

Yoneda S, Tanihara H, Kido N, Honda Y, Goto W, et al. 2001. Interleukin-1beta mediates ischemic injury in the rat retina. Exp Eye Res 73: 661-667.

Yoshida S, Yashar BM, Hiriyanna S, Swaroop A. 2002. Microarray analysis of gene expression in the aging human retina. Invest Ophthalmol Vis Sci 43: 2554-2560.

Young RW. 1976. Visual cells and the concept of renewal. Invest Ophthalmol Vis Sci 15: 700-725.

Young RW. 1987. Pathophysiology of age-related macular degeneration. Surv Ophthalmol 31: 291-306.

Young C, Tenkova T, Dikranian K, Olney JW. 2004. Excitotoxic versus apoptotic mechanisms of neuronal cell death in perinatal hypoxia/ischemia. Curr Mol Med 4: 77-85.

Yucel YH, Zhang Q, Weinreb RN, Kaufman PL, Gupta N. 2003. Effects of retinal ganglion cell loss on magno-, parvo-, koniocellular pathways in the lateral geniculate nucleus and visual cortex in glaucoma. Prog Retin Eye Res 22: 465-481.

Zeevalk GD, Nicklas WJ. 1990. Chemically induced hypoglycemia and anoxia: Relationship to glutamate receptor-mediated toxicity in retina. J Pharmacol Exp Ther 253: 1285-1292.

Zeevalk GD, Nicklas WJ. 1992. Evidence that the loss of the voltage-dependent Mg^{2+} block at the *N*-methyl-D-aspartate receptor underlies receptor activation during inhibition of neuronal metabolism. J Neurochem 59: 1211-1220.

Zhang X, Chaudhry A, Chintala SK. 2003. Inhibition of plasminogen activation protects against ganglion cell loss in a mouse model of retinal damage. Mol Vis 9: 238-248.

Zhang Y, Marcillat O, Giulivi C, Ernster L, Davies KJ. 1990. The oxidative inactivation of mitochondrial electron transport chain components and ATPase. J Biol Chem 265: 16330-16336.

Zhang K, Kniazeva M, Han M, Li W, Yu Z, et al. 2001. A 5-bp deletion in ELOVL4 is associated with two related forms of autosomal dominant macular dystrophy. Nat Genet 27: 89-93.

4 Cellular and Molecular Mechanisms of Mechanical Amplification in the Mammalian Cochlea

X. Wu · J. Zuo

Abstract: The mammalian cochlea is an amazing evolutionary achievement. The active mechanism housed in the cochlea, the cochlear amplifier, can amplify weak sound stimuli 100–1,000 times over a wide range of incoming frequencies. It confers remarkable sensitivity and exquisite frequency selectivity to the mammalian auditory system. The inner ear contains two types of sensory cells: inner hair cells, which convert the mechanical signal to an electrical signal, and outer hair cells (OHCs), which amplify the mechanical vibration of the cochlea's basilar membrane. OHCs can elongate and contract in response to an oscillation of membrane potential at frequencies of at least 70 kHz. This salient feature of OHCs, termed electromotility, forms the cellular basis of the sound amplification process. Electromotility originates in the motor protein particles embedded in the lateral plasma membrane of the OHCs. The motor protein—prestin—has been cloned. Several different molecular models of the hair cell motor complex have been proposed based on recent studies. This review aims to provide an update on the cellular and molecular mechanisms of cochlear amplification.

List of Abbreviations: ACh, acetylcholine; DPOAE, distortion product otoacoustic emission; IHCs, inner hair cells; MET, mechanoelectric transducel; NLC, nonlinear capacitance; OHCs, outer hair cells; PFAE, promyclocytic leukemia zinc finger protein; RC, resistance and capacitance; SLC, solute-carrier family; SSC, sub-surface cistern

1 Introduction

The delicacy of the mammalian auditory system is epitomized by its remarkable sensitivity, exquisite frequency resolution, and ability to operate over a wide frequency spectrum (Dallos, 1992; Hudspeth, 2000; Robles and Ruggero, 2001). These salient features depend mainly on the biomechanical properties of the cochlea, the sensory organ of the mammalian auditory system. The cochlea hosts a positive feedback loop—the cochlear amplifier—that can amplify an incoming sound (Gold, 1948; Davis, 1983). The mechanism by which the cochlear amplifier operates, unclear for several decades, is now being elucidated through the use of powerful tools in physiology and molecular genetics.

2 Structure of the Cochlea

2.1 General Anatomy

The snail-shaped cochlea, located in the temporal bone of the skull, contains a bony labyrinth and a membranous labyrinth. The bony labyrinth consists of the otic capsule (the external shell) and the modiolus (the internal axis). The membranous labyrinth, coiled inside the bony labyrinth, consists of three adjacent tubes: the scala vestibuli, the scala media, and the scala tympani (➲ *Figure 4-1*). The scala vestibuli and the scala media are separated by Reissner's membrane; the scala media and the scala tympani are separated by the basilar membrane and part of the osseous spiral lamina. The scala vestibuli and the scala tympani are filled with perilymph, a fluid whose ionic composition is similar to that of cerebrospinal fluid. The fluid sealed inside the scala media, the endolymph, contains a high concentration of potassium.

◘ Figure 4-1

Physiological anatomy of the inner ear. Incoming sound signals collected by the pinna stimulate the eardrum in the middle ear (a). Vibration of the eardrum is transferred to the oval window on the underside of the cochlea (b) via the three middle-ear ossicles. The vibration signal triggers an oscillation of pressure among the scala vestibuli, scala media, and scala tympani (c) that produces a shearing movement between the tectorial membrane and the cuticular plate (d). Deflection of the hair bundles as a result of the movement of the tectorial membrane causes the MET channels at stereocilia of hair cells to open. Then, potassium influx into the IHCs changes the IHCs' intracellular voltage. The resulting change in the IHCs' membrane potential releases glutamate into the synapses and, in turn, activates the afferent nerves, which transfer the electrical signal to the central nervous system. Adapted from *Figure 13-4* of Purves and group (2001). Permission is granted for reproduction in this review

Figure 4-1 (continued)

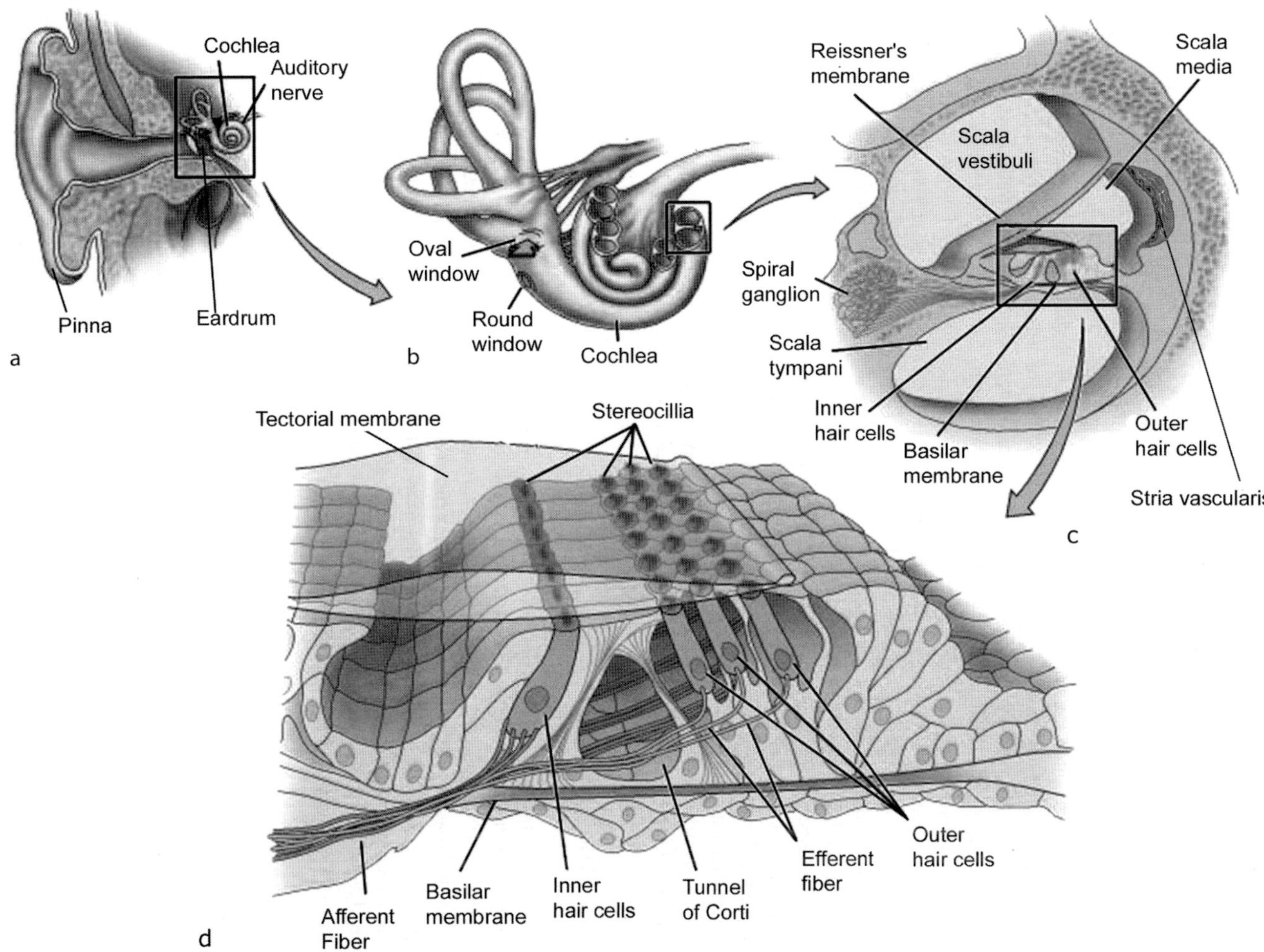

It is separated from the perilymph by virtue of tight junctions between adjacent cells that line the boundaries of the scala media (reviewed by Slepecky, 1996; *Figure 4-1 a–c*).

2.2 Organ of Corti

The organ of Corti resides in the scala media (*Figure 4-1 c, d*). It is a highly differentiated epithelium that rests on the basilar membrane and consists of different types of sensory cells ("hair" cells) and supporting cells. Each cochlea contains approximately 20,000 hair cells, depending on the species (Dallos, 1992; Hudspeth, 2000). Hair cells aligned in a tonotopic manner along the longitudinal axis of the spiral basilar membrane: hair cells in the apical turn respond to low-pitch sounds, whereas those in the basal turn respond to high-pitch sounds (Dallos, 1992; Hudspeth, 2000). Across the basilar membrane and emanating from it in a radial direction are four rows of hair cells. The innermost row, which consists of the inner hair cells (IHCs), is crowned with a crescent-shaped bundle of stereocilia. The three outer rows, which consist of the outer hair cells (OHCs), bear V- or W-shaped stereocilia bundles. The stereocilia are anchored in a matrix of dense filaments, the cuticular plate, on top of the hair cells. Overlying the hair cells is a collagenous-layered structure, the tectorial membrane. The OHCs' longest stereocilia are in contact with the bottom of the tectorial membrane. The stereocilia are deflected by a shearing movement between the tectorial membrane and the cuticular plate that results from a fluctuation in pressure between the scala vestibuli and the other two scalae. This pressure fluctuation is caused by the incoming sound waves, which are transferred to the oval window of the cochlea via the three small ossicles (malleus, incus, and stapes) of the middle ear. Deflection of the stereocilia opens mechanoelectric transducer (MET) channels at their tips to produce transduction currents and receptor potentials within the hair cells. These changes in receptor potentials cause the IHCs to release glutamate into synapses, an event that elicits action potentials in afferent auditory nerve fibers (see later) (Dallos, 1992; Hudspeth, 2000). Acoustic information is then transferred to the central nervous system via the ascending auditory pathway.

2.3 Stria Vascularis

On the lateral side of the scala media is the stria vascularis, a vascularized epithelium (*Figure 4-1 c*) composed of three cell layers: the marginal, basal, and intermediate cell layers. Large numbers of Na^+/K^+ ATPase and $Na^+/2Cl^-/K^+$ cotransporters are expressed on the surfaces of marginal cells facing the cochlear duct and basal cells facing the spiral ligament (Nakazawa et al., 1995b). Na^+/K^+ ATPase transports potassium into the scala media against its ionic gradient to maintain a high potassium concentration (approximately 140 mM) and positive endolymphatic potential (approximately +80 mV) in the scala media (Offner et al., 1987). The stria vascularis is considered the engine of ionic recycling in the cochlea. It provides energy for the operation of MET channels in hair cell stereocilia (Wangemann, 2002).

2.4 Innervation of Hair Cells

The nerve fibers in the cochlea can be classified as afferent fibers (toward the brain) or efferent fibers (toward the periphery). The afferent fibers that terminate on IHCs constitute approximately 90–95% of all afferent fibers in the cochlea. These afferent fibers originate from type I ganglion cells and are coated with a thick myelin sheath. Each fiber is connected to only one IHC, but each IHC is innervated by 2–10 individual afferent fibers. The principal neurotransmitter released by IHCs, glutamate, activates the afferent fibers in a quantal manner. The remaining 5–10% of afferent fibers are connected to OHCs and are unmyelinated. They originate from type II ganglion cells. Each afferent fiber is highly branched and is connected to 6–100 OHCs. OHCs are innervated mainly by efferent fibers, which are derived from the medial olivocochlear bundle. Efferent fibers release acetylcholine (ACh) as their principal neurotransmitter. Each OHC is in contact with 2–5 efferent synaptic boutons at its base. IHCs receive axodendritic efferent innervation onto afferent fibers from the lateral olivocochlear bundle (reviewed by Raphael and Altschuler, 2003).

3 Auditory Signaling in the Cochlea

3.1 Passive and Active Cochlea

Early studies of human cadavers led Von Bekesy to propose that the cochlea works as a wide-range frequency resonator (reviewed by Dallos, 1992; Robles and Ruggero, 2001) (➋ *Figure 4-2*). A stiffness gradient from the basal to apical basilar membrane defines the frequency resonance map. The basal portion of the basilar membrane is thicker, shorter, and stiffer than the apical portion. Therefore, the basal portion of basilar membrane resonates in response to high-pitch sounds whereas the apical basilar membrane resonates only in response to low-pitch sounds. However, in the living cochlea, a viscous force originating in the cochlear fluid and the organ of Corti itself can easily damp even the largest vibrations induced by passive sound resonance. Gold (1948) therefore predicted the existence of a positive electromechanical feedback mechanism in the inner ear that could boost the vibration of the basilar membrane and counteract the damping effect of the viscous force. This activity of the cochlea, later termed the "cochlear amplifier" by Davis in 1983, was shown by various lines of evidence to be intimately associated with the OHCs (Ryan and Dallos, 1975; Dallos and Harris, 1978; Harrison and Evans, 1979; Mountain, 1980).

Figure 4-2

The active cochlea. The straightened basilar membrane in the cochlea is similar to a wide-range frequency resonator. The basal part of the basilar membrane is narrow and thick, whereas the apical part is wide and thin. This decreasing stiffness and mass gradient from the basal to the apical part of the membrane corresponds to the tonotopic frequency gradient from high pitch to low pitch. However, the living cochlea requires a positive feedback loop to achieve the high sensitivity and frequency selectivity that are characteristic of mammalian hearing. The existence of such an active amplification process, termed cochlear amplification, has been confirmed by many lines of experimental evidence

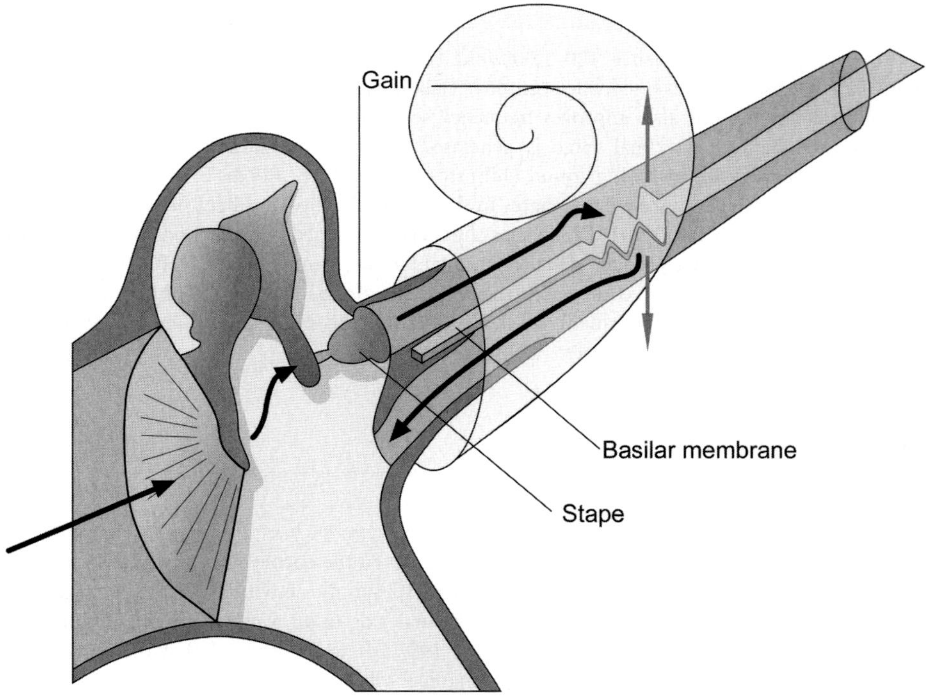

3.2 Otoacoustic Emission

The first experimental evidence supporting the concept of a cochlear amplifier was the discovery of otoacoustic emissions (Kemp, 1978), sound emissions produced by the active mechanical processes of the cochlea and propagated from the inner ear to the tympanum in the middle ear (Wilson, 1980). The connection between otoacoustic emissions and OHCs was suggested by the vulnerability of the otoacoustic emissions to electrical stimulation of the olivocochlear bundle and to OHC-specific ototoxic drugs (Mountain, 1980; Siegel and Kim, 1982; Lonsbury-Martin et al., 1987; Brown et al., 1989). Therefore, otoacoustic emissions have been widely used to examine the cochlear amplifier and OHC function. To experimentally examine an otoacoustic emission, two different frequencies (f_2 and f_1, with $f_2 > f_1$) can be used as input stimuli to generate a distortion product otoacoustic emission (DPOAE) (Kemp, 1979; Kim et al., 1980). Distorted sounds produced in DPOAE assays can be detected at frequencies that are different combinations of $n f_1 + m f_2$ (where n, m are any positive or negative integrals), of which the most prominent is at $2f_2 - f_1$ (Lonsbury-Martin and Martin, 1990). The DPOAE is frequently used to assess the function of the cochlea in humans and in laboratory animals because it is noninvasive and sensitive and it occurs over a relatively wide frequency range (Lonsbury-Martin and Martin, 1990; Jimenez et al., 1999).

3.3 The Cochlear Amplifier

The cochlear amplifier can enhance 100–1,000-fold (i.e., by 40–60 dB) the vibration of the basilar membrane in response to a weak acoustic stimulus (Dallos, 1992; Ashmore, 2002). It not only increases the sensitivity of hearing but also improves frequency selectivity by sharpening the tuning curve. This sharpening occurs because maximal force is generated only at the characteristic frequency position corresponding to that of the incoming sound (Johnstone et al., 1986; Robles et al., 1991; Olson, 2004). The cochlear amplifier's energy source originates in the electrochemical gradient between the endolymph and the OHC cytoplasm (Mills, 2004). Transducer channels are opened by deflections of the stereocilia on OHCs, and the resulting voltage gradient drives an influx of ionic current into the OHCs, thereby triggering them to generate force. This force increases the vibration of the basilar membrane (Dallos, 1992). Pharmacologic or pathologic impairment of the endolymphatic potential causes dysfunction of the cochlear amplifier (Ruggero and Rich, 1991; Mills and Schmiedt, 2004).

4 OHC Electromotility

4.1 Electromotility

The suggestion that an active amplification mechanism exists in the cochlea was particularly strengthened by the discovery that OHCs exhibit "electromotility," an OHC-specific activity in which the cylindrical body of the OHC elongates and shortens along its longitudinal axis upon a change in membrane potential (Brownell et al., 1985; Kachar et al., 1986; Ashmore, 1987). It was presumed that the contraction and elongation process feeds energy back into the mechanical vibration on a cycle-by-cycle basis and thus amplifies the movement of the basilar membrane (Dallos, 1992; Gitter et al., 1993). Electromotile responses by isolated OHCs can occur at a frequency of at least 70 kHz (Frank et al., 1999). This response occurs faster than do traditional biological force-generating processes, which normally require cell signaling and chemical cascades (Frank et al., 1999). Calcium is not directly involved in OHC electromotility (Ashmore, 1987), and treatment of OHCs with a nonhydrolyzable ATP analog that competes with endogenous ATP does not inhibit electromotility (Holley and Ashmore, 1988b). The latter finding suggested that electromotility is an ATP-independent property of OHCs (Holley and Ashmore, 1988b). The magnitude of the OHCs' electromotile response is, however, decreased by a change in membrane tension that results from reducing the intracellular turgor pressure (Santos-Sacchi, 1991; Sziklai and Dallos, 1997; Adachi et al., 2000). It was subsequently confirmed that the electromotility of OHCs originates in the plasma membrane, another feature of electromotility that is different from traditional cellular motility involving the cytoskeleton (Holley et al., 1992; Takahashi and Santos-Sacchi, 2001).

4.2 Nonlinear Capacitance

The electromotility of OHCs is commonly assessed by nonlinear capacitance (NLC) assay. NLC provides an independent indication of electromotility. A simple plasma membrane without any voltage-gated channels is normally modeled as a passive capacitor, because it is considered an insulating lipid bilayer flanked by fluids (intracellular cytoplasm and extracellular fluids) through which an electric current can flow. The passive capacitance of a plasma membrane is defined by the membrane's surface area and dielectric properties, and it displays linearity with respect to the membrane potential. Changes in membrane potential in plasma membranes that contain voltage-gated channels can cause conformational changes in those channels. In voltage-gated potassium channels, for example, an applied voltage causes the so-called S4 voltage sensor (normally the fourth transmembrane domain of the channel) to change its helix conformation and shift the position of its positively charged helix to open the channel (Bezanilla, 2000). Because the capacitance is also defined as the total electric charge divided by the voltage, membrane capacitance shows nonlinearity with respect to the voltage applied during the gating process. In 1990, Ashmore reported that for OHCs, a plot of NLC against membrane potential produces a bell-shaped curve. It is observed that the OHC's gating current is concurrent with the cell's motility (Ashmore, 1990; Santos-Sacchi, 1991). NLC represents the possibility of the charged particles being shifted between two opposite sides of the OHC plasma membrane (Dallos and Fakler, 2002). The peak of the bell-shaped NLC curve occurs at the membrane potential at which the charged particle can move across the plasma membrane at maximal efficiency (Dallos and Fakler, 2002).

4.3 The Lateral Wall of the OHC

Electromotility is generated along the OHC's lateral wall (Dallos et al., 1991; Kalinec et al., 1992; Huang and Santos-Sacchi, 1994), a highly specified structure composed of three distinct layers: the plasma membrane, the cortical lattice, and the subsurface cistern (SSC) (Slepecky, 1996) (➜ *Figure 4-3*). The SSC is a type of endoplasmic membrane structure akin to the canalicular reticulum, a structure specific to ion-transporting epithelia (Spicer et al., 1998). In human and mouse OHCs, the SSCs form a single layer of membranous sacs parallel to the lateral plasma membrane. Ca^{+} ATPase and CaM kinase IV are highly expressed in the SSC, which is therefore considered a Ca^{+} reservoir for OHCs (Schulte, 1993; Zine and Schweitzer, 1996; Koyama et al., 1999). The cortical lattice between the SSC and the plasma membrane is webbed with spectrin and actin filaments (Holley and Ashmore, 1988a). It is composed of many small individual microdomain sheets that adjoin each other at various angles (Holley et al., 1992). The cortical lattice and the plasma membrane are connected by an array of pillar-like structures (Arima et al., 1991). Freeze-fracture electron microscopy revealed a high density of protein particles in the lateral plasma membranes of OHCs (Kalinec et al., 1992; Gulley and Reese, 1977). The protein particles are estimated to have a diameter of 8–15 nm and to occur at a density of approximately 4,000 per μm^2 of membrane (Kalinec et al., 1992). Somatic motility of OHCs is thought to result from conformational alterations of these protein particles, and dynamic changes in the stiffness of the lateral wall are thought to be driven by oscillation of the membrane potential (Kalinec et al., 1992; Santos-Sacchi, 2002).

4.4 Modulation of Electromotility

OHCs undergo two categories of motility: electromotility, which is fast and occurs on a submillisecond time scale, and slow motility, which happens on a second-to-minute timescale (Brownell et al., 1985; Dulon and Schacht, 1992; Puschner and Schacht, 1997; Frank et al., 1999). Slow motility of OHCs is driven not by

Figure 4-3
Trilaminar structure of the OHC's lateral wall. The three distinct layers are the SSC layer, the cortical lattice, and the plasma membrane. The SSCs are membranous sacs akin to the endoplasmic reticulum. The space between the SSC layer and the plasma membrane is occupied by the cortical lattice, which is composed of actin and spectrin filaments connected to the plasma membrane by an array of pillar-like structures. Protein particles embedded at high density in the plasma membrane are considered the motor proteins that drive the OHC's electromotility. Adapted from *Figure 5-2* of Brownell and group (2001). Reprinted, with permission, from *Annual Review of Biomedical Engineering*, Volume 32001 by Annual Review www.annualreviews.org

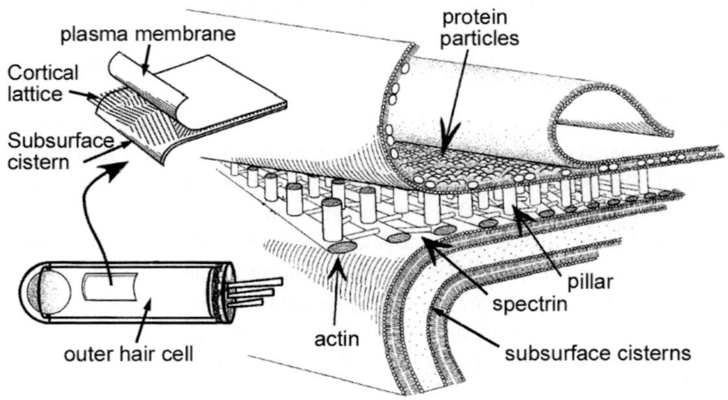

changes in membrane potential but by intracellular Ca^+ release or environmental challenges (e.g., high potassium bathing, high osmotic tension) (Dulon et al., 1988; Dulon and Schacht, 1992; Kakehata and Santos-Sacchi, 1995; Puschner and Schacht, 1997; Szonyi et al., 2001). The OHC's slow motility processes are involved in cytoskeleton reorganization mediated by phosphorylation of the actin–spectrin network (Zenner et al., 1985; Slepecky, 1989; Dulon and Schacht, 1992). The OHC's electromotility is modulated by ACh and other neurotransmitters via the efferent fibers that innervate them (Batta et al., 2004). ACh may modulate the electromotile output by affecting the OHC's slow motility and the stiffness of OHCs (Dulon and Schacht, 1992; Dallos et al., 1997; Frolenkov et al., 2000; He et al., 2003). ACh-based regulation is mediated by intracellular Rho-GTPase and calcium signaling (Dallos et al., 1997; Frolenkov et al., 2000; Kalinec et al., 2000; He et al., 2003; Zhang et al., 2003; Batta et al., 2004; Borko et al., 2005). The electromotile process in the OHC is accompanied by a change in the cell's axial stiffness (He and Dallos, 1999, 2000). The stiffness of the motor component of the OHC's plasma membrane also contributes notably to the overall axial stiffness of the OHC (He et al., 2003). Recently, cGMP-PKG-dependent phosphorylation of an OHC motor protein was shown to directly change the stiffness of the plasma membrane and modulate OHC electromotility (Szonyi et al., 1999; Deak et al., 2005). These modulations are believed to provide negative feedback to prevent acoustic overstimulation or injury to OHCs (Zenner, 1986).

4.5 Pharmacologic Inhibition of Electromotility

Salicylate, a well-known ototoxic drug, can reduce hearing sensitivity. A confluence of evidence has suggested that the main target of salicylate in the cochlea is the OHC (McFadden and Plattsmier, 1984; Santos-Sacchi, 1991; Shehata et al., 1991; Kakehata and Santos-Sacchi, 1996). Salicylate also reversibly inhibits OHC electromotility by competing with chloride to bind to the motor protein in the lateral membrane (Oliver et al., 2001). OHC electromotility can also be suppressed by the ototoxic drug quinine at high concentrations (Zheng et al., 2001b). Quinine also affects the OHC's resting potential, somatic stiffness, and efferent innervation (Lin et al., 1995; Yamamoto et al., 1997; Jarboe and Hallworth, 1999; Zheng et al., 2001b). Some ototoxic drugs can also affect protein–lipid interactions in the OHC plasma membrane (Oghalai et al., 2000). A good example is chlorpromazine, an ionic amphipathic drug that changes OHC motility by affecting membrane curvature and fluidity (Oghalai et al., 2000; Lue et al., 2001).

5 Molecular Identity of the OHC Motor

5.1 Motor Proteins of the OHC

The prevailing hypothesis is that most of the protein particles embedded in the lateral wall of the OHC are motor proteins (Kalinec et al., 1992; Santos-Sacchi, 2002). The primary reason for this hypothesis is that the OHC's gating current can be detected only along its lateral wall, where the force originates (Huang and Santos-Sacchi, 1993). The density of the charge movement during the gating of the OHC was estimated to be 8,000 per μm^2, which was several folds higher than the density of the protein particles in the lateral wall (Kalinec et al., 1992; Huang and Santos-Sacchi, 1993; Gale and Ashmore, 1997a). A unique feature of the OHC motor is that it can be driven by both the membrane potential and by mechanical stimulation bidirectionally (Dong et al., 2002). In one direction, OHC somatic motility can be driven by the change of the membrane potential; conversely, using a piezoelectric bimorph force to stretch and shorten an isolated OHC can generate a transient bidirectional current in OHC lateral wall (Gale and Ashmore, 1997b).

Using NLC as an indicator, it has been shown that several experimental parameters can affect the function of the motor protein. First, an increase in turgor pressure in the OHC shifts the NLC curve toward the depolarization side (Kakehata and Santos-Sacchi, 1995; Adachi et al., 2000). Second, stepwise elevations (from +40 to −120 mV) of the preconditioning membrane potential of the OHC also shift the NLC curve to the depolarization side (Santos-Sacchi et al., 1998). Finally, the peak voltage of NLC measured in isolated OHCs also shifts 20–25 mV rightward (depolarizing) when the environmental temperature is increased by 10°C (Santos-Sacchi and Huang, 1998).

5.2 GLUT5 as the Motor Protein Candidate and the Possible Model

In 1999, Geleoc and group proposed GLUT5 as a candidate motor protein for OHCs. Previous studies using immunostaining, immunogold labeling, or in situ hybridization had demonstrated that GLUT5 is expressed in the lateral wall of OHCs from different species (Nakazawa et al., 1995a; Geleoc et al., 1999; Belyantseva et al., 2000). GLUT5 belongs to a family of facilitative hexose transporters. Its specific substrate is fructose (Inukai et al., 1995; Corpe et al., 2002). Applying fructose to OHCs results in their shortening and a shift in the NLC (Geleoc et al., 1999; Ashmore et al., 2000). The rate of shortening induced by fructose depends on the preconditioning membrane potential (Geleoc et al., 1999; Ashmore et al., 2000). A charge dipole in the second transmembrane helix of GLUT5 could be the voltage sensor (Ashmore et al., 2000), which, in the model proposed by Ashmore and colleagues (➋ *Figure 4-4 a*), is within the motor protein itself and is involved in the conformational change in GLUT5 (Ashmore et al., 2000; Meech and Holley, 2001). Depolarization of the membrane potential triggers a conformational change in the channel and releases a neutral molecule such as fructose from its bound state in the motor protein (Ashmore et al., 2000; Meech and Holley, 2001).

5.3 Prestin as the Motor Protein Candidate and the Possible Model

Using subtractive hybridization, Peter Dallos' group (Zheng et al., 2000) enriched for transcripts expressed in OHCs but not in nonmotile IHCs. They identified a novel gene, prestin, which they proposed encodes the OHC motor protein. It was named after the musical term "presto" meaning fast tempo. The prestin gene encodes an 81-kDa protein containing 744 amino acids, which is a member of the anion transporter, solute-carrier family (SLC) 26 (Zheng et al., 2000). Prestin is predicted to consist of 10–12 transmembrane domains and to have both its carboxyl (C) and amino (N) termini intracellularly located

Figure 4-4

Models of the OHC's motor complex. (a). Ashmore and his colleagues (2000) proposed that fructose (F) can bind to a transporter that has charged side groups as its intrinsic voltage sensor. Hyperpolarization in membrane potential translocates the voltage sensor and exposes intracellular fructose-binding sites. The binding of the sugar to the intracellular site causes a conformational change in the transporter. (b). Oliver and group (2001) proposed that Cl^- is the motor protein's extrinsic voltage sensor. During hyperpolarization, Cl^- binding to an internal site of the motor translocates the sensor without reaching the external surface of the channel and enlarges the motor. This enlargement subsequently leads to the expanding of the membrane surface area. (c). Santos-Sacchi (2003) proposed that the motor protein has its own voltage sensor, and the binding of Cl^- mainly allows the motor protein to remain in its contracted conformation

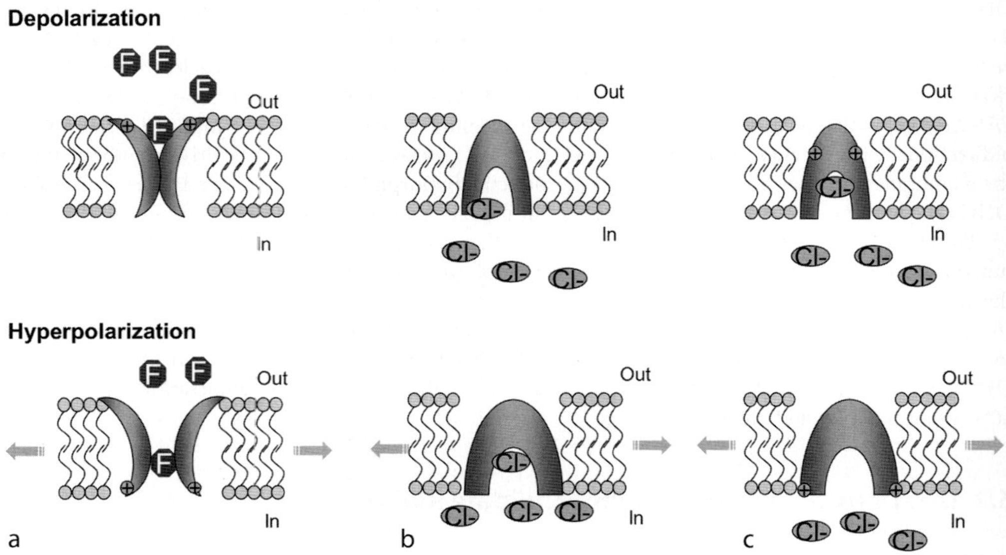

(Zheng et al., 2001a; Navaratnam et al., 2005). The C-terminal portion of prestin, composed of 245 amino acids, contains a conserved sequence domain named the sulfate transporter and antisigma factor (STAS) domain (Zheng et al., 2001a). Immunostaining demonstrated that prestin is highly expressed along the lateral plasma membrane of the OHC (Zheng et al., 2000). The developmental expression profile of prestin is tightly correlated with the OHC's acquisition of electromotility (Belyantseva et al., 2000). Transfection of prestin into human embryonic kidney (HEK) 293T cells conferred those cells with some OHC-specific features, such as NLC and electromotility up to 20 kHz (Zheng et al., 2000). In addition, electromotility in the prestin-transfected HEK 293T cells showed reciprocal sensitivity to mechanical tension and voltage changes (Ludwig, et al., 2001) and could be modulated by treatment with salicylate (Zheng, et al., 2000). We demonstrated in vivo that prestin is indeed required for OHC electromotility and cochlear amplification by generating and characterizing prestin-null mice (Liberman et al., 2002). In these mice, OHC electromotility is abolished and hearing thresholds are elevated by 40–60 dB, as expected (Liberman et al., 2002). Furthermore, frequency selectivity, the remarkable property of mammalian cochleae, is abolished in prestin-knockout mice, further demonstrating the contribution of prestin to the cochlear amplifier (Cheatham, et al., 2004).

Anions such as chloride and bicarbonate are required for the normal function of prestin (Oliver et al., 2001). Chloride has been proposed to serve as an extrinsic voltage sensor that acts to change the conformation of prestin by binding to certain sites during voltage changes (Oliver et al., 2001) (*Figure 4-4 b*). However, another study showed that replacing chloride with sulfate does not change the gating valence, despite shifting the NLC (Rybalchenko and Santos-Sacchi, 2003). In addition, an increase in the intracellular chloride concentration shifts the peak NLC toward more negative potentials

(Rybalchenko and Santos-Sacchi, 2003). Santos-Sacchi proposed that the anion serves only as an allosteric modulator rather than an extrinsic voltage sensor (Santos-Sacchi, 2003) (➲ *Figure 4-4 c*). The intrinsic charged residues of prestin sense the polarization of the membrane, whereas the binding of the anion increases the probability of prestin remaining in the contracted state, which leads to shortening of the hair cell (Santos-Sacchi, 2003). During hyperpolarization, the release of the anion restores prestin to its original conformation (Santos-Sacchi, 2003).

5.4 A Motor Complex Containing More Than One Molecule?

The size (8–14 μm in diameter) of the membrane particles in OHCs is approximately 2–3 times as large as one molecule of either prestin or GLUT5. The change in surface area during charge transfer in a native OHC is also 3–4 times as large as that in prestin-transfected cells (Dong and Iwasa, 2004), although these cells exhibit similar charge displacement (approximately 0.8 *e*) during conformational transitions (Dong and Iwasa, 2004). In prestin-transfected HEK 293T cells, the sensitivity of the NLC curve to changes in environmental temperature and preconditioning voltage is lower than that of native OHCs (Santos-Sacchi et al., 2001; Dong and Iwasa, 2004). These results suggest that additional prestin-interacting proteins are required to achieve the efficient prestin-mediated amplification exhibited by native OHCs. To date, only one putative prestin-interacting protein, promyelocytic leukemia zinc-finger protein (PFAE), has been identified by yeast two-hybrid screening (Nagy et al., 2005).

6 Somatic Electromotility Versus Hair Bundle Motility

6.1 Uncertainty About the Somatic Electromotility Hypothesis

Although compelling evidence supports somatic electromotility of OHCs as the cellular basis for the cochlear amplifier, some uncertainty remains. In a microchamber system, OHC electromotility could produce a consistent oscillating force of at least 79 kHz (Frank et al., 1999). This range covers almost the entire range of acoustic frequencies detected by mammals (Frank et al., 1999). However, in vivo, the resistance and capacitance (RC) of the OHC membranes create a low-pass filter that attenuates the rapid change in receptor potential driven by transduction currents (Santos-Sacchi, 1992; Gale and Ashmore, 1997a). Therefore, the receptor potential is too small to drive the somatic motility when the stimulus frequency is higher than several kHz (Santos-Sacchi, 1992; Gale and Ashmore, 1997a). One possible solution to overcome this problem is that the OHC motor could be driven, without RC-mediated attenuation, by an extracellular potential change generated by the neighboring OHCs within the organ of Corti (Dallos and Evans, 1995). This hypothesis is supported by the results of a recent study (Fridberger et al., 2004) showing that in the basilar membrane of guinea pig cochlea, the extracellular potential does indeed drive the OHC motor at high frequencies. In an alternative "local resonating" model described by Ospeck and coworkers (2003), the frequency limit of the cochlear amplifier (estimated at 10–13 kHz) depends on the OHC's membrane capacitance and local viscous drag but does not depend directly on the RC.

6.2 Active Amplification by Hair Bundles

An alternative view of the source of cochlear amplification is that it arises in the OHC's bundles of stereocilia. The stereocilia bundle is composed of many actin filaments, and the kinocilium containing mainly axonemal microtubules. At least three types of actin-dependent molecular motor proteins are expressed in the stereocilia and are involved in adaptation of the hair cell MET channels: myosin 15A (Rzadzinska et al., 2004), myosin 7A (Kros et al., 2002), and myosin 1C (Gillespie and Cyr, 2004). Many studies have shown that fast adaptation of MET channels can amplify the oscillation of hair bundles in the nonmammalian auditory system (Crawford and Fettiplace, 1985; Howard and Hudspeth, 1987; Ricci et al., 2000; Manley et al., 2001). The opening of MET channels by deflection of hair bundles to a positive phase causes an influx of calcium (Ricci et al., 2000). Calcium is then thought to bind to the MET channels and cause the hair bundle to move further in the positive direction (Ricci et al., 2000). Theoretically, this active hair bundle movement can occur at a frequency of at least 20 kHz without the interference of the RC filter (Manley, 2001). Recent results from mammalian hair cell studies have further strengthened this hypothesis (Chan and Hudspeth, 2005; Kennedy et al., 2005). Kennedy and colleagues (2005) found that a change in the mechanical force generated in the hair bundle of isolated rat OHCs can occur on a submillisecond timescale. Hudspeth and colleagues took a cochlear turn from a gerbil and, to mimic the in vivo condition, made a two-chamber system in which they demonstrated that movement of the hair bundle can generate enough force for cochlear amplification (Chan and Hudspeth, 2005). This active movement of the hair bundles could be diminished by blocking the ionic transduction current, but it was retained when calcium influx alone occurred (Chan and Hudspeth, 2005). They proposed that the hair bundle's calcium-dependent force generator alone can drive cochlear amplification and that somatic electromotility is responsible only for keeping the hair bundles in their optimal position (Chan and Hudspeth, 2005). A follow-up review of these two papers therefore suggested that prestin be renamed "andantino," after the musical term *andante*, meaning moderate tempo (Kros, 2005).

6.3 Hair Bundle Movement Dependent on Somatic Electromotility?

However, new support for the somatic electromotility model of cochlear amplification has been provided by Jia and He (2005), who demonstrated, using a coil preparation of the gerbil cochlea, that active hair bundle movement depends on the developmental onset of somatic electromotility. At postnatal day 4 (P4), when hair bundle MET channels are functional but the OHCs are not yet electromotile, no hair bundle motility could be detected from the apical turn of the coil in response to a sinusoidal voltage command (Jia and He, 2005). The hair bundle movement appeared when somatic electromotility developed. This observation is supported by findings from studies of the cochlea of prestin knockout mice, in which active movement of the hair bundle is not detected despite the presence of functional MET channels (Jia and He, 2005). In addition, the amplitude of the motility-associated hair bundle movement is 20 dB greater than that of the transduction-channel-based hair bundle motion. They propose that somatic motility in conjunction with bundle motion is responsible for the cochlear amplifier (Jia and He, 2005).

7 Summary

The evolutionary process rarely selects one cellular mechanism over a competing mechanism in an "all-or-none" stepwise fashion. Instead, the transition is generally gradual. Both hair bundle movement and somatic electromotility may be important for cochlear amplification in mammals. Somatic electromotility may contribute more than just an "andantino" to cochlear amplification (Kros, 2005). Perhaps the first mechanical amplification system to develop was that in nonmammalian hair cells, which is based on hair bundle movement. More demanding selection pressures from the environment during the evolution of the system might have led to the development of somatic electromotility as a more efficient way to amplify sound. The newly acquired somatic-based amplification system then maintained a close association with the ancient hair bundle based amplification system throughout evolution (Manley, 2000). Active movement of hair bundles coupled with somatic electromotility may therefore reflect an evolutionary transition. However, the hair bundle movement observed by Jia and He (2005) is clearly not the same as that observed by Chan and Hudspeth (2005) and Kennedy and group (2005), because the former was not affected by changes in calcium concentration and not inhibited by streptomycin, a MET channel blocker. Therefore, the next challenging questions confronting auditory scientists are these: What are the relative contributions of the two different types of hair bundle movement to the cochlear amplifier? How is somatic electromotility related directly to hair bundle movement?

Acknowledgments

We would like to thank Drs. David He and Janet Davies for comments. This work was supported in part by NIH grants DC06471, DC05168, DC04761, and Cancer Center Core Grant CA21765; by the American Lebanese Syrian Associated Charities (ALSAC).

References

Adachi M, Sugawara M, Iwasa KH. 2000. Effect of turgor pressure on outer hair cell motility. J Acoust Soc Am 108: 2299-2306.

Arima T, Kuraoka A, Toriya R, Shibata Y, Uemura T. 1991. Quick-freeze, deep-etch visualization of the 'cytoskeletal spring' of cochlear outer hair cells. Cell Tissue Res 263: 91-97.

Ashmore J. 2002. Biophysics of the cochlea—biomechanics and ion channelopathies. Br Med Bull 63: 59-72.

Ashmore JF. 1987. A fast motile response in guinea-pig outer hair cells: The cellular basis of the cochlear amplifier. J Physiol 388: 323-347.

Ashmore JF. 1990. Forward and reverse transduction in the mammalian cochlea. Neurosci Res Suppl 12: S39-50.

Ashmore JF, Geleoc GS, Harbott L. 2000. Molecular mechanisms of sound amplification in the mammalian cochlea. Proc Natl Acad Sci USA 97: 11759-11764.

Batta TJ, Panyi G, Szucs A, Sziklai I. 2004. Regulation of the lateral wall stiffness by acetylcholine and GABA in the outer hair cells of the guinea pig. Eur J Neurosci 20: 3364-3370.

Belyantseva IA, Adler HJ, Curi R, Frolenkov GI, Kachar B. 2000. Expression and localization of prestin and the sugar transporter GLUT-5 during development of electromotility in cochlear outer hair cells. J Neurosci 20:RC116.

Bezanilla F. 2000. The voltage sensor in voltage-dependent ion channels. Physiol Rev 80: 555-592.

Borko R, Batta TJ, Sziklai I. 2005. Slow motility, electromotility and lateral wall stiffness in the isolated outer hair cells. Hear Res 207: 68-75.

Brown AM, McDowell B, Forge A. 1989. Acoustic distortion products can be used to monitor the effects of chronic gentamicin treatment. Hear Res 42: 143-156.

Brownell WE, Bader CR, Bertrand D, de Ribaupierre Y. 1985. Evoked mechanical responses of isolated cochlear outer hair cells. Science 227: 194-196.

Brownell WE, Spector AA, Raphael RM, Popel AS. 2001. Micro- and nanomechanics of the cochlear outer hair cell. Annu Rev Biomed Eng 3: 169-194.

Chan DK, Hudspeth AJ. 2005. $Ca2^{+}$ current-driven nonlinear amplification by the mammalian cochlea in vitro. Nat Neurosci 8 149-155.

Cheatham MA, Huynh KH, Gao J, Zuo J, Dallos P. 2004. Cochlear function in Prestin knockout mice. J Physiol 560: 821-830.

Corpe CP, Bovelander FJ, Munoz CM, Hoekstra JH, Simpson IA, et al. 2002. Cloning and functional characterization of the mouse fructose transporter, GLUT5. Biochim Biophys Acta 1576: 191-197.

Crawford AC, Fettiplace R. 1985. The mechanical properties of ciliary bundles of turtle cochlear hair cells. J Physiol 364: 359-379.

Dallos P. 1992. The active cochlea. J Neurosci 12: 4575-4585.

Dallos P, Evans BN. 1995. High-frequency motility of outer hair cells and the cochlear amplifier. Science 267: 2006-2009.

Dallos P, Evans BN, Hallworth R. 1991. Nature of the motor element in electrokinetic shape changes of cochlear outer hair cells. Nature 350: 155-157.

Dallos P, Fakler B. 2002. Prestin, a new type of motor protein. Nat Rev Mol Cell Biol 3: 104-111.

Dallos P, Harris D. 1978. Properties of auditory nerve responses in absence of outer hair cells. J Neurophysiol 41: 365-383.

Dallos P, He DZ, Lin X, Sziklai I, Mehta S, et al. 1997. Acetylcholine, outer hair cell electromotility, and the cochlear amplifier. J Neurosci 17: 2212-2226.

Davis H. 1983. An active process in cochlear mechanics. Hear Res 9: 79-90.

Deak L, Zheng J, Orem A, Du GG, Aguinaga S, et al. 2005. Effects of cyclic nucleotides on the function of prestin. J Physiol 563: 483-496.

Dong XX, Iwasa KH. 2004. Tension sensitivity of prestin: Comparison with the membrane motor in outer hair cells. Biophys J 86: 1201-1208.

Dong XX, Ospeck M, Iwasa KH. 2002. Piezoelectric reciprocal relationship of the membrane motor in the cochlear outer hair cell. Biophys J 82: 1254-1259.

Dulon D, Aran JM, Schacht J. 1988. Potassium-depolarization induces motility in isolated outer hair cells by an osmotic mechanism. Hear Res 32: 123-129.

Dulon D, Schacht J. 1992. Motility of cochlear outer hair cells. Am J Otol 13: 108-112.

Frank G, Hemmert W, Gummer AW. 1999. Limiting dynamics of high-frequency electromechanical transduction of outer hair cells. Proc Natl Acad Sci USA 96: 4420-4425.

Fridberger A, de Monvel JB, Zheng J, Hu N, Zou Y, et al. 2004. Organ of Corti potentials and the motion of the basilar membrane. J Neurosci 24: 10057-10063.

Frolenkov GI, Mammano F, Belyantseva IA, Coling D, Kachar B. 2000. Two distinct $Ca^{(2+)}$-dependent signaling pathways regulate the motor output of cochlear outer hair cells. J Neurosci 20: 5940-5948.

Gale JE, Ashmore JF. 1997a. An intrinsic frequency limit to the cochlear amplifier. Nature 389: 63-66.

Gale JE, Ashmore JF. 1997b. The outer hair cell motor in membrane patches. Pflugers Arch 434: 267-271.

Geleoc GS, Casalotti SO, Forge A, Ashmore JF. 1999. A sugar transporter as a candidate for the outer hair cell motor. Nat Neurosci 2: 713-719.

Gillespie PG, Cyr JL. 2004. Myosin-1c, the hair cell's adaptation motor. Annu Rev Physiol 66: 521-545.

Gitter AH, Rudert M, Zenner HP. 1993. Forces involved in length changes of cochlear outer hair cells. Pflugers Arch 424: 9-14.

Gold T. 1948. Hearing. II. The physical basis of the action of the cochlea. Proc R Soc B 135: 492-498.

Gulley RL, Reese TS. 1977. Regional specialization of the hair cell plasmalemma in the organ of corti. Anat Rec 189: 109-123.

Harrison RV, Evans EF. 1979. Cochlear fibre responses in guinea pigs with well defined cochlear lesions. Scand Audiol Suppl 9: 83-92.

He DZ, Dallos P. 1999. Somatic stiffness of cochlear outer hair cells is voltage-dependent. Proc Natl Acad Sci USA 96: 8223-8228.

He DZ, Dallos P. 2000. Properties of voltage-dependent somatic stiffness of cochlear outer hair cells. J Assoc Res Otolaryngol 1: 64-81.

He DZ, Jia S, Dallos P. 2003. Prestin and the dynamic stiffness of cochlear outer hair cells. J Neurosci 23: 9089-9096.

Holley MC, Ashmore JF. 1988a. A cytoskeletal spring in cochlear outer hair cells. Nature 335: 635-637.

Holley MC, Ashmore JF. 1988b. On the mechanism of a high-frequency force generator in outer hair cells isolated from the guinea pig cochlea. Proc R Soc Lond B Biol Sci 232: 413-429.

Holley MC, Kalinec F, Kachar B. 1992. Structure of the cortical cytoskeleton in mammalian outer hair cells. J Cell Sci 102(3): 569-580.

Howard J, Hudspeth AJ. 1987. Mechanical relaxation of the hair bundle mediates adaptation in mechanoelectrical transduction by the bullfrog's saccular hair cell. Proc Natl Acad Sci USA 84: 3064-3068.

Huang G, Santos-Sacchi J. 1993. Mapping the distribution of the outer hair cell motility voltage sensor by electrical amputation. Biophys J 65: 2228-2236.

Huang G, Santos-Sacchi J. 1994. Motility voltage sensor of the outer hair cell resides within the lateral plasma membrane. Proc Natl Acad Sci USA 91: 12268-12272.

Hudspeth AJ. 2000. Hearing, In Principles of Neural Science. Kandel ER, Schwartz JH, Jessell TM, editors. New York: McGraw-Hill (Health Professions Division), pp. 590-614.

Inukai K, Katagiri H, Takata K, Asano T, Anai M, et al. 1995. Characterization of rat GLUT5 and functional analysis of chimeric proteins of GLUT1 glucose transporter and GLUT5 fructose transporter. Endocrinology 136: 4850-4857.

Jarboe JK, Hallworth R. 1999. The effect of quinine on outer hair cell shape, compliance and force. Hear Res 132: 43-50.

Jia S, He DZ. 2005. Motility-associated hair-bundle motion in mammalian outer hair cells. Nat Neurosci 8: 1028-1034.

Jimenez AM, Stagner BB, Martin GK, Lonsbury-Martin BL. 1999. Age-related loss of distortion product otoacoustic emissions in four mouse strains. Hear Res 138: 91-105.

Johnstone BM, Patuzzi R, Yates GK. 1986. Basilar membrane measurements and the travelling wave. Hear Res 22: 147-153.

Kachar B, Brownell WE, Altschuler R, Fex J. 1986. Electrokinetic shape changes of cochlear outer hair cells. Nature 322: 365-368.

Kakehata S, Santos-Sacchi J. 1995. Membrane tension directly shifts voltage dependence of outer hair cell motility and associated gating charge. Biophys J 68: 2190-2197.

Kakehata S, Santos-Sacchi J. 1996. Effects of salicylate and lanthanides on outer hair cell motility and associated gating charge. J Neurosci 16: 4881-4889.

Kalinec F, Holley MC, Iwasa KH, Lim DJ, Kachar B. 1992. A membrane-based force generation mechanism in auditory sensory cells. Proc Natl Acad Sci USA 89: 8671-8675.

Kalinec F, Zhang M, Urrutia R, Kalinec G. 2000. Rho GTPases mediate the regulation of cochlear outer hair cell motility by acetylcholine. J Biol Chem 275: 28000-28005.

Kemp DT. 1978. Stimulated acoustic emissions from within the human auditory system. J Acoust Soc Am 64: 1386-1391.

Kemp DT. 1979. Evidence of mechanical nonlinearity and frequency selective wave amplification in the cochlea. Arch Otorhinolaryngol 224: 37-45.

Kennedy HJ, Crawford AC, Fettiplace R. 2005. Force generation by mammalian hair bundles supports a role in cochlear amplification. Nature 433: 880-883.

Kim DO, Molnar CE, Matthews JW. 1980. Cochlear mechanics: Nonlinear behavior in two-tone responses as reflected in cochlear-nerve-fiber responses and in ear-canal sound pressure. J Acoust Soc Am 67: 1704-1721.

Koyama M, Spicer SS, Schulte BA. 1999. Immunohistochemical localization of Ca^{2+}/Calmodulin-dependent protein kinase IV in outer hair cells. J Histochem Cytochem 47: 7-12.

Kros C. 2005. Hearing: Aid from hair force. Nature 433: 810-811.

Kros CJ, Marcotti W, van Netten SM, Self TJ, Libby RT, et al. 2002. Reduced climbing and increased slipping adaptation in cochlear hair cells of mice with Myo7a mutations. Nat Neurosci 5: 41-47.

Liberman MC, Gao J, He DZ, Wu X, Jia S, et al. 2002. Prestin is required for electromotility of the outer hair cell and for the cochlear amplifier. Nature 419: 300-304.

Lin X, Hume RI, Nuttall AL. 1995. Dihydropyridines and verapamil inhibit voltage-dependent K^+ current in isolated outer hair cells of the guinea pig. Hear Res 88: 36-46.

Lonsbury-Martin BL, Martin GK. 1990. The clinical utility of distortion-product otoacoustic emissions. Ear Hear 11: 144-154.

Lonsbury-Martin BL, Martin GK, Probst R, Coats AC. 1987. Acoustic distortion products in rabbit ear canal. I. Basic features and physiological vulnerability. Hear Res 28: 173-189.

Ludwig J, Oliver D, Frank G, Klocker N, Gummer AW, et al. 2001. Reciprocal electromechanical properties of rat prestin: The motor molecule from rat outer hair cells. Proc Natl Acad Sci USA 98: 4178-4183.

Lue AJ, Zhao HB, Brownell WE. 2001. Chlorpromazine alters outer hair cell electromotility. Otolaryngol Head Neck Surg 125: 71-76.

Manley GA. 2000. Cochlear mechanisms from a phylogenetic viewpoint. Proc Natl Acad Sci USA 97: 11736-11743.

Manley GA. 2001. Evidence for an active process and a cochlear amplifier in nonmammals. J Neurophysiol 86: 541-549.

Manley GA, Kirk DL, Koppl C, Yates GK. 2001. In vivo evidence for a cochlear amplifier in the hair-cell bundle of lizards. Proc Natl Acad Sci USA 98: 2826-2831.

McFadden D, Plattsmier HS. 1984. Aspirin abolishes spontaneous oto-acoustic emissions. J Acoust Soc Am 76: 443-448.

Meech R, Holley M. 2001. Ion-age molecular motors. Nat Neurosci 4: 771-773.

Mills DM. 2004. Relationship of neural and otoacoustic emission thresholds during endocochlear potential development in the gerbil. J Acoust Soc Am 116: 1035-1043.

Mills DM, Schmiedt RA. 2004. Metabolic presbycusis: Differential changes in auditory brainstem and otoacoustic emission responses with chronic furosemide application in the gerbil. J Assoc Res Otolaryngol 5: 1-10.

Mountain DC. 1980. Changes in endolymphatic potential and crossed olivocochlear bundle stimulation alter cochlear mechanics. Science 210: 71-72.

Nagy I, Bodmer M, Schmid S, Bodmer D. 2005. Promyelocytic leukemia zinc finger protein localizes to the cochlear outer hair cells and interacts with prestin, the outer hair cell motor protein. Hear Res 204: 216-222.

Nakazawa K, Spicer SS, Schulte BA. 1995a. Postnatal expression of the facilitated glucose transporter, GLUT 5, in gerbil outer hair cells. Hear Res 82: 93-99.

Nakazawa K, Spicer SS, Schulte BA. 1995b. Ultrastructural localization of Na,K-ATPase in the gerbil cochlea. J Histochem Cytochem 43: 981-991.

Navaratnam D, Bai JP, Samaranayake H, Santos-Sacchi J. 2005. N-terminal mediated homo-multimerization of prestin, the outer hair cell motor protein. Biophys J.

Offner FF, Dallos P, Cheatham MA. 1987. Positive endocochlear potential: Mechanism of production by marginal cells of stria vascularis. Hear Res 29: 117-124.

Oghalai JS, Zhao HB, Kutz JW, Brownell WE. 2000. Voltage- and tension-dependent lipid mobility in the outer hair cell plasma membrane. Science 287: 658-661.

Oliver D, He DZ, Klocker N, Ludwig J, Schulte U, et al. 2001. Intracellular anions as the voltage sensor of prestin, the outer hair cell motor protein. Science 292: 2340-2343.

Olson ES. 2004. Harmonic distortion in intracochlear pressure and its analysis to explore the cochlear amplifier. J Acoust Soc Am 115: 1230-1241.

Ospeck M, Dong XX, Iwasa KH. 2003. Limiting frequency of the cochlear amplifier based on electromotility of outer hair cells. Biophys J 84: 739-749.

Purves D, Augustine GJ, Fitzpatrick D, Katz LC, La Mantia A-S, et al. 2001. Neuroscience, 3rd edn (Sunderland: Sinauer Associates, Inc.).

Puschner B, Schacht J. 1997. Calmodulin-dependent protein kinases mediate calcium-induced slow motility of mammalian outer hair cells. Hear Res 110: 251-258.

Raphael Y, Altschuler RA. 2003. Structure and innervation of the cochlea. Brain Res Bull 60: 397-422.

Ricci AJ, Crawford AC, Fettiplace R. 2000. Active hair bundle motion linked to fast transducer adaptation in auditory hair cells. J Neurosci 20: 7131-7142.

Robles L, Ruggero MA. 2001. Mechanics of the mammalian cochlea. Physiol Rev 81: 1305-1352.

Robles L, Ruggero MA, Rich NC. 1991. Two-tone distortion in the basilar membrane of the cochlea. Nature 349: 413-414.

Ruggero MA, Rich NC. 1991. Application of a commercially-manufactured Doppler-shift laser velocimeter to the measurement of basilar-membrane vibration. Hear Res 51: 215-230.

Ryan A, Dallos P. 1975. Effect of absence of cochlear outer hair cells on behavioural auditory threshold. Nature 253: 44-46.

Rybalchenko V, Santos-Sacchi J. 2003. Cl- flux through a non-selective, stretch-sensitive conductance influences the outer hair cell motor of the guinea-pig. J Physiol 547: 873-891.

Rzadzinska AK, Schneider ME, Davies C, Riordan GP, Kachar B. 2004. An actin molecular treadmill and myosins maintain stereocilia functional architecture and self-renewal. J Cell Biol 164: 887-897.

Santos-Sacchi J. 1991. Reversible inhibition of voltage-dependent outer hair cell motility and capacitance. J Neurosci 11: 3096-3110.

Santos-Sacchi J. 1992. On the frequency limit and phase of outer hair cell motility: Effects of the membrane filter. J Neurosci 12: 1906-1916.

Santos-Sacchi J. 2002. Functional motor microdomains of the outer hair cell lateral membrane. Pflugers Arch 445: 331-336.

Santos-Sacchi J. 2003. New tunes from Corti's organ: The outer hair cell boogie rules. Curr Opin Neurobiol 13: 459-468.

Santos-Sacchi J, Huang G. 1998. Temperature dependence of outer hair cell nonlinear capacitance. Hear Res 116: 99-106.

Santos-Sacchi J, Kakehata S, Takahashi S. 1998. Effects of membrane potential on the voltage dependence of motility-related charge in outer hair cells of the guinea-pig. J Physiol 510(1): 225-235.

Santos-Sacchi J, Shen W, Zheng J, Dallos P. 2001. Effects of membrane potential and tension on prestin, the outer hair cell lateral membrane motor protein. J Physiol 531: 661-666.

Schulte BA. 1993. Immunohistochemical localization of intracellular Ca-ATPase in outer hair cells, neurons and fibrocytes in the adult and developing inner ear. Hear Res 65: 262-273.

Shehata WE, Brownell WE, Dieler R. 1991. Effects of salicylate on shape, electromotility and membrane characteristics of isolated outer hair cells from guinea pig cochlea. Acta Otolaryngol 111: 707-718.

Siegel JH, Kim DO. 1982. Efferent neural control of cochlear mechanics? Olivocochlear bundle stimulation affects cochlear biomechanical nonlinearity. Hear Res 6: 171-182.

Slepecky N. 1989. Cytoplasmic actin and cochlear outer hair cell motility. Cell Tissue Res 257: 69-75.

Slepecky NB. 1996. Structure of the mammalian cochlea. In The Cochlea, Dallos P, Popper AN, Richard RF, editors. New York: Springer-Verlag; pp. 44-129.

Spicer SS, Thomopoulos GN, Schulte BA. 1998. Cytologic evidence for mechanisms of K^+ transport and genesis of Hensen bodies and subsurface cisternae in outer hair cells. Anat Rec 251: 97-113.

Sziklai I, Dallos P. 1997. Hyposmotic swelling induces magnitude and gain change in the electromotile performance of isolated outer hair cells. Acta Otolaryngol 117: 222-225.

Szonyi M, He DZ, Ribari O, Sziklai I, Dallos P. 1999. Cyclic GMP and outer hair cell electromotility. Hear Res 137: 29-42.

Szonyi M, He DZ, Ribari O, Sziklai I, Dallos P. 2001. Intracellular calcium and outer hair cell electromotility. Brain Res 922: 65-70.

Takahashi S, Santos-Sacchi J. 2001. Non-uniform mapping of stress-induced, motility-related charge movement in the outer hair cell plasma membrane. Pflugers Arch 441: 506-513.

Wangemann P. 2002. K^+ cycling and the endocochlear potential. Hear Res 165: 1-9.

Wilson JP. 1980. Evidence for a cochlear origin for acoustic re-emissions, threshold fine-structure and tonal tinnitus. Hear Res 2: 233-252.

Yamamoto T, Kakehata S, Yamada T, Saito T, Saito H, et al. 1997. Effects of potassium channel blockers on the acetylcholine-induced currents in dissociated outer hair cells of guinea pig cochlea. Neurosci Lett 236: 79-82.

Zenner HP. 1986. Motile responses in outer hair cells. Hear Res 22: 83-90.

Zenner HP, Zimmermann U, Schmitt U. 1985. Reversible contraction of isolated mammalian cochlear hair cells. Hear Res 18: 127-133.

Zhang M, Kalinec GM, Urrutia R, Billadeau DD, Kalinec F. 2003. ROCK-dependent and ROCK-independent control of cochlear outer hair cell electromotility. J Biol Chem 278: 35644-35650.

Zheng J, Long KB, Shen W, Madison LD, Dallos P. 2001a. Prestin topology: Localization of protein epitopes in relation to the plasma membrane. Neuroreport 12: 1929-1935.

Zheng J, Ren T, Parthasarathi A, Nuttall AL. 2001b. Quinine-induced alterations of electrically evoked otoacoustic emissions and cochlear potentials in guinea pigs. Hear Res 154: 124-134.

Zheng J, Shen W, He DZ, Long KB, Madison LD, et al. 2000. Prestin is the motor protein of cochlear outer hair cells. Nature 405: 149-155.

Zine A, Schweitzer L. 1996. Development of intracellular Ca-ATPase in the gerbil outer hair cell lateral wall. Brain Res 721: 1-10.

5 Neurochemistry of the Gustatory System

D. V. Smith · J. D. Boughter Jr

Abstract: The sense of taste is mediated by receptor mechanisms that are distributed on modified epithelial cells within structures called taste buds, which are located on the tongue and other parts of the oral cavity. These cells are innervated by branches of one of three cranial nerves, including the chorda tympani and greater superficial petrosal branches of VII, the lingual-tonsillar branch of IX, and the superior laryngeal branch of X. These nerves project into the rostral portion of the nucleus of the solitary tract (NST) of the medulla. Gustatory information is carried from there to oral motor circuits within the brainstem and to the parabrachial nuclei (PbN) of the pons, from which pathways arise to thalamus and insular cortex and also into areas of the limbic forebrain, including the lateral hypothalamus (LH), central nucleus of the amygdala (CeA), and the bed nucleus of the stria terminalis (BST). Taste buds contain cells that can be classified into types on ultrastructural grounds and these in turn have been shown to exhibit expression of a wide array of molecules, many of which are characteristic of a particular cell type. The transduction of chemical stimuli by taste receptor cells is mediated by several different mechanisms. Salts and acids interact directly with ion channels to depolarize receptor cells, whereas sweet- and bitter-tasting stimuli and amino acids interact with G-protein-coupled receptors of the T1R and T2R families, linked to second-messenger pathways. Although some of these receptors appear to be segregated into different cells, the electrophysiological and calcium imaging data show that taste bud cells are often responsive to stimuli of more than one quality. This multiple sensitivity is evident to an even greater degree in first-order nerve fibers and in central gustatory neurons, due to convergence of afferent input onto higher-order neurons. Afferent input into the NST appears to be mediated by glutamate and most second-order neurons are maintained under tonic GABAergic inhibition. Some of the neurons in the NST are excited by substance P and some are inhibited by met-enkephalin. Areas of the forebrain that receive gustatory input, including the insular cortex, LH, CeA, and BST, provide descending modulatory control over taste neurons in both the NST and PbN.

1 Anatomy of the Mammalian Taste System

The gustatory system in mammals provides sensory input that is critical for the regulation of ingestive behavior and the avoidance of toxic substances. Taste is unique among sensory systems to the extent that it interfaces with neural substrates of reward and motivation (Pfaffmann, 1964). For example, sweet-tasting and bitter-tasting stimuli produce inherent preference and avoidance, respectively, and sweet tastes can serve as reinforcing stimuli. The anatomy of this system reflects the dual role of taste as both a discriminative system, designed to determine subtle differences in taste quality and intensity, and a motivational one, which underlies the acceptance and rejection of potential foods. Anatomically, the taste system sits at the interface between the external environment and the internal milieu, making it in a sense a rostral extension of the visceral afferent system (Norgren, 1985).

1.1 Distribution of Taste Buds and Taste Receptors

Taste begins with molecular events at the surface membranes of modified epithelial cells, which share many characteristics with neurons. Taste receptor cells are organized within specialized structures—taste buds—on the tongue and other parts of the oral cavity [for reviews see Kinnamon (1987), Reutter and Witt (1993)]. As described later, taste buds contain cells of different types, which can be distinguished on the basis of their ultrastructure and their molecular phenotypes.

Taste buds on the tongue are located within three different kinds of papillae. On the anterior two-thirds of the tongue, numerous fungiform papillae contain taste buds on their dorsal surface. In humans, each of these papillae contains about three to five taste buds, whereas in rodents there is typically a single taste bud on each fungiform papilla. On the posterior portion of the tongue, taste buds are contained in two kinds of papillae: the foliate on the sides of the tongue and the circumvallate along the posterior border. In humans, there are about 12 circumvallate papillae, whereas in rodents there is only one located on the midline, typically termed the vallate papilla. In addition to these papillae on the tongue, taste buds are distributed within the epithelium of the soft palate, the nasopharynx, the larynx, and the upper reaches of the esophagus. Counts of taste buds in rodents show that about 25% of the taste buds are located on the anterior portion of the tongue, about 50% on the posterior part of the tongue, and about 12% on the soft palate, with the remainder scattered within these other areas (Miller and Smith, 1984).

Cells within the taste bud are in a constant state of renewal. New cells arise throughout life from cell division of basal cells within the taste bud and the adjacent epithelium. The lifespan of taste bud cells in the rat has been estimated at about 9–10 days (Beidler and Smallman, 1965; Farbman, 1980), based on both tritiated thymidine and bromodeoxyuridine studies. Although we know that brain-derived neurotrophic factor (BDNF) is critical for the initial development of taste papillae and taste buds and their successful innervation by gustatory axons (Nosrat et al., 1997), the molecular mechanisms important for cell turnover during adulthood are not well understood. It is known that some taste receptor cells in adult rodents express BDNF (Yee et al., 2003) and/or the neural cell adhesion molecule (NCAM; Nelson and Finger, 1993; Smith et al., 1993; Yee et al., 2001), either of which could play a role in nerve terminal growth and/or synaptogenesis, but there are no experimental data to support this idea.

Nothing is really known about how individual nerve fibers make synaptic contact with particular taste receptor cells during turnover, although the expression of a number of cell-surface epitopes has been demonstrated on taste bud cells (see later). This process is clearly important for our understanding about how any neural code for taste might be maintained while the receptor cells are being constantly replaced throughout life. It is known that branches of single chorda tympani (CT) axons innervating several fungiform papillae have similar taste sensitivities, showing that axons tend to make multiple synaptic contacts with similar receptor cells (Oakley, 1975). Cross-reinnervation experiments show that both the sensitivities of the axons (Oakley, 1974) and the molecular expression of taste bud cells (Smith et al., 1999) are determined by the epithelium itself and not by the nerve (chorda tympani or glossopharyngeal) that innervates it. Thus, a role exists for axon-guidance molecules in this process, but so far, nothing is known about how such mechanisms function within the taste bud.

1.2 Afferent Innervation of Taste Buds

Taste receptor cells are not neurons, but modified epithelial cells. They make synaptic connections with nerve fibers of one of three cranial nerves: VII, IX, or X. Separate branches of the facial (VII) nerve innervate taste buds of the fungiform papillae and the soft palate (Whiteside, 1927; Miller and Spangler, 1982). The CT nerve carries gustatory afferent information from the fungiform taste buds, whereas the greater superficial petrosal (GSP) branch of VII carries taste information from the palate. The cell bodies of both the CT and GSP neurons are located in the geniculate ganglion of the facial nerve. The CT also provides innervation for taste buds in the more rostral foliate papillae. However, the lingual-tonsillar branch of the glossopharyngeal (IX) nerve innervates the majority of the foliate taste buds and all of those of the circumvallate taste buds (Whiteside, 1927; Guth, 1957). The cell bodies of these IX nerve afferent fibers are located within the petrosal ganglion of the glossopharyngeal nerve. The superior laryngeal nerve (SLN), a branch of the vagus (X) nerve, innervates taste buds distributed on the laryngeal surface of the epiglottis (Khaisman, 1976; Belecky and Smith, 1990); its cell bodies lie in the nodose ganglion. Both the IX and the X nerves are probably involved in the innervation of taste buds on the nasopharynx and esophagus, although the exact distribution of this innervation in either humans or rodents is not well understood.

1.3 Brainstem Projections of Gustatory Nerves

Afferent fibers of the VII, IX, and X nerves terminate within the medulla in the nucleus of the solitary tract (NST) in an overlapping rostral to caudal distribution (Contreras et al., 1982; Hamilton and Norgren, 1984; Hanamori and Smith, 1989). The most rostral pole of the NST receives afferent input from both the anterior tongue and the palate via the VII nerve. Input from the IX nerve overlaps with the more caudal portion of the VII nerve distribution within the NST, although it extends to more caudal levels. Similarly, the gustatory input from the X nerve, although overlapping somewhat with that of the VII and IX, is more caudally distributed still. The most caudal aspects of the NST receive vagal afferent input from the viscera. A schematic diagram of the ascending and descending connections of the rodent gustatory system is shown in ➲ *Figure 5-1*, which depicts the afferent innervation of the NST by cranial nerves VII, IX, and X.

◘ Figure 5-1
Schematic diagram of the gustatory pathway in rodents. Taste receptor cells are innervated by one of three cranial nerves (VII, IX, or X), which project topographically into the rostral portion of nucleus of the solitary tract (NST). Cells within the NST send projections into the reticular formation (RF), through which connections are made to oral motor nuclei: V, VII, and XII and the nucleus ambiguous (NA). Ascending fibers connect to the parabrachial nuclei (PbN) of the pons, from which two parallel pathways emerge. One pathway carries taste information to the insular cortex (IC) via the ventral posterior medial nucleus, parvicellularis (VPMpc), of the thalamus. The other pathway projects into areas of the limbic forebrain involved in food and water regulation, reinforcement, reward, and stress, including the lateral hypothalamus (LH), the central nucleus of the amygdala (CeA), and the bed nucleus of the stria terminalis (BST). These areas and the IC are interconnected and send descending projections back to both the PbN and NST

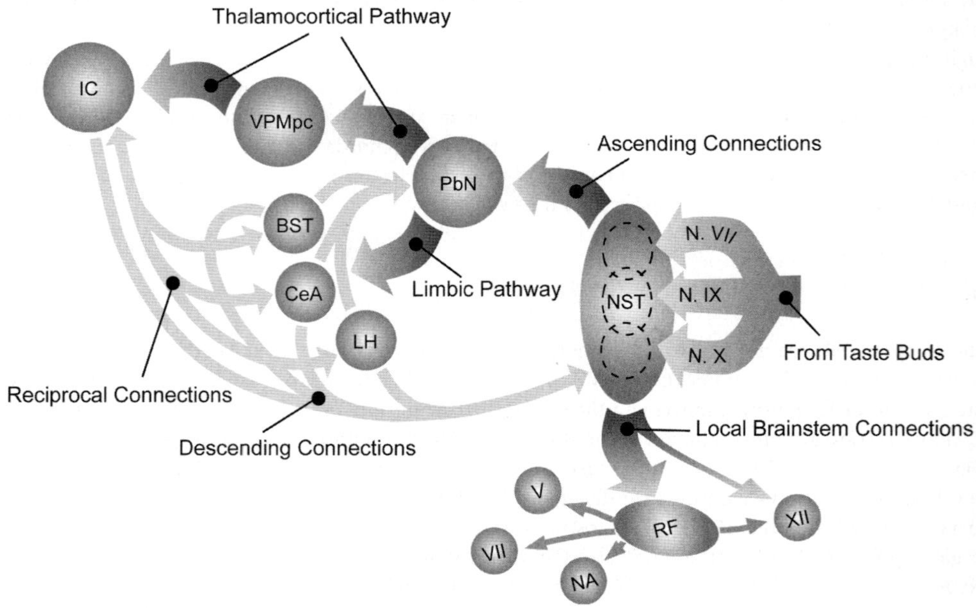

Electrophysiology studies have shown that there is convergence of gustatory input onto neurons of the NST, both within a receptive field (e.g., anterior tongue; Vogt and Mistretta, 1990) and across receptive fields (e.g., anterior tongue and palate; Travers and Norgren, 1991). That is, single fungiform papillae converge onto CT nerve fibers, which converge even further as those afferent fibers innervate neurons of the NST. In addition, there is evidence that axons from different peripheral nerves also converge onto single neurons of the NST. This convergence serves to increase the breadth of responsiveness of NST neurons over that seen in peripheral gustatory axons (Travers and Smith, 1979).

Gustatory neurons in the NST of rodents send second-order projections to the parabrachial nuclei (PbN) of the pons (Norgren and Leonard, 1973), which is an obligatory relay in the taste system of most mammals (➲ *Figure 5-1*). In these species, the gustatory projection parallels that of the visceral afferent system. In primates, including humans, there is a direct projection from the rostral NST to the ventral posterior medial nucleus of the thalamus (Pritchard, 1991); only the visceral portions of the NST appear to have an obligatory relay in the primate PbN. Electrophysiological evidence in rodents suggests further convergence of afferent taste information onto cells of the PbN (Van Buskirk and Smith, 1981).

1.4 Forebrain Connections of the Gustatory System

The close association between taste and motivational systems is most clearly seen in the anatomy of the forebrain projections in rodents. From the PbN, the gustatory projection follows two routes: a thalamocortical pathway and a limbic forebrain pathway (➲ *Figure 5-1*). Like other sensory systems, taste projects into a thalamic nucleus, specifically the ventral posterior medial nucleus, parvicellularis (VPMpc), adjacent to the tongue somatosensory projection (Emmers et al., 1962). From the VPMpc, a cortical projection goes to the dysgranular and agranular insular cortex (Kosar et al., 1986). In addition to this thalamocortical projection, there is a substantial distribution of taste information into several areas of the limbic forebrain, including the lateral hypothalamus (LH), the central nucleus of the amygdala (CeA), the substantia innominata, and the bed nucleus of the stria terminalis (BST) (Norgren, 1974, 1976; Halsell, 1992). These areas are important in food and water regulation, responses to stress, conditioned taste aversion learning, and mechanisms of reinforcement, reward, and drug addiction. Electrophysiology studies have shown the existence of taste-responsive neurons in several of these areas, including the LH (Norgren, 1970; Yamamoto et al., 1989) and the CeA (Nishijo et al., 1998).

2 Characteristics of Taste Receptor Cells

Taste begins with transduction events in receptor cells of the taste buds. There has been a lot of recent progress in our understanding of gustatory receptor mechanisms and transduction. The challenge is to relate molecular mechanisms that are evident at the receptor level to the extensive literature on gustatory neurophysiology and taste-mediated behavior.

2.1 Ultrastructure of Taste Buds and Cell Types

For many years, it has been recognized that there are different types of cells within the taste bud (Kinnamon, 1987; Reutter and Witt, 1993). A schematic diagram of a mammalian taste bud is shown in ➲ *Figure 5-2a*. This figure indicates different cell types within the taste bud, most of which extend apical microvilli into the taste pore, which opens into the oral cavity. Taste bud cells are innervated by distal branches of primary neurons of the VII, IX, or X cranial nerve, with which they make synaptic contact. Taste bud cell types were first delineated on ultrastructural grounds, based on a number of criteria. The simplest distinction was between cells that appeared translucent in electron micrographs (light cells) and those that were more electron dense (dark cells). However, this distinction is dependent to a large extent on methods of fixation and tissue processing (Kinnamon et al., 1985; Cottler-Fox et al., 1987), which can make the distinction between light and dark cells problematic. Other criteria were used to classify cells as Type I, Type II, Type III, or basal cells (Type IV), based largely on the extensive work of Murray (1971) on rabbit taste buds. Type I cells are typically electron dense (i.e., "dark"), have numerous microvilli on their apical end, often have nuclei with prominent heterochromatin and membranous invaginations, and have dense-core vesicles near their apex. Type I cells extend cytoplasmic and plasma membrane processes as irregularly shaped lamellae, which surround neighboring cells, as evident in the immunocytochemical expression of the H blood group antigen (➲ *Figure 5-2c*, green) or in electronmicrographs of transversely cut taste buds (➲ *Figure 5-2e*). Type II cells are electron lucent ("light"), have large round nuclei, and a single microvillus on their apical end; they are also triangular in longitudinal section. Another type of "light" cell is the Type III cell, which differs from the Type II in having numerous microvilli on its apical end and often contains clearly recognizable synapses in contact with first-order nerve fibers. In addition, Type III cells are typically long and clearly spindle-shaped in longitudinal sections through the taste bud.

Figure 5-2

Cell types in mammalian taste buds. (a) Schematic diagram of a mammalian taste bud, which is an onion-shaped structure containing different cell types, including basal cells, dark cells, and light cells. Some of these epithelial cells make synaptic contact with distal processes of cranial nerves VII, IX, or X, the cell bodies of which lie within cranial nerve ganglia. Microvilli of the taste bud cells project into an opening in the epithelium, the taste pore, where they make contact with gustatory stimuli dissolved in saliva. (b) The characteristic spindle shape of taste receptor cells is revealed when a subset of light cells in the rat vallate papilla is immunoreacted to an antibody against α-gustducin, a gustatory G protein subunit. (c) When sectioned transversely, light cells appear round in cross section, as shown by α-gustducin immunoreactivity in mouse taste cells (*red*), whereas the characteristic shape of dark cells produced by their thin cytoplasmic projections enveloping neighboring light cells is revealed with an antibody against the H blood group antigen (*green*). (d) Electronmicrograph through a rat vallate taste bud showing the appearance of light (*L*) and dark (*D*) cells. E: At a higher magnification, the thin cytoplasmic extensions of dark cells can be seen surrounding neighboring light cells [a–c from Smith and Shepherd (2003)]

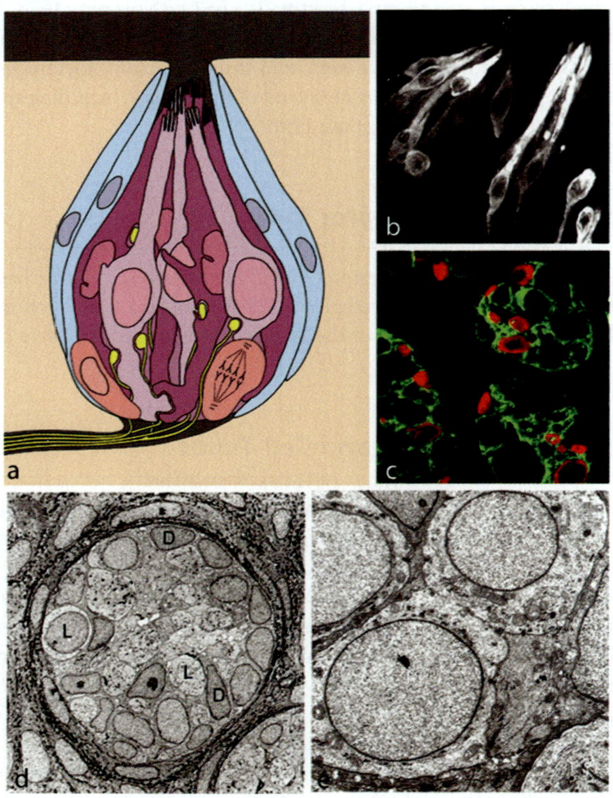

In sections cut horizontally through the taste bud, Type I cells are quite distinctive in their shapes. These cells typically have a smaller soma than either Type II or Type III cells and they have sheet-like cytoplasmic projections that extend for some distance and envelop neighboring "light" cells (Royer and Kinnamon, 1988; Pumplin et al., 1997). Often, the cytoplasmic extensions of different Type I cells interdigitate with one another to completely envelop the neighboring cells. In contrast, Type II and Type III cells are relatively round in cross section and considerably larger than the Type I cells. This structural arrangement has led some investigators to suggest a supporting role for the Type I cells (Reutter and Witt, 1993; Finger, 2005), a view supported by the presence of the glutamate-aspartate transporter (GLAST) in Type I cells (Lawton et al., 2000). The existence of synapses on the Type III cells (Murray, 1986; Royer and Kinnamon, 1991)

suggests strongly that these are receptor cells, although the role of Type II cells is confusing in that regard, given their molecular profiles (see later) and apparent lack of synaptic connections, at least in species other than mice (Kinnamon et al., 1985).

2.2 Immunocytochemical Diversity Among Taste Bud Cells

There has been a considerable amount of research on the molecular expression of cells within the taste bud, a good deal of which relates to the classification of cells into types based on ultrastructural criteria. Some of the data on molecular expression within the taste bud relates to known mechanisms of transduction or synaptic function, but much of it reflects the expression of molecules for which a functional role is not apparent. For example, NCAM is expressed on a subset of taste bud cells and on the nerve fibers innervating the taste buds (➋ *Figure 5-3i, j, l*; Nelson and Finger, 1993; Smith et al., 1993); these cells appear to be Type III cells (Yee et al., 2001). Although it is interesting to speculate that NCAM could play a role in cell–cell recognition during turnover and synaptogenesis, there is no experimental evidence to support this idea. Taste bud cells also express a number of other neural-related molecules, including neuron-specific enolase (NSE), ubiquitin carboxyl terminal hydrolase (protein gene-related product 9.5; PGP 9.5; ➋ *Figure 5-3k*), and serotonin (5-HT; Yee et al., 2001). On the other hand, taste bud cells also express several molecular markers of epithelial cells, including the human blood group antigens A (➋ *Figure 5-3k*), H (➋ *Figure 5-2c, d*), B and Lewis^{b} (Smith et al., 1994), and a number of cytokeratins (➋ *Figure 5-3b, c, e–h, i, l*; Knapp et al., 1995; Zhang and Oakley, 1996). Molecules associated with synaptic function are also expressed by cells in the taste bud, including synaptophysin, synaptobrevin-2 (Stb-2), SNAP-25, and synaptotagmin (Pumplin and Getschman, 2000; Yang et al., 2000, 2004). Similarly, several receptor molecules, G proteins, and other transduction elements, such as the inositol triphospate receptor 3 (IP_3R3) and phospholipase Cβ2 (PLCβ2) have been described in taste bud cells, as delineated in the section later on gustatory transduction (Clapp et al., 2001, 2004).

Figure 5-3

Immunocytochemical expression in rat taste bud cells. (a) Horizontal section through the entire vallate papilla, showing immunoreactivity to α-gustducin in a subset of Type II cells. (b) Same horizontal section as in a, showing the expression of cytokeratin 7. (c) Double-labeling in a transverse section through the vallate papilla showing immunoreactivity to cytokeratin 7 (*green*) and α-gustducin (*red*), which are expressed in different cells. Cytokeratin 7 is a cytoplasmic marker of Type I cells and α-gustducin is expressed in a subset of Type II cells. (d) Transverse section showing immunoreactivity to the H blood group antigen (*green*) and α-gustducin (*red*). The H antigen is a cell-surface marker for Type I cells. (e) Double-labeling of a taste bud for α-gustducin (*red*) and cytokeratin 18 (*green*). All gustducin-positive cells (Type II) express cytokeratin 18 (*yellow*), but there are other cytokeratin 18-positive cells that do not express gustducin (*green*). (f) Transverse section through the vallate papilla showing double labeling for cytokeratin 18 (*green*) and α-gustducin (*red*); again, all gustducin-positive cells double-label for cytokeratin 18 (*yellow*). Both are expressed by light cells, which are recognizable by their relatively round shapes in transverse section. (g) Double-labeling for cytokeratin 20 (*green*) and α-gustducin (*red*); these markers are on separate cells. (h) Transverse section through vallate taste buds showing immunocytochemical expression of cytokeratin 19, a cytoplasmic marker of Type I cells. (i) Double-labeling of vallate taste buds for cytokeratin 18 (*green*) and NCAM (*red*); NCAM is expressed on Type III cells, but not by cells expressing cytokeratin 18. (j) Image of a coronal section of an entire vallate trench, lined with taste buds expressing NCAM (*green*), which also stains innervating fibers of the IX nerve, and α-gustducin (*red*). The gustducin- and NCAM-positive cells are different subsets; one reflecting Type II cells (gustducin) and the other Type III (NCAM). (k) Double-labeling for PGP 9.5 (*red*), which stains nerve fibers and Type III cells, and the A blood group antigen (*green*), which does not colocalize with PGP 9.5. (l) Longitudinal section through vallate taste buds showing separate subsets of cells expressing NCAM (*red*) and cytokeratin 18 (*green*). Lack of coexpression of cytokeratin 18 with a Type III cell marker suggests that it may be limited to Type II cells

Figure 5-3 (continued)

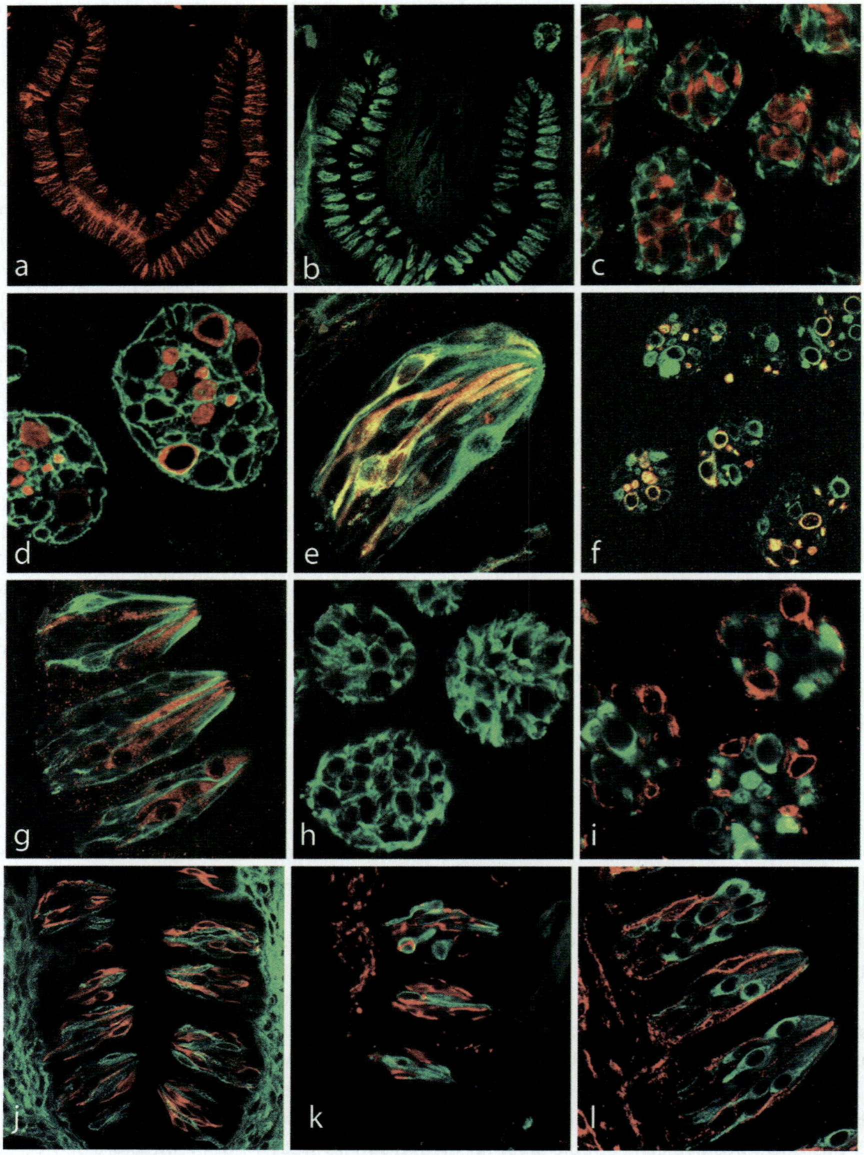

Although it is possible to relate the expression of particular molecules to ultrastructural cell type, the role of most of these markers in taste bud cell function remains elusive, as does the role of the cell types themselves. The known distributions of several of these molecules with the types of cells found in mammalian taste buds are depicted in *Figure 5-4*. Type I (dark) cells of the rat express a number of molecules, including GLAST (Lawton et al., 2000), the H blood group antigen (*Figure 5-3d*; Pumplin et al., 1997), the B blood group antigen, cytokeratin 7 (*Figure 5-3c*), cytokeratin 19 (*Figure 5-3h*), and the carbohydrate marker 2B8 (Pumplin et al., 1997). This expression is most evident in transverse sections through the taste bud, in which the unique shape of Type I cells is clear (*Figure 5-2c*, *Figure 5-3c, d, h*). As far as can be determined, all Type I cells express these markers, with no evidence of subsets with different molecular profiles. On the other hand, light cells (Types II and III) are considerably more diverse in their molecular expression. For example, a subset of Type II cells expresses the gustatory G-protein subunit

Figure 5-4

Schematic diagram indicating distribution of the immunocytochemical expression of several markers by ultrastructurally identifiable cell types in rat taste buds. Type I cells, which in transverse section are seen to surround neighboring light cells, express several markers, including the H and B blood group antigens, cytokeratins 7 and 19, the 2B8 carbohydrate antigen, and the glutamate-aspartate transporter (GLAST). These appear to be on all Type I cells; in transverse section, the cytoplasmic extensions of Type I cells are apparent (*Figures 5-2c* and *5-3c, d, h*). All Type II cells express synaptobrevin-2 (Stb-2), the type III IP_3 receptor (IP_3R3), and phospholipase Cβ2 (PCLβ2). The colocalization of cytokeratin 18 with α-gustducin and its nonoverlap with the Type III cell marker NCAM (*Figure 5-3i, l*) suggest that it is also a marker of Type II cells. Within the Type II cells are recognizable subsets of neurons, one expressing α-gustducin and the other ubiquitin carboxyl terminal hydrolase (PGP 9.5), neuron-specific enolase (NSE), and brain-derived neurotrophic factor (BDNF). A subset of cells expresses the A blood group antigen and some of these cells also express gustducin; cells expressing A do not express PGP 9.5 (*Figure 5-3k*). A subset of the gustducin-expressing cells shows immunoreactivity to the Lewis^b blood group antigen, but these cells are separate from the cells expressing the A antigen. All Type III cells appear to express Stb-2, SNAP-25, BDNF, and NCAM. Within the Type III cells are subsets expressing either serotonin (5-HT) or PGP 9.5 and NSE. A subset of the 5-HT cells also expresses IP_3R3 and PLCβ2

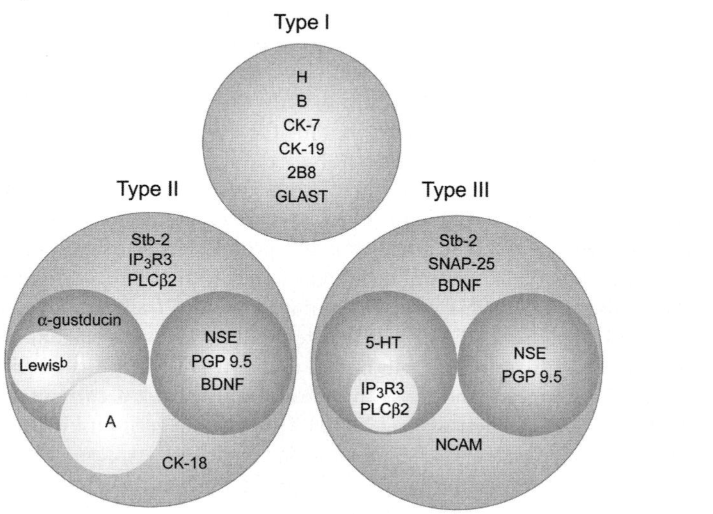

α-gustducin (Yang et al., 2000), whereas another subset expresses the ubiquitin protein PGP 9.5 and NSE (Yee et al., 2001, 2003). Among the gustducin-expressing cells, another subset expresses the Lewis^b blood group antigen and a separate subset the A blood group antigen (Pumplin et al., 1999); some cells express the A antigen and not gustducin, but they do not express PGP 9.5 (*Figure 5-3k*). All Type II cells appear to express synaptobrevin, PLCβ2, and IP_3R3 (Clapp et al., 2004; Yang et al., 2004). The Type II cells that exhibit PGP 9.5 staining also express BDNF (Yee et al., 2003), but they do not express the A antigen (*Figure 5-3k*). Immunocytochemical studies with cytokeratins have shown that cells expressing cytokeratin 18 have the morphological characteristics of "light" cells (*Figure 5-3k, i*) and that a subset of these express α-gustducin (*Figure 5-3e, f*), so it is possible that all Type II cells express this epithelial marker.

Type III cells are also molecularly diverse. They all appear to express synaptobrevin-2, SNAP-25, BDNF, and NCAM (Yee et al., 2001; Yang et al., 2004). Within this cell type are two subsets, one that expresses PGP 9.5 and NSE and another that exhibits 5-HT staining after animals have been pretreated with the 5-HT precursor, 5-hydoxy-L-tryptophan (Yee et al., 2001). Many of these 5-HT cells also appear to express PCLβ2 and IP_3R3 (Clapp et al., 2004). The lack of colocalization between cytokeratin 18 and NCAM (*Figure 5-3i, l*) suggests that Type III cells do not express cytokeratin 18. The A blood group antigen, seen on a subset of Type II cells (Pumplin et al., 1999), does not colocalize with either PGP 9.5 (*Figure 5-3k*) or NCAM, suggesting that it is also not expressed by Type III cells.

There are a number of other molecules that have been identified in taste buds, although no particular cell types have been identified. These include the peptides cholecystokinin (CCK; Herness et al., 2002) and vasoactive intestinal polypeptide (VIP; Herness, 1989), leptin receptors (Shigemura et al., 2003), and the synaptic proteins syntaxin-1, and synaptophysin (Pumplin and Getschman, 2000). Colocalization studies involving recently cloned taste receptors have demonstrated that the bitter receptors (T2Rs) are found in cells expressing α-gustducin (Adler et al., 2000), whereas the T1Rs (sweet and amino acid receptors) are not always found with gustducin (Kim et al., 2003). However, there are no data indicating which cell types express the T1Rs. Both T1Rs and T2Rs appear to colocalize with IP_3R3 and PLCβ2 (Miyoshi et al., 2001), which are expressed by all Type II and a subset of Type III cells (➲ *Figure 5-4*). These various data indicate that the cells within taste buds exhibit a complex array of phenotypes. As we learn more about the role of these molecules in taste bud function, we will begin to gain insight into this complexity.

3 Transduction of Chemical Stimuli

3.1 Ion Channels Function in Taste Cell Excitability

Taste receptor cells are compartmentalized into apical and basolateral portions; taste stimuli interact with the apical membrane but tight junctions generally prohibit chemicals from reaching the basolateral membrane. The interaction between a taste stimulus and a taste receptor or apical channel produces depolarization of the taste receptor cell, either directly or indirectly. This receptor potential leads to activation of voltage-dependent ion channels within the cell and/or to neurotransmitter release onto afferent gustatory nerve fibers. Voltage-dependent Na^+ and K^+ channels in taste cells underlie the generation of action potentials, although the role of these in transduction and taste cell signaling is not completely understood. About 75% of rat vallate and foliate taste receptor cells are capable of generating action potentials, which may encode stimulus intensity and/or produce depolarization necessary to open voltage-activated Ca^{2+} channels on the basolateral membrane (Chen et al., 1996). Taste cells express several types of potassium channels, including delayed rectifier, inward rectifier, transient, leak, and Ca^{2+}-activated K^+ current (Herness and Gilbertson, 1999). Cells that respond in situ to taste stimuli can be separated into two primary subsets: those that possess both inward and outward voltage-gated currents (due to voltage-activated Na^+ and K^+ channels), and those that possess only outward currents (due predominantly to delayed rectifier K^+ channels; Gilbertson et al., 2001). However, the relationship of morphological cell type as defined by immunocytochemical markers to function as defined by current profile is not well understood. For example, cells expressing the gustatory G protein α-gustducin express both inward and outward currents (➲ *Figure 5-5a-c*; see also Medler et al., 2003). Often, these cells also express taste receptors for either sweet or bitter stimuli (Adler et al., 2000; Kim et al., 2003), suggesting that they are receptor cells capable of generating action potentials in response to taste stimuli. In comparison, Type II cells that express CK18 are found both with and without inward current (e.g., ➲ *Figure 5-5d-f*). A subset of cells that express the A blood group antigen and all NCAM-expressing cells was found to possess large inward Na^+ and outward K^+ currents, as well as voltage-gated Ca^{2+} currents (Medler et al., 2003). In the same study, putative Type I cells that express the H blood group antigen were found both with and without small inward Na^+ current. These physiology studies, together with calcium imaging experiments (Caicedo et al., 2003), provide evidence that Type II taste cells involved directly in transduction of sweet- and bitter-tasting stimuli possess the current types necessary for an excitatory response; the relationships of other cell types to current profile is less clear.

3.2 Neurotransmitters in Taste Cells

Subsets of mammalian taste receptor cells possess proteins involved in synaptic machinery, such as synaptobrevin and SNAP-25, and form synaptic contacts with afferent nerve processes (e.g., Yang et al., 2004); these appear to be primarily Type III cells. Additionally, taste cells express a number of neurotransmitters and

Figure 5-5

Electrophysiological responses recorded from identified rat taste receptor cells in intact fungiform taste buds. (a) Taste bud cell filled with Lucifer yellow dye after patch recording. (b) Same taste cell after immunostaining for α-gustducin, showing that the recorded cell expressed this G protein subunit. (c) Inward Na^+ and outward K^+ currents recorded from this cell following a voltage-step protocol (−80 to +40mV). (d) Another taste cell filled with Lucifer yellow. (e) Immunoreactivity to cytokeratin 18, showing that the recorded cell contained both Lucifer yellow and cytokeratin 18. (f) Only outward K^+ currents were seen in this cell

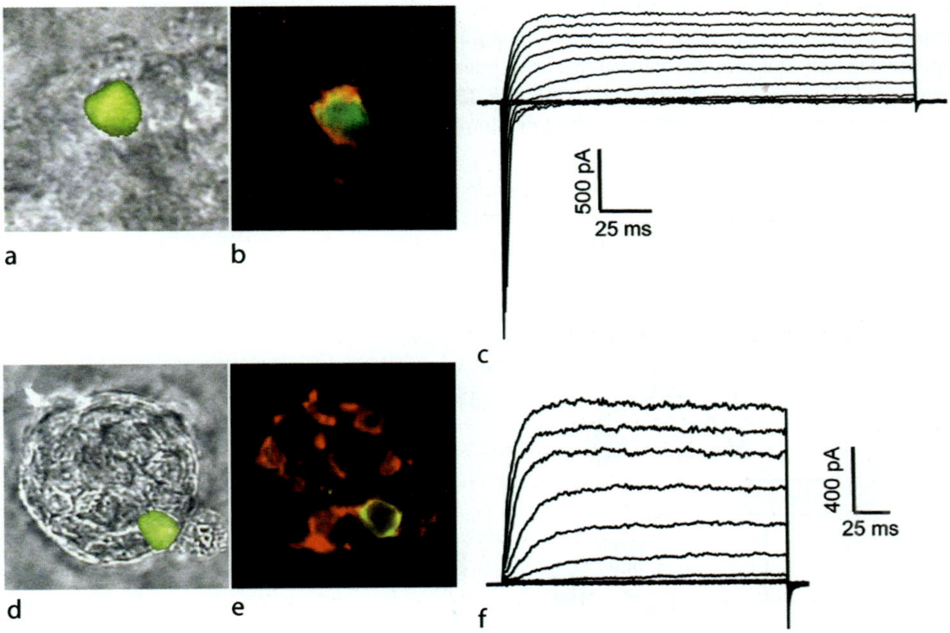

neuropeptides, including serotonin, glutamate, γ-aminobutyric acid (GABA), noradrenalin, vasoactive intestinal peptide (VIP), and CCK (e.g., Kaya et al., 2004; Shen et al., 2005). Recent evidence suggests that a likely candidate for the taste cell—nerve fiber synapse is serotonin, as gustatory stimulation of taste cells causes serotonin release (Huang et al., 2005). There is also ample evidence that some transmitters may play a paracrine role within taste buds. Nonoverlapping subsets of taste cells express either serotonin or serotonergic receptors (Kaya et al., 2004), and serotonin causes inhibition of K^+ current in taste cells (Herness and Chen, 1997). Similarly, some taste cells respond to adrenergic stimulation, and subsets of taste cells express subtypes of adrenergic receptors (Herness et al., 2002). The brain-gut peptide CCK is expressed in subsets of taste cells and may play a signaling role within the taste bud, possibly modifying responses to gustatory stimuli (Herness et al., 2002; Lu et al., 2003).

3.3 Salt and Acid Stimuli are Transduced by Ion Channels

Ion channels located in the apical membrane of taste receptor cells respond directly to taste stimuli in the oral cavity and can be thought of as taste receptors for stimuli having a sour or salty taste (➋ *Figure 5-6*). The predominate group of ion channels involved in transduction of ionic stimuli are members of the degenerin (DEG)/epithelial sodium channel (ENaC) superfamily. These channels are strongly Na^+-selective, but are also somewhat permeable to other cations, such as K^+ and H^+. During gustatory stimulation with sodium salts, Na^+ ions diffuse through open ENaC channels and directly depolarize taste receptor cells. This diffusion is reversibly blocked by application of the diuretic drug amiloride to the

Figure 5-6
Schematic diagram of cellular elements involved in gustatory transduction. Sodium salts stimulate taste bud cells by passing through open ENaC channels on the epical membrane; these channels also admit H^+ and K^+ ions to some extent. ENaC channels are blocked by the diuretic drug amiloride; a nonamiloride-sensitive transduction pathway for Na^+ also exists and may involve passage of ions through the tight junctions separating the taste cells. Acidic stimuli (H^+) engage a number of ion channels, including the two-pore domain K^+ channels (K_2P) depicted here. Both salts and acids depolarize taste cells by ionic entry into the cell. Bitter stimuli, on the other hand, engage a family of G-protein-coupled receptors (GPCRs), the T2Rs. Binding of bitter substances to these proteins leads to activation of PLCβ2 and opening of TRPM5 channels, which admit Ca^{+2} to depolarize the cell and produce neurotransmitter release. Sweet stimuli and amino acids (umami stimuli) engage the T1R receptors and activate PLCβ2 and TRPM5 in a similar fashion. Other second-messenger components, such as IP_3 and cAMP have also been implicated in sweet and bitter transduction. See text for more detailed description

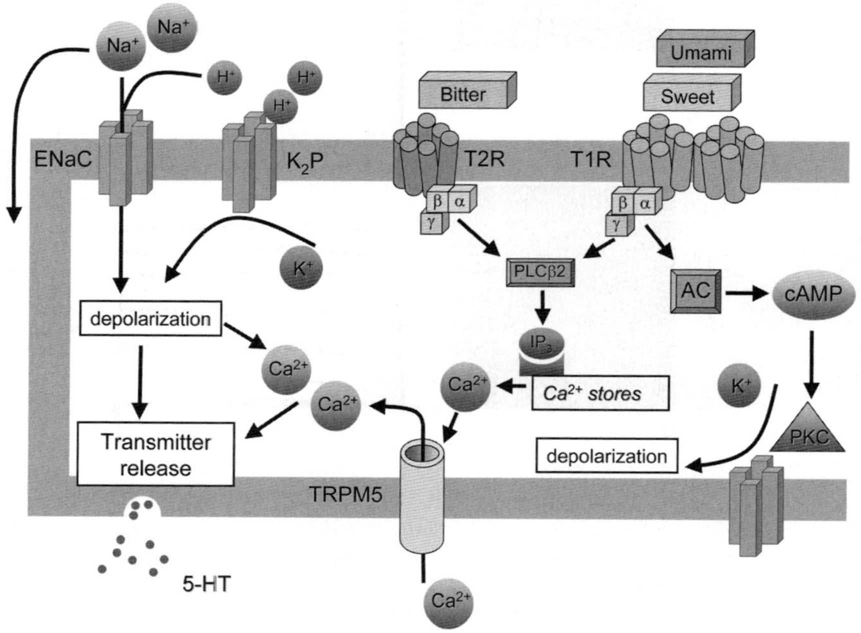

tongue surface, which reduces the response to NaCl stimulation in taste receptor cells, gustatory afferent nerves, and gustatory neurons in the central nervous system (CNS; Boughter and Gilbertson, 1999). ENaCs in taste cells have a high homology with, and share many characteristics of, those found in other transporting epithelia in the body. DEG/ENaC channels have two transmembrane domains, a large extracellular loop, and a small pore-forming domain, and they function as heteroligomeric complexes (Benos and Stranton, 1999). These channels also conduct H^+, which can lead to taste cell depolarization, and appear to play some role in acid taste (Gilbertson et al., 1993). Other ion channels have been proposed to function in acid taste (*Figure 5-6*), including acid-sensing ion channels (ASICs; Lin et al., 2002) and two-pore domain K^+ channels (K_2P; Lin et al., 2004; Richter et al., 2004). These channels are expressed in taste tissue and are sensitive to changes in extracellular pH.

Not all of the gustatory response to salts is mediated by ENaCs. A second, "amiloride-insensitive" mechanism exists, which does not discriminate among Na^+, K^+, or NH_4^+. This mechanism has been described as a paracellular pathway, within which ions pass through the tight junctions between cells to interact with basolateral ion channels, which are not blocked by lingual application of amiloride (Ye et al., 1991). Amiloride-sensitive and -insensitive inputs to the CNS are segregated into different subsets of gustatory neurons, an organization that apparently underlies behavioral discrimination of Na^+ versus nonNa^+ salts (Spector et al., 1996; St John and Smith, 2000). It has been recently suggested that the amiloride-insensitive salt taste receptor may be a variant of a vanilloid-type 1 (VR-1) receptor (Lyall et al., 2004).

3.4 Sweet, Umami, and Bitter Tastes are Transduced by GPCRs

Three genes have been discovered in mammals that account for much of the behavioral sensitivity toward stimuli characterized by humans as tasting sweet or savory (umami). These genes, the *TAS1Rs*, encode functional G-protein-coupled receptors (GPCRs; ➋ *Figure 5-6*) that have been demonstrated in vitro to bind sweet stimuli such as sugars, artificial sweeteners, and sweet proteins, or umami stimuli such as L-amino acids (e.g., Bachmanov et al., 2001; Kitagawa et al., 2001; Max et al., 2001; Montmayeur et al., 2001; Nelson et al., 2001; Sainz et al., 2001; Li et al., 2002). "Umami" is a Japanese word describing the savory taste imparted by certain amino acids, such as monosodium glutamate (MSG). The T1Rs are ClassC GPCRs, characterized by a large N-terminal domain, which contains the binding site for some sweet ligands (Xu et al., 2004). The receptors T1R1 and T1R2 are coexpressed with T1R3, but not with each other, in different subsets of taste receptor cells (Nelson et al., 2001). Heterologous expression studies suggest that T1R1 and T1R3 together form a functional receptor for umami-tasting stimuli but are not responsive to sweet-tasting stimuli, whereas T1R2 and T1R3 together comprise a broad-spectrum receptor for sweet-tasting stimuli, including both natural and artificial sweeteners (Li et al., 2002; Nelson et al., 2001, 2002). Recent studies have suggested that these subunits may also function as homomeric receptors, with relatively weak ligand affinity (Zhao et al., 2003).

A second class of taste receptors, the T2Rs (➋ *Figure 5-6*), function as receptors for bitter-tasting compounds (Adler et al., 2000; Chandrashekar et al., 2000). To date, approximately 25 human and 33 mouse functional bitter taste receptor genes (*TAS2Rs)* have been identified (Fischer et al., 2005). T2Rs are Class A GPCRs, characterized by a short N-terminal domain. They are expressed in taste receptors cells throughout the oral cavity; individual cells tend to express multiple T2Rs (Adler et al., 2000). The ligands for some of the T2R receptors have been discovered using either heterologous expression studies or genetic linkage with known taste sensitivity phenotypes. The mouse receptor mT2R5 was found to respond only to cycloheximide out of a panel of 12 stimuli in a heterologous cell assay (Chandrashekar et al., 2000). Naturally occurring strain variation in taste sensitivity for several other bitter-tasting compounds has been linked to the cluster of 24 *Tas2r* genes on distal mouse chromosome 6 (Bachmanov et al., 2001; Lush et al., 1995; Nelson et al., 2005). Much recent progress has been made in elucidating the specific function of human bitter taste receptors, including the identification of the taste receptor gene hTAS2R38, which underlies the ability to taste phenylthiocarbamide (Kim et al., 2003). Functional expression studies have indicated that various human T2R receptors often respond to a single structural class of bitter stimuli such as β-glucopyranosides (Bufe et al., 2002), or to relatively small subsets of stimuli such as denatonium and 6-nitrosaccharin (Pronin et al., 2004), or the artificial sweeteners saccharin and acesulfame-K, which possess a bitter taste at higher concentrations (Kuhn et al., 2004). In contrast, hT2R14 was shown to be broadly responsive to multiple bitter compounds (Behrens et al., 2004). Recent evidence shows that human *TAS2Rs* are extremely diverse, with as many as 151 known haplotypes of the 25 human T2R genes (Kim et al., 2005).

3.5 Transduction Pathways

Taste receptor cells express diverse elements of cellular transduction machinery and multiple second-messenger pathways have been described, especially for transduction of bitter- and sweet-tasting stimuli (➋ *Figure 5-6*). Taste receptor cells express many second-messenger components that are common to GPCR-coupled cascades such as cyclic nucleotide and phosphoinositide-signaling systems (Herness and Gilbertson, 1999). Increases in intracellular calcium are crucial for both sweet- and bitter-taste transduction, as demonstrated in situ in mouse vallate taste cells that respond to gustatory stimulation (Caicedo et al., 2002). Two salient elements in the receptor-dependent transduction of sweet-, umami-, and bitter-tasting stimuli are a taste cell-specific phospholipase C (PLCβ2) and a transient receptor potential-related channel (TRPM5; Perez et al., 2003; Zhang et al., 2003). Ligand binding to one of these GPCRs activates a G protein, such as the taste cell-specific G protein α-gustducin, which in turn causes the activation of PLCβ2. This leads to activation of TRPM5, which mediates calcium entry into the taste receptor cell. The influx of Ca^{2+} via TRPM5 may contribute to the receptor potential and/or mediate neurotransmitter release onto afferent nerve fibers. The dependency of sweet, umami, and bitter transduction on this pathway is indicated by studies with knockout mice: deletion of the genes for either PLCβ2 or TRPM5 causes a severe decrement in the behavioral and neural response to sweet, umami, or bitter stimuli (Zhang et al., 2003). Other transduction elements have been shown to be involved in bitter and sweet taste, such as cyclic AMP (Cummings et al., 1996) or inositol triphosphate (IP_3; Herness and Gilbertson, 1999); these may play a modulatory role in taste cell signaling.

4 Distribution of Gustatory Sensitivities

As noted earlier, taste transduction involves a variety of mechanisms, including direct permeation or block of ion channels and activation of metabotropic and ionotropic receptors [for reviews, see Herness and Gilbertson (1999), Lindemann (1996)]. There is less known, however, about how these mechanisms are distributed within and across taste receptor cells.

4.1 Sensitivity of Taste Receptor Cells

Intracellular recording experiments conducted as long as 30 years ago suggested that taste cells are broadly responsive to stimuli representing different taste qualities (Ozeki and Sato, 1972; Sato and Beidler, 1997; Tonosaki and Funakoshi, 1984). However, because of their relatively small membrane potentials and the possibility of leak currents associated with penetrating such small cells with sharp electrodes, many investigators viewed these earlier intracellular experiments with skepticism (e.g., Herness and Gilbertson, 1999; Kinnamon, 1988). Many experiments since then have employed patch-clamp recording methods on isolated taste receptor cells (e.g., Avenet and Lindemann, 1987; Kinnamon and Roper, 1988; Cummings et al., 1996), but the range of stimuli that can be applied to an isolated cell preparation is limited, making this approach unfeasible for addressing the question of how the cells are tuned across qualities.

Gilbertson et al. (2001) recorded whole-cell responses of 120 taste cells of the rat fungiform papillae and soft palate maintained within an intact epithelium in a modified Ussing (MU) chamber, which permitted the flow of tastants across the apical membrane while monitoring the activity of cells in intact taste buds with their basolateral membranes bathed in Tyrode's solution. Taste stimuli were: 0.1 *M* sucrose, KCl, and NH_4Cl, 0.032 *M* NaCl, and 3.2 m*M* HCl and quinine hydrochloride (QHCl). When cells were held at their resting potentials, taste stimulation resulted in conductance changes; reversible currents greater than 5 pA were considered reliable responses. Sucrose and QHCl produced a decrease in outward current and membrane conductance, whereas NaCl, KCl, NH_4Cl, and HCl elicited inward currents accompanied by increased conductance. Combinations of responses to pairs of the four basic stimuli (sucrose, NaCl, HCl, and QHCl) across the 71–84 cells tested with each pair were predictable from the probabilities of responses to individual stimuli, indicating an independent distribution of sensitivities. Of 62 cells tested with all four basic stimuli, 59 responded to at least one of the stimuli; 16 of these (27.1%) responded to only one, 20 (33.9%) to two, 15 (25.4%) to three, and 8 (13.6%) to all of the basic stimuli. Cells with both inward (Na^+) and outward (K^+) voltage-activated currents were significantly more broadly tuned to gustatory stimuli than those with only inward currents. These data confirmed the earlier observations with intracellular recording that individual taste receptor cells are relatively broadly responsive to stimuli of differing taste quality (Ozeki and Sato, 1972). Similar results were found using calcium imaging in lingual slice preparations with stimulation restricted to the apical membranes (Caicedo et al., 2002). Thus, several laboratories using highly divergent methods have all shown that taste receptor cells are sensitive to stimuli of more than one quality.

In contrast, expression studies have suggested that the genes for the T1R and T2R receptors, which are involved in responses to sweet and bitter stimuli, respectively, are not colocalized within receptor cells

(Zhang et al., 2003). These authors and others (e.g., Scott, 2004) have suggested, therefore, that taste quality is represented by different cells in a "labeled-line" fashion. However, we have known for many years that sweet and bitter stimuli are predominantly transduced and coded by different cells in the gustatory pathway (e.g., Travers and Smith, 1979). Further, these cells respond also to salts and acids (see later), making the response of any one cell ambiguous with respect to sensory quality, since firing rate is also dependent on stimulus concentration (Smith and Scott, 2003). Even though sweet and bitter stimuli are largely represented by different cells, this is not evident that these cells comprise "labeled lines" for coding sweetness and bitterness, as these cells, particularly in the CNS, respond readily to other classes of stimuli.

The results of one type of experiment in particular have been used to postulate a labeled-line code for taste. Expressing an opiate receptor within taste cells normally expressing the T1R receptors resulted in mice showing behavioral preference for an opiate agonist (Zhao et al., 2003), which the authors interpreted as evidence for a "labeled line" for sweet taste. A similar result was obtained when a T2R receptor was expressed in cells normally expressing T1R receptors: mice then preferred a bitter ligand (Mueller et al., 2005). That is, when the receptor cells normally responding to sugar were activated by an opiate or a bitter taste, mice showed a behavioral preference for these stimuli as though they tasted sweet, showing that these receptor cells are "hardwired" to signal the sweet taste (or to lead to preference and intake). However, does "hard-wired" necessarily mean "labeled line?" Placing an opiate receptor or a bitter receptor in cells that normally respond to sweet stimuli means that these agonists would then produce a pattern of input to the CNS, which mimics that normally produced by sugar. These experiments, however, say nothing about how that information is represented by activity of gustatory neurons, which is the essential question of gustatory neural coding. The input from these receptors to their central targets could take the form of either a specific labeled line or an across-neuron pattern (i.e., an ensemble code); these molecular substitution studies do not differentiate between these possibilities.

4.2 Responsiveness of Peripheral Gustatory Neurons

Since the earliest electrophysiology studies of single peripheral taste fibers, it has been recognized that individual gustatory neurons are often responsive to stimuli representing more than one of the four taste qualities (Pfaffmann, 1955), which led to the initial suggestion that taste quality is represented by the pattern of activity across responsive neurons (Pfaffmann, 1959). Nevertheless, neurons within the taste pathway can be grouped into classes based on their relative sensitivities. For example, when the stimuli 0.1 *M* sucrose (sweet), 0.03 *M* NaCl (salty), 3 m*M* HCl (sour), and 1 m*M* QHCl (bitter) are ordered hedonically from most to least preferred, the response profiles of hamster CT nerve fibers show a single peak as sucrose-, NaCl-, or HCl-best fibers (Frank, 1973). A QHCl-best class of fibers is evident in the IX nerve (Hanamori et al., 1988). Thus, peripheral taste fibers have an organization to their sensitivities; there are four neuron classes in the peripheral nervous system defined by their sensitivities to four basic stimuli: sucrose-, NaCl-, HCl-, and QHCl-best fibers. Sensitivities to sucrose and QHCl are largely contained in different neurons, as one might expect from the distribution of T1R and T2R receptors, but the responses of each of these fiber types, although relatively narrowly tuned, are not absolutely specific. Sucrose-best neurons typically respond somewhat to NaCl and QHCl-best to HCl, whereas the NaCl- and HCl-best fibers are relatively more broadly tuned. Cell classifications based on each neuron's best stimulus are also seen when the relative response profiles of these peripheral fibers are subjected to a hierarchical cluster analysis, which groups the fibers into similar classes based on their responses to all four or even more stimuli (Frank et al., 1988; Hanamori et al., 1988).

5 Central Gustatory Nuclei

5.1 Electrophysiology of Central Taste Neurons

Neurons in the central gustatory pathway are typically more broadly tuned to taste stimuli than peripheral fibers and often respond to tactile and temperature stimulation as well (Travers, 1993; Smith and Scott, 2003). Taste-responsive neurons of the NST receive converging input from peripheral gustatory axons

(Vogt and Mistretta, 1990) and this convergence can occur from separate taste bud populations, e.g., the anterior tongue and palate (Travers et al., 1986; Travers and Norgren, 1991). This convergence produces greater breadth of tuning than in peripheral fibers, so that medullary taste cells are quite broadly responsive (Smith and Scott, 2003). Nevertheless, the cells can still be classified according to their response properties into sucrose-, NaCl-, HCl-, or QHCl-best neurons. Multivariate analyses of the similarities in response profiles indicate that cells within these various groups are relatively similar to one another, but they respond to a range of stimuli (e.g., Smith et al., 1983). The responses of an NST neuron of the rat to several taste stimuli, including an array of bitter compounds, are shown in ❯ *Figure 5-7*. As predicted by the coexpression of T2Rs in taste receptor cells, the responses to bitter stimuli are highly correlated in central gustatory neurons (Lemon and Smith, 2005), although these same neurons respond to stimuli representing other taste qualities as well, such as NaCl and HCl (❯ *Figure 5-7*.)

5.2 Neurotransmitters and Neuromodulators

Responses of brainstem cells to gustatory stimulation are subject to several modulatory influences. For example, glucose, insulin, and pancreatic glucagon, when systemically administered, alter the responses of cells in the rat NST to tongue stimulation with glucose (Giza and Scott, 1983, 1987; Giza et al., 1993). Conditioned taste aversion learning shifts the patterns of response to taste stimuli recorded from the rat NST (Chang and Scott, 1984) and PbN (Tokita et al., 2004). The precise mechanisms underlying these effects are unknown. There are direct descending projections from gustatory areas of the ventral forebrain to both PbN and NST (❯ *Figure 5-1*) and input from the CeA appears to be critical for the effects of such learning on PbN neurons (Tokita et al., 2004). Inputs from the gustatory cortex, the CeA, the lateral hypothalamus, and the BST have all been shown to excite or inhibit the activity of taste-responsive neurons of the rostral NST (Smith and Li, 2000; Cho et al., 2002, 2003; Li et al., 2002; Smith et al., 2005).

Recent studies have begun to reveal mechanisms of synaptic transmission within the gustatory region of the NST. Electrophysiological recordings from cells in the rostral NST from both rats and hamsters have shown that γ-amino butyric acid (GABA) produces inhibition of activity in these cells, which is mediated predominantly by the $GABA_A$ receptor subtype (Liu et al., 1993; Wang and Bradley, 1993). These studies show that the gustatory portion of the NST is under the influence of a tonic GABAergic inhibitory network. One of the roles of GABA is to regulate the breadth of tuning of these cells, which respond to more taste stimuli when GABA activity is blocked by bicuculline (Smith and Li, 1998). That is, GABA inhibition serves to regulate the breadth of responsiveness of gustatory neurons in the NST, a role similar to that seen in other sensory systems (Fuzessery and Hall, 1996; Kyriazi et al., 1996; Sato et al., 1996; Suga et al., 1997).

Glutamate antagonists block the responses of cells in the gustatory zone of the NST. Excitatory postsynaptic potentials (EPSPs) recorded from rat NST cells in vitro in response to electrical stimulation of the solitary tract are reduced by both CNQX and APV, antagonists to AMPA-kainate and NMDA glutamate receptors, respectively (Wang and Bradley, 1995). Both of these agents also reversibly block or reduce the responses to chemical stimulation of the anterior tongue in hamster NST cells recorded in vivo (Li and Smith, 1997). All cells responsive to taste stimulation are blocked by CNQX, regardless of their profiles of sensitivity; there is no evidence that the neurotransmitter is different for cells with different responsiveness (i.e., sucrose- versus NaCl-best). Therefore, it is likely that an excitatory amino acid such as glutamate acts as a neurotransmitter between gustatory afferent fibers and taste-responsive cells in the NST.

Figure 5-7

Trains of action potentials recorded extracellularly from a neuron in the rostral portion of the nucleus of the solitary tract (NST) of a rat. About 10s of activity is shown for each stimulus, the application of which is indicated by the arrowheads. This cell did not respond to the sweet-tasting stimuli (sucrose and fructose), but showed robust responses to sodium salts, nonsodium salts, acids, and all the 10 bitter stimuli applied to the tongue and palate. Arrowheads indicated the time of stimulus application. The interspike interval histogram shown at the lower left indicates that no spikes fell within the neuron's refractory period, demonstrating that the spikes are recorded from a single neuron. Data from Lemon and Smith (2005)

Figure 5-7 (continued)

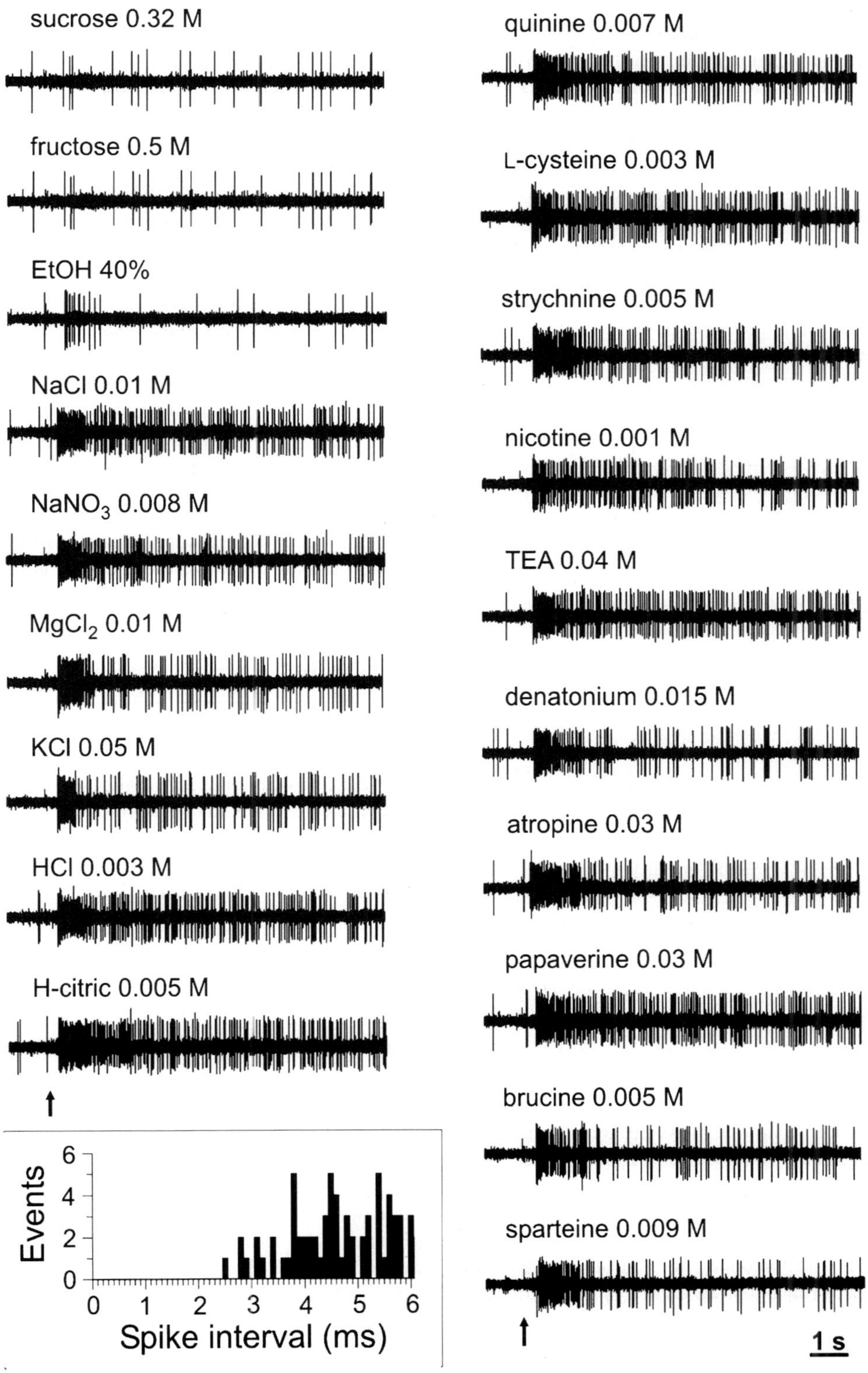

Evidence also indicates that taste-responsive cells in the NST are excited by substance P (SP) and inhibited by met-enkephalin. Immunocytochemical studies have shown that tachykinin- and opioid-containing neurons are present within the gustatory zone of the NST (Davis and Kream, 1993) and that SP-containing fibers enter this nucleus from a number of yet unknown sources (Duncan et al., 1990). In vitro experiments on rat brainstem slices have shown that bath application of SP excites a number of cells in the gustatory zone of the NST (King et al., 1993). A role for these SP-responsive NST cells in gustatory processing has been demonstrated in the hamster NST in vivo, where responses to anterior tongue stimulation with NaCl or sucrose are enhanced by local microinjection of SP (Davis and Smith, 1997). In vivo experiments also show that a subset of taste-responsive neurons in the NST is inhibited by met-enkephalin (Li et al., 2003), showing a role for endogenous opiates in regulating gustatory processing. Recent in vitro recording from the rostral portion of the rat solitary nucleus demonstrates a functional role for cholinergic receptors (AChRs), including α7 nicotinic, nonα7 nicotinic, and muscarinic AChRs (V.V. Uteshev and Smith, 2006). Although the region of the NST in which afferent terminals of the VII and IX nerves are most dense is heavily immunoreactive for the presence of acetylcholinesterase, (AChE; (Barry et al., 1993, 2001) it is not yet clear if or how cholinergic mechanisms modulate taste activity.

Several investigators have reported neurotransmitter actions within the gustatory cortex of rats similar to those seen in the medulla. For example, Ogawa and colleagues have shown that glutamate receptor antagonists excited about 80% of taste-responsive neurons in the rat gustatory cortex, whereas glutamate receptor antagonists blocked this excitatory effect (Otawa et al., 1995). On the other hand, GABA decreased the activity of about 60% of taste-responsive cortical neurons and this effect was antagonized by bicuculline methiodide, a $GABA_A$ receptor antagonist (Ogawa et al., 1998). Both SP and CGRP appear to increase firing in cortical gustatory neurons and CGRP immunoreactivity in the cortical gustatory area is increased by aversive stimuli such as quinine and by palatable stimuli after rats are trained to avoid them (Yamamoto et al., 1990).

6 Forebrain Interactions with the Taste System

6.1 Forebrain Targets of the Gustatory System

From the NST, ascending gustatory fibers project to third-order cells within the PbN of the pons (Norgren, 1978; Travers, 1988) and in turn to multiple forebrain nuclei, including the thalamus, insular cortex (IC), LH, CeA, and the BST (Norgren, 1976; Saper and Loewy, 1980; Halsell, 1992). These forebrain targets (➋ *Figure 5-1*) send centrifugal axons to both the PbN and NST (van der Kooy et al., 1984; Allen et al., 1991; Halsell, 1998). All of these forebrain regions of the gustatory system have been shown to produce modulation of brainstem taste activity in the NST and/or the PbN.

6.2 Forebrain Modulation of Brainstem Neural Activity

Earlier research, involving either decerebration or electrical stimulation, suggested that descending pathways from the forebrain could modulate brainstem gustatory activity (Matsuo et al., 1984; Murzi et al., 1986; Mark et al., 1988). A recent series of experiments has examined systematically the centrifugal influence of various forebrain areas on taste-responsive neurons in the hamster NST and PbN (Smith and Li, 2000; Cho et al., 2002, 2003; Li et al., 2002, 2005; Smith et al., 2005) and similar data are available for some areas of the rat forebrain (Lundy and Norgren, 2001, 2004). The experiments in hamsters employed bilateral stimulating electrodes and drug-injection pipettes in forebrain nuclei and extracellular recording of neural activity from NST neurons in vivo. Brief electrical pulses (<0.1 mA, 0.5 ms) were applied repeatedly (1/3 Hz) to forebrain sites and peri-stimulus time histograms (PSTHs) were accumulated over 100–200 stimulus trials to reveal excitatory or inhibitory modulation of NST or PbN neuronal activity. Once such a connection was confirmed, electrical stimulus trains (100 Hz, 0.2 ms, at 0.9 × orthodromic threshold) were applied for 15 s before and during taste stimulation trials to determine the effects of descending modulation on taste-evoked activity. The forebrain was also stimulated with DL-homocysteic acid (DLH) to limit stimulation to neuronal somata and not fibers of passage. Taste stimuli were 32 m*M* sucrose (S), 32 m*M* NaCl (N), 3.2 m*M* citric acid (C), and 32 m*M* quinine hydrochloride (Q), applied to the anterior tongue. Each NST cell was classified as S-, N-, C-, or Q-best based on its response to these stimuli. The effects of stimulation of each forebrain target on NST activity are summarized in the following paragraphs.

6.2.1 Insular Cortex

Previous experiments demonstrated that blocking the activity of the IC with procaine resulted in either increased or decreased responses to taste stimulation in rat NST (Di Lorenzo and Monroe, 1995). In an initial experiment, Smith and Li (2000) used multibarrel glass micropipettes to record the activity of NST neurons extracellularly and to apply the $GABA_A$ antagonist bicuculline methiodide (BICM) into the vicinity of the cell. The ipsilateral IC was stimulated while the spontaneous activity of each NST cell was recorded. The baseline activity of 17/50 cells (34%) was modulated by cortical stimulation: eight cells were inhibited and nine were excited. BICM microinjected into the NST blocked the cortical-induced inhibition. Although the excitatory effects were distributed across S-, N-, and C-best neurons, the inhibitory effects of cortical stimulation were significantly more common in N-best cells. These data suggest that corticofugal input to the NST may differentially inhibit gustatory afferent activity through GABAergic mechanisms. A more recent experiment (unpublished observations), in which electrodes were implanted in the IC bilaterally, has demonstrated a slightly greater influence from the contralateral than the ipsilateral IC on NST taste cells: 16/50 cells (32%) were modulated ipsilaterally, whereas 20/50 (40%) responded to contralateral stimulation. Eleven of these neurons received converging modulation from both sides of the cortex: three were excited and eight inhibited bilaterally.

6.2.2 Lateral Hypothalamus

Stimulation of the LH produces increases in food intake and alterations in taste preference behavior (Frank et al., 1982; Vasudev et al., 1985), whereas damage to this area has opposite effects (Grossman et al., 1978; Touzani and Velley, 1990). Half of the taste-responsive cells in the NST (49/99) were modulated by LH stimulation (Cho et al., 2002). Contralateral stimulation was more often effective (41 cells) than ipsilateral (13 cells) and always excitatory; 10 cells were excited bilaterally. Six cells were inhibited by ipsilateral stimulation. A subset of these cells ($n = 13$) was examined for the effects of microinjection of DLH into the LH. The effects of electrical stimulation were completely imitated by DLH, indicating that cell somata in and around the LH are responsible for these effects. Other cells ($n = 14$) were tested for the effects of electrical stimulation of the LH on the responses to stimulation of the tongue with standard tastants (S, N, C, and Q). Responses to taste stimuli were more than doubled by the excitatory influence of the LH, with no alteration of spontaneous activity. These effects would enhance taste discriminability by increasing the signal-to-noise ratio of taste-evoked activity. Thus, in addition to its role in feeding and metabolism, the LH exerts descending control over the processing of gustatory information through the brainstem. Specifically, when the LH is active, neurons of the NST may be capable of finer taste discriminations.

6.2.3 Central Nucleus of the Amygdala

The CeA contains neurons that respond differentially to hedonically positive and negative taste stimuli (Nishijo et al., 1998) and both the CeA and basolateral amygdala are involved in conditioned taste aversion learning (Yamamoto et al., 1994). Extracellular action potentials were recorded from 109 taste-responsive cells in the NST and analyzed for a change in excitability following electrical and chemical stimulation of the CeA (Li et al., 2002). An orthodromic excitatory response was observed in 33 of 109 taste-responsive cells (30.3%). An initial decrease in excitability, suggestive of postsynaptic inhibition, was observed in 3 of the

109 cells (2.7%). Many of these neurons were under the influence of the contralateral CeA (28/36 = 77.8%) as well as the ipsilateral (22/36 = 61.1%); 14 (38.9%) were excited bilaterally. Latencies for excitation were longer following ipsilateral than after contralateral CeA stimulation. The effect of microinjection of DLH into the CeA closely resembled the effect of electrical stimulation on each of the nine cells tested: DLH excited eight and inhibited one of these electrically activated NST cells. Application of subthreshold electrical stimulation to the CeA during taste trials enhanced the taste responses of every CeA-responsive NST cell tested with this protocol, as shown for four CeA neurons in ➲ *Figure 5-8*. NST cells modulated by the CeA were significantly less responsive to taste stimuli than cells that were not. Given that local-circuit neurons in the NST are less responsive to taste stimuli than cells that project to the PbN (Cho et al., 2002), this finding suggests that the CeA may preferentially target neurons involved in brainstem motor pathways, as may occur following taste aversion learning. In another study (Cho et al., 2003), electrodes were implanted bilaterally in both the LH and CeA and 113/215 cells (52.6%) were modulated by either one or both of these forebrain sites. Both the LH and CeA altered the responses of 52 of these cells, showing that these sites produce converging modulation of medullary taste neurons. Studies of both the rat (Lundy and Norgren, 2001, 2004) and hamster (Li et al., 2005) PbN show that the LH and the CeA also provide descending modulation of pontine taste neurons, but that this influence is largely inhibitory, in contrast to the excitatory effect on NST neurons.

6.2.4 Bed Nucleus of the Stria Terminalis

Electrical stimulation of the ipsilateral BST inhibited the activity of 29 of 121 NST taste cells (24.0%); two were excited (Smith et al., 2005). Stimulation of the contralateral BST inhibited 14 neurons and excited six. Seven cells were inhibited bilaterally and two were excited by contralateral stimulation and inhibited by ipsilateral. In all, 43 of 121 NST cells (35.5%) were modulated by stimulation of the BST. These results demonstrated that most of the BST influence on NST taste cells was inhibitory. This inhibition, like the excitation produced by LH and CeA stimulation, was distributed across all cell types in the NST (S-, N-, C, and Q-best).

This series of experiments demonstrates extensive centrifugal modulation of brainstem gustatory activity for cells in both the NST and PbN. Essentially every forebrain target of the gustatory system, including the IC, LH, CeA, and BST, plays a descending modulatory role in the processing of taste information. The LH and CeA have predominantly an excitatory effect on the NST and an inhibitory effect on the PbN. The IC and particularly the BST produce significant inhibition of medullary taste responses. This extensive neural substrate no doubt underlies the modulation of taste activity by physiological and experiential factors. Further research should be directed toward determining how these pathways are engaged by alterations in blood glucose, gastric distension, conditioned taste aversion learning, and other physiological conditions known to alter taste sensitivity and to examination of the neurotransmitter systems underlying these effects.

Figure 5-8

Taste response profiles of several NST neurons to taste stimulation before and during subthreshold CeA stimulation. (a) This PSTH depicts the response of neuron CeA-NST-15 to single-pulse electrical stimulation of the CeA, showing a brief excitatory response. (b) The control response to each of the four basic stimuli (sucrose, NaCl, citric acid, and QHCl) are shown first for the same NST neuron as in a. The responses (impulse/s) during the 10-s stimulus are shown as black bars and the 5-s prestimulus and poststimulus water rinse periods are shown as open bars; each histogram reflects 20 s of activity. After the four taste stimuli are the responses to the two most effective stimuli in the presence of electrical stimulus (ES) trains delivered to the CeA. The responses to both QHCl and citric acid were enhanced by CeA stimulation. (c–e) Control responses and responses to the two most effective taste stimuli during ES for three additional NST neurons. For each of these cells, CeA stimulation markedly enhanced the response to taste stimulation, but had no effect on the firing rate during the water prerinse. (f) Mean (+SEM) firing rates to all taste stimuli tested before (taste) and

Figure 5-8 (continued)

during (taste + ES) electrical stimulation of the CeA. Activity during the 5-s prerinse period (*open bars*) was unaffected by ES, whereas the response to taste stimuli (*shaded bars*) was significantly (*) increased. From Li et al. (2002)

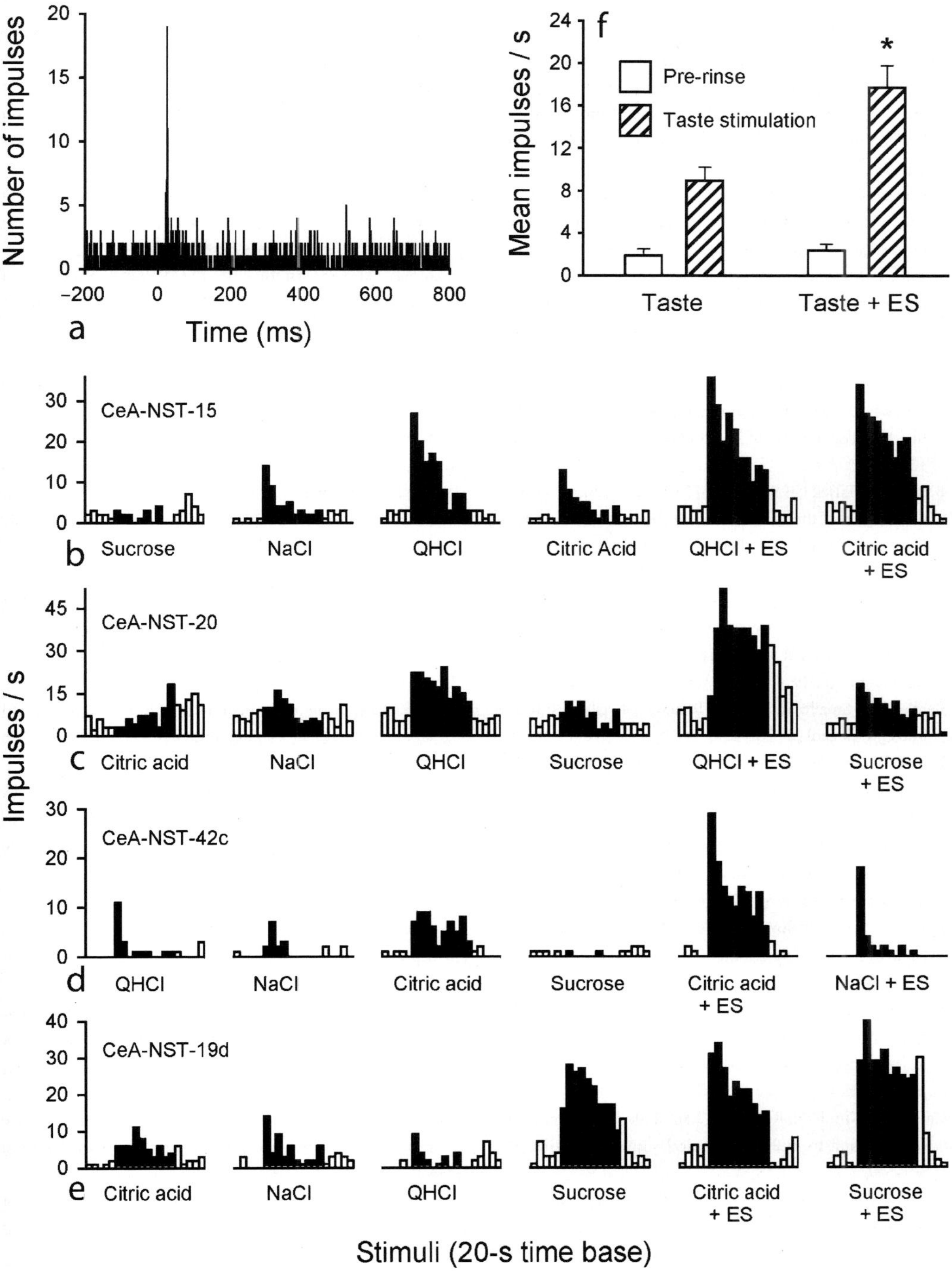

References

Adler E, Hoon MA, Mueller KL, Chandrashekar J, Ryba NJP, et al. 2000. A novel family of mammalian taste receptors. Cell 100: 693-702.

Allen GV, Saper CB, Hurley JM, Cechetto DF. 1991. Organization of visceral and limbic connections in the insular cortex of the rat. J Comp Neurol 311: 1-23.

Avenet P, Lindemann B. 1987. Patch-clamp study of isolated taste receptor cells of the frog. J Membr Biol 97: 223-240.

Bachmanov AA, Li X, Li S, Neira M, Beauchamp GK, et al. 2001. High-resolution genetic mapping of the sucrose octaacetate taste aversion (SoA) locus on mouse Chromosome 6. Mamm Genome 12: 695-699.

Bachmanov AA, Li X, Reed DR, Ohmen JD, Li S, et al. 2001. Positional cloning of the mouse saccharin preference (Sac) locus. Chem Senses 26: 925-933.

Barry MA, Haglund S, Savoy LD. 2001. Association of extracellular acetylcholinesterase with gustatory nerve terminal fibers in the nucleus of the solitary tract. Brain Res 921: 12-20.

Barry MA, Halsell CB, Whitehead MC. 1993. Organization of the nucleus of the solitary tract in the hamster: Acetylcholinesterase, NADH dehydrogenase and cytochrome oxidase histochemistry. Microsc Res Tech 26: 231-244.

Behrens M, Brockhoff A, Kuhn C, Bufe B, Winnig M, et al. 2004. The human taste receptor hTAS2R14 responds to a variety of different bitter compounds. Biochem Biophys Res Commun 319: 479-485.

Beidler LM, Smallman R. 1965. Renewal of cells within taste buds. J Cell Biol 27: 263-272.

Belecky TL, Smith DV. 1990. Postnatal development of palatal and laryngeal taste buds in the hamster. J Comp Neurol 293: 646-654.

Benos DJ, Stranton BA. 1999. Functional domains within the degenerin/epithelial sodium channel (Deg/ENaC) superfamily of ion channels. J Physiol 520: 631-644.

Boughter JD Jr, Gilbertson TA. 1999. From channels to behavior: An integrative model of NaCl taste. Neuron 22: 213-215.

Bufe B, Hofmann T, Krautwurst D, Raguse JD, Meyerhof W. 2002. The human TAS2R16 receptor mediates bitter taste in response to beta-glucopyranosides. Nat Genet 32: 397-401.

Caicedo A, Kim K-N, Roper SD. 2002. Individual mouse taste cells respond to multiple chemical stimuli. J Physiol 544: 501-509.

Caicedo A, Pereira E, Margolskee RF, Roper SD. 2003. Role of the G-protein subunit α-gustducin in taste cell responses to bitter stimuli. J Neurosci 23: 9947-9952.

Chandrashekar J, Mueller KL, Hoon MA, Adler E, Feng L, et al. 2000. T2Rs function as bitter taste receptors. Cell 100: 703-711.

Chang F-CT, Scott TR. 1984. Conditioned taste aversions modify neural responses in the rat nucleus tractus solitarius. J Neurosci 4: 1850-1862.

Chen Y, Sun X-D, Herness MS. 1996. Characteristics of action potentials and their underlying outward currents in rat taste receptor cells. J Neurophysiol 75: 820-831.

Cho YK, Li C-S, Smith DV. 2002a. Gustatory projections from the nucleus of the solitary tract to the parabrachial nuclei in the hamster. Chem Senses 27: 81-90.

Cho YK, Li C-S, Smith DV. 2002b. Taste responses of neurons of the hamster solitary nucleus are enhanced by lateral hypothalamic stimulation. J Neurophysiol 87: 1981-1992.

Cho YK, Li C-S, Smith DV. 2003. Descending influences from the lateral hypothalamus and amygdala converge onto medullary taste neurons. Chem Senses 28: 155-171.

Clapp TR, Stone LM, Margolskee RF, Kinnamon SC. 2001. Immunocytochemical evidence for co-expression of Type III IP_3 receptor with signaling components of bitter taste transduction. BMC Neurosci 2: 6.

Clapp TR, Yang R, Stoick CL, Kinnamon SC, Kinnamon JC. 2004. Morphological characterization of rat taste receptor cells that express components of the phospholipase C signaling pathway. J Comp Neurol 468: 311-321.

Contreras RJ, Beckstead RM, Norgren R. 1982. The central projections of the trigeminal, facial, glossopharyngeal and vagus nerves: An autoradiographic study in the rat. J Auton Nerv Syst 6: 303-322.

Cottler-Fox M, Arvidson K, Hammarlune E, Friberg U. 1987. Fixation and occurrence of dark and light cells in taste buds of fungiform papillae. Scand J Dental Res 95: 417-427.

Cummings TA, Daniels C, Kinnamon SC. 1996. Sweet taste transduction in hamster: Sweeteners and cyclic nucleotides depolarize taste cells by reducing a K^+ current. J Neurophysiol 75: 1256-1263.

Davis BJ, Kream RM. 1993. Distribution of tachykinin- and opioid-expressing neurons in the hamster solitary nucleus: An immuno- and in situ hybridization histochemical study. Brain Res 616: 6-16.

Davis BJ, Smith DV. 1997. Substance P modulates taste responses in the nucleus of the solitary tract of the hamster. Neuroreport 8: 1723-1727.

Di Lorenzo PM, Monroe S. 1995. Corticofugal influence on taste responses in the nucleus of the solitary tract in the rat. J Neurophysiol 74: 258-272.

Duncan HJ, Brining SK, Smith DV. 1990. Distribution of substance P-immunoreactive nerve terminals in gustatory regions of the hamster solitary nucleus. Chem Senses 15: 568.

Emmers R, Benjamin RM, Blomquist AJ. 1962. Thalamic localization of afferents from the tongue in albino rats. J Comp Neurol 118: 43-48.

Farbman AI. 1980. Renewal of taste bud cells in rat circumvallate papillae. Cell Tiss Kinetics 13: 349-357.

Finger TE. 2005. Cell types and lineages in taste buds. Chem Senses 30 (Suppl. 1): i54-i55.

Fischer A, Gilad Y, man O, Paabo S. 2005. Evolution of bitter taste receptors in humans and apes. Mol Biol Evol 22: 432-436.

Frank M. 1973. An analysis of hamster afferent taste nerve response functions. J Gen Physiol 61: 588-618.

Frank ME, Bieber SL, Smith DV. 1988. The organization of taste sensibilities in hamster chorda tympani nerve fibers. J Gen Physiol 91: 861-896.

Frank RA, Preshaw RL, Stutz RM, Valenstein ES. 1982. Lateral hypothalamic stimulation: Stimulus-bound eating and self-deprivation. Physiol Behav 29: 17-21.

Fuzessery ZM, Hall JC. 1996. Role of GABA in shaping frequency tuning and creating FM sweep selectivity in the inferior colliculus. J Neurophysiol 76: 1059-1073.

Gilbertson TA, Roper SD, Kinnamon SC. 1993. Proton currents through amiloride-sensitive Na^+ channels in isolated hamster taste cells: Enhancement by vasopressin and cAMP. Neuron 10: 931-942.

Gilbertson TA, Boughter JD Jr, Zhang H, Smith DV. 2001. Distribution of gustatory sensitivities in rat taste cells: Whole-cell responses to apical chemical stimulation. J Neurosci 21: 4931-4941.

Giza BK, Scott TR. 1983. Blood glucose selectively affects taste-evoked activity in rat nucleus tractus solitarius. Physiol Behav 31: 643-650.

Giza BK, Scott TR. 1987. Intravenous insulin infusions in rats decrease gustatory-evoked responses to sugars. Am J Physiol 252: R994-R1002.

Giza BK, Deems RO, Vanderweele DA, Scott TR. 1993. Pancreatic glucagon suppresses gustatory responsiveness to glucose. Am J Physiol 265: R1231-R1237.

Grossman SP, Dacey D, Halaris AE, Collier T, Routtenberg A. 1978. Aphagia and adipsia after preferential destruction of nerve cell bodies in hypothalamus. Science 202: 537-539.

Guth L. 1957. The effects of glossopharyngeal nerve transection on the circumvallate papilla of the rat. Anat Rec 128: 715-731.

Halsell CB. 1992. Organization of parabrachial nucleus efferents to the thalamus and amygdala in the golden hamster. J Comp Neurol 317: 57-78.

Halsell CB. 1998. Differential distribution of amygdaloid input across rostral solitary nucleus subdivisions in rat. Ann NY Acad Sci 855: 482-485.

Hamilton RB, Norgren R. 1984. Central projections of gustatory nerves in the rat. J Comp Neurol 222: 560-577.

Hanamori T, Smith DV. 1989. Gustatory innervation in the rabbit: Central distribution of sensory and motor components of the chorda tympani, glossopharyngeal and superior laryngeal nerves. J Comp Neurol 282: 1-14.

HanamoriT, Miller IJ Jr, Smith DV. 1988. Gustatory responsiveness of fibers in the hamster glossopharyngeal nerve. J Neurophysiol 60: 478-498.

Herness MS. 1989. Vasoactive intestinal peptide-like immunoreactivity in rodent taste cells. Neuroscience 33: 411-419.

Herness MS, Gilbertson TA. 1999. Cellular mechanisms of taste transduction. Ann Rev Physiol 61: 873-900.

HernessMS, ZhaoF-L, Lu S-G, Kaya N, Shen T. 2002. Expression and physiological actions of cholecystokinin in rat taste receptor cells. J Neurosci 22: 10018-10029.

Herness S, Chen Y. 1997. Serotonin inhibits calcium-activated K^+ current in rat taste receptor cells. Neuroreport 8: 3257-3261.

Herness S, Zhao FL, Kaya N, Lu SG, Shen T, et al. 2002. Adrenergic signaling between rat taste receptor cells. J Physiol 543: 601-614.

Huang YJ, Maruyama Y, Lu KS, Pereira E, Plonsky I, et al. 2005. Mouse taste buds use serotonin as a neurotransmitter. J Neurosci 25: 843-847.

Kaya N, Shen T, Lu SG, Zhao FL, Herness S. 2004. A paracrine signaling role for serotonin in rat taste buds: Expression and localization of serotonin receptor subtypes. Am J Physiol 286: R649-R658.

Khaisman EB. 1976. Particular features of the innervation of taste buds of the epiglottis in monkeys. Acta Anat 95: 101-115.

KimM-R, Kusakabe Y, Miura H, Shindo Y, Ninomiya Y, et al. 2003. Regional expression patterns of taste receptors and gustducin in the mouse tongue. Biochem Biophys Res Commun 312: 500-506.

Kim U, Wooding S, Ricci D, Jorde LB, Drayna D. 2005. Worldwide haplotype diversity and coding sequence variation at human bitter taste receptor loci. Hum Mutat 26: 199-204.

Kim UK, Jorgenson E, Coon H, Leppert M, Risch N, et al. 2003. Positional cloning of the human quantitative trait locus underlying taste sensitivity to phenylthiocarbamide. Science 299: 1221-1225.

King MS, Wang L, Bradley RM. 1993. Substance P excites neurons in the gustatory zone of the rat nucleus tractus solitarius. Brain Res 619: 120-130.

Kinnamon JC. 1987. Organization and innervation of taste buds. Neurobiology of taste and smell. Finger TE, Silver WL, editors. New York: Wiley; pp. 277-297.

Kinnamon JC, Taylor BJ, Delay RJ, Roper SD. 1985. Ultrastructure of mouse vallate taste buds. I. Taste cells and their associated synapses. J Comp Neurol 235: 48-60.

Kinnamon S. 1988. Taste transduction: A diversity of mechanisms. Trends Neurosci 11: 491-496.

Kinnamon SC, Roper SD. 1988. Membrane properties of isolated mudpuppy taste cells. J Gen Physiol 91: 351-371.

Kitagawa M, Kusakabe Y, Miura H, Ninomiya Y, Hino A. 2001. Molecular genetic indentification of a candidate receptor gene for sweet taste. Biochem Biophys Res Commun 283: 236-242.

Knapp L, Lawton A, Oakley B, Wong L, Zhang C. 1995. Keratins as markers of differentiated taste cells of the rat. Differentiation 58: 341-349.

Kosar E, Grill HJ, Norgren R. 1986. Gustatory cortex in the rat. II. Thalamocortical projections. Brain Res 379: 342-352.

Kuhn C, Bufe B, Winnig M, Hofmann T, Frank O, et al. 2004. Bitter taste receptors for saccharin and acesulfame K. J Neurosci 24: 10260-10265.

Kyriazi HT, Carvell GE, Brumberg JC, Simons DJ. 1996. Quantitative effects of GABA and bicuculline methiodide on receptive-field properties of neurons in real and simulated whisker barrels. J Neurophysiol 75: 547-560.

Lawton DM, Furness DN, Lindemann B, Hackney CM. 2000. Localization of the glutamate-aspartate transporter, GLAST, in rat taste buds. Eur J Neurosci 12: 3163-3171.

Lemon CH, Smith DV. 2005. Neural representation of bitter taste in the nucleus of the solitary tract. J Neurophysiol 94: 3719-3729.

Li C-S, Smith DV. 1997. Glutamate receptor antagonists block gustatory afferent input to the nucleus of the solitary tract. J Neurophysiol 77: 1514-1525.

Li C-S, Cho YK, Smith DV. 2002. Taste responses of neurons in the hamster solitary nucleus are modulated by the central nucleus of the amygdala. J Neurophysiol 88: 2979-2992.

Li C-S, Cho YC, Smith DV. 2005. Modulation of parabrachial taste neurons by electrical and chemical stimulation of the lateral hypothalamus and amygdala. J Neurophysiol 93: 1183-1196.

Li C-S, Davis BJ, Smith DV. 2003. Opioid modulation of taste responses in the nucleus of the solitary tract. Brain Res 965: 21-34.

Li X, Staszewski L, Xu H, Durick K, Zoller M, et al. 2002. Human receptors for sweet and umami taste. Proc Natl Acad Sci USA 99: 4692-4696.

Lin W, Ogura T, Kinnamon SC. 2002. Acid-activated cation currents in rat vallate taste receptor cells. J Neurophysiol 88: 133-141.

Lin W, Burks CA, Hansen DR, Kinnamon SC, Gilbertson TA. 2004. Taste receptor cells express pH-sensitive leak K^+ channels. J Neurophysiol 92: 2909-2919.

Lindemann B. 1996. Taste reception. Physiol Rev 76: 719-766.

Liu H, Behbehani MM, Smith DV. 1993. The influence of GABA on cells in the gustatory region of the hamster solitary nucleus. Chem Senses 18: 285-305.

Lu SG, Zhao FL, Herness S. 2003. Physiological phenotyping of cholecystokinin-responsive rat taste receptor cells. Neurosci Lett 351: 157-160.

Lundy RF, Norgren R. 2004. Activity in the hypothalamus, amygdala, and cortex generates bilateral and convergent modulation of pontine gustatory neurons. J Neurophysiol 91: 1143-1157.

Lundy RF Jr, Norgren R. 2001. Pontine gustatory activity is altered by electrical stimulation in the central nucleus of the amygdala. J Neurophysiol 85: 770-783.

Lush IE, Hornigold N, King P, Stoye JP. 1995. The genetics of tasting in mice. VII. Glycine revisited, and the chromosomal location of Sac and Soa. Genet Res 66: 167-174.

Lyall V, Heck GL, Vinnikova AK, Ghosh S, Phan TH, et al. 2004. The mammalian amiloride-insensitive non-specific salt taste receptor is a vanilloid receptor-1 variant. J Physiol 558: 147-159.

Mark GP, Scott TR, Chang F-CT, Grill HJ. 1988. Taste responses in the nucleus tractus solitarius of the chronic decerebrate rat. Brain Res 443: 137-148.

Matsuo R, Shimizu N, Kusano K. 1984. Lateral hypothalamic modulation of oral sensory afferent activity in nucleus tractus solitarius neurons of rats. J Neurosci 4: 1201-1207.

Max M, Shanker YG, Huang L, Rong M, Liu Z, et al. 2001. Tas1r3, encoding a new candidate taste receptor, is allelic to the sweet responsiveness locus Sac. Nat Genet 28: 58-63.

Medler KF, Margolskee RF, Kinnamon SC. 2003. Electrophysiological characterization of voltage-gated currents in defined taste cell types in mice. J Neurosci 23: 2608-2617.

Miller IJ Jr, Smith DV. 1984. Quantitative taste bud distribution in the hamster. Physiol Behav 32: 275-285.

Miller IJ Jr, Spangler KM. 1982. Taste bud distribution and innervation on the palate of the rat. Chem Senses 7: 99-108.

Miyoshi M, Abe K, Emori Y. 2001. IP(3) receptor type 3 and PLCbeta2 are co-expressed with taste receptors T1R and T2R in rat taste bud cells. Chem Senses 26: 259-265.

Montmayeur JP, Liberles SD, Matsunami H, Buck LB. 2001. A candidate taste receptor gene near a sweet taste locus. Nat Neurosci 4: 492-498.

Mueller KL, Hoon MA, Erlenbach I, Chandrashekar J, Zuker CS, et al. 2005. The receptors and coding logic for bitter taste. Nature 434: 225-229.

Murray RG. 1971. Ultrastructure of taste receptors. Handbook of sensory physiology, Vol. IV. Chemical Senses.

Part 2. Taste. Beidler LM, editor. Berlin: Springer-Verlag; pp. 31-50.

Murray RG. 1986. The mammalian taste bud type III cell: a critical analysis. J Ultrastruct Molec Struct Res 95: 175-188.

Murzi E, Hernandez L, Baptista T. 1986. Lateral hypothalamic sites eliciting eating affect medullary taste neurons in rats. Physiol Behav 36: 829-834.

Nelson G, Chandrashekar J, Hoon MA, Feng L, Zhao G, et al. 2002. An amino-acid taste receptor. Nature 416: 199-202.

Nelson G, Hoon MA, Chandrashekar J, Zhang Y, Ryba NJP, et al. 2001. Mammalian sweet taste receptors. Cell 106: 381-390.

Nelson GM, Finger TE. 1993. Immunolocalization of different forms of neural cell adhesion molecule (NCAM) in rat taste buds. J Comp Neurol 336: 507-516.

Nelson TM, Munger SD, Boughter JD Jr, 2005. Haplotypes at the Tas2r locus on distal chromosome 6 vary with quinine taste sensitivity in inbred mice. BMC Genet 6: 32.

Nishijo H, Uwano T, Tamura R, Ono T. 1998. Gustatory and multimodal neuronal responses in the amygdala during licking and discrimination of sensory stimuli in awake rats. J Neurophysiol 79: 21-36.

Norgren R. 1970. Gustatory responses in the hypothalamus. Brain Res 21: 63-71.

Norgren R. 1974. Gustatory afferents to ventral forebrain. Brain Res 81: 285-295.

Norgren R. 1976. Taste pathways to hypothalamus and amygdala. J Comp Neurol 166: 17-30.

Norgren R. 1978. Projections from the nucleus of the solitary tract in the rat. Neuroscience 3: 207-218.

Norgren R. 1985. Taste and the autonomic nervous system. Chem Senses 10: 143-161.

Norgren R, Leonard CM. 1973. Ascending central gustatory pathways. J Comp Neurol 150: 217-238.

Nosrat CA, Blomlof J, El Shamy WM, Ernfors P, Olson L. 1997. Lingual deficits in BDNF and NT3 mutant mice leading to gustatory and somatosensory disturbances, respectively. Development 124: 1333-1342.

Oakley B. 1974. On the specification of taste neurons in the rat tongue. Brain Res 75: 85-96.

Oakley B. 1975. Receptive fields of cat taste fibers. Chem Senses Flav 1: 431-442.

Ogawa H, Hasegawa K, Otawa S, Ikeda I. 1998. GABAergic inhibition and modifications of taste responses in the cortical taste area in rats. Neurosci Res 32: 85-95.

Otawa S, Takagi K, Ogawa H. 1995. NMDA and non-NMDA receptors mediate taste afferent inputs to cortical taste neurons in rats. Exp Brain Res 106: 391-402.

Ozeki M, Sato M. 1972. Responses of gustatory cells in the tongue of rat to stimuli representing four taste qualities. Comp Biochem Physiol A 41: 391-407.

Perez CA, Margolskee RF, Kinnamon SC, Ogura T. 2003. Making sense with TRP channels: Store-operated calcium entry and the ion channel Trpm5 in taste receptor cells. Cell Calcium 33: 541-549.

Pfaffmann C. 1955. Gustatory nerve impulses in rat, cat and rabbit. J Neurophysiol 18: 429-440.

Pfaffmann C. 1959. The afferent code for sensory quality. Am Psychol 14: 226-232.

Pfaffmann C. 1964. Taste, its sensory and motivating properties. Am Sci 52: 187-206.

Pritchard T. 1991. The primate gustatory system. Smell and taste in health and disease. Getchell TV, Doty RL, Bartoshuk LM, Snow JB Jr, editors. New York: Raven; pp. 109-125.

Pronin AN, Tang H, Connor J, Keung W. 2004. Identification of ligands for two human bitter T2R receptors. Chem Senses 29: 583-593.

Pumplin DW, Getschman E. 2000. Synaptic proteins in rat taste bud cells: Appearance in the Golgi apparatus and relationship to α-gustducin and the Lewis^{b} and A antigens. J Comp Neurol 427: 171-184.

Pumplin DW, Yu C, Smith DV. 1997. Light and dark cells of rat vallate taste buds are morphologically distinct cell types. J Comp Neurol 378: 389-410.

Pumplin DW, Getschman E, Boughter JD Jr, Yu C, Smith DV. 1999. Differential expression of carbohydrate antigens on rat taste bud cells: Relation to the functional marker α-gustducin. J Comp Neurol 415: 230-239.

Reutter K, Witt M. 1993. Morphology of vertebrate taste organs and their nerve supply. Mechanisms of taste transduction. Simon SA, Roper SD, editors. Boca Raton, FL: CRC Press; pp. 29-82.

Richter TA, Dvoryanchikov GA, Chaudhari N, Roper SD. 2004. Acid-sensitive two-pore domain potassium (K2P) channels in mouse taste buds. J Neurophysiol 92: 1928-1936.

Royer SM, Kinnamon JC. 1988. Ultrastructure of mouse foliate taste buds: Synaptic and nonsynaptic interactions between taste cells and nerve fibers. J Comp Neurol 270: 11-24.

Royer SM, Kinnamon JC. 1991. HVEM serial-section analysis of rabbit foliate taste buds: I. Type III cells and their synapses. J Comp Neurol 306: 49-72.

Sainz E, Korley JN, Battey JF, Sullivan SL. 2001. Identification of a novel member of the T1R family of putative taste receptors. J Neurochem 77: 896-903.

Saper CB, Loewy AD. 1980. Efferent connections of the parabrachial nucleus in the rat. Brain Res 197: 291-317.

Sato T, Beidler LM. 1997. Broad tuning of rat taste cells to four basic taste stimuli. Chem Senses 22: 287-293.

Sato H, Katsuyama N, Tamura H, Hata Y, Tsumoto T. 1996. Mechanisms underlying orientation selectivity of neurons in the primary visual cortex of the macaque. J Physiol 494: 757-771.

Scott K. 2004. The sweet and the bitter of mammalian taste. Curr Opin Neurobiol 14: 423-427.

Shen T, Kaya N, Zhao FL, Lu SG, Cao Y, et al. 2005. Co-expression patterns of the neuropeptides vasoactive intestinal peptide and cholecystokinin with the transduction molecules alpha-gustducin and T1R2 in rat taste receptor cells. Neuroscience 130: 229-238.

Shigemura N, Miura H, Kusakabe Y, Hino A, Ninomiya Y. 2003. Expression of leptin receptor (Ob-R) isoforms and signal transducers and activators of transcription (STATs) mRNAs in the mouse taste buds. Arch Histol Cytol 66: 253-260.

Smith DV, Li C-S. 1998. Tonic GABAergic inhibition of taste-responsive neurons in the nucleus of the solitary tract. Chem Senses 23: 159-169.

Smith DV, Li C-S. 2000. GABA-mediated corticofugal inhibition of taste-responsive neurons in the nucleus of the solitary tract. Brain Res 858: 408-415.

Smith DV, Scott TR. 2003. Gustatory neural coding. Handbook of olfaction and gustation, 2nd Edition. Doty RL, editor. New York: Marcel Dekker; pp. 731-758.

Smith DV, Shepherd GM. 2003. Chemical senses: Taste and olfaction. Fundamental neuroscience. Squire LR, Bloom FE, McConnell S, Roberts J, Spitzer N, Zigmond MJ, editors. New York: Academic; pp. 631-667.

Smith DV, Akeson RA, Shipley MT. 1993. NCAM expression by subsets of taste cells is dependent upon innervation. J Comp Neurol 336: 493-506.

Smith DV, Ye M-K, Li C-S. 2005. Medullary taste responses are modulated by the bed nucleus of the stria terminalis. Chem Senses 30: 421-434.

Smith DV, Klevitsky R, Akeson R, Shipley MT. 1994. Taste bud expression of human blood group antigens. J Comp Neurol 343: 130-142.

Smith DV, Van Buskirk RL, Travers JB, Bieber SL. 1983. Gustatory neuron types in hamster brain stem. J Neurophysiol 50: 522-540.

SmithDV, Som J, Boughter JD Jr, St. John SJ, Yu C, et al. 1999. Cellular expression of α-gustducin and the A blood group antigen in rat fungiform taste buds cross-reinnervated by the IXth nerve. J Comp Neurol 409: 118-130.

Spector AC, Guagliardo NA, St. John SJ. 1996. Amiloride disrupts NaCl versus KCl discrimination performance: Implications for salt taste coding in rats. J Neurosci 16: 8115-8122.

St. John SJ, Smith DV. 2000. Neural representation of salts in the rat solitary nucleus: Brainstem correlates of taste discrimination. J Neurophysiol 84: 628-638.

Suga N, Zhang YF, Yan J. 1997. Sharpening of frequency tuning by inhibition in the thalamic auditory nucleus of the moustached bat. J Neurophysiol 77: 2098-2114.

Tokita K, Karadi Z, Shimura T, Yamamoto T. 2004. Centrifugal inputs modulate taste aversion learning associated parabrachial neuronal activities. J Neurophysiol 92: 265-279.

Tonosaki K, Funakoshi M. 1984. Intracellular taste cell responses of mouse. Comp Biochem Physiol A 78: 651-656.

Touzani K, Velley L. 1990. Ibotenic acid lesion of the lateral hypothalamus increases preference and aversion thresholds for saccharin and alters the morphine modulation of taste. Pharmacol Biochem Behav 36: 585-591.

Travers JB. 1988. Efferent projections from the anterior nucleus of the solitary tract of the hamster. Brain Res 455: 283-294.

Travers JB, Smith DV. 1979. Gustatory sensitivities in neurons of the hamster nucleus tractus solitarius. Sens Processes 3: 1-26.

Travers SP. 1993. Orosensory processing in neural systems of the nucleus of the solitary tract. Mechanisms of Taste Transduction. Simon SA, Roper SD, editors. Boca Raton FL: CRC Press; pp. 339-394.

Travers SP, Norgren R. 1991. Coding the sweet taste in the nucleus of the solitary tract: Differential roles for anterior tongue and nasoincisor duct gustatory receptors in the rat. J Neurophysiol 65: 1372-1380.

Travers SP, Pfaffmann C, Norgren R. 1986. Convergence of lingual and palatal gustatory neural activity in the nucleus of the solitary tract. Brain Res 365: 305-320.

Van Buskirk RL, Smith DV. 1981. Taste sensitivity of hamster parabrachial pontine neurons. J Neurophysiol 45: 144-171.

Kooy van der D, Koda LY, McGinty JF, Gerfen CR, Bloom FE. 1984. The organization of projections from the cortex, amygdala, and hypothalamus to the nucleus of the solitary tract in rat. J Comp Neurol 224: 1-24.

Uteshev VV, Smith DV. 2006. Cholinergic modulation of neurons in the gustatory region of the nucleus of the solitary tract. Brain Research 1084: 38-53.

Vasudev R, Gentil CG, Covian MR. 1985. Effect of electrical and chemical stimulation of the lateral hypothalamus on taste preferences. Braz J Med Biol Res 18: 3-14.

Vogt MB, Mistretta CM. 1990. Convergence in mammalian nucleus of solitary tract during development and functional differentiation of salt taste circuits. J Neurosci 10: 3148-3157.

Wang L, Bradley RM. 1993. Influence of GABA on neurons of the gustatory zone of the rat nucleus of the solitary tract. Brain Res 616: 144-153.

Wang L, Bradley RM. 1995. In vitro study of afferent synaptic transmission in the rostral gustatory zone of the rat nucleus of the solitary tract. Brain Res 702: 188-198.

Whiteside B. 1927. Nerve overlap in the gustatory apparatus of the rat. J Comp Neurol 44: 363-377.

Xu H, Staszewski L, Tang H, Adler E, Zoller M, et al. 2004. Different functional roles of T1R subunits in the heteromeric taste receptors. Proc Natl Acad Sci USA 101: 14258-14263.

Yamamoto T, Matsuo R, Kiyomitsu Y, Kitamura R. 1989. Response properties of lateral hypothalamic neurons during ingestive behavior with special reference to licking of various taste solutions. Brain Res 481: 286-297.

Yamamoto T, Shimura T, Saiko N, Yasoshima Y, Sakai N. 1994. Neural substrates for conditioned taste aversion in the rat. Behav Brain Res 65: 123-137.

Yamamoto T, Matsuo R, Ichikawa H, Wakisaka S, Akai M, et al. 1990. Aversive taste stimuli increase CGRP levels in the gustatory insular cortex of the rat. Neurosci Lett 112: 167-172.

Yang R, Stoick CL, Kinnamon JC. 2004. Synaptobrevin-2-like immunoreactivity is associated with vesicles at synapses in rat circumvallate taste buds. J Comp Neurol 471: 59-71.

Yang R, Crowley HH, Rock ME, Kinnamon JC. 2000. Taste cells with synapses in rat circumvallate papillae display SNAP-25-like immunoreactivity. J Comp Neurol 424: 205-215.

Yang R, Tabata S, Crowley HH, Margolskee RF, Kinnamon JC. 2000. Ultrastructural localization of gustducin immunoreactivity in micovilli of type II taste cells in the rat. J Comp Neurol 425: 139-151.

Ye Q, Heck GL, De Simone JA. 1991. The anion paradox in sodium taste reception: Resolution by voltage-clamp studies. Science 254: 724-726.

Yee C, Jones KR, Finger TE. 2003. Brain-derived neurotrophic factor is present in adult mouse taste cells with synapses. J Comp Neurol 459: 15-24.

Yee C, Yang R, Bottger B, Finger TE, Kinnamon JC. 2001. "Type III" cells of rat taste buds: Immunohistochemical and ultrastructural studies of neuron-specific enolase, protein gene product 9.5, and serotonin. J Comp Neurol 440: 97-108.

Zhang C, Oakley B. 1996. The distribution and origin of keratin 20-containing taste buds in rat and human. Differentiation 61: 121-127.

Zhang Y, Hoon MA, Chandrashekar J, Mueller KL, Cook B, et al. 2003. Coding of sweet, bitter, and umami tastes: Different receptor cells sharing similar signalling pathways. Cell 112: 293-301.

Zhao GQ, Shang Y, Hoon MA, Chandrashekar J, Erlenbach I, et al. 2003. The receptors for mammalian sweet and umami taste. Cell 115: 255-266.

6 Neurochemistry of the Main Olfactory System

M. Ennis · K. A. Hamilton · A. Hayar

Abstract: Many aspects of animal and human behavior are guided by, or dependent on, the sense of smell-olfaction. This chapter summarizes the major facts of the anatomy, neurochemistry, molecular biology and physiology of the olfactory system. We emphasize the mammalian olfactory system, particularly rodents, because of their widespread use as a laboratory model and the rich database related to this species.

List of Abbreviations: 5-HT, serotonin; 6-OHDA, 6-hydroxydopamine; AChE, acetylcholinesterase; ACIII, adenylate cyclase type III; AMPA, alpha-amino-3-hydroxy-5-methyl-4-isoxazole propionic acid; AOB, accessory olfactory bulb; AON, anterior olfactory nucleus; AP4, 2-amino-4-phosphonobutyric acid; BAPTA, bis (2-aminophenoxy)ethane-N,N,N,N-tetra-acetate; BDNF, brain-derived neurotrophic factor; CaM, calmodulin; cAMP, cyclic adenosine monophosphate; CB, calbindin; CCK, cholecystokinin; ChAT, choline acetyltransferase; CNGs, cyclic nucleotide gated channels; CR, calretinin; CREB, cAMP-response element-binding protein; CRF, corticotropin-releasing factor; DA, dopamine; DB, diagonal band; DBH, dopamine-β-hydroxylase; EM, electron microscopy; ENK, enkephalin; EPL, external plexiform layer; ET, external tufted; GABA, gamma-aminobutyric acid; GAD, glutamic acid decarboxylase; GC, granule cell; GCL, granule cell layer; GL, glomerular layer; IAS, intrabulbar association system; IP_3, inositol trisphosphate; IPL, internal plexiform layer; IR, insulin receptor; JG, juxtaglomerular; LC, locus coeruleus; LHRH, leutenizing hormone releasing hormone; LOT, lateral olfactory tract; M, muscarinic; MAPK/ERK, mitogen-activated protein kinase/extracellular signal-related kinase; MCL, mitral cell layer; mGluR, metabotropic glutamate receptor; MOB, main olfactory bulb; NADPH, nicotinamide adenine dinucleotide phosphate; NAG, N-acetyl-aspartyl-glutamate; NDB, nucleus of the horizontal limb of the diagonal band; NE, norepinephrine; NMDA, *N*-methyl D-Aspartate; NPY, neuropeptide Y; OCNC1, olfactory cyclic nucleotide gated channel subunit-1; OMP, olfactory marker protein; ON, olfactory nerve; ONL, olfactory nerve layer; ORN, olfactory receptor neurons; PC, piriform cortex; PG, periglomerular; PKC, protein kinase C; POC, primary olfactory cortex; PV, parvalbumin; RMS, rostral migratory stream; RNA, ribonucleic acid; SA, short axon; SP, substance P; SSA, superficial short axon; TH, tyrosine hydroxylase; TRH, thyrotropin-releasing hormone; Trk, tropomyosin related kinase; VG, Van Gehuchten; VIP, vasoactive intestinal polypeptide

1 Introduction

The olfactory system consists of two parallel systems, the main olfactory system and the accessory olfactory system. This chapter focuses on the main olfactory system in mammals, specifically on the main olfactory bulb (MOB) and primary olfactory cortex (POC), and it emphasizes the neuroanatomy, neurochemistry, and neurophysiology of these regions in rodents. The sense of olfaction is critically important for food consumption, emotional responses, aggression, maternal and reproductive functions, neuroendocrine regulation, and the recognition of conspecifics, predators, and prey. In many species, olfaction plays a more pivotal role in these functions than in humans, with olfactory cues exceeding visual or auditory cues in importance. The olfactory system has long been an attractive model to study cellular mechanisms underlying the encoding, transfer, processing, and decoding of sensory information. Recent interest in this area has been sparked by a series of dramatic breakthroughs over the past decade in our understanding of the organization and function of the peripheral olfactory system, cloning of the olfactory receptors, and identification of the olfactory transduction machinery (Breer, 2003). These advances, together with a wealth of accumulated knowledge about the anatomy and connectivity of the MOB (Shipley et al., 1996, 2004; Shepherd et al., 2004), have set the stage for experiments aimed at unraveling the mechanisms of early sensory processing by bulbar and cortical circuits.

2 The Olfactory Epithelium

2.1 Organization and Cell Types

Odor molecules that enter the nasal cavities and bind to first-order neurons, olfactory receptor neurons (ORNs), which are contained in a neuroepithelial sheet lining the upper and caudal aspects of the nasal septum and the cribriform plate region of the nasal cavities. ORNs are located in a specific portion of the nasal epithelium, the olfactory neuroepithelium, which is thicker than the surrounding respiratory epithelium and contains a number of distinct cell types (➋ *Figure 6-1*).

◘ Figure 6-1
The olfactory epithelium showing the organization of the olfactory receptor neurons (ORNs) and other cell types

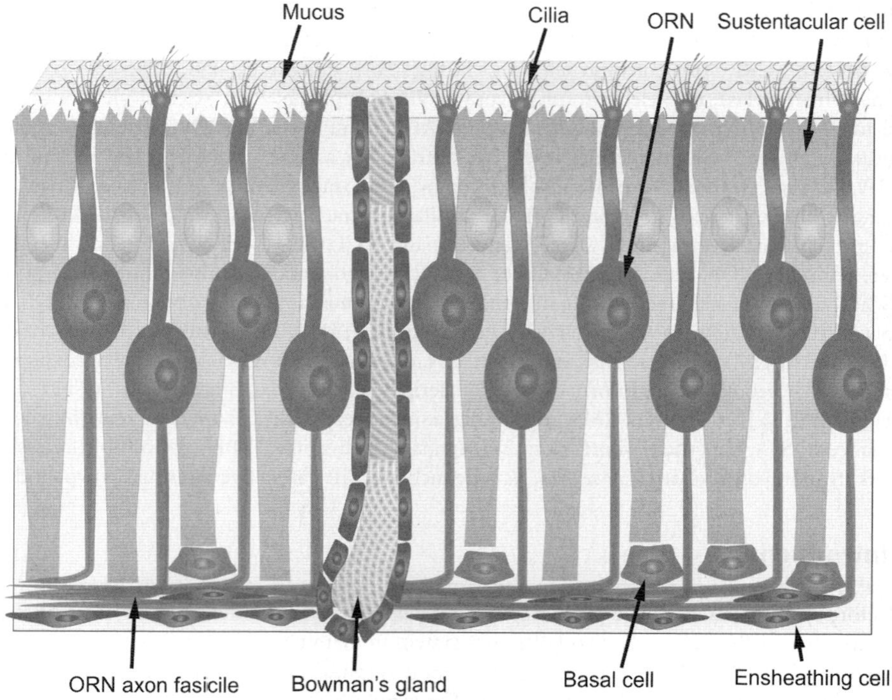

2.1.1 Microvillar Cells and Sustentacular Cells

As the name suggests, microvillar cells (not shown) have superficial microvilli in contact with the mucus (Moran et al., 1982; Muller and Marc, 1984; Rowley et al., 1989; Morrison and Costanzo, 1990). They lie deep within the neuroepithelium and extend an "axon-like" projection, which can reach the MOB, as suggested by tract-tracing studies (Rowley et al., 1989). Sustentacular cells separate and partially wrap the ORNs. Like the microvillar cells, their surface contains microvilli that project into the mucus layer. They may function in mucus composition regulation (Getchell et al., 1984; Getchell and Getchell, 1992). As they express molecules of the P450 enzyme systems (Hadley and Dahl, 1982), they may play a role in detoxification.

2.1.2 Basal Cells

The basal cells are deeply located globose- and horizontal-type stem cells for the replacement of the ORNs (Cuschieri and Bannister, 1975a, b; Graziadei and Graziadei, 1979), which in rodents have a life span of approximately 40 days.

2.1.3 Olfactory Receptor Neurons

ORNs are bipolar sensory neurons with both an apical dendrite and a deep or basal axon (Cajal, 1911a, b). The dendrites extend superficially, forming an olfactory knob with multiple cilia extending into the mucus (Menco, 1984; Menco and Farbman, 1985a, b). Odor transduction takes place in the cilia. Basally, the ORNs give rise to axons—the olfactory nerve (ON) fibers. The axons are ensheathed by specialized Schwann cells, the ensheathing cells (DeLorenzo, 1957). These fibers form bundles, which then collect as groups of fascicles, pass through the cribriform plate, and synapse in the MOB.

2.1.3.1 Neurochemistry of ORNs

Olfactory Marker Protein ORNs express high levels of olfactory marker protein (OMP), which is unique in the olfactory system to ORNs (Margolis, 1972; Keller and Margolis, 1975). OMP is found in a number of mammalian species and it appears to be expressed in all mature ORNs. Studies in mice containing a null mutation for OMP suggest that this protein may play a role in ORN adaptation to odors (Ivic et al., 2000), signal amplification, and transduction (Youngentob et al., 2004).

Glutamate, Carnosine, Copper, and Zinc ORNs contain glutamate and glutamate antibodies stain ON fibers and axon terminals in the glomerular layer (GL) (Liu et al., 1989; Sassoè-Pognetto et al., 1993). Carnosine, a soluble dipeptide, is uniquely expressed in high concentrations in mammalian ORNs, and it is present in ON axon terminals in the GL (Ferriero and Margolis, 1975; Margolis, 1980; Biffo et al., 1990). Carnosine colocalizes with glutamate in the ON axon terminals (Sassoè-Pognetto et al., 1993), and it satisfies criteria for neurotransmitter candidacy, including (1) carnosine synthetic and degradative enzymes are present in ORNs, (2) the peptide is released by depolarization in a Ca^{2+}-dependent manner in ON synaptosomes, and (3) high affinity binding sites for carnosine are present in the GL (Ferriero and Margolis, 1975; Margolis, 1980; Burd et al., 1982; Rochel and Margolis, 1982; Margolis et al., 1983, 1985, 1987; Margolis and Grillo, 1984; Biffo et al., 1990). Zinc and copper are also present in high concentrations in ON axon terminals (Biffo et al., 1990). The potential neuromodulatory roles of carnosine, zinc, and copper are discussed later.

2.2 Odor Receptors, Transduction, and Physiology of ORNs

The binding of odors to receptors located on the cilia of ORNs elicits electrical signals (see Reed, 1992 for review). Olfactory receptors comprise a family of approximately 1,000 G-protein coupled receptors with 7 transmembrane domains that correspond to roughly 1,000 genes. Each mammalian ORN seems to express only a single receptor gene and approximately 10,000 ORNs express the same receptor, although there may be some exceptions (Rawson et al., 2000). The ~10,000 ORNs expressing the same receptor gene are scattered across one of the four epithelial expression zones (Buck and Axel, 1991; Ressler et al., 1993; Vassar et al., 1993; Strotmann et al., 1994; Sullivan et al., 1995). As discussed in *Section 2.3*, collections of ORNs expressing the same receptor project focally onto very few glomeruli in the MOB. Biochemical and electrophysiological studies suggest that the cAMP signal transduction pathway predominates in mammalian ORNs (Schild and Restrepo, 1998; Firestein, 2001; Zufall and Munger, 2001). Odors activate a G-protein, G_{olf}, which activates adenylate cyclase type III (ACIII), leading to a rise in cyclic nucleotide levels (i.e., cAMP). cAMP binds to, and opens, cyclic nucleotide gated channels (CNGs) (Zufall et al., 1991). These channels are permeable to Na^+, K^+, and Ca^{2+} (Frings et al., 1995). The channels are preferentially permeable to Ca^{2+}, and their activation increases intracellular Ca^{2+} within the cilia (Leinders-Zufall et al., 1997, 1998). Increased Ca^{2+} leads, in turn, to activation of a Ca^{2+}-activated chloride conductance (Kleene and Gesteland, 1991; Lowe and Gold, 1993), which further depolarizes the cell (due to high intracellular Cl^- levels relative to the mucus), leading to the generation of action potentials that propagate down the axon to the MOB. Genetic null mutations for G_{olf} (Belluscio et al., 1998), CNGA2 subunits (also called OCNC1) (Brunet et al., 1996), and ACIII (Wong et al., 2000) firmly establish the essential role for these molecules in odor transduction. Mice with null mutations for any of these three transduction elements are functionally anosmic. The transduction process is negatively regulated via several mechanisms, allowing for relatively rapid adaptation of odor responses. Increased intracellular Ca^{2+} triggers adaptation, as demonstrated by reduced adaptation when intracellular Ca^{2+} is chelated by BAPTA (Leinders-Zufall et al., 1998, 1999). The rise in intracellular Ca^{2+} activates calmodulin, which binds to CNGs, thereby decreasing their affinity for cAMP. The CNGA4 subunits are especially important for this mechanism of adaptation (Bradley et al., 2001; Munger et al., 2001). Ca^{2+}-calmodulin has also been shown to activate CaM kinase II, leading to phosphorylation of ACIII and reduced production of cAMP. CaM kinase II phosphorylates phosphodiesterase, thus increasing inactivation of cAMP. Finally, there is tonic inhibition of CNGs by extracellular Ca^{2+}, which is quite high in mucus, perhaps serving to increase the ORN signal-to-noise ratio (Zufall and Firestein, 1993).

2.3 Topography of ORN Projections to MOB

The ORNs of each olfactory epithelium give rise to several million axons that coalesce in the ON layer (ONL) of the ipsilateral MOB and synapse within the glomeruli (Meisami, 1989). Because of the widespread epithelial distribution of ORNs expressing the same receptor, each glomerulus in the MOB receives input from widely dispersed ORNs. In the same manner, adjacent ORNs may project to spatially distant glomeruli (Astic and Saucier, 1986; Saucier and Astic, 1986; Astic et al., 1987; Clancy et al., 1994; Schoenfeld et al., 1994). Early mapping studies showed that there is a crude topography such that ORNs located in the four expression zones project to homologous regions of MOB (Astic and Saucier, 1986; Saucier and Astic, 1986; Clancy et al., 1994; Schoenfeld et al., 1994). The mRNA for the receptor genes is orthogradely transported along ORN axons to their axon terminals. This property led to the discovery that despite the scattered distribution of ORNs expressing the same receptor within the epithelial zones, there is a remarkably precise topographic projection of ORNs expressing the same receptor. Specifically, the ORNs expressing the same receptor project to one or two glomeruli located on the medial and lateral side of each MOB (➔ *Figure 6-2*) (Ressler et al., 1993, 1994; Vassar et al., 1994; Mombaerts et al., 1996; Wang et al., 1998; Potter et al., 2001; Treloar et al., 2002). Studies in transgenic animals showed that this projection pattern is topographically fixed across animals. That is, the same glomeruli identified in different mice receive inputs from the same restricted population of ORNs bearing the same receptor (Mombaerts et al., 1996; Wang etal., 1998; Potter et al., 2001; Treloar et al., 2002). Rough calculations confirm an approximately 1:2 ratio between the number of different types of receptors (~1,000) and the total number of glomeruli (~1,800) in mice.

The implications of these receptor localization and expression studies for the odor specificity of individual ORNs are still not fully understood. However, experiments suggest that a receptor may recognize a specific ligand or epitope expressed in common by a number of odors, and that the same receptor may respond to several similar epitopes (Krautwurst et al., 1998; Zhao et al., 1998). In turn, this suggests that an ORN expressing the same receptor responds differentially to a family of odorants containing a common epitope, as well as to other odorants containing different, but structurally similar epitopes. In support of this notion, a variety of electrophysiological studies demonstrate that individual ORNs respond with different degrees of excitation to a spectrum of odors (Malnic et al., 1999; Duchamp-Viret et al., 1999; Hamana et al., 2003). Together, these findings imply that an individual odor will activate a distinct pattern of glomerular activity that depends on the number of specific ORN-receptor gene groups activated.

The results of odor mapping studies in the bulb are consistent with the preceding model. Functional studies utilizing 2-deoxyglucose (Jourdan et al., 1980; Astic and Saucier, 1981; Benson et al., 1985; Johnson and Leon, 2000a, b; Johnson et al., 1998), c-fos (Onoda, 1992; Guthrie and Gall, 1995a, b), magnetic resonance imaging (Yang et al., 1998), Ca^{2+} or voltage-sensitive dye imaging (Cinelli et al., 1995; Friedrich and Korshung, 1997; Wachowiak and Cohen, 1999, 2001; Wachowiak et al., 2000), and intrinsic imaging (Rubin and Katz, 1999; Uchida et al., 2000; Belluscio and Katz, 2001; Rubin and Katz, 2001) suggest that low concentrations of single odorants activate a restricted number of glomeruli. Often, several responsive glomeruli are clustered. These studies suggest that different odors activate unique sets of glomeruli that are topographically fixed from animal to animal. Higher odorant concentrations activate larger numbers of glomeruli, probably resulting from activation of additional ORNs that express a different receptor that has a lower affinity for that odorant. These mapping studies suggest that glomeruli are functional modules that

Figure 6-2
The projections of ORNs to the glomerular layer (GL) of the MOB. Note that ORNs expressing different odorant receptor genes (shown as blue, red, or green cells) are interspersed and widely distributed, yet the axons of ORN expressing the same odorant receptor gene converge onto the same glomerulus (or pairs of medial and lateral glomeruli) in the GL (represented as blue, red, or green glomeruli). EPL, external plexiform layer; MCL, mitral cell layer; ONL, olfactory nerve layer

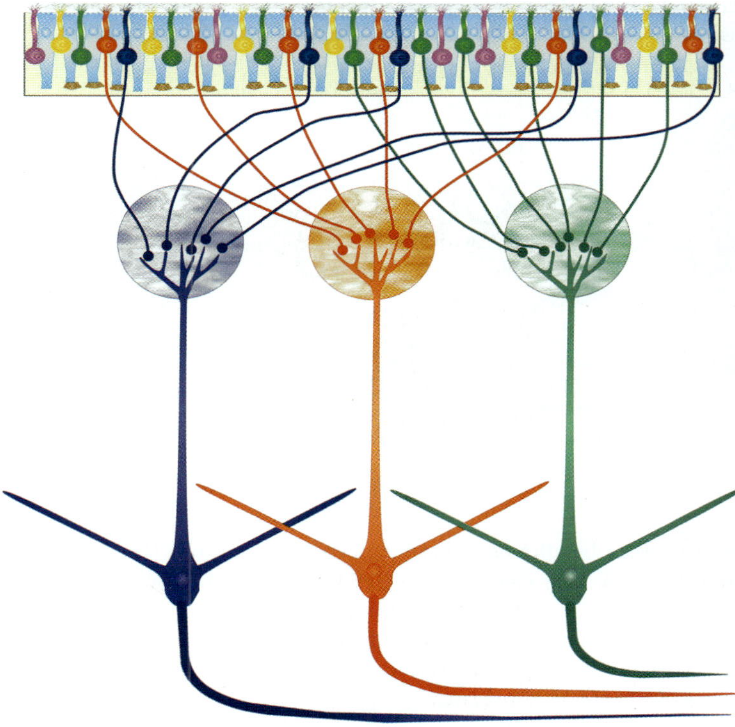

represent a map of the activity of ORNs. As odors typically activate groups of glomeruli, the olfactory network must extract information from such patterns to recognize and distinguish different odors. The MOB is the first component in the olfactory system that performs this neural computation.

3 Main Olfactory Bulb

The MOB in rodents is situated at the rostral pole of the cranial cavity and it is connected to the frontal cortex by a slender peduncle. The bulb can be thought of as an elongated onion composed of distinct layers or laminae that are organized as concentric circles. These layers, from superficial to deep, are the ONL, GL, external plexiform layer (EPL), mitral cell layer (MCL), internal plexiform layer (IPL), GC layer (GCL), and the ependymal layer (*Figures 6-3* and *6-4*).

3.1 Olfactory Nerve Layer

The outer and most superficial MOB layer, the olfactory nerve layer (ONL), consists of ON axons and glial cells (Cajal, 1911a, b; Pinching and Powell, 1971b; Doucette, 1989). The glial or ensheathing cells envelop the ON axons and express the Schwann cell marker S100. The deepest third of the ONL also contains

Figure 6-3
Architecture of the MOB (A, B) and primary olfactory cortex (POC, C). A and B: Coronal section (Nissl stain) of the rat MOB at low (A) and high (B) magnifications. C: Coronal section through the rat brain showing several structures of the POC. Abbreviations: ac, anterior commissure; Acb, nucleus accumbens; AONvp, anterior olfactory nucleus, ventroposterior division; DP, dorsal peduncular cortex; IL, infralimbic cortex; Pir, piriform cortex; SEL, subependymal layer; Scale bar in A and C = 1 mm. Modified from Handbook of Chem. Neuroanat. Integrated Sys. CNS, Vol. 12, Part III, Chapter III, The Olfactory System, M. Shipley etal., pp. 469–573, 1996, with permission from Elsevier

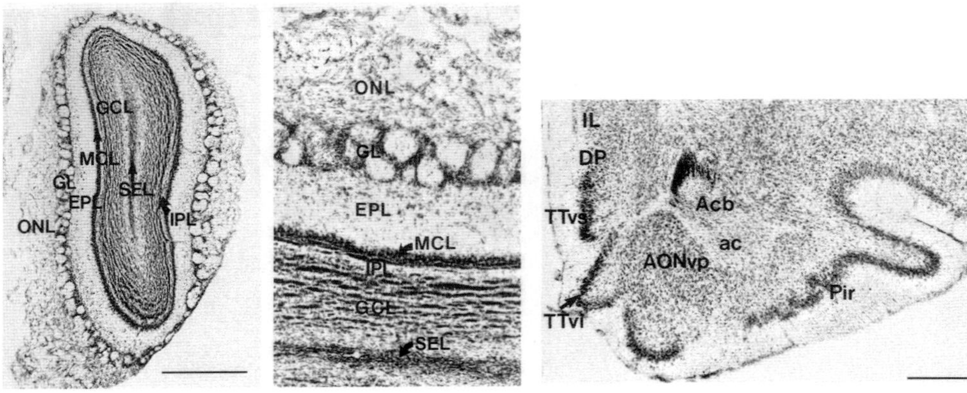

astrocytes that reside between bundles of ensheathing cell-wrapped axons (Doucette, 1984; Bailey and Shipley, 1993). Astrocytes in this same region express QPRT, the degradative enzyme for quinolinic acid, an NMDA receptor agonist (Bailey and Shipley, 1993). Quinolinic acid, like glutamate, may modulate glutamatergic transmission between the ON axon terminals and postsynaptic targets. The roles of other neuroactive substances of ORN axons (glutamate, zinc, carnosine), as well as receptors expressed by the ON axon terminals (D2, $GABA_B$), are discussed later.

3.2 Glomerular Layer

Deep to the ONL, the GL comprises neuropil-rich ovoid structures—the glomeruli—each of which is surrounded by a shell of small neurons and glia. Within the glomeruli, ON axons synapse with mitral and tufted cells, as well as with the intrinsic neurons of the GL. Glomerular diameters are 80–160 μm in rat (Meisami and Safari, 1981; Meisami and Sendera, 1993) and 50–100 μm in mouse (Royet et al., 1988), and the number of glomeruli varies among species: 2000–3000 glomeruli/bulb for rabbits (Allison, 1949); 1800–2000 for mice (Allison, 1953; White, 1972; Royet et al., 1988); 3000 for rats (Meisami and Safari, 1981). More recent stereological studies yielded higher numbers: 6300 glomeruli in rabbit and 4200 in rat (Royet et al., 1998). Adjacent glomeruli are somewhat isolated from each other by astrocytes residing in the glomerular shell. Wedge-shaped astrocytes with somata in the glomerular shell send branched processes into the glomerular core (Bailey and Shipley, 1993; Chao et al., 1997). The astrocytes are restricted to a single glomerulus and they appear to cordon off adjacent glomeruli. This observation provides additional evidence that each glomerulus functions as a discrete unit.

3.2.1 Neuron Types of the GL

The neurons of the GL are classified as three cell types, which include: (1) periglomerular (PG) cells, (2) external tufted (ET) cells, and (3) short axon (SA) cells (Golgi, 1875; Van Gehuchten and Martin, 1891;

Figure 6-4
The basic circuitry of the MOB. Axons of ORNs travel in the ONL and synapse in the GL on the dendrites of mitral cells (MC), tufted cells (external tufted cell, ET; middle tufted cell, MT), and generic juxtaglomerular (JG) neurons, which include periglomerular cells (PG), ET cells, and short axon cells (SA). SA cells interconnect different glomeruli. There are serial and reciprocal synapses between the apical dendrites of mitral/tufted cells and the processes of JG neurons. Superficial tufted cells (ST) are located in the superficial EPL or at the GL–EPL border. The lateral dendrites of mitral/tufted cells form serial and reciprocal synapses with the apical dendrites of granule cells (GC) in the EPL. GCs are located in the GCL and the MCL. The axons of mitral/tufted cells project locally to GCs (not shown) and also to primary olfactory cortex via the lateral olfactory tract (LOT). The bulb also contains other populations of interneurons neurons, including the van Gehuchten cells (VG) within the EPL

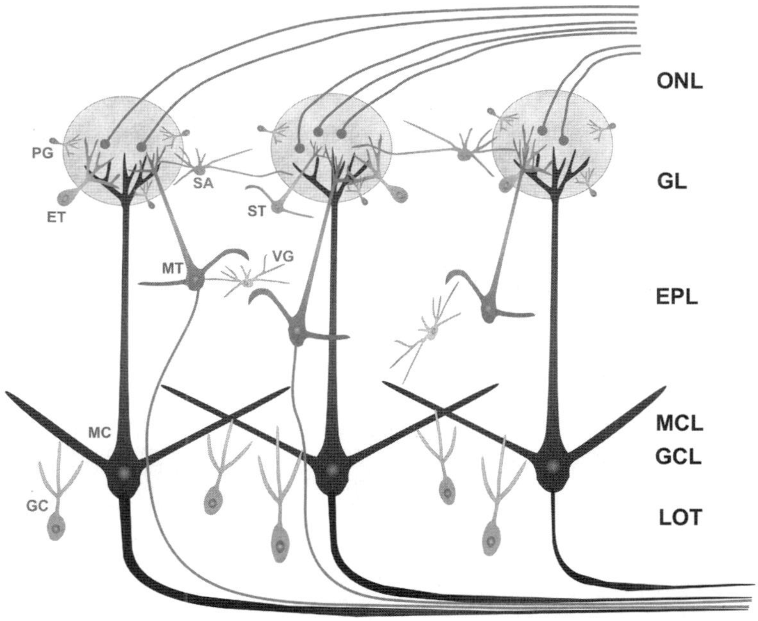

Blanes, 1898; Cajal, 1911a, b; Pinching and Powell, 1971a, b, c; Pinching and Powell, 1972b, c). Collectively, the intrinsic neurons of the GL are referred to as juxtaglomerular (JG) cells. The term JG is also used here with regard to the cited studies in which the subtype of glomerular neuron was not specified. The morphology and features of these cells are illustrated in *Figure 6-5* and are only briefly reviewed here as more detailed descriptions are available (Hayar et al., 2004a, b).

3.2.1.1 ET Cells These are the largest (10–15 μm) cells in the GL, and electrophysiological recordings have confirmed that they are excitatory. They are dispersed in the juxtaglomerular regions, surrounding/deep to the glomeruli. Most have one apical dendrite that arborizes extensively throughout one glomerulus (Pinching and Powell, 1971a; Hayar et al., 2004a, b). Rarely, ET cells have two or three apical dendrites that ramify in different glomeruli. Most ET cells have secondary or lateral dendrites that extend in the superficial EPL. Some ET cells have axons that appear to synapse with PG cells or SA cells, or more infrequently, project out of MOB (Pinching and Powell, 1971a). ET cells are somewhat similar to tufted cells of the EPL and mitral cells, but growing evidence suggests that all tufted cell subtypes exhibit distinct anatomical and physiological properties.

3.2.1.2 PG Cells These cells are the most numerous ones in the GL, and they are thought to be inhibitory in nature. They are small (5–8 μm), spherical or ovoid, and are distributed in the periglomerular regions

Figure 6-5

Neuron types in the MOB. All neurons were labeled in MOB slices by intracellular filling with biocytin during electrophysiological recordings. A–C: Reconstruction of three cells representing the three major types of JG cells in the GL: external tufted cell (A); periglomerular cell (B); short axon cell (C). D–H: Photomicrographs of biocytin-filled superficial tufted (D), EPL interneuron (E), middle tufted cell (F) in the EPL; mitral cell (G, soma in the MCL); granule cell (H, soma in the GCL). Calibration bars in A–H = 100 μm. Figures D–F reprinted from K.A. Hamilton et al., Neuroscience 133: 819–829, 2005, with permission form Elsevier, Ltd

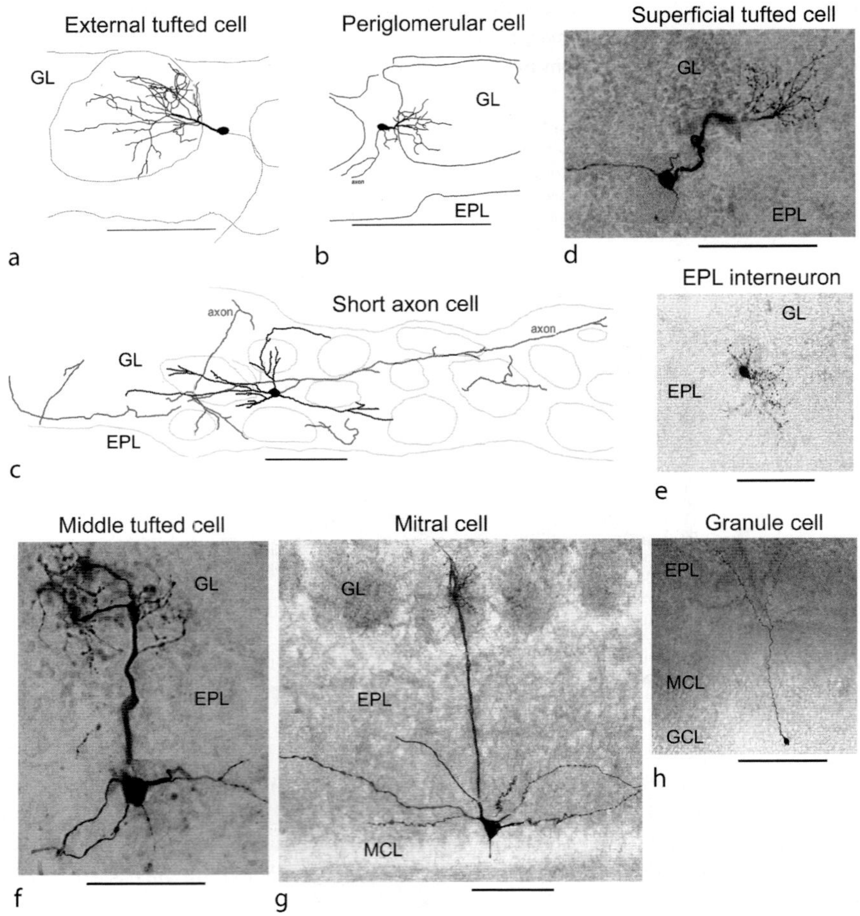

surrounding the glomeruli. Their dendrites are typically restricted to a small subregion of a glomerulus (Pinching and Powell, 1971a; Hayar et al., 2004a, b). These dendrites most commonly receive synapses from axon terminals containing spherical vesicles that have been thought to be collateral branches of ET cell axons and centrifugal fibers (Pinching and Powell, 1971c, 1972c). Other synapses along PG dendrites include asymmetrical (morphologically excitatory) synapses from the ON terminals and from mitral/tufted cell dendrites. Some of the mitral/tufted cell synapses are paired with reciprocal symmetrical (morphologically inhibitory) synapses back onto the parent mitral/tufted cell dendrites (Pinching and Powell, 1971b; Kasowski et al., 1999; Toida et al., 1998, 2000). Physiological recordings indicate that PG cells also receive monosynaptic dendrodendritic excitatory input from ET cells (Hayar et al., 2004b). PG cell axons appear to be rare (Pinching and Powell, 1971a; Hayar et al., 2004a), but they have been reported to extend over distances equivalent to four to five glomeruli. They appear to form symmetrical (morphologically inhibitory) synapses onto mitral/tufted cell dendrites and ET cells and other JG cells (Pinching and Powell, 1971c).

3.2.1.3 SA Cells These cells are roughly of the same size (8–12 μm) as the ET cells. They are distinguished by multiple dendrites that seem to harvest information from multiple glomeruli (Hayar et al., 2004b). The dendrites may receive synaptic inputs from ET cell dendrites (Hayar et al., 2004b), tufted cell collaterals, or from other SA cells. SA cells have axons that can extend up to 1–2 mm within the GL (Aungst et al., 2003). The axons appear to synapse onto the dendrites of PG cells (Pinching and Powell, 1971a). They do not receive direct ON input (Pinching and Powell, 1971c; Hayar et al., 2004b).

3.2.2 Neurochemistry of JG Neurons

As illustrated in *Tables 6-1* and *6-2*, GL neurons are rich in neuroactive substances and mapping studies have revealed that PG cells in particular are neurochemically heterogeneous (Halász, 1990).

Table 6-1
Neurotransmitters in MOB neurons

Transmitter	Cell type (layer)	References
Aspartate	Mitral	Fuller and Price (1988), Halász (1987), Watanabe and Kawana (1984)
CCK	JG, ET, Tufted, Cajal	Bonnemann et al. (1989), Matsutani et al. (1988), Seroogy et al. (1985)
CRF	Mitral, Tufted	Bassett et al. (1992), Imaki et al. (1989)
DA	JG	Baker et al. (1983,1984,1988), Baker (1986), Gall et al. (1987), Halász et al. (1981), McLean and Shipley (1988)
ENK	PG, Granule, EPL Int,	Bogan et al. (1982), Davis et al. (1982), Kosaka et al. (1995)
GABA	PG, Granule, EPL Int.	Gall et al. (1987), Kosaka et al. (1987d), Mugnaini et al. (1984a, b)
GABA + DA	PG	Kosaka et al. (1985,1988)
GABA + Parvalbumin	Sup. Tufted (EPL)	Kosaka et al. (1987d)
GABA + ENK	Granule	Kosaka et al. (1987d)
GABA and SP and DA	PG	Davis et al. (1982)
Glutamate	JG, Mitral, Tufted	Liu et al. (1989)
NAG	Mitral	Blakely et al. (1987), Ffrench-Mullen et al. (1985)
NADPH-diaphorase	SA (GL), EPL Int.	Alonso et al. (1995), Davis (1991), Scott et al. (1987), Villalba et al. (1989)
Neurotensin	EPL Int.	Matsutani et al. (1988)
NPY	PG, GCL, EPL Int., SA	Briñón et al. (1992), Gall et al. (1986), Nakajima et al. (1996), Ohm et al. (1988), Sanides-Kohlrausch and Wahle (1990a), Scott et al. (1987), Seroogy et al. (1989)
Somatostatin	JG, SA (GL), Deep SA, EPL Int.	Davis et al. (1982), Matsutani et al. (1988), Scott et al. (1987), Seroogy et al. (1989)
SP	JG, Sup. Tufted (EPL), EPL Int.	Baker (1986), Davis et al. (1982), Kream et al. (1984), Wahle et al. (1990)
TRH	PG	Merchenthaler et al. (1988), Tsuruo et al. (1988)
VIP	PG, Sup. Tufted (EPL), Van Gehuchten (EPL)	Crespo et al. (2002), Gall et al. (1986), López-Mascaraque et al. (1989), Sanides-Kohlrausch and Wahle (1990b)

EPL Int., EPL interneuron; Sup. Tufted, superficial tufted cell. Adapted from Handbook of Chem. Neuroanat. Integrated Sys. CNS, Vol. 12, Part III, Chapter III, The Olfactory System, M. Shipley et al., pp. 469–573, 1996, Elsevier, Ltd

Table 6-2
Relative frequency of neurons expressing calcium binding proteins in a 25-μm-thick rat MOB section

	Calcium Binding Protein			
Cell Type (Layer/s)	Neurocalcin	Cabindin D-28k	Calretinin	Parvalbumin
Periglomerular (GL)	+++++	+++++	+++++	++
Sup. Short Axon (GL, EPL)	−	++	+++	++++
Sup. Tufted (EPL)	+++++	−	−	−
Van Gehuchten (EPL)	+	+	++	+++
Mitral (MCL)	−	−	+++	−
Deep Short Axon (MCL/IPL, IPL, GCL)	+	+	+	++
Granule (MCL, IPL, GCL)	++++	−	+++++	+

+++++, more than 500 cells per section; ++++, 250–500 cells per section; +++, 100–250 cells per section; ++, 25–100 cells per section; +, less than 25 cells per section; −, immunonegative. MCL/IPL, IPL border with MCL. Adapted from Briñón etal., J. Comp. Neurol. 407: 404–414 (1999), with permission from John Wiley & Sons, Inc

Electrophysiologically, ET cells appear to be glutamatergic (Hayar et al., 2004b) and they may correspond to JG cells stained by glutamate antibodies (Liu et al., 1989). The GL also contains the largest population of dopamine (DA) containing cells in the brain. The majority of DA containing cells are PG cells, but some ET cells are also dopaminergic (Halász et al., 1977; Davis and Macrides, 1983; Halász et al., 1985; Gall et al., 1986; McLean and Shipley, 1988). Another major neurotransmitter of JG neurons is GABA, which appears to be predominantly contained in PG cells (Ribak et al., 1977; Mugnaini et al., 1984a; Kosaka et al., 1985, 1987a, b, c, d, 1988). Some JG cells that send axons to the deeper GCL contain nicotinamide adenine dinucleotide phosphate (NADPH)-diaphorase, neuropeptide-Y (NPY), and somatostatin, which Scott et al. (1987) suggested may provide a direct route for PG cells to influence granule cells (GCs). Davis (1991) concluded that NADPH in the GL is primarily, if not exclusively, contained in PG cells. The NADPH PG cells are distinct from cells that express calbindin immunoreactivity (Alonso et al., 1993). Other neurotransmitter and neurochemical markers of JG cells include: (1) enkephalin (ENK) (Bogan et al., 1982; Davis et al., 1982; Kosaka et al., 1987d), (2) thyrotropin-releasing hormone (TRH), (Merchenthaler et al., 1988; Tsuruo et al., 1988), (3) acetylcholinesterase (AChE), a marker for cholinoceptive neurons (Nickell and Shipley, 1988b), (4) cholecystokinin (CCK) (Seroogy et al., 1985; Matsutani et al., 1988), (5) aspartic acid (Watanabe and Kawana, 1984; Halász, 1987; Fuller and Price, 1988), (6) vasoactive intestinal polypeptide (VIP) (Gall et al., 1986; Sanides-Kohlrausch and Wahle, 1990a), (7) protein kinase C (Saito et al., 1988) and two Ca^{2+} binding proteins, (8) calretinin (CR) (Jacobowitz and Winsky, 1991; Rogers, 1992), and (9) calbindin-d28k (CB) (Baimbridge and Miller, 1982; Celio, 1990; Briñón et al., 1992).

Three-dimensional analyses using serial-sectioning/mirror-imaging and immunocytochemical double-labeling techniques revealed that about half of the PG cell population can be classified into at least three cell types, including: (1) GABA-positive, (2) CR-positive, and (3) CB-positive (Kosaka et al., 1995, 1998). Moreover, the DA, TRH, and ENK neurons were identified as further subpopulations. About 80% of the DA cells contained GABA in the rat MOB, and in turn, corresponded to about 50% of GABAergic cells (Kosaka et al., 1985, 1987a, c, 1988, 1995; Gall et al., 1987). Therefore, the DA PG neurons are regarded as a subpopulation of GABAergic neurons. Clearly, the neurochemical diversity of JG neurons is remarkable and at present, the role of these neuronal subtypes and their neurochemical constituents in olfactory processing is unclear.

3.2.3 Synaptic and Bicompartmental Organization of the GL

3.2.3.1 Excitatory Systems in the GL In addition to ON axons, each glomerulus contains the apical-dendritic tufts of about 20 mitral cells, 200 tufted cells and 1500–2000 JG cells (Shipley et al., 1996).

Sensory transmission from ON axon terminals is mediated by glutamate acting at AMPA and NMDA ionotropic glutamate receptor subtypes (Bardoni et al., 1996; Ennis et al., 1996, 2001; Aroniadou-Anderjaska et al., 1997; Chen and Shepherd, 1997; Keller et al., 1998). The distribution of these receptors in MOB is discussed later (*Section 3.8* and *Table 6-3*). Electrophysiological studies, described later,

Table 6-3
Neurotransmitter receptors in the MOB

Receptor	GL	EPL	MCL	IPL	GCL	References
Cholinergic						Fonseca et al. (1991a, b), Le Jeune et al. (1996), Rotter et al. (1979), Sahin et al. (1992), Spencer et al. (1986)
M1	–	++	–	++	++	
M2	++	++	++	++	++	
M3	–	++	–	++	–	
M4	–	++	–	++	–	
Nicotinic	++	++	++	–	++	
Noradrenergic						Booze et al. (1989), Day et al. (1997), Domyancic and Morilak (1997), McCune et al. (1993), Nicholas et al. (1993), Palacios and Kuhar (1980), Pieribone et al. (1994), Rosin et al. (1996), Sargent-Jones et al. (1985), Talley et al. (1996), Wanaka et al. (1989), Winzer-Serhan et al. (1997a), Woo and Leon (1995), Young and Kuhar (1980a), Yuan et al. (2003)
$\alpha1$	++	++	++	++	++	
$\alpha2$	++	++	++	++	++	
$\beta1$	++	++	++	–	++	
$\beta2$	++	–	–	++	++	
Dopaminergic						Coronas et al. (1997), Koster et al. (1999), Mansour et al. (1990a, b), Nickell et al. (1991)
D1	++	–	–	–	–	
D2	++	–	–	–	++	
Serotonergic						Clemett et al. (2000), Cornea-Hebert et al. (1999), Hamada et al. (1998), McLean et al. (1995), Pompeiano et al. (1992, 1994), Tecott et al. (1993), Whitaker-Azmitia et al. (1993), Wright et al. (1995), Yuan et al. (2003)
5-HT_{1A}	++	++	++	++	++	
5-HT_{1C}	–	–	–	–	++	
5-$HT_{2A/C}$	++	++	++	NR	++	
5-HT_3	++	–	–	–	–	
Glutamatergic						Duvoisin et al. (1995), Gall et al. (1990), Giustetto et al. (1997), Hamilton and Coppola (2003), Kinoshita et al. (1996, 1998), Kinzie et al. (1995, 1997), Martin et al. (1992, 1993), Masu et al. (1991), Miller et al. (1990), Molnar et al. (1993), Monaghan and Cotman (1982), Montague and Greer (1999), Monyer et al. (1994), Ohishi et al. (1993a, b, 1995, 1998), Petralia and Wenthold (1992), Petralia et al. (1994a, b), Romano et al. (1995), Sahara et al. (2001), Sassoè-Pognetto and Ottersen (2000), Saugstad et al. (1997), Shigemoto et al. (1992, 1993), Sun et al. (2000), Tanabe et al. (1992), Van Den Pol (1995), Wada et al. (1998), Watanabe et al. (1993), Wisden and Seeburg (1993)

Table 6-3 (continued)

Receptor	GL	EPL	MCL	IPL	GCL	References
Kainate	–	++	++	++	–	
NMDA	++	++	++	++	++	
AMPA	++	++	++	++	++	
mGlu	++	++	++	++	++	
GABAergic						Bonino et al. (1999), Bowery et al. (1987), Chu et al. (1990), Fritschy et al. (1992), Laurie et al. (1992), Margeta-Mitrovic et al. (1999), Persohn et al. (1992), Richards et al. (1987)
$GABA_A$	++	++	++	–	++	
$GABA_B$	++	–	++	NR	++	
CRF	NR	++	++	NR	++	Chen et al. (2005), Van Pett et al. (2000)

++, receptors present; –, receptors absent; NR, not reported. Adapted from Handbook of Chem. Neuroanat. Integrated Sys. CNS, Vol. 12, Part III, Chapter III, The Olfactory System, M. Shipley, pp. 469–573, 1996, Elsevier, Ltd

have shown that mitral/tufted cells and ET cells are excited by these synapses. ON axon terminals synapse with some subclasses of PG cells, but not with others, and that they do not synapse with SA cells (Hayar et al., 2004a, b; for review see Kosaka et al., 1998). As will also be discussed later, mitral/tufted cells are glutamatergic, and dendritic release of glutamate from these cells has been reported to produce synaptically and nonsynaptically mediated excitation of neighboring cells (i.e., glutamate spillover).

Recent in vitro electrophysiological findings have also provided a new insight into the roles of the different types of JG neurons in olfactory processing (Hayar et al., 2004a, b). Notably, these studies have shown that the spontaneous and sensory-evoked activity patterns of ET cells differ markedly from those of SA and PG cells. ET cells spontaneously generate rhythmic spike bursts and receive monosynaptic ON input, as has also been reported for JG cells in vivo (Getchell and Shepherd, 1975; Wellis and Scott, 1990). SA and PG cells, by contrast, have relatively low levels of spontaneous spike activity, do not possess the intrinsic capacity to generate spike bursts, and mostly respond di- or polysynaptically to ON stimuli. Some dopaminergic PG cells exhibit intrinsic pacemaking activity however (Puopolo et al., 2005). Thus, most SA and PG cells do not appear to receive direct ON input, perhaps because their dendrites ramify in glomerular compartments devoid of ON terminals. Anatomical studies have revealed that the glomeruli have a bicompartmental organization. Each glomerulus has several interdigitating compartments, one of which is rich in ON terminals and the second devoid of ON input (Kosaka et al., 1997; Kasowski et al., 1999). Calbindin-positive JG neurons extend their dendrites only into glomerular compartments devoid of ON terminals, suggesting that they do not receive direct sensory innervation (Toida et al., 1998, 2000). It is reasonable to speculate, therefore, that the SA and PG cells that do not receive direct input (Hayar et al., 2004b) may correspond to these calbindin-positive glomerular interneurons. If most PG cells and all SA cells are devoid of ON input, what circuit mediates their polysynaptic responses to ON stimulation? Recent studies have shown that ET cells, which respond monosynaptically to ON stimulation, in turn relay this input to PG and SA cells (Hayar et al., 2004b). These studies confirmed that ET cells provide direct excitatory glutamatergic input to PG and SA cells. Notably, this microcircuit is intra- versus interglomerular in nature. Mitral/tufted cell dendrites are also likely to provide intraglomerular excitatory input to the PG and SA cells based on anatomical grounds, but this has not been confirmed electrophysiologically (*Figures 6-6*).

3.2.3.2 Inhibitory Systems in the GL

Dendrodendritic Inhibition As noted earlier, PG cells contain GABA and they form inhibitory synapses onto mitral/tufted cell dendrites in the glomeruli. In the periglomerular spaces, their axons also form symmetrical synapses onto the mitral/tufted cell dendrites and PG and SA cells (Pinching and Powell, 1971c). Physiological studies support the notion that PG cells receive excitatory input from mitral/tufted cells and ET cells, and in return, make feedforward and feedback inhibitory GABAergic synapses onto these

Figure 6-6
Intra- and interglomerular circuitry. Glutamatergic ET cells rhythmically burst at sniffing frequencies and receive monosynaptic ON input. ET cells of the same glomerulus burst synchronously and trigger monosynaptic bursts of EPSPs in other JG interneurons (PG and SA cells), most of which do not receive direct ON input. ET cells are synchronized by ON input, glutamate spillover, and interconnections via gap junctions (ovals). ET cells are also synchronized by IPSP bursts from the same and different sets of PG cells. PG cell dendrites ramify in a restricted portion of a single glomerulus and provide local intraglomerular inhibition of ET cells via dendro-dendritic interactions, and also exert presynaptic inhibition of ON terminals. By contrast, SA cells have dendrites and axons extending throughout several glomeruli, and thus might subserve interglomerular interactions (i.e., lateral inhibition). The synchronously bursting ET cell glomerular "ensemble" may constitute an oscillating rhythm generator that monosynaptically synchronizes the activity of PG and SA cells within the same glomerulus, and perhaps also coordinates the activity of mitral/tufted cells via glutamate spillover

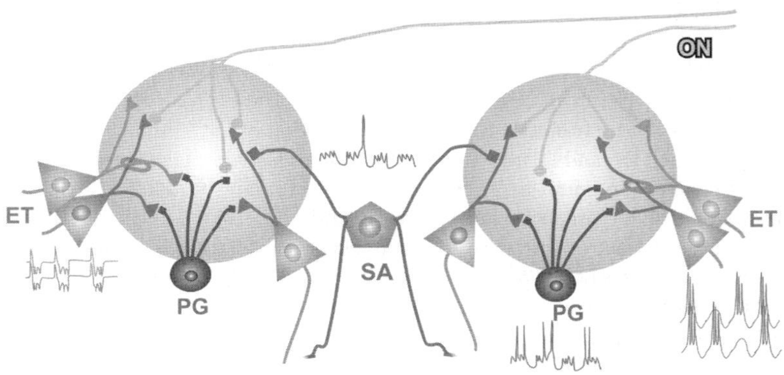

cells (Shepherd and Greer, 1998; Hayar et al., 2004b). Such inhibition is thought to be primarily mediated by $GABA_A$ receptors (see *Section 3.8.3*). Because PG cell dendrites ramify within a restricted portion of a glomerulus, their inhibition is presumably localized to microdomains of the extensive mitral/tufted cell dendritic arbors or to nearby JG cells (Kasowski et al., 1999). By contrast to PG cells, SA cells extend dendrites and axons across multiple glomeruli, suggesting that these cells mediate interglomerular functions, such as lateral inhibition of output neurons in neighboring glomeruli (Aungst et al., 2003). The latter study suggests the intriguing possibility that SA cells are excitatory. The possible interactions among ET, PG, and SA cells are illustrated in a simplified network model in *Figure 6-6*. Detailed EM-immunocytochemical work from the Kosaka and Toida laboratories has deduced the synaptic interactions of subclasses of GABA-, DA-, and CB-positive PG neurons with mitral/tufted cell dendrites. A review of this work is beyond the scope of this chapter and interested readers are referred to the original publications (Toida et al., 1998, 2000; Kosaka et al., 2001). PG cells and ET cells also express $GABA_A$ receptors (see *Section 3.8.3*), and there is electrophysiological evidence that some PG cells can release GABA onto themselves, and perhaps neighboring PG cells, in certain circumstances. This occurs through Ca^{2+}-dependent release of GABA (Smith and Jahr, 2002). GABA has been reported to depolarize PG cells at their resting potential, probably due to the elevated intracellular chloride concentrations (Siklos et al., 1995; Smith and Jahr, 2002). It was suggested that GABA inhibits PG cells by activating a chloride conductance that reduces the neuronal input resistance and shunts excitatory inputs. GABA released from PG cells was recently shown to inhibit other PG cells in the same glomerulus via $GABA_A$ receptors (Murphy et al., 2005).

Presynaptic Inhibition of ON Axon Terminals

Dopamine and D2 receptors The GL contains several hundred thousand DA neurons, but the MOB receives no known extrinsic DA input. In mammals, D1 receptor mRNA is expressed in the GL and GCL (Coronas et al., 1997). Immunocytochemical localization of D1 receptors is faint and primarily in the GCL (Levey et al., 1993). By contrast, D1-like ligand binding is present at very low levels in all layers of the MOB with the exception of the ONL (Mansour et al., 1990a; Nickell et al., 1991; Coronas et al., 1997). The functional significance of D1 receptors in MOB remains unclear. By contrast, only the ONL and GL have high densities of D2 receptors in rats and mice (Mansour et al., 1990a; Nickell et al., 1991; Levey et al., 1993; Coronas et al., 1997; Koster et al., 1999) and to a lesser extent in guinea pigs, but not in cats or monkeys (Camps et al., 1990). In the GL, the JG neurons express D2 receptors (Mansour et al., 1990a). Some immunocytochemical labeling for D2 receptors, as well as in situ hybridization, has been reported in the GCL and EPL, however D2 binding sites are consistently restricted to the ONL and GL (Mansour et al., 1990a; Levey et al., 1993; Coronas et al., 1997). Other anatomical evidence indicates that most, if not all, of the D2 receptors in the GL occur on ON axon terminals. ORNs express D2 receptors and bulbectomy, a manipulation that causes retrograde degeneration of ORNs, eliminates D2 receptor mRNA in the olfactory epithelium (Koster et al., 1999). Taken together, these findings indicate that DA released from JG neurons may presynaptically modulate ON terminals via activation of D2 receptors.

In agreement with this, DA and D2 receptor agonists reduced spontaneous and ON-evoked activity in mitral cells, as well as odor-evoked activity in the GL and odor detection performance, in a variety of species (Nowycky et al., 1983; Doty and Risser, 1989; Sallaz and Jourdan, 1992; Wachowiak and Cohen, 1999; Hsia et al., 1999; Berkowicz and Trombley, 2000; Ennis et al., 2001). These effects are mediated by presynaptic suppression of glutamate release from ON terminals via inhibition of Ca^{2+} influx (Wachowiak and Cohen, 1999). In a similar manner, DA and D2 receptor agonists suppressed spontaneous and ON-evoked activity in JG cells, but had no effect on mitral-to-JG cell transmission (Ennis et al., 2001). The inhibitory effects of DA were abolished in D2 receptor knockout mice (Ennis et al., 2001). Intriguingly, this presynaptic regulation occurs in the absence of conventional anatomical synapses from JG cells to ON terminals. The close proximity of the dendrites of DA-positive and GABA-positive PG cells to ON terminals may facilitate this "nonsynaptic" presynaptic regulation. Further, synaptic vesicles are present in some dendrites opposed to ON axon terminals (Bonino et al., 1999). What role might such presynaptic inhibition serve? One possibility is that presynaptic inhibition of ON terminals by DA provides a mechanism for increasing the range of concentrations that can be processed by MOB neurons: as activity increases in ON terminals, dopaminergic JG cells are more strongly excited. This in turn provides negative feedback to ON terminals reducing the release of glutamate. Such a scheme would effectively increase the dynamic range of information transfer from ORNs to MOB neurons. Interestingly, systemic administration of D2 receptor agonists has been reported to prevent odorant specific 2-dexyglucose patterns in MOB and to reduce odorant detectability (Doty and Risser, 1989; Sallaz and Jourdan, 1992). Related to this question of how DA participates in odor processing is the degree to which these receptors are tonically active in vivo? If, for example, ON terminals are tonically inhibited by DA via D2 receptors, this might serve to filter out weak signals ("noise"). This might sharpen the spatial pattern of active glomeruli, and facilitate the detection of predominant odors. There is experimental support for this possibility. Blockade of D2-like receptors by systemic administration of spiperone increased the number of mitral cells that responded to single or multiple odorants (Wilson and Sullivan, 1995). One interpretation of this study is that reduced D2 presynaptic inhibition of ON terminals increases the odor responsiveness of mitral cells, but does so at the cost of reduced odorant discrimination. Other effects of both D1 and D2 receptor activation have been reported (Duchamp-Viret et al., 1997; Brünig et al., 1999; Davison et al., 2004).

GABA, $GABA_B$ Receptors, and Taurine $GABA_B$ receptors play a presynaptic inhibitory role, apparently very similar to that described for D2 receptors. As noted earlier, GABAergic PG cells represent a large population of GL interneurons. In the rat MOB, the glomeruli have the highest concentration of $GABA_B$ receptors as determined by radioligand binding (Bowery et al., 1987; Chu et al., 1990) and immunohistochemical localization of $GABA_B$ receptor subunits (Margeta-Mitrovic et al., 1999). EM-immunohistochemistry revealed that the dense labeling in the GL is due to the presence of $GABA_B$ receptors on ON terminals and the somata of PG cells (➋ *Table 6-4*) (Bonino et al., 1999). A variety of imaging and electrophysiological studies have provided solid evidence that GABA released from PG neurons presynaptically inhibits glutamate release from ON terminals via activation of these $GABA_B$ receptors (Keller et al., 1998; Wachowiak and Cohen, 1999; Aroniadou-Anderjaska et al., 2000; Palouzier-Paulignan et al., 2002; Murphy et al., 2005). The MOB also contains the highest levels of the putative inhibitory transmitter

Table 6-4
Gap junctions and connexins in the MOB

Layer/Cell Type	Gap Junctions[a]	Connexin mRNA[b]	Connexin Protein[c]	References
MOB			C×36	Rash et al. (2000)
ONL			C×36, 43	Paternostro et al. (1995), Reyher et al. (1991), Teubner et al. (2000), Zhang and Restrepo (2003)
GL			C×36, 43	Belluardo et al. (2000), Christie et al. (2005), Paternostro et al. (1995), Theis et al. (2003), Teubner et al. (2000), Zhang and Restrepo (2003)
JG cells		C×36		Christie et al. (2005)
PG cells		C×36		Kosaka and Kosaka (2003, 2004), Landis et al. (1974)
M/T–PG dendrites	TEM			Kosaka and Kosaka (2003, 2005)
M/T–M/T dendrites	TEM			Christie et al. (2005), Kosaka and Kosaka (2004, 2005), Rash et al. (2005)
M/T-unidentified dendrites	TEM			Kosaka and Kosaka (2003, 2005), Kosaka et al. (2005)
EPL			C×43	Reyher et al. (1991)
Mitral cell proximal dendrites			C×45	Teubner et al. (2000)
Tufted cells (and/or interneurons?)		C×36, 43		Condorelli et al. (1998, 2000), Miragall et al. (1996)
Unidentified cells			C×43	Theis et al. (2003)
Tufted cells-Interneurons	TEM			Kosaka and Kosaka (2003)
M/T-Interneuron dendrites	TEM			Kosaka and Kosaka (2003)
M/T-Unidentified dendrites	TEM			Kosaka and Kosaka (2003)
M/T-GC dendrites	FF			Landis et al. (1974)
EPL/MCL		C×36	C×36, 43	Belluardo et al. (2000)
MCL		C×36	C×36, 43	Belluardo et al. (2000), Christie et al. (2005), Condorelli et al. (2000), Reyher et al. (1991)
Mitral cells	FF	C×36, 43, 45	C×45	Belluardo et al. (2000), Condorelli et al. (1998), Miragall et al. (1996), Paternostro et al. (1995), Theis et al. (2003), Teubner et al. (2000), Zhang and Restrepo (2002)
IPL			C×43	Reyher et al. (1991), Kosaka et al. (2005), Rash et al. (2005)
GCL				
GCs	TEM, FF	C×36	C×43	Christie et al., (2005), Condorelli et al. (1998), Reyher et al. (1991), Kosaka et al. (2005)
Unidentified cells			C×43	Theis (2003)

[a]Standard transmission electron microscopy (TEM) showing junctions between two cell types or freeze-fracture (FF) replica analysis or intramembranous particles

[b]In situ hybridization or nuclease protection assay

[c]Western blots (entire MOB, presumably including AOB), immunocytochemistry (layers) or immunohistochemistry (cells)

Abbreviations: M/T, mitral/tufted

taurine, exceeding concentrations of GABA and glutamate (Collins, 1974; Margolis, 1974; Banay-Schwartz et al., 1989a, b; Ross et al., 1995; Kamisaki et al., 1996). Taurine is found in ON axons, in various neurons, and in astrocytes (Kratskin et al., 2000; Kratskin and Belluzzi, 2002; Pow et al., 2002). In the ON terminals and some postsynaptic dendrites, taurine is colocalized with glutamate (Didier et al., 1994). Observation of spontaneous taurine release from MOB synaptosomes suggests that taurine may be abundantly released (Kamisaki et al., 1996). In electrophysiological recordings, taurine directly activated presynaptic $GABA_B$ receptors and inhibited ON terminals, and also induced Cl^- currents in mitral/tufted cells. Surprisingly, taurine had no direct effect on PG cells (Belluzzi et al., 2003).

3.2.3.3 Neuromodulation in the GL Neuromodulatory systems in the GL include carnosine and certain heavy metals (copper and zinc), metabotropic glutamate receptors (mGluRs), and neuromodulatory inputs from centrifugal afferents. mGluRs and centrifugal inputs are discussed later (➋ *Sections 3.8.2* and ➋ *3.9*). Carnosine, a dipeptide synthesized by ORNs, is localized in ON terminals in the GL and fulfils many criteria for neurotransmitter candidacy. However, no direct postsynaptic actions of carnosine have been revealed to date (MacLeod and Straughan, 1979; Nicoll and Alger, 1980; Frosch and Dichter, 1982; Trombley et al., 1998). Carnosine did not affect currents evoked by glutamate, GABA, or glycine in cultured MOB neurons (Trombley et al., 1998). Because carnosine is a chelator of both zinc and copper, it has been suggested that it might modulate transmission at ON synapses by regulating zinc and copper. Depending on the concentration, both zinc and copper can augment or block responses mediated by NMDA and GABA receptors. Both zinc and copper inhibit NMDA and GABA receptor-mediated currents and synaptic transmission in MOB neurons (Trombley and Shepherd, 1996; Trombley et al., 1998). Carnosine prevented the action of copper and reduced the effect of zinc. These results suggest that carnosine may indirectly influence glutamate actions on MOB neurons by modulating the effects of synaptically released zinc and copper. What roles might carnosine, zinc and copper play in olfactory processing? At present, there are no clear answers to this question. Is carnosine, like other peptide transmitters, preferentially released by high frequencies of ON activity? If so, carnosine may be preferentially released during intense odor stimulation. Zinc is known to be preferentially released during high frequency neural activity. There are neurotoxic effects of zinc and copper, and Trombley (Horning etal., 2000) has speculated that carnosine, by preventing the actions of these metals, may serve an important neuroprotective function, perhaps to protect MOB neurons during intense or high frequency activity.

3.3 External Plexiform Layer

The EPL lies beneath or deep to the glomeruli, and it primarily consists of dense neuropil formed by the dendrites of mitral cells and GCs that ascend from the MCL and GCL, respectively. Relative to other MOB layers, the EPL has a low cell density. In Nissl-stained sections, however, it can nevertheless be seen to contain significant numbers of neurons (➋ *Figure 6-3*). These include several subtypes of tufted cells and intrinsic interneurons, which are described later. Because tufted cells are in many aspects similar to mitral cells, and as mitral and tufted cell dendrites cannot be distinguished ultrastructurally, the term mitral/tufted cell is often used when generalizing to these two cell populations. The dominant feature of the EPL is nevertheless the extensive dendrodendritic synapses between mitral/tufted cells and GCs.

3.3.1 Neuron Types of the EPL

3.3.1.1 Tufted Cells Tufted cells are the most numerous cells of the EPL. Like ET cells and mitral cells, tufted cells of the mammalian EPL have one (or rarely several) apical (primary) dendrite(s) ending in a glomerular tuft, hence the name (Cajal, 1890) (➋ *Figure 6-5*). The size of the tufted cell bodies gradually increases from the superficial to the deep EPL (Cajal, 1890; Pinching and Powell, 1971a; Macrides and Schneider, 1982; Orona et al., 1984). Golgi staining and dye injection studies (➋ *Figure 6-5*) have shown that superficial tufted cells tend to have asymmetrical branching patterns, extending sparsely branched lateral dendrites obliquely to the projection path of the primary dendrites (Macrides and Schneider, 1982; Orona et al., 1984). Middle tufted cells, deep tufted cells, and mitral cells exhibit more symmetrical branching patterns (i.e., shaped like a bishop's miter), with one apical dendrite and tangentially arrayed lateral dendrites (➲ *Figure 6-5*). Tufted cells that closely resemble mitral cells in size have also been referred to as displaced mitral cells (Blanes, 1898), but this term is now generally reserved for the deepest tufted cells located near the MCL (Mori et al., 1983; Kiyoshi et al., 1984; Mori, 1987a, b). Tufted cells utilize glutamate as their principle transmitter (Liu et al., 1989; Christie et al., 2001), but they also stain for a number of other substances (➲ *Table 6-1*). Many superficial tufted cells contain CCK (Seroogy et al., 1985), and they form a topographically organized, reciprocal network (Schoenfeld et al., 1985), which is described in ➲ *Section 3.3.2.4*. In the Chinese hamster, many superficial and middle tufted cells are immunoreactive for substance P. The rat has some of these cells, but in other rodents, few if any occur (Baker et al., 1986; Matsutani et al., 1988). The deep EPL of the Syrian hamster has a few NADPH diaphorase-positive tufted cells (Davis, 1991), and it also has a relatively large population of dopaminergic tufted cells (Halász et al., 1981; Baker, 1986; Gall et al., 1987). The latter cells are relatively rare in other rodent species, but are observed in primates (Smith et al., 1991) and amphibians (Boyd and Delaney, 2002). TH-immunoreactive tufted cells do not coexpress GABA, unlike TH-positive PG cells (Gall et al., 1987). As discussed in ➲ *Section 3.4*, many mitral/tufted cells are also immunoreactive for corticotropin-releasing factor (CRF) (Imaki et al., 1989; Bassett et al., 1992), and the rodent EPL exhibits a high density of CRF receptors (De Souza et al., 1985).

Axonal projections of middle and deep tufted cells are similar, but not identical, to those of mitral cells (Schoenfeld and Macrides, 1984; Schoenfeld et al., 1985). The local axon collaterals of tufted cells course mainly in the IPL. The projections of tufted cells beyond the MOB terminate densely in the AON and, to a lesser extent, in other rostral olfactory cortical structures (Schoenfeld et al., 1985; Scott, 1986). Few tufted cell axons project into more caudal POC regions.

3.3.1.2 EPL Interneurons The EPL contains anatomically and neurochemically heterogeneous subtypes of intrinsic interneurons. From a functional standpoint, these cells are not well understood. More information on their synaptic interactions is clearly needed to understand the role of these cells in olfactory processing. Their interactions with tufted cells may be particularly important, because tufted cells are thought to comprise a parallel "channel" for transmitting most of the activity from the MOB, separately from the mitral cells (Shepherd et al., 2004). With Golgi staining methods, the EPL was originally described in the cat as containing multipolar neurons having multiple dendrites originating from two poles, but no axons (Van Gehuchten and Martin, 1891). Other interneuron subtypes were subsequently described in other species (for a review, see Halász, 1990). These subtypes include bipolar interneurons with long processes arrayed tangentially to the MCL (horizontal cells) (Valverde, 1965) and multipolar neurons with varicose dendrites, referred to as Van Gehuchten (VG) cells, which occur together with superficial short axon (SSA) cells resembling those of the GL (➲ *Figure 6-5*) (Schneider and Macrides, 1978). VG cells are characterized by several relatively thick primary dendrites that remain in the EPL and axons terminating around mitral and tufted cells. VG cells with smooth dendrites, horizontal cells thought to be equivalent to SSA cells, and small satellite cells with varicose dendrites that envelop small segments of mitral cell dendrites, have been described (López-Mascaraque et al., 1990). Other VG and SA cells typically have varicose dendrites that branch within either the superficial or deep half of the EPL. With parvalbumin (PV) immunostaining however, interneurons with dendrites that are either varicose or relatively smooth have been observed, and some of these interneurons exhibit intermediate branching patterns (Kosaka et al., 1994a, b). Many EPL interneurons stain for GABA (Mugnaini et al., 1984a; Gall et al., 1987; Kosaka et al., 1987d), including some of those identified as VG and SSA cells (Ohm et al., 1990), and are therefore presumed to be inhibitory. Some of the GABA interneurons are also immunoreactive for PV (Kosaka et al., 1987d). As summarized in ➲ *Tables 6-1* and ➲ *6-2*, EPL interneuron subtypes are neurochemically diverse and stain for (1) NADPH diaphorase (Scott et al., 1987; Villalba et al., 1989; Alonso et al., 1995), (2) AChE (Le Jeune and Jourdan, 1994), (3) SP (Baker, 1986; Wahle et al., 1990), (4) ENK (Bogan et al., 1982; Davis et al., 1982), (5) NPY (Gall et al., 1986; Scott et al., 1987; Seroogy et al., 1989; Sanides-Kohlrausch and Wahle, 1990a), (6 and 7) neurotensin and somatostatin (Matsutani et al., 1988), (8) VIP (Gall et al., 1986; López-Mascaraque et al., 1989;

Sanides-Kohlrausch and Wahle, 1990b; Nakajima et al., 1996), (9) PV (Cielo, 1990; Ohm et al., 1990; Kosaka et al., 1994a, b; Alonso et al., 1995, 1998; Kakuta et al., 1998; Briñón et al., 1999, 2001; Jia and Halpern, 2004), (10) CB, (11) neurocalcin, and (12) CR (Ohm et al., 1991; Briñón et al., 1992, 1999, 2001; Alonso et al., 1995, 1998; Jia and Halpern, 2004).

3.3.2 Circuitry, Synaptology, and Sublaminar Organization of the EPL

3.3.2.1 Overview of EPL Circuitry and Reciprocal Mitral/Tufted Cell—Granule Cell Dendrodendritic Interactions The neuropil of the EPL contains the vertically oriented apical dendrites of mitral/tufted cells ascending to receive monosynaptic ON input in the GL, and laterally oriented secondary dendrites of the mitral/tufted cells. The majority of the synapses of the EPL are between the lateral dendrites of mitral/tufted cells and the dendrites of GCs. Most of the synapses are: (1) asymmetrical synapses from the mitral/tufted cell lateral dendrites onto GC dendrites, and (2) symmetrical synapses from the spines ("gemmules") of GC dendrites onto the mitral/tufted cell lateral dendrites (Price and Powell, 1970b–d). These synapses are mostly reciprocal (Hirata, 1964; Rall et al., 1966; Price and Powell, 1970d; Woolf et al., 1991), and therefore occur in roughly equal proportion (Jackowski et al., 1978). Similar synapses are formed by mitral/tufted cell bodies and GC dendrites. The mitral/tufted cell lateral dendrites are tangentially arrayed relative to the MOB layers and can extend up to ~1–2 mm from the soma (Mori et al., 1983). One mitral/tufted cell can therefore receive feedback inhibition and lateral inhibition mediated by GCs that are excited by other mitral/tufted cells. As detailed knowledge of the functions of these synapses has been derived from studies of mitral cells, they are described in greater detail in ➋ *Section 3.4.1.1*. The mitral/tufted cell dendrites also form synapses with intrinsic EPL interneurons. Through the mitral/tufted cells, these EPL interneurons may therefore indirectly receive input related to the overlying glomeruli (Hamilton et al., 2005). The third element of the EPL is an intrabulbar association system (IAS), and the fourth element is the centrifugal afferent input to the EPL from the POC. The interneuron synapses, IAS, and centrifugal inputs are also described separately later.

3.3.2.2 Synaptology of EPL Tufted Cells Tufted cells are, in many respects, similar to mitral cells in terms of synaptology. The dendritic tufts receive monosynaptic input from ON axon terminals and the lateral dendrites synapse with JG cells in the GL and with GC dendrites in the EPL (Shepherd, 1972; Shepherd et al., 2004). Like mitral cells (*see* ➋ *Section 3.4*), tufted-to-GC synaptic interactions involve a strong NMDA receptor component (Christie et al., 2001). The lateral extent of tufted cell inhibition is limited to several glomerular widths, however, ~400 μm as opposed to 750 μm for mitral cells in rats (Christie et al., 2001). The weaker lateral inhibition of tufted cells could merely be due to the fact that their lateral dendrites are relatively short (Mori et al., 1983; Orona et al., 1984). However, recent evidence suggests that inhibition from the intrinsic EPL interneurons, not from GCs, tunes the responses of middle tufted cells to odor stimulation (Nagayama et al., 2004).

3.3.2.3 Synaptology of EPL Interneurons The EPL interneurons were originally thought to interact chiefly with GC dendrites (Schneider and Macrides, 1978). However, more recent studies have suggested that they instead interact with mitral/tufted cell dendrites (López-Mascaraque et al., 1990; Nagayama et al., 2004), a view that is supported by ultrastructural studies. Ultrastructural studies have shown that the PV-IR interneurons of the rat EPL receive asymmetrical synapses from mitral/tufted cell bodies and dendrites, and form symmetrical synapses onto the mitral/tufted cells, many of which are reciprocal (Toida et al., 1994, 1996; Crespo et al., 2001). Surprisingly, although the PV-IR interneurons were observed to form frequent contacts with each other, no synapses were observed at the contact points. However, symmetrical synapses from unidentified PV-immunonegative axons were observed on the PV-immunoreactive cell bodies, where they would be expected to shunt the more distal mitral/tufted cell inputs. The PV-immunoreactive interneurons also receive asymmetrical synapses from unidentified axons (Toida et al., 1996), which could be centrifugal afferents. Synapses from centrifugal fibers have been observed on SA cells (Pinching and Powell, 1972a–c). Combined morphological and electrophysiological analyses showed that multipolar

EPL interneurons with highly varicose dendrites are excited polysynaptically by ON stimulation, most likely via input from mitral/tufted cells (Hamilton et al., 2005). The multipolar interneurons exhibit robust spontaneous excitatory synaptic activity mediated by AMPA/kainate receptors, consistent with immunocytochemical staining studies showing that mouse EPL neurons are strongly reactive to the GluR1 AMPA receptor subunit (Petralia and Wenthold, 1992; Giustetto et al., 1997; Montague and Greer, 1999; Hamilton and Coppola, 2003). Similar electrophysiological recordings were obtained from varicose multipolar interneurons in slices from normal mice and from fluorescent cells exhibiting the same morphology in slices from transgenic mice labeled to reveal GABAergic interneurons. The dendrites of the interneurons that were located in the superficial EPL bridged the space below several adjacent glomeruli (➋ *Figure 6-5*). Interneurons located in the deeper zones bridged an equivalent EPL area, which was approximately equal to the glomerular width (Hamilton et al., 2005). This suggests that multipolar EPL interneurons are excited by mitral/tufted cells and may in turn inhibit mitral/tufted cells within EPL domains that are topographically related to pairs of overlying glomeruli.

3.3.2.4 Sublaminar Organization of the EPL As noted earlier, superficial, middle, and deep tufted cells are distinguished by cell body location and size (Cajal, 1890; Pinching and Powell, 1971a; Macrides and Schneider, 1982; Orona et al., 1983). The lateral dendrites of the superficial, middle, and deep tufted cells course within superficial, intermediate, and deep EPL zones, respectively (Scott and Harrison, 1987). Type I and II mitral cell subtypes have also been identified according to the distribution of their lateral dendrites within the deep and intermediate zones, respectively (Kiyoshi et al., 1982; Orona et al., 1984). Electrophysiological recordings have shown that the sensitivity of these tufted cell and mitral cell subtypes to ON stimulation is correlated with the depth of the lateral dendrites within the EPL. The more superficial tufted cells are more easily excited by ON stimulation than the deeper tufted cells, and the type II mitral cells are more easily excited than type I mitral cells (Schneider and Scott, 1983; Wellis et al., 1989; Ezeh et al., 1993). In response to odor stimulation, superficially located tufted cells also exhibit more prolonged excitation than mitral cells (Luo and Katz, 2001). Thus, anatomically distinct mitral/tufted cell subtypes appear to be functionally distinct.

A differential sublaminar distribution is also observed for the dendrites of superficial and deep GCs. The apical dendrites of superficial GCs have very dense spines, and they terminate within both the superficial and deep portions of the EPL. By contrast, apical dendrites of deeper GCs terminate exclusively within the deep EPL (Orona et al., 1983; Greer, 1987). The dendrites of a third GC subtype produce spines predominantly within the superficial EPL (Mori, 1987a, b). Perhaps as a result of these morphological differences, the superficial EPL stains more strongly for glutamic acid decarboxylase (GAD) (Mugnaini et al., 1984b), for the $\alpha 3$ $GABA_A$ receptor subunit (Panzanelli et al., 2005), and for the AMPA receptor subunit, GluR4 (Montague and Greer, 1999). Both the superficial and deep EPL stain faintly for cytochrome oxidase activity, whereas the intermediate EPL stains darkly (Mouradian and Scott, 1988). Together, these observations suggest that morphologically and neurochemically distinct subsets of mitral/tufted cells might synapse with different subsets of GCs and intrinsic interneurons within the superficial and deep EPL sublaminae (Macrides et al., 1985; Mori, 1987a, b).

3.3.2.5 Intrabulbar Association System Many superficial tufted cells do not appear to project beyond the bulb. The axons of these cells contain CCK (Seroogy et al., 1985), and they form a topographically organized, reciprocal network, the IAS, connecting lateral and medial regions of each MOB (Schoenfeld et al., 1985). The axons of the IAS travel through the EPL and MCL to the IPL, where they coalesce into tracts that travel to the opposite side of the bulb and terminate on the apical dendrites of GCs coursing through the IPL en route to the EPL (Liu and Shipley, 1994). The IAS projection exhibits a high degree of point-to-point, topographical organization. As noted in ➋ *Section 2.3* earlier, the ORNs which express the same receptor project to one glomerulus in the lateral bulb and to a second glomerulus in the medial bulb (Ressler et al., 1993; Vassar et al., 1994; Mombaerts et al., 1996; Wang et al., 1998). The IAS projections from a single glomerulus preferentially target the IPL deep to the second glomerulus (Belluscio et al., 2002; Lodovichi et al., 2003). Because CCK typically causes membrane depolarization, activation of the CCK tufted cells presumably depolarizes GCs located on the opposite side of the bulb. Thus, the IAS appears to modulate the activities of subsets of MOB neurons receiving input from ORNs expressing the same receptor, on opposite sides of the MOB.

3.4 Mitral Cell Layer, Mitral/Tufted Cell Synapses, and Neurochemistry

Deep in the EPL is the MCL, a thin layer that contains the somata of mitral cells (25–35 μm diameter, rat) as well as numerous GCs (Cajal, 1911a, b). There are ~40,000 mitral cells (Meisami, 1989) and ~100,000 GCs in the MCL (Frazier and Brunjes, 1988). Thus, mitral cells make up less than 50% of the cells in this layer. Together with tufted cells, mitral cells are the major class of output cells of the bulb. They extend a single apical dendrite through the EPL to the GL, where it arborizes extensively throughout much of a single glomerulus (➋ *Figure 6-5*). There are about 25 mitral cells (and 50 tufted cells) associated with a single glomerulus (Cajal, 1911a, b; Allison, 1953). The apical dendrites are synaptically contacted by ON terminals (Price and Powell, 1970a; Shepherd, 1972). The secondary dendrites of mitral cells may extend upto 2 mm in the EPL and are oriented tangentially, i.e., parallel to the surface of the bulb. Mitral cells have been subdivided into two classes based on the organization of the lateral dendrites as reviewed in detail elsewhere (Orona et al., 1984; Shipley et al., 1996). These lateral dendrites participate in dendrodendritic synapses with dendrites of GCs, as reviewed earlier and later. In addition, they may receive centrifugal axon inputs and inputs from VG EPL interneurons (Jackowski et al., 1978; Toida et al., 1996). Mitral cells give off axon collateral which terminate, within the bulb, in the IPL and GCL (Mori et al., 1983; Price and Powell, 1970c), or exit the MOB and innervate a number of olfactory-related brain regions collectively known as the POC. Projections to POC are discussed later.

3.4.1 Neurochemistry of Mitral Cells

Mitral (and tufted) cells are glutamatergic (Liu et al., 1989; Christie et al., 2001). A few small mitral cells appear to contain aspartate and project to the largest component of POC, piriform cortex (PC) (Fuller and Price, 1988). Another proposed transmitter for mitral cells is N-acetyl-aspartyl-glutamate (NAG) (➋ *Table 6-1*) based on anatomical grounds (Ffrench-Mullen et al., 1985; Blakely et al., 1987), but there has been little neurophysiological support for this notion (Whittemore and Koerner, 1989; see ➋ *Section 5.3.1.1*). The neuropeptide CRF has been demonstrated in mitral and some tufted cells using both immunocytochemistry and in situ hybridization in the rat (Imaki et al., 1989; Bassett et al., 1992). CRF fibers were also observed in layer Ia of the PC, consistent with the dense axonal projections of mitral cells to layer Ia of the PC. Mitral cells express type 1 CRF receptor mRNA (Van Pett et al., 2000), and mouse mitral cells also label for a soluble splice variant of the type 2 CRF receptor (Chen et al., 2005). SP mRNA transcripts have been detected in up to one-half of the rat mitral cells (Warden and Young, 1988), but to date no studies using immunocytochemistry have detected SP in mitral cells of any species (Inagaki et al., 1982; Shults et al., 1984; Baker, 1986). Finally, CR, a Ca^{2+}-binding protein, has been shown by immunohistochemistry to be localized in mitral cells (➋ *Table 6-2*) (Jacobowitz and Winsky, 1991). Transmitter candidates for mitral and tufted cells are discussed further in ➋ *Section 5.3.1.1.*

3.4.1.1 Glutamate

ON Input As noted earlier, mitral/tufted cell apical dendrites respond to glutamate released from the ON via AMPA and NMDA receptors (Ennis et al., 1996, 2001; Aroniadou-Anderjaska et al., 1997; Chen and Shepherd 1997). The NMDA receptor-mediated response component is of unusually long duration, leading to a late spiking component in response to ON input (Ennis et al., 1996, 2001; Aroniadou-Anderjaska et al., 1997). An mGluR-mediated component of ON-evoked currents in mitral cells also occurs but is typically small under normal conditions (De Saint Jan and Westbrook, 2005; Ennis et al., 2006), yet, as will be discussed later, blockade of these receptors significantly alters ON-evoked spiking.

Self And Neighbor Excitation Electrophysiological studies indicate that mitral and tufted cells have functional ionotropic glutamate autoreceptors that mediate recurrent excitatory interactions among their apical and lateral dendrites (Isaacson, 1999; Aroniadou-Anderjaska et al., 1999b; Carlson et al., 2000; Schoppa and Westbrook, 2001; Didier et al., 2001; Salin et al., 2001; Urban and Sakmann, 2002). Mitral cell bodies and lateral dendrites have NMDA (Isaacson, 1999; Isaacson and Murphy, 2001), AMPA, and kainate autoreceptors (Lowe, 2003). Ultrastructural studies indicate that NMDA receptors of the lateral dendrites are extrasynaptic (Sassoè-Pognetto et al., 2003). Mitral cell primary dendrites have AMPA autoreceptors (Salin et al., 2001). NMDA autoreceptors on mitral cells increase the firing frequency during prolonged discharges (Friedman and Strowbridge, 2000). Thus, glutamate released from mitral/tufted cell dendrites appears to spillover and excite glutamate receptors on the same cell and on neighboring mitral/tufted cell dendrites. Such excitatory interactions among mitral tufted cell apical dendrites in the glomeruli are thought to synchronize the discharge of mitral cells associated with the same glomerulus (Carlson et al., 2000; Schoppa and Westbrook, 2001).

mGluRs In dissociated rat and frog, MOB culture preparations, Group I mGluRs increased Ca^{2+} release from internal stores in mitral/tufted cells as well as in MOB interneurons (Geiling and Schild, 1996; Carlson et al., 1997), or it depolarized and increased the frequency of miniature excitatory postsynaptic currents (Schoppa and Westbrook, 1997). Other studies indicate that activation of Group III mGluRs with AP4 inhibits Ca^{2+} currents in mitral cells and presynaptically decreases mitral cell-to-GC synaptic transmission (Trombley and Westbrook, 1992). More recent studies in rat and mouse MOB slices demonstrate that activation of mGluR1 directly depolarizes and increases the firing of MCs, and that these effects persist in the presence of blockers of fast synaptic transmission (Heinbockel et al., 2004). The same study showed that mGluR1 induces a voltage-dependent inward current consisting of multiple components sensitive to K^{+} and Ca^{2+} channel blockade and intracellular Ca^{2+} chelation. mGluR1 antagonists also altered mitral cell membrane potential bistability, increasing the duration of the up- and downstates, and substantially attenuated ON-evoked spikes. These findings suggest that endogenous glutamate tonically modulates MC excitability and responsiveness to ON input via activation of mGluR1.

Mitral/Tufted-to-JG and Mitral-to-GC Excitation Glutamate released from mitral/tufted cells, acting at ionotropic receptor subtypes, mediates dendrodendritic transmission at synapses with JG cells and with GCs (Bardoni et al., 1996; Isaacson and Strowbridge, 1998; Schoppa et al., 1998; Aroniadou-Anderjaska et al., 1999a; Chen et al., 2000; Christie et al., 2001). The mitral/tufted-to-GC cell excitation triggers GABA release from the GC dendrites (Isaacson and Strowbridge, 1998; Schoppa et al., 1998; Chen et al., 2000), which inhibits mitral cells via $GABA_A$ receptors (Shepherd, 1972; Jahr and Nicoll, 1982; Schoppa et al., 1998; Chen et al., 2000; see ➲ *Section 3.8.3*). Ca^{2+} influx via the NMDA receptor channel is sufficient to trigger GABA release from GCs, and thus, to drive dendrodendritic inhibition of mitral cells (Chen et al., 2000; Halabisky et al., 2000); however, NMDA-evoked depolarization can also trigger GABA release from these cells via activation of voltage-dependent Ca^{2+} channels.

Modulation of Mitral/Tufted-to-Granule Cell Dendrodendritic Interactions mGluRs may also occur at the mitral-to-granule synapses, although their role is unclear. As noted earlier, activation of Group III receptors was reported presynaptically to decrease mitral-to-GC synaptic transmission (Trombley and Westbrook, 1992), and activation of $GABA_B$ receptors has been observed to reduce dendrodendritic inhibition via suppression of GABA release from GCs due to reduction of high voltage-activated Ca^{2+} currents (Isaacson and Vitten, 2003). Although DA has no direct effect on resting membrane properties of mitral/tufted cells (Ennis et al., 2001), activation of D2 receptors was found to reduce glutamate release onto GCs via inhibition of N and/or P/Q high voltage-activated Ca^{2+} channels (Davila et al., 2003; Davison et al., 2004).

3.5 Internal Plexiform Layer and Granule Cell Layer

The IPL is the relatively thin layer that lies deep to the MCL. It has a low density of cells, composed mostly of dendrites derived from GCs and axons derived from mitral/tufted cells and centrifugal sources (discussed later). As noted earlier, the IPL contains a network of CCK-positive axons and terminals, which derive from CCK-containing tufted cells as part of the IAS (Liu and Shipley, 1994). The major intrinsic cell type in the IPL appears to be the horizontal cell. Horizontal cell dendrites are oriented parallel to the MCL and horizontal cell axons extend into the EPL (Price and Powell, 1970c). Neurocalcin (Bastianelli et al., 1993; Porteros et al., 1996), CB (Briñón et al., 1992), NADPH diaphorase (Alonso et al., 1995), and ENK (Bogan et al., 1982) have been found in these cells. The IPL also contains a few multipolar neurons, which are larger than GCs and express AChE (Carson and Burd, 1980; Nickell and Shipley, 1988b; Le Jeune and Jourdan, 1994).

The GCL is the deepest neuronal layer in the bulb, and it contains the largest number of cells. Most of the neurons of the GCL are the GCs, but there are also small numbers of Golgi cells, Cajal cells, and Blanes cells. As discussed earlier, the GCs are inhibitory GABAergic cells that form dendrodendritic synapses with mitral/tufted cells in the EPL.

3.5.1 Neuron Types of the GCL

3.5.1.1 GCs GCs are axon-less cells with small cell bodies that are mostly tightly packed into row-like aggregates of 3–9 somata (➲ *Figure 6-5*) (Reyher et al., 1991). GC bodies are also found mixed with mitral cell bodies within the MCL, however. Golgi studies indicate that most GCs have a thicker, longer apical dendrite that ramifies within the EPL and shorter basal dendrites that ramify within the GCL. As described earlier, several types of GCs occur, which have different dendritic ramifications within the superficial and deep portions of the EPL. Not all GCs follow this pattern, however. The exceptions include GCs with basal dendrites that project deeper, toward the center of the MOB, and GCs with apical dendrites that do not enter the EPL (Schneider and Macrides, 1978). The superficial and deep GCs can also be distinguished using a variety of staining methods. For example, superficial GCs have more dense spines on their apical dendrites and they stain for the transcription factor Er81 (Stenman et al., 2003), whereas deep GCs stain for Ca^{2+}/calmodulin-dependent kinase IV (Baker et al., 2001). The GC spines are also immunoreactive for Ca^{2+}-camodulin-dependent kinase II, expression of which is developmentally regulated. Because MOB interneurons develop postnatally and are added throughout life, the differential staining of cell bodies in the superficial versus deep GCL could be related to the migration and maturation of interneuron progenitors (see later). One function of the cholinergic input to the MOB appears to be related to the survival of these cells (Cooper-Kuhn et al., 2004), which is influenced by nicotinic receptors containing the β_2 subunit (Mechawar et al., 2004). Subpopulations of GCs are either intensely or lightly immunoreactive for CR (Jacobowitz and Winsky, 1991). GCs and some SA cells are also immunoreactive for methionine-enkephalin (Bogan et al., 1982; Davis et al., 1982), whereas deep SA cells are immunoreactive for somatostatin (Davis et al., 1982). Other possible transmitters of GCs are summarized in ➲ *Table 6-1.*

As discussed earlier, GCs are excited by mitral/tufted cells at asymmetrical synapses in the EPL. Most GCs contain GABA (Ribak et al., 1977), which, via $GABA_A$ receptors, inhibits mitral/tufted cells (Shepherd, 1972; Jahr and Nicoll, 1982; Schoppa et al., 1998; Chen et al., 2000) at symmetrical synapses within the EPL. Within both the EPL and GCL, the GCs also receive asymmetrical synapses from a variety of centrifugal afferents, however, which modulate these synaptic interactions. Some of the centrifugal afferents have restricted terminations on the GC bodies and dendritic arbors, suggesting that different sources of centrifugal input could differentially modulate GC synaptic interactions with mitral/tufted cells (Price and Powell, 1970c). In large part, these centrifugal fibers arise from neurons in POC (e.g., PC and AON), which form the bulk of synaptic contacts onto GC somata and proximal dendrites within the GCL (Price and Powell, 1970c). Deep to the EPL, GCs also receive synapses from the collateral branches of mitral and tufted cell axons. During mitral cell excitation, coincident, γ-frequency stimulation of GC proximal dendrites relieves the Mg^{2+} blockade of their NMDA receptors and facilitates recurrent mitral cell inhibition (Halabisky and Strowbridge, 2003), presumably via activation of these axon collaterals. In addition, GCs receive minor inputs from Golgi, Cajal, and Blanes cells (Price and Powell, 1970c) and GCs also receive symmetrical synapses thought to be from the axons of deep SA cells, described later. The SA cells could be a source of GC inhibition as noted earlier. GABAergic centrifugal afferents from the nucleus of the horizontal limb of the diagonal band are another likely source of this inhibition (Kunze et al., 1992).

GCs express high levels of the transient outward potassium current activated by depolarization—the A-type current (Schoppa and Westbrook, 1999). This current plays an important role in dendrodendritic transmission with mitral cells in the sense that it dampens the AMPA-mediated synaptic depolarization of GCs and thereby allows NMDA receptor-mediated depolarization to play a relatively greater role. A-currents may also negatively modulate backpropagation of spikes in GC lateral dendrites (Christie and Westbrook, 2003). As noted earlier, activation of $GABA_B$ receptors on GCs has been reported to reduce GABA release from these cells via inhibition of high voltage-activated Ca^{2+} channels (Isaacson and Vitten, 2003). GCs also receive excitatory glutamatergic projections arising from POC (see ➋ *Section 3.7.2.1*). GCs express type 1 and 2 CRF receptor mRNA (Van Pett et al., 2000; Chen et al., 2005).

3.5.1.2 Other Deep Interneurons The deep MOB layers contain several subtypes of SA cells (Price and Powell, 1970b; Schneider and Macrides, 1978; López-Mascaraque et al., 1990). The somata of these cells mostly occur within the IPL and GCL, and they project multiple dendrites and axons that ramify within the EPL, MCL, and GCL. The deep SA cells include horizontal cells, similar to those seen in the EPL, and Cajal cells (Cajal, 1890; Van Gehuchten and Martin, 1891; Blanes, 1898). The Cajal cells, which are also known as vertical cells of Cajal, have fusiform cell bodies that project dendrites perpendicularly rather than tangentially to the MCL. The apical dendrites extend through the MCL into the EPL, the basal dendrites extend into the GCL, and the axons project out of the GCL into the EPL (Cajal, 1911a, b). By contrast, the Blanes cells have numerous dendrites emerging from all sides of the soma. The axons from these cells can extend considerable distances, but they remain within the GCL (Cajal, 1911a, b). Larger Golgi cells (Golgi, 1875) and Blanes cells (Blanes, 1898) occur in the deeper layers. Golgi cell morphology varies considerably, but the dendrites are generally radial and the axons remain within the GCL (Cajal, 1911a, b). The dendrites of all four of these deep interneuron subtypes are varicose, but the Blanes cell dendrites are also densely spiny (Schneider and Macrides, 1978). For a detailed discussion of early literature concerning these cells, see Halász (1990).

Very little is known about the functions of the deep interneurons, but they are presumed to be inhibitory. Their axon terminals have been thought to be a source of the numerous symmetrical synapses observed on the cell bodies, basal dendrites, and spines of GCs (Price and Powell, 1970d), and some types have been shown to be immunoreactive for GABA (Gracia-Llanes et al., 2003). Recent ultrastructural evidence indicates that the VIP-containing deep short-axon cells form symmetrical synapses onto other deep interneurons containing VIP, calbindin, or NPY, but not onto GCs (Gracia-Llanes et al., 2003). Deep interneurons have also been shown to receive asymmetrical synapses that are presumed to be from mitral/tufted cell axon collaterals (Price and Powell, 1970b). It therefore appears that at least some of these interneurons inhibit other deep interneurons in response to their excitation by mitral/tufted cells. As with the more superficial interneurons, however, the deep interneurons are immunoreactive for a rich variety of substances, most notably neuropeptides (➋ *Table 6-1*) and Ca^{2+} binding proteins (➋ *Table 6-2*), suggesting they could serve a variety of functions. Some GCL interneurons contain NPY (Gall et al., 1986; Ohm et al., 1988). These are probably Golgi or Cajal cells, giving rise to axons that ramify in the more superficial layers such as the EPL and GL. Cajal cells stain for NADPH diaphorase, and they are immunoreactive for neurocalcin, CB, and potentially, NPY (Gall et al., 1986; Briñón et al., 1992; Bastianelli et al., 1993; Alonso et al., 1995; Porteros et al., 1996). Blanes cells also stain for NADPH diaphorase, and immunoreactivity for NPY has been observed in a population of "Blanes-like" cells near the ventricular layer (Scott et al., 1987), although the exact identity of these NADPH diaphorase/NPY cells has yet to be confirmed. Recent work shows that Blanes cells mediate persistent monosynaptic GABAergic inhibition of GCS (Pressler and Strowbridge, 2006).

3.6 Subependymal Layer and Rostral Migratory Stream

The deepest layer in the MOB is the subependymal layer (also called subventricular zone), which is a cell-poor region lining the ventricle (if present) in adults. Most MOB interneurons originate postnatally from progenitor cells within this layer (Hinds, 1968; Altman, 1969; Bayer, 1983). In adults, interneurons (primarily GCs and PG cells) are also continually generated from these progenitor cells and their offspring generated en route migrate to the MOB within the rostral migratory stream (RMS) (Luskin, 1993; Lois and Alvarez-Buylla, 1994; Smith and Luskin, 1998; Wichterle et al., 2001). The RMS progenitors express the ETS transcription factor Er81, and they originate from the lateral ganglionic eminence (Wichterle et al., 2001; Stenman et al., 2003). They continue to migrate and divide in the absence of the bulb (Kirschenbaum et al., 1999), but they only reach their final destinations and mature if the GL and MCL+plexiform layers are present (Liu and Rao, 2003). Recent evidence shows that the extracellular matrix glycoprotein tenascin-R is important to this process (Saghatelyan et al., 2004). Many of the offspring nevertheless die (Kaplan et al., 1985; Brunjes and Armstrong, 1996; Kato et al., 2000), with olfactory stimulation apparently being critical to the survival of new interneurons during development (Frazier-Cierpial and Brunjes, 1989; Corotto et al., 1994; Najbauer and Leon, 1995; Fiske and Brunjes, 2001) and of mature, presumably synaptically connected, interneurons in adults (Petreanu and Alvarez-Buylla, 2002; Hack et al., 2005). Two subsets of new interneurons have been identified, which express either TH, which is required for DA synthesis, or Ca^{2+}-calmodulin-dependent protein kinase IV during migration (Baker et al., 2001). The transcription factors Pax6 and Olig2 play important roles in determining the dopaminergic phenotype of a periglomerular subset of these interneurons (Hack et al., 2005).

Both the progenitors and new interneurons express functional $GABA_A$ receptors. Electrophysiological studies indicate that the new interneurons subsequently express functional AMPA receptors, then NMDA receptors, before they exhibit spiking activity, responses to ON stimulation, and spontaneous glutamatergic EPSCs (Belluzzi et al., 2003; Carleton et al., 2003). These electrophysiological studies provide evidence that the new interneurons become functionally integrated into the MOB circuitry. Neural cell adhesion protein-deficient mice, which exhibit defective migration of new interneurons into the GCL resulting in reduced MOB size, have been shown to exhibit impaired odor discrimination (Gheusi et al., 2000). Although threshold detection and short-term odor memory were normal in these mice, in normal mice both olfactory memory and survival of the new interneurons were improved following rearing in an odor-enriched environment (Rochefort et al., 2002). Thus, the neurons of the subependymal layer and RMS appear to be important both to MOB development and to certain olfactory functions during adulthood.

3.7 Outputs and Inputs of the MOB

3.7.1 Outputs of MOB

The output neurons of the MOB are mitral and tufted cells. These cells project to a number of structures collectively designated as the POC (de Olmos et al., 1978). The features of POC, including input from MOB, are described later (➋ *Figure 6-7*).

Figure 6-7
Major connections of the main olfactory system. Axons of MOB mitral/tufted cells (circles in the EPL and MCL, respectively) project as the LOT to synapse in a number of structures collectively referred to as primary olfactory cortex (POC). Centrifugal inputs to MOB include feedback projections from POC as well as inputs from subcortical forebrain and brainstem neuromodulatory cell groups. Abbreviations: AON, anterior olfactory nucleus; DP, dorsal peduncular cortex; Ent, entorhinal cortex; IG-AHC, indusium griseum-anterior hippocampal continuation; LC, locus coeruleus; NDB, nucleus of the diagonal band; PeCo, periamygdaloid cortex; PC, piriform cortex; RN, raphe nuclei (dorsal and median raphe); TT, taenia tecta; Tu, olfactory tubercle

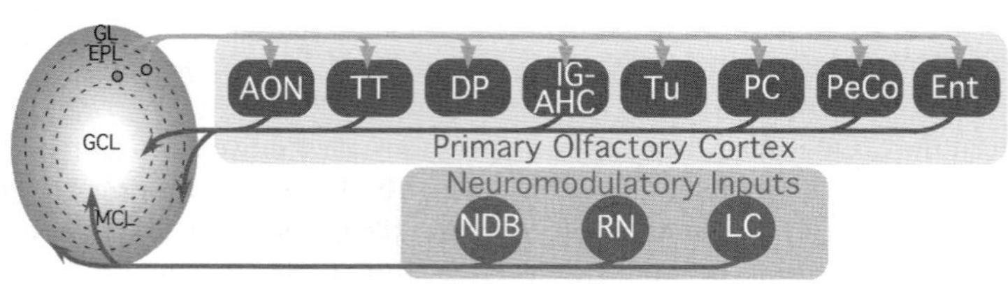

3.7.2 Extrinsic Inputs to MOB

Extrinsic afferent input to MOB, also referred to as centrifugal fibers, can be subdivided into two classes: (1) inputs arising from olfactory-related structures, in particular those arising from POC, and (2) inputs arising from non-olfactory, so-called neuromodulatory transmitter systems including ACh, NE, and 5-HT. The modulatory transmitter systems are considerable and are described in ➒ *Section 3.9.*

3.7.2.1 Neurochemistry of POC inputs to MOB Before departing to the MOB, it is germane to consider the neurochemistry of the input to the MOB from POC. As described later, feedback projections to the MOB from POC arise predominantly from glutamatergic, pyramidal neurons in layer II and III of PC, as well as other POC structures. These projections massively target GCs, where they form asymmetrical synapses on the cell bodies, basal dendrites, and spines of GCs (Price and Powell, 1970a). Activation of these feedback projections produces a negative field-potential recorded in the GCL (Walsh, 1959; Nakashima et al., 1978), as expected, if excitatory currents are flowing into GCs. Similar stimulation elicits IPSPs in mitral cells (Yamamoto et al., 1963; Nicoll, 1971; Mori and Takagi, 1978), due to GC excitation and subsequent GABA release (Halász and Shepherd, 1983). The transmitter of these feedback projections to GCs is glutamate. Activation of POC input to MOB excites GCs as measured by voltage-sensitive dye and field-potential recordings in vitro. This excitation is mediated by glutamate acting at both AMPA and NMDA receptors (Laaris and Ennis, 2002). A major function of these projections is to provide an inhibitory regulation of the firing rate and excitability state of mitral/tufted cells. Activation of these inputs modifies (inhibits) odor responses in MOB (Kerr and Hagbarth, 1955; von Baumgarten et al., 1962).

3.8 Amino Acid Receptor and Gap Junction Distribution in MOB

The following sections review the distributions of glutamate and GABA receptors and gap junction proteins in the MOB. In some cases, additional details about the neurophysiology and functional significance of the receptors are provided in the layer-specific sections earlier.

3.8.1 Ionotropic Glutamate Receptors

The three classes of ionotropic glutamate receptors- NMDA, AMPA, and kainate- are distributed extensively throughout the MOB. NMDA receptors are found in every layer of MOB except the IPL (Watanabe et al., 1993; Monyer et al., 1994; Petralia et al., 1994b). AMPA receptors are found in all layers except the ONL and subependymal layer (Molnar et al., 1993; Martin et al., 1993; Petralia and Wenthold, 1992; van den Pol, 1995). Kainate receptors are found in the MCL, EPL, and IPL (Monaghan and Cotman, 1982; Gall et al., 1990; Miller et al., 1990; Wisden and Seeburg, 1993).

3.8.1.1 Distribution in the GL Immunoreactivity for the NR1 NMDA receptor subunit is present at synapses between ON terminals and profiles that mostly appear to be mitral and tufted cell dendrites (Giustetto et al., 1997). JG neurons, including PG cells, also express NMDA NR1 subunit, which is also present at the dendrodendritic synapses with mitral/tufted cells in the glomeruli, as well as at synapses from centrifugal fibers and/or tufted cell collaterals in the periglomerular regions (Giustetto et al., 1997; Montague and Greer, 1999). Immunocytochemical studies show that AMPA receptors are also localized to processes and cell bodies of mitral, tufted, and PG cells (Petralia and Wenthold, 1992; Molnar et al., 1993). JG neurons, including PG cells, express GluR1, GluR2/3, and GluR4 AMPA receptor subunits (Giustetto et al., 1997; Montague and Greer, 1999); GluR4 staining is also associated with glia in the GL. Like NMDA receptors, AMPA receptors are present postsynaptically at synapses formed by ON terminals and at mitral/tufted-to-PG cell dendrodendritic synapses. Thus, NMDA and AMPA subunits are localized to ON-to-mitral/tufted cell synapses and to dendrodendritic synapses in the glomeruli (Giustetto et al., 1997).

3.8.1.2 Distribution in the Deeper Layers In the deeper layers, both mitral and tufted cells express NR1, GluR1, GluR2/3, and GluR4 AMPA receptor subunits, and kainate GluR5/6/7receptor subunits, as indicated by immunohistochemical staining (Giustetto et al., 1997; Montague and Greer, 1999). Additionally, the interneurons of the EPL express GluR1, as do GCs (Giustetto et al., 1997; Montague and Greer, 1999; Hamilton and Coppola, 2003). GCs also express GluR2/3 subunits, but they are weakly immunoreactive for GluR1 and 4 receptor subunits (Montague and Greer, 1999). At the mitral/tufted-to-GC synapses, GluR2/3 and/or GluR1 AMPA receptor subunits are colocalized with NR1 NMDA receptor subunits. The GluR2 and 4 subunits occur in equal proportion as both flip and flop splice variants, whereas GluR1 and 3 subunits occur mostly as flip variants (Horning et al., 2004). In primary cell cultures, AMPA receptors of presumptive interneurons (most likely GCs), desensitize more rapidly and completely than AMPA receptors of presumptive mitral/tufted cells. This could be due to the occurrence of a higher proportion of flip subunits (Blakemore and Trombley, 2003). In dissociated preparations, AMPA receptors of presumptive GCs have also been reported to exhibit limited Ca^{2+} permeability, suggesting that GC AMPA receptors must include GluR2 subunits, which regulate Ca^{2+} permeability (Jardemark et al., 1997). The AMPA receptors of mitral cells and JG cells appear to exhibit Ca^{2+} permeability, however (Blakemore and Trombley, 2005; Ma and Lowe, 2005).

3.8.2 Metabotropic Glutamate Receptors

Neuroanatomical studies demonstrate that the MOB expresses high levels of mGluRs, suggesting that these receptors play important roles in olfactory processing. The eight mGluRs identified to date are subdivided into three groups based on sequence homology, signal transduction mechanisms, and pharmacology (Conn and Pin, 1997): Group I mGluRs (mGluR1, mGluR5), Group II mGluRs (mGluR2, mGluR3), and Group III mGluRs (mGluR4, mGluR6-8) (➜ *Figure 6-8*).

3.8.2.1 Distribution in the GL JG cells express several types of mGluRs. Deciphering mGluR expression in the GL from the literature is difficult because mGluR expression has not been systematically examined in the different JG cell types. ET cells appear to have the same mGluR makeup as the deeper mitral/tufted cells; i.e., they express high levels of mGluR1 and lower levels of mGluR7 and mGluR8 (Ohishi et al., 1995; Kinzie et al., 1995; Saugstad et al., 1997). Other, as of yet unidentified JG cells have been reported to express low-to-moderate levels of Group I (mGluR5), Group II (mGluR2), and Group III (mGluR4/7/8) receptors (Ohishi et al., 1993a, 1998; Romano et al., 1995; Saugstad et al., 1997).

3.8.2.2 Distribution in Mitral/Tufted Cells Mitral/tufted cells express high levels of mGluR1 (Masu et al., 1991; Martin et al., 1992; Shigemoto et al., 1992; van den Pol, 1995; Sahara et al., 2001). Electron microscopy studies demonstrated that mGluR1 is present on the somata and apical and lateral dendrites of mitral/tufted cells (van den Pol, 1995). This expression pattern suggests that mGluR1 could mediate mitral/tufted cell responses to glutamatergic inputs from ON terminals, and/or could function as an auto- or heteroreceptor for glutamate released from apical or lateral dendrites of mitral/tufted cells. Mitral/tufted cells also express mRNA for mGluR7 and mGluR8 (Kinzie et al., 1995; Saugstad et al., 1997), although immunocytochemical studies indicate that these receptors are present on mitral/tufted cell axon terminals in the GCL and PC (Kinoshita et al., 1998; Wada et al., 1998). Mitral/tufted cells do not express Group II receptors (mGluR2/3), which contrast markedly with the strong expression of mGluR2 in mitral cells in the Accessory olfactory Bulb (AOB) (Ohishi et al., 1993b, 1998).

3.8.2.3 Distribution in the GCL GCs express the highest levels of mGluR5 (Group I) in the brain (Romano et al., 1995). EM studies have shown that mGluR5 is localized to portions of GC dendrites in the EPL apposed to presynaptic glutamatergic synapses from mitral/tufted cell lateral dendrites (van den Pol, 1995). This suggests that mGluR5 may mediate, at least in part, responses of GCs to glutamatergic inputs from mitral/tufted cells. GCs also express low-to-moderate levels of mGluR2 (Group II) (Ohishi et al., 1993a, 1998), and low levels of mGluR4 and mGluR7 (Group III) (Kinzie et al., 1995; Ohishi et al., 1995; Saugstad

Figure 6-8
The distribution of Group I, Group II, and Group III mGluRs in the MOB. See *Section 3.8.2* **for details**

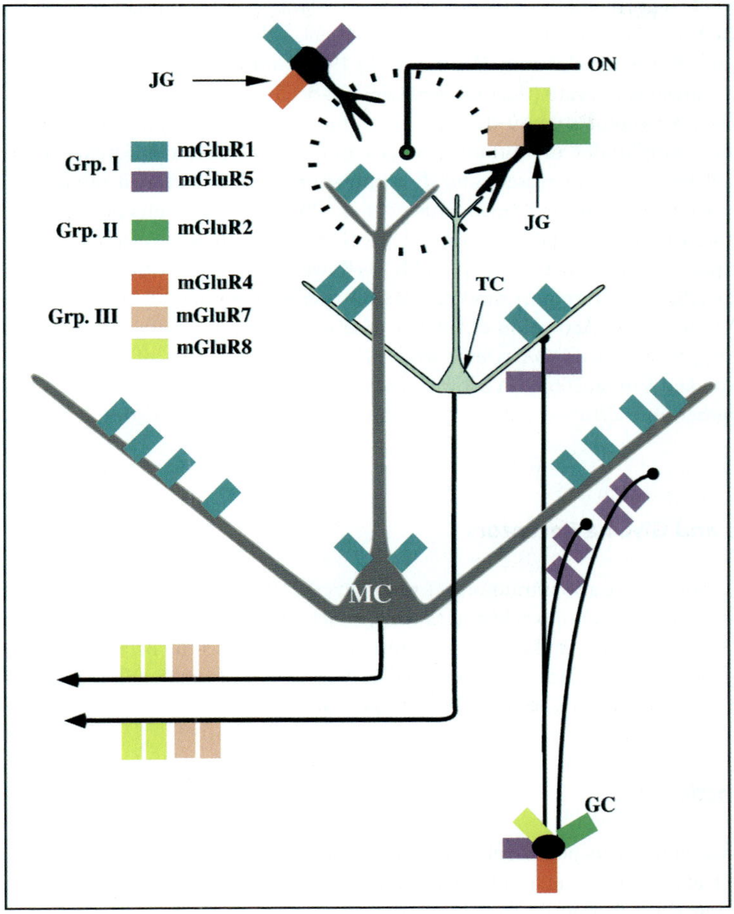

et al., 1997; Wada et al., 1998). The precise cellular localization of these Group II/III mGluRs on GCs is not known. Recent experiments have shown that activation of mGluR5 directly and potently activates GCs. This action is mediated by an apparent inward current that involves, at least in part, a closure of K^+ channels (Heinbockel and Ennis, 2003).

3.8.3 GABA Receptors

$GABA_A$ receptors are present in every layer of MOB except for the IPL (Palacios et al., 1981b; Bowery et al., 1987; Richards et al., 1987; Chu et al., 1990; Zhang et al., 1991; Persohn et al., 1992; Fritschy et al., 1992; Laurie et al., 1992). PG cells moderately express mRNAs for the $\beta 2$ and $\beta 3$ $GABA_A$ receptor subunits (Laurie et al., 1992), and several ET cell subpopulations occur, which are differentially immunoreactive for the $\alpha 1$, $\alpha 3$, or both $\alpha 1$ and $\alpha 3$ receptor subunits (Panzanelli et al., 2005). Putative SA cells of the GCL express $\alpha 1$ and $\beta 2$ receptor subunits (Laurie et al., 1992). At the granule-to-mitral/tufted cell synapses of the EPL, $\alpha 1$, $\beta 2$, $\beta 3$, and $\gamma 2$ $GABA_A$ receptor subunits occur (Fritschy et al., 1992; Giustetto et al., 1998; Sassoè-Pognetto et al., 2000). Mitral/tufted cells also express β_1 $GABA_A$ receptor subunits (Laurie et al., 1992). In addition, some mitral cells, middle tufted cells, and deep tufted cells express the $\alpha3$ $GABA_A$ receptor subunit, and many of the mitral cells coexpress CR (Panzanelli et al., 2005). The $\alpha1$ $GABA_A$ receptor subunits are perisynaptic, which suggests that GABA overflow from nearby synapses might regulate glutamate release from mitral/tufted cell dendrites (Panzanelli et al., 2004). GCs also express $GABA_A$ receptors, including the $\alpha2$, $\alpha4$, $\alpha5$, $\beta3$, and δ receptor subunits (Laurie et al., 1992) and they exhibit inhibition (Mori and Kishi, 1982), which in a lower vertebrate (tiger salamander) has been shown to be GABAergic (Wellis and Kauer, 1994). In $GABA_A$ $\beta3$ receptor subunit-deficient mice, functional expression of GC $GABA_A$ receptors was almost eliminated, amplitudes of mitral/tufted cell miniature IPSPs, theta-frequency oscillations, and γ-frequency oscillations were increased, and discrimination of closely related mixtures of alcohols after training was poor relative to normal mice (Nusser et al., 2001). These results provide functional evidence for the importance of both GC and mitral/tufted cell $GABA_A$ receptors in olfaction.

$GABA_B$ receptors are restricted to the GL in MOB (Bowery et al., 1987; Chu et al., 1990). However, there are reports that $GABA_B$ autoreceptors modulate GABA release at granule-to-mitral cell reciprocal synapses (Isaacson and Vitten, 2003). Activation of $GABA_B$ receptors has been observed to reduce dendrodendritic inhibition via suppression of GABA release from GCs due to reduction of high voltage-activated Ca^{2+} currents (Isaacson and Vitten, 2003). In a lower vertebrate (frog), $GABA_B$ and dopamine D_2 receptors both appear to strengthen mitral/tufted cell odor signaling by reducing spontaneous activity (Duchamp-Viret et al., 2000).

3.8.4 Glycine and Glycine Receptors

Cells in the EPL and GCs are also immunoreactive for glycine (Pourcho et al., 1992), and the MOB contains significant levels of both this amino acid and taurine, which inhibit cultured olfactory bulb cells (Trombley and Shepherd, 1994; Trombley et al., 1999) and mitral/tufted cells in situ (Belluzzi et al., 2004). The MOB exhibits strong labeling for glycine receptors (van den Pol and Gorcs, 1988), and the β receptor subunit has been localized to mitral/tufted cell synapses (Malioso et al., 1991).

3.8.5 Gap Junctions

Gap junctions are thought to play a role in synchronization of MOB activity (Schoppa and Westbrook, 2002; Christie et al., 2005; Hayar et al., 2005). Gap junction inhibitors block γ-frequency (20–70) Hz oscillations generated by chemically and electrically coupled networks of mitral and GCs (Friedman and Strowbridge, 2003), suggesting a role for both types of synapses. ➲ *Table 6-3* summarizes the distribution of gap junctions and the gap junction proteins, connexins, in the MOB. Gap junctions have been observed on mitral, tufted, and GC bodies (Reyher et al., 1991; Paternostro et al., 1995; Miragall et al., 1996; Kosaka and Kosaka, 2003). In the GL, mitral/tufted cell dendrites have also been observed to form gap junctions with each other, with PG cells, and with processes of unidentified cells (Kosaka and Kosaka, 2003, 2004, 2005; Kosaka et al., 2005; Rash et al., 2005). The junctions tend to occur as mixed synapses with glutamatergic synapses. Some of the unidentified processes involved in these synapses are not immunoreactive for either DA or GABA and could be mitral/tufted cell dendrites (Kosaka and Kosaka, 2003). In support of this, mitral/tufted cell dendrites of a lower vertebrate (tiger salamander) have been observed to form asymmetrical synapses with each other in the GL (Allen and Hamilton, 2000). In addition to their GL synapses, mammalian mitral/tufted cell dendrites form some gap junctions with dendrites of GCs (Landis et al., 1974; Kosaka et al., 2005) and other interneurons (Kosaka and Kosaka, 2003) in the EPL.

The predominant gap junction protein of MOB neurons appears to be connexin36 (Condorelli et al., 1998, 2000; Belluardo et al., 2000; Rash et al., 2000, 2005), although mitral/tufted cells and unidentified interneurons in the EPL and GCL have been reported to express other connexin mRNAs (connexin43, Miragall et al., 1996; connexin45, Zhang and Restrepo, 2002) or proteins (connexin43, Theis et al., 2003) that are mainly expressed by glial cells. It is noteworthy that expression of connexin36 mRNA is particularly high within (but not uniform among) the glomeruli (Teubner et al., 2000; Zhang and Restrepo, 2003), and that expression of this connexin remains high into adulthood (Hormuzdi et al., 2001). This is in contrast to the hippocampus, where connexin36 mRNA expression peaks during early postnatal life and subsequently declines (Rozental et al., 2000). In addition, γ-frequency electrical signaling is disrupted in connexin36 knockout mice (Hormuzdi et al., 2001), providing evidence that this connexin is involved in oscillatory signaling. Gap junction channels formed by connexin36 exhibit unitary conductances of $\sim$15 pS, weak transjunctional voltage dependence, and are also permeable to cAMP and IP_3 (Rozental et al., 2000; Teubner et al., 2000). Thus, gap junctions connecting MOB neurons could participate in both fast oscillatory signaling and slower, second messenger-mediated signaling, even in adults.

3.9 Neuromodulatory Inputs to MOB

MOB and most components of POC receive extrinsic inputs from cholinergic, noradrenergic, serotonergic, and dopaminergic cell groups in the brainstem and basal forebrain.

3.9.1 Cholinergic Inputs to MOB

3.9.1.1 Cholinergic Innervation Pattern and Receptor Distribution In the mouse, about 3.5% of all neurons that project to the bulb originate in the nucleus of the horizontal limb of the diagonal band (NDB); far fewer originate in the vertical limb of DB (Carson, 1984; Shipley and Adamek, 1984). At least two distinct transmitter-specific populations of NDB neurons project to the MOB (Zaborszky et al., 1986). About 20% of the NDB neurons that project to the bulb are cholinergic; most of these cells are concentrated in the rostromedial portion of the horizontal limb of NDB. Many NDB-MOB projection neurons are GABAergic and occur mainly in the caudo-lateral aspect of NDB (Zaborszky et al., 1986). AChE is concentrated in the IPL, GCL, inner third of the EPL, and the GL. Some glomeruli are more intensely stained for AChE and correspond to regions of leutenizing hormone releasing hormone (LHRH) innervation (Zheng et al., 1988), and may include the modified glomerular complex as defined by Greer and colleagues (Teicher et al., 1980; Greer et al., 1982). There are also AChE-positive neurons in the bulb (Nickell and Shipley, 1988b). Choline acetyltransferase (ChAT), the biosynthetic enzyme for ACh, is located in axons distributed in the same layers as described for AChE. As shown in ➋ *Figure 6-9*, cholinergic fibers are distributed throughout the MOB, but are especially heavy in the GL and IPL. The GABAergic projection from NDB (Zaborszky et al., 1986) is more difficult to characterize than the cholinergic projection, because the intrinsic GABAergic PG cells and GCs provide intrinsic GABAergic innervation.

Cholinergic receptor localization in the MOB is in close agreement with terminal staining from the NDB (➋ *Table 6-4*). Muscarinic receptors are found in all layers while nicotinic receptors are found in all layers except the IPL (see ➋ *Table 6-4*). The EPL exhibits intense staining for muscarinic-1 (M_1), M_3, and M_4 receptors (Rotter et al., 1979; Spencer et al., 1986; Buckley et al., 1988; Fonseca et al., 1991b). M_2 receptors are immunocytochemically localized to PG cells in the GL and tufted cells in the EPL (Fonseca et al., 1991a). M_2 receptors are also present in the IPL and GCL. Nicotinic receptors show a different regional distribution throughout the MOB and are concentrated in the GL and EPL (Hunt and Schmidt, 1978; Sahin et al., 1992).

3.9.1.2 Cholinergic Actions in MOB Only limited information is available about cholinergic actions in MOB. Electrical activation of NDB has been reported to depress (Nickell and Shipley, 1988a) or increase (Kunze et al., 1991, 1992) mitral cell activity indirectly via primary effects on GABAergic GCs. NDB stimulation also reduced the field potential in the MOB caused by stimulation of the anterior commissure (Nickell and Shipley, 1993), an effect mediated by presynaptic inhibition of anterior commissure terminals via muscarinic receptors. One interpretation of these results is that cholinergic input to MOB may function to modulate interhemispheric transmission of olfactory information. In this regard, it is noteworthy that anterior commissural fibers are required for access and recall of olfactory memories between the two hemispheres. Infusion of ACh into MOB was reported to reduce paired-pulse depression of lateral olfactory tract (LOT)-evoked field potentials recorded in the GCL. This effect was attributed to

Figure 6-9
Neuromodulatory transmitter inputs to MOB. Darkfield photomicrographs showing the distribution of cholinergic (a), noradrenergic (b), and serotonergic (c) fibers revealed respectively with immunohistochemistry for choline acetyltransferase (ChAT), dopamine-β-hydroxylase (DBH), and serotonin (5-HT). Reprinted from Handbook of Chem. Neuroanat. Integrated Sys. CNS, Vol. 12, Part III, Chapter III, The Olfactory System, M. Shipley et al., pp. 469–573, 1996, with permission from Elsevier, Ltd

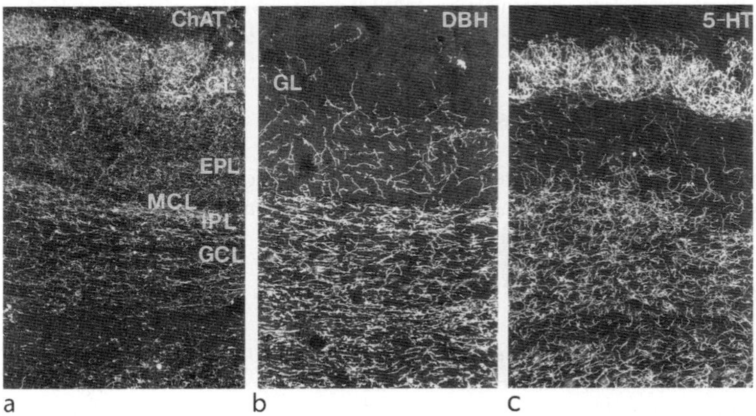

muscarinic receptor-mediated inhibition of GABA release from GCs (Elaagouby et al., 1991). In slice preparations, nicotinic but not muscarinic receptor agonists directly excited mitral cells, and this effect appeared to be due to an inward current with a reversal potential of –5 to +10 mV (Castillo et al., 1999). In slices, muscarinic receptor agonists inhibited GCs (Castillo et al., 1999), and paradoxically, also appeared to increase GABA release from these cells. The same study reported that in the GL only bipolar PG cells were sensitive to nicotine (Castillo et al., 1999). The morphological identity of these cells is unclear.

3.9.2 Dopaminergic (DA) Input to MOB

There is no known extrinsic DA innervation of the MOB. As noted earlier, however, the MOB contains several hundred thousand intrinsic JG DA neurons that mediate potent presynaptic inhibition of ON terminals via activation of D2 receptors. This presynaptic inhibition as well as DA receptor localization is discussed in *Section 3.2.3.2.*

3.9.3 Noradrenergic (NE) Input to MOB

3.9.3.1 NE Innervation Pattern and Receptor Distribution A significant modulatory input to the MOB is from the pontine nucleus, locus coeruleus (LC). In the rat, all LC neurons contain the neurotransmitter, norepinephrine (NE); LC contains the largest population of NE neurons in the brain. It has been estimated that up to 40% of LC neurons (400–600 of a total of 1,600 LC neurons) project to the bulb in the rat (Shipley et al., 1985). A subset of LC neurons projecting to MOB contain NPY (Bouna et al., 1994). NE fibers terminate in MOB with laminar specificity. LC axons project mainly to the subglomerular layers of the bulb, particularly the IPL and GCL (McLean et al., 1989). The EPL and MCL are moderately innervated, whereas the GL is nearly devoid of NE input.

NE receptors occur in multiple layers of the MOB and they are expressed by multiple cell types, in general consistent with the pattern of NE fiber innervation (*Figure 6-10* and *Table 6-4*). Receptor

Figure 6-10
The distribution of NE receptors in the MOB. See *Section 3.9.3* **for details. CFF, centrifugal fibers**

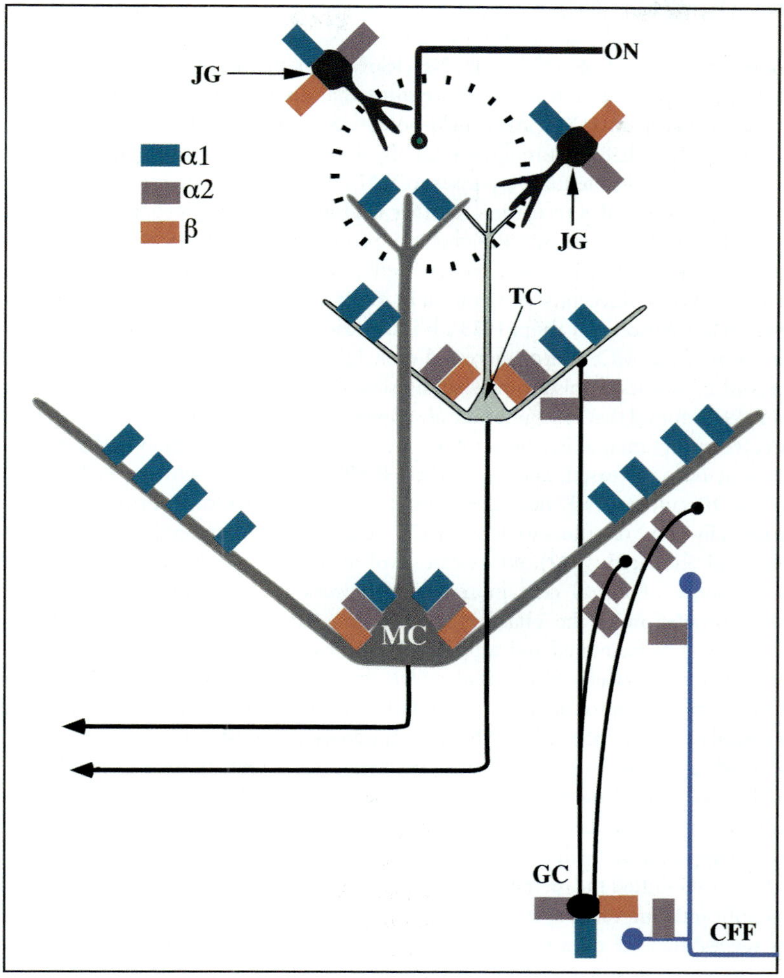

localization studies also suggest that individual MOB cells express multiple NE receptor subtypes. α receptors are abundantly expressed in MOB. Binding sites for $\alpha 1$ receptors are particularly dense in the EPL, and moderate in the MCL and IPL/GCL (Young and Kuhar 1980a; Sargent-Jones et al., 1985); the density of $\alpha 1$ receptors in the EPL was the highest in the brain (Young and Kuhar, 1980a). Consistent with these findings are observations that $\alpha 1$ receptor mRNA is expressed by mitral cells ($\alpha 1_A$, $\alpha 1_B$, $\alpha 1_D$ subtypes) and GCs ($\alpha 1_A$, $\alpha 1_D$ subtypes) (McCune et al., 1993; Pieribone et al., 1994; Domyancic and Morilak, 1997; Day et al., 1997). The MCL, IPL, and GCL express moderate-to-dense levels of $\alpha 2$ receptors (Young and Kuhar, 1979; Nicholas et al., 1993; Rosin et al., 1996; Talley et al., 1996; Winzer-Serhan et al., 1997a, b), and $\alpha 2_C$ receptor mRNA is expressed by both mitral cells and GCs (Nicholas et al., 1993; Rosin et al., 1996; Winzer-Serhan et al., 1997a, b). The high density of $\alpha 2$ receptor binding sites and NE fibers in the IPL/GCL suggests the possibility that these receptors may be present presynaptically on axon terminals of NE fibers themselves or on other centrifugal inputs to the soma and/or proximal dendrites of GCs (*Figure 6-10*). Unidentified JG neurons have been reported to express $\alpha 1$ and $\alpha 2$ receptors (Rosin et al., 1996; Winzer-Serhan et al., 1997a, b). By contrast, the level of β receptors is relatively low in the MOB, and they appear to be localized preferentially to the GL, IPL, and GCL (Palacios and Kuhar, 1982; Booze et al., 1989; Wanaka et al., 1989; Woo and Leon, 1995). JG, mitral, and tufted cells, as well as subsets of GCs express the $\beta 1$ receptor subtype (Yuan et al., 2003).

3.9.3.2 Physiological Actions of NE Although, NE clearly plays significant roles in olfactory function, the effects of NE at the cellular and network levels are somewhat discrepant. For example, LC stimulation was reported to have no effect on LOT-evoked field-potentials recorded in the GCL (Perez et al., 1987). A subsequent study reported that LC stimulation initially decreased and then subsequently increased paired-pulse depression of GC field-potential responses to LOT stimulation (Okutani et al., 1998). These effects were attributed to activation of β receptors. Another field-potential study reported that NE infusion into MOB, acting at $\alpha 1$ receptors, increased the depolarization of GC dendrites elicited by LOT stimulation. Mitral cell responses to antidromic shocks were not affected, suggesting that NE excites GC (Mouly et al., 1995). In neonatal animals, β receptor stimulation in MOB decreased LOT-evoked, paired-pulse inhibition of GC field potentials (Wilson and Leon, 1988). It is unclear if this was mediated by presynaptic inhibition of transmitter release from MCs and/or increased excitability of GCs.

Cellular recording studies are also somewhat discrepant. In the rabbit and cat, Salmoiraghi et al. (1964) and McLennan (1971) found that iontophoretically applied NE inhibited mitral cells. This effect was blocked by the $GABA_A$ receptor antagonist, bicuculline. In the isolated turtle bulb (Jahr and Nicoll, 1982), mitral cell spike activity increased, and GABAergic IPSPS decreased, following bath application of NE. In dissociated MOB cultures, NE decreased mitral cell-to-GC dendrodendritic synaptic transmission acting presynaptically at $\alpha 2$ receptors to decrease Ca^{2+} currents in both granule and mitral cells (Trombley, 1992; Trombley and Shepherd, 1992). NE release, evoked by selective chemical activation of LC in vivo, enhanced the response of mitral cells in response to weak (i.e., perithreshold) but not strong (i.e., suprathreshold) stimulation of the olfactory epithelium (Jiang et al., 1996). Interestingly, NE release from LC axon terminals is facilitated and suppressed by activation of presynaptic nicotinic and muscarinic cholinergic receptors, respectively (El-Etri et al., 1999). In rat MOB slices, application of NE or $\alpha 1$ receptor agonists, but not $\alpha 2$ or β receptor agonists, also selectively increased mitral cell responses to perithreshold intensity ON stimulation (Ciombor et al., 1999). Noradrenergic agonists had no effect on ON-evoked field-potentials recorded in the GL, or on ON-evoked postsynaptic currents in mitral cells (Hayar et al., 2001). This suggests that NE-evoked modulation of ON-evoked mitral cell spiking is mediated by postsynaptic actions on bulb neurons. In voltage clamp recordings, NE or $\alpha 1$ agonists directly evoked an inward current in mitral cells that appeared to be due to closure of K^+ channels. In current clamp recordings from bistable mitral cells, $\alpha 1$ agonists shifted the membrane potential from the downstate (–52 mV) toward the upstate (–40 mV), and significantly increased spike generation in response to perithreshold ON input. Taken together, these findings suggest that NE release directly alters mitral cell excitability in a manner that could increase their sensitivity to weak ON input, perhaps to improve the detection of weak odorants.

NE inputs to the bulb are critical to olfactory function. Olfactory cues increase the discharge of LC neurons in behaving animals (Aston-Jones and Bloom, 1981) and increase NE levels in the MOB (Chanse and Kopin, 1968; Rosser and Keverne, 1985; Brennan et al., 1990). LC projections to the main and accessory olfactory bulb are pivotal to the formation and/or recall of specific olfactory memories, pheromonal regulation of pregnancy and postpartum maternal behavior (Pissonnier et al., 1985; Kaba and Keverne, 1988; Brennan et al., 1990; Kendrick et al., 1992). NE plays an important role in the so-called Bruce effect in mice: when impregnated female mice are exposed to the odor of a strange male, they abort; if exposed to the odor of the impregnating male they do not abort (Kaba et al., 1989). Systemic administration of adrenergic receptor antagonists or 6-OHDA lesions, selectively destroying only the NE inputs to the MOB, cause the female to abort when presented with the odor of the impregnating male (Rosser and Keverne, 1985; Kaba and Keverne, 1988; Brennan et al., 1990). Finally, NE has been shown to play a critical role in olfactory learning in young animals. In neonatal rats, NE release via tactile stimulation leads to a preference for an odor associatively paired with this stimulation (Sullivan et al., 1989). The conditioned preference is associated with odor-specific metabolic changes in the bulb (Coopersmith and Leon, 1984). Following the conditioning, there is an increased inhibition of mitral cells by the odor (Sullivan et al., 1989). Such conditioning is abolished by eliminating NE input to the bulb or via β receptor antagonists (Sullivan et al., 1989, 1992, 2000; Wilson and Sullivan, 1991; Moriceau and Sullivan, 2004). Recent studies from McLean's laboratory demonstrate that this β receptor-dependent neonatal learning involves activation of intracellular cAMP and CREB pathways (see McLean and Harley, 2004 for review).

3.9.4 Serotonergic (5-HT) Inputs to MOB

3.9.4.1 5-HT Innervation Pattern, Receptor Distribution, and Physiological Actions The midbrain dorsal and median raphe provides strong inputs to the MOB. In the rat, about 1,000 dorsal and 300 median raphe neurons project to the bulb. These neurons are serotonergic and they do not contain TH (McLean and Shipley, 1987a, b) or substance P (Zaborszky et al., 1986). As shown in ➋ *Figure 6-9*, 5-HT fibers are present in all layers of MOB, but with varying densities. Input to the GL is especially dense, while the EPL contains very low density. [^{3}H]5-HT-labeled boutons have been observed in close proximity to probable SA cells in the EPL (Halász et al., 1978). The MCL, IPL, and GCL have a fairly heavy and uniform innervation, but not as dense as that of the GL. Thick serotonergic fibers preferentially innervate the glomeruli of MOB, whereas thinner serotonergic axons preferentially innervate inframitral layers (McLean and Shipley, 1987a, b). In neocortex, thick axons arise from the median raphe and thin axons arise from dorsal raphe (McLean and Shipley, 1987a, b) and the same segregation occurs in MOB.

In agreement with the 5-HT fiber distribution, 5-HT receptors are localized in most layers of the MOB (➋ *Table 6-4*). The 5-HT$_{1A}$ subtype is present on processes in the EPL, MCL, and minimally in the GCL; mitral cells and GCs appear to express the 5-HT$_{1A}$ receptor (Pompeiano et al., 1992). 5-HT$_{1C}$ is present in the GCL. It is possible that many of the cells possessing 5-HT receptors are glia (Whitaker-Azmitia et al., 1993). The 5-HT$_{2}$ receptors are present in the GL, EPL, and MCL. The 5-HT$_{2A}$ receptor subtype is found in the EPL and in mitral and tufted cells (McLean et al., 1993, 1995; McLean, 1994; Pompeiano et al., 1994; Hamada et al., 1998; Cornea-Hebert et al., 1999). In mitral and tufted cells, 5-HT$_{2A}$ receptors colocalize with the β1 receptor (Yuan et al., 2003). The 5-HT$_{2A}$ receptors are also present in unidentified JG cells in the GL (Hamada et al., 1998; Cornea-Hebert et al., 1999). The 5-HT$_{2C}$ receptor subtype is expressed at fairly high levels by GCs and at relatively low levels by unidentified JG neurons (Clemett et al., 2000). JG cells have also been reported to express 5-HT$_{3}$ receptors (Tecott et al., 1993).

At present, there is only limited data about the actions of serotonin in the MOB. In the GL, it was recently reported that serotonin depolarized 34% of JG cells in vitro via activation of 5HT$_{2C}$ receptors (Hardy et al., 2005). The serotonin-induced depolarization was due to activation of a nonselective cation current with a reversal potential of –44 mV. The heterogeneous electrophysiological properties of 5HT-responsive JG cells suggested that several types of JG cells could be targeted by 5HT centrifugal fibers. A subset of mitral cells was also depolarized by serotonin acting at 5HT$_{2A}$ receptors. In contrast with these results, another subset of mitral cells was hyperpolarized by serotonin, an action that was indirectly mediated by GCs as it was blocked by GABA$_{A}$ receptor antagonists (Hardy et al., 2005). This effect of serotonin was also thought to be mediated by 5HT$_{2A}$ receptors. Behavioral studies indicated that lesion of serotonergic fibers reversed conditioned olfactory learning (Morizumi et al., 1994), and also induced glomerular atrophy. Behavioral work on neonates by McLean et al. (1996) showed that serotonin depletion or 5-HT$_{2}$ receptor antagonism compromised olfactory learning and that serotonin release or 5-HT$_{2}$ receptor activation promoted odor conditioning. 5-HT release appeared to facilitate β receptor-mediated, NE-induced olfactory learning by facilitating cAMP mechanisms (see McLean and Harley, 2004 for review).

4 Olfactory Input Regulates Neurochemistry of Specific MOB Neurons

ORNs turn over throughout life, most MOB interneurons are added after birth, and new interneurons are continually added during adulthood. The MOB is consequently a highly plastic neural system that is readily affected by both sensory experience and deprivation. A variety of studies showed that occlusion of one naris during early postnatal life markedly affected development of the ipsilateral MOB, resulting in numerous cellular changes and ~25% smaller MOB (Brunjes, 1994). The studies also showed that naris occlusion resulted in reduced DA levels (Baker et al., 1983; Brunjes et al., 1985; Wilson and Wood, 1992). During development, the maturing dopaminergic JG neurons only exhibited TH immunoreactivity after they reached the GL, and they failed to do so after naris occlusion (McLean and Shipley, 1988; Baker, 1990; Baker and Farbman, 1993; Baker et al., 1993). In adults, both naris occlusion and destruction of the mature olfactory sensory neurons by nasal irrigation with $ZnSO_4$ or detergent, or by severing the ON, reduced transneuronal regulation of DA levels and TH expression in rats (Nadi et al., 1981; Kawano and Margolis, 1982; Baker et al., 1983, 1993; Erlich et al., 1990; Cho et al., 1996; Couper Leo et al., 2000), hamsters (Kream et al., 1984) and dogs (Nadi et al., 1981). In concert with the reduced TH expression, dopamine D2 receptor density was increased in the ONL and GL (Guthrie et al., 1991). TH expression was also downregulated in mice homozygous for a null mutation in the olfactory cyclic nucleotide gated channel subunit-1 (OCNC1), which rendered them functionally anosmic (Baker et al., 1999). The loss of TH expression following naris occlusion and deafferentation was not due to cell death, because JG neurons could be detected with antibodies to other DA enzymes (Baker et al., 1984). Moreover, both GAD and GABA, which are coexpressed by many dopaminergic JG neurons, were still detected (Kosaka et al., 1987a, b, c; Baker et al., 1988, 1993; Stone et al., 1990). Depolarizing stimuli could induce TH expression in vitro and in vivo (McMillian et al., 1994; Philpot et al., 1998), and the in vitro studies showed that the induction required Ca^{2+} influx into the bulb neurons (Cigola et al., 1998). In an MOB neuron/olfactory epithelium coculture, TH expression was abolished by NMDA receptor blockade (Puche and Shipley, 1999), suggesting that glutamate released by ORNs stimulates the dopaminergic JG neurons, resulting in Ca^{2+} influx and regulation of TH expression. Thus, both the developmental induction and maintenance of the DA phenotype appear to be transneuronally regulated and to depend on the presence and normal functioning of the ON. Although the intracellular mechanisms remain to be elucidated, the distributions of c-fos immediate early gene mRNA and Fos protein immunoreactivity were found partially to overlap with TH distribution and were downregulated following naris occlusion (Guthrie and Gall, 1995a, b; Jin et al., 1996). Fos-B was implicated in the regulation of TH expression through interactions with the AP-1 motif (Liu et al., 1999).

The expression of other neurotransmitters and receptors is also influenced by the afferent sensory input to the MOB. In the hamster, many JG neurons express substance P, and this peptide was found to be downregulated following chemical deafferentation (Kream et al., 1984). By contrast in rats, glomerular mGluR1 immunoreactivity was increased. Initially, the mGluR1a mRNA level increased, but it then declined (like TH expression), presumably due to product inhibition (Ferraris et al., 1997; Casabona et al., 1998). As with chemical deafferentation, neonatal naris occlusion of rats and mice had both negative and positive effects. It reduced the density of β1- and β2-adrenergic receptors as determined by quantitative autoradiography (Woo and Leon, 1995), although NE levels were not reduced following either naris occlusion (Brunjes et al., 1985) or chemical deafferentation (Nadi et al., 1981). Similarly, although levels of GluR1 were not reduced following chemical deafferentation (Ferraris et al., 1997), immunoreactivity of cell bodies in the EPL for both GluR1 (Hamilton and Coppola, 2003) and for PV were greatly reduced, as was immunoreactivity of nondopaminergic JG neurons for calbindin (Philpot et al., 1997). Focal surgical deafferentation had similar effects on PV and calbindin immunoreactivity (Couper Leo et al., 2000). CR immunoreactivity of other interneurons was not affected, however. In addition, although the number of synapses between mitral cell bodies and GC dendrites were reduced following naris occlusion, mitral/tufted cell inhibition appeared to be increased, possibly via enhanced NMDA receptor-mediated excitation of the GCs by the mitral/tufted cells at their reciprocal synapses (Wilson, 1995). Expression of voltage-sensitive Na^+ channel αII and β1 subunits was also downregulated in tufted cells, but not in mitral cells (Sashihara et al., 1997). The afferent sensory input to MOB therefore appears to influence the expression of a variety of neurotransmitters, receptors, and ion channel proteins, with different effects observed on different cell types.

Neurotrophic factors, receptors, and kinases were also shown to be affected by naris occlusion. These include insulin receptor (IR) kinase, brain-derived neurotrophic factor (BDNF), neurotrophin receptors TrkB and TrkC, and mitogen-activated protein kinase/extracellular signal-related kinase (MAPK/ERK)

(Mackay-Sim and Chuah, 2000). Expression of IR kinase was reduced, as was the insulin-induced suppression of current through the Kv1.3 channel, which is typically tyrosine phosphorylated in the N and C termini by IR kinase (Fadool et al., 2000). The Kv1.3 channel carries most of the outward current of cultured MOB neurons (Fadool and Levitan, 1998). Moreover, BDNF levels initially increased in vivo and subsequently decreased, without concomitant effects on TrkB levels (McLean et al., 2001; Tucker and Fadool, 2002). In cultured MOB neurons isolated from the occluded side, however, the BDNF-induced phosphorylation of Kv1.3 was increased, indicating that tyrosine kinase activity was increased. Activation of ERK pathway was also downregulated (Mirich et al., 2004). Levels of other kinases did not appear to be altered, however, at least as examined by immunocytochemical staining (Liu, 2000). These results suggest that altered levels of neurotrophic factors and altered kinase expression or activity may contribute to the developmental changes observed in MOB neurons following naris occlusion.

5 Primary Olfactory Cortex: Piriform Cortex

The MOB projects to a collection of structures referred to as POC (de Olmos et al., 1978). These structures may be divided into three groups: (1) the AON, (2) medial olfactory cortex (indusium griseum, anterior hippocampal continuation, taenia tecta, and the olfactory tubercle), and (3) lateral olfactory cortex comprising, from rostral to caudal, PC and the periamygdaloid, transitional and entorhinal cortices (*Figure 6-7*). At least from an olfactory perspective PC is perhaps the most extensively studied and well understood member of POC. A thorough presentation of the cytoarchitecture, connections, and neurochemistry of each of the components of POC is beyond the scope of the present report, and readers are directed to more complete reviews of these subjects (Shipley et al., 1996, 2004). Here, we focus on PC (*Figures 6-3* and *6-7*) as an example of POC.

5.1 Organization and Architecture

PC is the largest component of the lateral olfactory cortical areas. It extends rostrocaudally and is located in the cortical mantle ventral to the rhinal sulcus (*Figure 6-3*). Haberly and Price (1978a) divided PC into three layers that were further subdivided on the basis of cytoarchitecture and afferent connections. Layer I, the superficial plexiform layer, is divided into Ia and Ib, which receive different afferents: layer Ia, from the ipsilateral MOB, and layer Ib, association fibers from AON and from other parts of POC. Layer II, the superficial compact cell layer, is divided into two zones; the more superficial zone has a lower cell density and the deeper zone has a higher cell density. Layer III is the thickest cell layer. The endopiriform nucleus lies deep to the PC (Loo, 1931; Haberly and Price, 1978a). Behan and Haberly (1999) reported that this nucleus, while similar in many regards to PC proper, has unique intrinsic and extrinsic connections that differ from PC. Additionally, electrophysiological properties of endopiriform neurons, such as certain K^+ currents (e.g., A-current), differ from those of PC neurons (Banks et al., 1996). As reviewed later, there is considerable rostrocaudal heterogeneity in the structure and connections of anterior versus posterior PC (Haberly and Price, 1978a; Luskin and Price, 1983). There is evidence for distinct subdivisions within the anterior PC based on cytoarchitecture, connections, and neurochemistry (Ekstrand et al., 2001b). Candidate transmitters and neurotransmitter receptors in PC are summarized in *Tables 6-5* and *6-6*.

5.2 PC Cell Types

Neurons in PC can be divided into two main classes: (1) principal or output neurons—the pyramidal cells and (2) intrinsic interneurons.

Table 6-5
Neuroactive substances in PC neurons

Neurotramitter	Cell Location	Cell Type	References
Calbindin/ Calretinin	Mainly layer II	Pyramidal, basket, bi-multipolar	Celio (1990), Morales and Bloom (1997)
CCK	Mainly layer II and III	Pyramidal, multipolar, basket	Roberts et al. (1982), Westenbroek et al. (1987)
Dyn B	Layer II and III	Nonpyramidal fusiform, multipolar	Fallon and Leslie (1986)
ENK	Mainly layer II	Fusiform, pyramidal, multipolar	Harlan et al. (1987)
GABA/GAD	All layers	Diverse interneurons	Ekstrand et al. (2001a), Haberly et al. (1987), Morales and Bloom (1997)
Glutamate	Layers II and III	Pyramidal cells	See ➋ *Section 3.8.1*
Neurotensin	NR	NR	Hara et al. (1982)
Parvalbumin	Mainly layers II and III	Multipolar, basket	Celio (1990)
VIP	Mainly layers II and III	Bipolar, basket	Ekstrand et al. (2001a), Roberts et al. (1982), Sanides-Kohlrausch and Wahle (1990b)

NR, not reported. Adapted from Handbook of Chem. Neuroanat. Integrated Sys. CNS, Vol. 12, Part III, Chapter III, The Olfactory System, M. Shipley et al., pp. 469–573, 1996, Elsevier, Ltd

5.2.1 Pyramidal Cells

PC has two principal layers of pyramidal cells, layers II and III, corresponding to superficial and deep pyramidal cells. Both superficial and deep pyramidal neurons share characteristic features similar to pyramidal cells in the hippocampus and other cortical areas, including: (1) a primary apical-dendritic trunk that extends radially toward the pial surface and arborizes in layer Ia and Ib (Haberly, 1983), (2) secondary or basal dendrites that extend from the soma into deeper parts of PC. Both the apical and basal dendrites are heavily invested with spines and varicosities, and (3) a myelinated axon that typically extends deep to the soma that terminates on other local pyramidal cells and interneurons (see later), or projects back to MOB. Reconstructions of intracellularly filled pyramidal cells show that they have extensive axonal projections covering almost an entire cerebral hemisphere, including local connections to anterior and posterior PC as well as arborizations in orbital cortex, insular cortex, olfactory tubercle, perirhinal cortex, entorhinal cortex, and amygdaloid cortex (Johnson et al., 2000). The collaterals of deep pyramidal axons travel into the endopiriform nucleus (Tseng and Haberly, 1989), whereas those of superficial pyramidal neurons do not appear to enter this structure. The apical dendritic tree of superficial pyramidal cells generally branches more extensively and is more densely invested with spines and varicosities than that of deep pyramidal cells. Pyramidal cells contain glutamate and participate in intrinsic (intra-PC) and extrinsic excitatory projections as discussed later.

5.2.2 Interneurons

PC possesses a variety of interneurons as evidenced by morphological and neurochemical heterogeneity (Haberly and Presto, 1986; Haberly et al., 1987; Ekstrand et al., 2001a). A thorough consideration of PC

Table 6-6
Neurotransmitter receptors in the PC

Receptor	Layer	References
Cholinergic		Buckley et al. (1988), Hill et al. (1993), Levey et al. (1991), Rotter et al. (1979), Sahin et al. (1992), Seguela et al. (1993), Spencer et al. (1986)
mAChR1	I, II	
mAChR2	II	
mAChR3	II	
mAChR4	II	
nAChR	II, III	
Noradrenergic		Day et al. (1997), Domyancic and Morilak (1997), Nicholas et al. (1993), Palacios and Kuhar (1980), Pieribone et al. (1994), Rosin et al. (1996), Sargent-Jones et al. (1985), Talley et al. (1996), Unnerstall et al. (1984), Wanaka et al. (1989), Winzer-Serhan et al. (1997a, b), Young and Kuhar (1979, 1980a)
$\alpha1$	II, III	
$\alpha2$	I, II, III	
$\beta1$	II	
$\beta2$	II	
Dopaminergic		Fremeau et al. (1991), Huang et al. (1992), Mansour et al. (1990a, b)
D_1	II	
D_2	NR	
Serotonergic		Clemett et al. (2000), Cornea-Hebert et al. (1999), Hamada et al. (1998), Hoffman and Mezey (1989), Mengod et al. (1990a, b), Morales and Bloom (1997), Pompeiano et al. (1992, 1994), Tecott et al. (1993), Wright et al. (1995)
5-HT_{1A}	II	
5-HT_{1C}	II	
5-$HT_{2A/C}$	I, II, III	
5-HT_3	NR	
Glutamatergic		Gall et al. (1990), Kinoshita et al. (1998), Kinzie et al. (1995), Monaghan et al. (1985), Ohishi et al. (1993a, b, 1995, 1998), Petralia and Wenthold (1992), Petralia et al. (1994a, b), Romano et al. (1995), Saugstad et al. (1997), Shigemoto et al. (1992), Sun et al. (2000), Wada et al. (1998), Wisden and Seeburg (1993)
KA	II	
NMDA	Ia, II	
AMPA	II	
mGluR		
Group I mGluR (mGluR1/5)	II, III	
Group II mGluR (mGluR2,3)	NR	
Group III mGluR (mGluR4,7,8)	Ia	
GABAergic		Bowery et al. (1987), Margeta-Mitrovic et al. (1999), Palacios et al. (1981b), Young and Kuhar (1980b)
$GABA_A$	I, III	
$GABA_B$		

NR, not reported

Adapted from Handbook of Chem. Neuroanat. Integrated Sys. CNS, Vol. 12, Part III, Chapter III, The Olfactory System, M. Shipley et al., pp. 469–573, 1996, with permission from Elsevier, Ltd interneurons is beyond the scope of this review and detailed descriptions of these cells are available elsewhere (Shipley et al., 1996). Here, we present the distribution of the main interneuron types and some of their major neurochemical features. Interneurons are distributed throughout all layers and regions of this structure, and most, if not all, contain GABA (Haberly et al., 1987). Major interneuronal subtypes include the following:

1. Horizontal cells with large fusiform somata, found only in the superficial part of layer I (Haberly, 1983). These GABAergic cells produce dendrites that ramify within layer Ia and to a lesser extent within layer Ib (Haberly et al., 1987). Layer I interneurons are thought to function in feedforward inhibitory systems.
2. Spiny cells, found in layers I and III. Spiny actually refers to several distinct morphological subtypes (e.g., those with spherical or fusiform somata) that exhibit dendritic spines (Haberly, 1983). The dendrites of layer I cells extend into all parts of layer I and occasionally into deeper layers. Spiny cells give rise to unmyelinated axons that are horizontally oriented. Giant cells, another subcategory of spiny cell, have the largest soma size in PC. These cells are concentrated in the ventral parts of layer III.
3. Smooth cells, also found in layers I and III. Those in layer I exhibit a variety of cell shapes and sizes and they are distinguished by a lack of spines on the cell body or dendrites and by the presence of beaded varicosities on the distal dendrites. Smooth cells are the most numerous nonpyramidal cells in layer III, and they are preferentially distributed in the middle and deep parts of layer III. The dendrites emerge from numerous sites on the soma, usually branch once or twice near the soma, and then extend radially in all directions, with few subsequent bifurcations. The dendritic tree typically respects the boundaries of layer III.
4. Neurogliaform cells, found in layers I and II, have spherical somata and they are the most common cell type in layer I. The dendritic trees of these cells are usually restricted to a single sublamina in layer I. The axons are unmyelinated and branch extensively. Neurogliaform cells in layer II are essentially similar to those in layer I.
5. Semilunar cells, located in layer IIa, lack a basal dendritic tree, but they have several apical dendrites that arborize in layer I and axons that extend deep to the somata.
6. Multipolar cells, located throughout layer III and the endopiriform nucleus (Tseng and Haberly, 1989), have dendrites that are confined to layer III. The axons of these cells collateralize and synapse extensively within layer III and also extend to layer I. Many multipolar cells are GABAergic (Haberly et al., 1987).

In parallel with their diverse anatomical makeup, interneurons in PC are neurochemically heterogeneous, and they are thought to subserve multiple functions. CR and CCK also colocalize in GABAergic interneurons in PC (Morales and Bloom, 1997). Many GABAergic interneurons, described as "basket cells," colocalize with Ca^{2+} binding proteins (PV, calbindin), VIP, or CCK (Kubota and Jones, 1993; Ekstrand et al., 2001a). Basket cells are primarily distributed in layers II/III and exhibit diverse molecular markers and morphological characteristics. These cells are thought to predominantly form axo-somatic or proximal axo-dendritic synapses with neurons in layers II and III, and to participate both in feedback and feedforward inhibitory circuits.

5.3 PC Circuitry

5.3.1 Input from MOB

The output of MOB is through mitral cells and tufted cells. Their main axons gather at the caudo-lateral extent of MOB forming the LOT. These caudally directed axons give off collaterals in the AON and in other regions of POC (➋ *Figure 6-7*). The MOB sends a projection to the entire extent of PC, terminating in the superficial half of layer I, which is layer Ia. Within layer Ia, mitral/tufted cell axons synapse on the apical dendrites of pyramidal cells and with certain types of interneurons (Haberly, 1983). At the ultrastructural level, mitral/tufted cell axon terminals form asymmetrical synapses with dendritic spines. Some of the outputs of the MOB have a modest degree of topographical organization. For example, neurons in the dorsolateral quadrant of MOB project to the dorsal part of the external subdivision of AON, whereas output cells of the ventral half of MOB project to the lateral subdivision (Schoenfeld and Macrides, 1984). Intracellular HRP injections into mitral cells show that their axons form collateral terminal arbors within AON and PC (Ojima et al., 1984). The terminal arbors have a patchy anterior–posterior distribution in layer Ia of AON and PC in rabbit (Ojima et al., 1984). Some mitral cells branch and project to both PC and the olfactory tubercle. Mitral cells that are close together are reported to have similar patterns of axonal projections to the olfactory cortex (Buonviso et al., 1991). These findings, as well as observation that ORNs bearing the same odor receptor converge on single glomeruli, suggested that mitral cells of the same glomerulus might terminate with topographic specificity in PC. Studies by Buck and colleagues (Zou et al., 2001), using a genetic transneuronal tract tracing approach, reported that mitral/tufted cells associated with the same glomerulus projected to patches in PC. Such findings suggest that the representation of glomerular input to PC may have a higher degree of topographical organization than previously suspected. However, odor responses in PC, as determined by Fos mapping, do not show discrete patches, perhaps indicating that a restricted afferent input is distributed by intracortical processing (Illig and Haberly, 2003).

5.3.1.1 Neurochemistry of MOB Inputs The MOB projection to PC, via the mitral/tufted cells, is glutamatergic. Bulbectomy reduces evoked release of aspartate and NAG in PC (Collins and Probett, 1981a; Ffrench-Mullen et al., 1985) and stimulation of LOT induces glutamate and aspartate release in PC (Collins and Probett, 1981b). Receptor localization studies indicate that ionotropic glutamate and mGluRs are present in PC (➋ *Table 6-6*). Stimulation of mitral/tufted cell axons in the LOT produces monosynaptic depolarization and spiking in superficial pyramidal cells. LOT-evoked excitation of PC is also blocked by DNQX (Collins and Buckley, 1989), a potent postsynaptic antagonist of kainate and AMPA receptors. By contrast, selective NMDA receptor antagonists do not reduce LOT-evoked monosynaptic excitation of PC cells. LOT-evoked responses in PC are attenuated by 2-amino-4-phosphonobutyric acid (AP4) (Collins, 1982; Hori et al., 1982; Ffrench-Mullen et al., 1985, 1986; Hasselmo and Bower, 1991), an agonist of Group III mGluRs. The AP4-induced suppression appears to be mediated by presynaptic inhibition of glutamate release from mitral/tufted cell axon terminals (Hasselmo and Bower, 1991), consistent with anatomical evidence that these cells express Group III mGluRs on their axon terminals. Collins and Richards (1990) reported that protein kinase inhibitors reduce LOT-evoked monosynaptic excitation of PC, further supporting a role for a metabotropic, second messenger-mediated regulation of excitatory transmission at this synapse.

5.3.2 Intrinsic and Association Connections

PC has extensive connections, including (1) intrinsic, local, short translaminar connections between neurons in different layers of the PC, and (2) associative, longer-range, rostrocaudally directed connections with different parts of the PC.

5.3.2.1 Intrinsic or Local Connections There are extensive translaminar connections from superficial to deeper layers and vice versa. Layer II pyramidal cell axon collaterals synapse with deeper layer III pyramidal cells and with local inhibitory interneurons in layers I and II. Local collaterals from deeper pyramidal cells synapse with local interneurons or with more superficial pyramidal cells. Local projections from pyramidal cells to GABAergic interneurons play an important role in regulating PC neuronal excitability, including the expression of LTP and seizures, via feedback and feedforward inhibition (Ekstrand et al., 2001a).

5.3.2.2 Association Connections Cortico-cortical projections within PC are extensive and exhibit laminar and regional organization (Haberly and Price, 1978a, b; Luskin and Price, 1983). Axons from pyramidal cells of layer IIb are primarily directed at more caudal sites in PC, whereas pyramidal cells in layer III project predominantly to rostral parts of PC. Overall, the rostral-to-caudal associational connections are heavier than the caudal-to-rostral directed counterparts. Both systems terminate more heavily within the lateral than the medial parts of PC (Datiche et al., 1996). The contralateral PC also receives commissural fibers that arise from layer IIb of the anterior parts of PC. All of these association fiber systems terminate in a highly laminar fashion in layer Ib, immediately below the zone that contains the afferent input from the MOB; a lighter projection terminates in layer III.

5.3.2.3 Features and Neurochemistry of a Simple Model of PC Network Operation The orderly laminar structure of the PC and the segregation of MOB and association fiber inputs in layer I facilitate the analysis of PC's physiological organization. Haberly and others have provided a detailed analysis of the site and sequence of synaptic responses mediated by afferent inputs (i.e., LOT) and associative circuits (Haberly and Shepherd, 1973; Satou et al., 1983a, b, c; Haberly and Bower, 1984; Rodriguez and Haberly, 1989). LOT-shocks, or shocks applied to layer Ib to activate association fibers, result in a volley of activity that sweeps sequentially through PC. Field-potential recording studies suggest that there is little spatial topography to the pattern of activity following LOT or association fiber shocks. Thus, activity spreads diffusely from the site of activation, both rostrally and caudally. Single LOT shocks elicit a field potential exhibiting monosynaptic (A1) and disynaptic (B1) excitatory components (Haberly and Shepherd, 1973; Rodriguez and Haberly, 1989). The A1 component reflects monosynaptic excitation of pyramidal cell apical dendrites by LOT fibers. The B1 component reflects disynaptic excitation of pyramidal cells mediated by association fibers. The A1 and B1, mono- and disynaptic components are mediated by glutamate acting at both AMPA/kainate and NMDA receptors (Collins, 1982; Collins and Buckley, 1989). Baclofen, a $GABA_B$ receptor agonist, inhibits responses elicited by association fibers, but not LOT fibers, via a presynaptic mechanism (Tang and Hasselmo, 1994). Feedback and feedforward inhibitory circuits within PC suppress repetitive firing from pyramidal cells, and hence, reactivation of intracortical excitatory circuits. Fast ($GABA_A$) and slow ($GABA_B$) feedforward inhibitory systems independently regulate the excitability of the pyramidal cell somata and apical dendrites (Tseng and Haberly, 1988; Kanter et al., 1996; Kapur et al., 1997). Selective blockade of the apical-dendritic inhibition enhances NMDA receptor-dependent afferent synaptic responses in pyramidal cells (Kanter et al., 1996). Computational modeling suggests that inhibitory circuits in PC operate to reiterate patterned oscillatory activity from the bulb. Slow $GABA_B$ receptor-mediated inhibition is thought to tune PC activity to the 3–10 Hz θ range while fast $GABA_A$ receptor-mediated inhibition provides tuning in the 40–50 Hz γ range (Wilson and Bower, 1992). Thus, at least part of the oscillatory activity emerges from intrinsic properties of intracortical inhibitory circuits.

5.4 Extrinsic Outputs of PC

Outputs of PC can be categorized into three main classes, which include: (1) MOB, (2) cortical structures, and (3) subcortical structures (➋ *Figure 6-7*).

5.4.1 Feedback to the MOB

As noted earlier, PC and other structures of POC project heavily back to the MOB (➋ *Figure 6-7*). These feedback projections are heavier from the rostral than the caudal parts of POC (Shipley and Adamek, 1984) and arise mainly from pyramidal neurons in layer II and to a lesser extent in layer III.

5.4.2 Cortical Projections

There are also direct projections from PC to insular and orbital cortex. Insular and orbital cortices are also the primary cortical targets of ascending pathways arising in the nucleus of the solitary tract in the medulla and appear to contain the primary cortical representations for both gustatory and visceral sensations. Thus, olfactory projections to insular and orbital cortex may form a part of the circuitry that integrates olfactory and gustatory signals to generate the integrated perception of flavor.

5.4.3 Subcortical Projections

There are direct olfactory projections to the hypothalamus from neurons in the deepest layers of PC that terminate most heavily in the lateral hypothalamic area. Other components of PC project to medial and anterior parts of the hypothalamus and to the thalamus (Benjamin et al., 1982; Price and Slotnick, 1983).

5.5 Amino Acid Receptor Distribution in PC

5.5.1 Glutamate Receptors

Receptor localization studies indicate that layer II of PC contains an extensive amount of AMPA and kainate receptor subtypes, whereas layers Ia and II stain for NMDA receptors (Monaghan et al., 1985; Gall et al., 1990; Petralia and Wenthold, 1992; Wisden and Seeburg, 1993; Molnar et al., 1993; Petralia et al., 1994a, b). Receptors for mGluR1, mGluR2/3, mGluR5, mGluR7, and mGluR8 are present in PC (Shigemoto et al., 1992; Kinzie et al., 1995, 1997; Romano et al., 1995; Saugstad et al., 1997; Kinoshita et al., 1998; Wada et al., 1998). Pyramidal cells express mGluR1 (Shigemoto et al., 1992). Consistent with evidence that mGluR7-8 are presynaptically located on mitral/tufted cell axons, these receptors are densely expressed in layer Ia where these axons terminate (Kinzie et al., 1997; Kinoshita et al., 1998; Wada et al., 1998), and expression of mGluR7-8 was markedly decreased after bulbectomy or LOT transection. At the EM level, mGluR7 and 8 were found to be colocalized on axon terminals in layer Ia (Wada et al., 1998) (➲ *Table 6-6*).

5.5.2 GABA Receptors

$GABA_A$ receptors are located in layers I and III of PC (Young and Kuhar, 1980b; Palacios et al., 1981b; Bowery et al., 1987). Although studies have shown intense staining for $GABA_A$ receptors, there is only a very weak signal $GABA_B$ receptors (Bowery et al., 1987).

5.6 Neuromodulatory Inputs to PC

5.6.1 Cholinergic Inputs to PC

5.6.1.1 Cholinergic Innervation Pattern and Receptor Distribution Cholinergic inputs to PC arise from NDB cells that are codistributed among, but are distinct from, those that project to MOB. Thus, the cholinergic inputs to MOB and PC originate from separate populations of NDB neurons. The distribution of cholinergic inputs along the rostrocaudal axis of PC is fairly homogenous (Lysakowski et al., 1989). Layer I receives a sparse cholinergic innervation, whereas layers II and III receive a moderate and fairly uniform cholinergic innervation; the density of fibers is somewhat heavier in layer II than III. All four muscarinic subtypes (M_1, M_2, M_3, and M_4) are present in layer II (Rotter et al., 1979; Spencer et al., 1986; Buckley et al., 1988). The M_1 receptor subtype has also been localized on dendritic spines in layer I, but the cell type was not specified (Levey et al., 1991). Nicotinic cholinergic receptors are located in layers II and III, but as with the muscarinic receptors, the cellular and dendritic locations of these receptors are not known (Sahin et al., 1992; Hill et al., 1993; Seguella et al., 1993) (➲ *Table 6-6*).

5.6.1.2 Cholinergic Actions In PC slices, ACh and muscarinic cholinergic agonists suppress intrinsic (layer Ib) fiber transmission without affecting transmission at afferent (layer Ia) fiber synapses (Hasselmo and

Bower, 1992; Hasselmo et al., 1992b; Hasselmo et al., 1997). This suppression is presynaptically mediated by the M1 muscarinic subtype. Cholinergic agonists also appear to directly increase the excitability of pyramidal neurons, increasing neuronal bursts induced by intracellular depolarization. This effect is mediated, in part, by blockade of the slow afterhyperpolarization mediated by a Ca^{2+}-dependent K^+ current. ACh, acting at muscarinic receptors, has been shown to reduce the M-current in PC pyramidal neurons (Constanti and Galvan, 1983), and to depolarize and increase the firing rate of pyramidal cells (Libri et al., 1994; Postlewaithe et al., 1998). In the later study, muscarinic receptor stimulation was also reported to induce a slow poststimulus afterdepolarization and to depress EPSPs elicited by association fibers. Activation of the NDB, which is the source of cholinergic projections to PC, in vivo increased the spontaneous firing rate of PC cells, and it also increased the disynaptic excitatory (B1), and decreased the disynaptic inhibitory (P2), field-potential components evoked by LOT stimulation (Zimmer et al., 1999). NDB stimulation decreased the P2 component following activation of association fibers in caudal PC, and also reduced the paired-pulse inhibition of the P2 component following LOT and caudal PC shocks. These effects were reversed by scopolamine, suggesting the involvement of muscarinic receptors. These results suggest that activation of cholinergic inputs to PC increases the excitability of pyramidal cells, probably by a disinhibitory mechanism. A subsequent in vivo study reported similar findings: NDB stimulation enhanced the late, disynaptic component of the evoked potential elicited by LOT stimulation, but caused a suppression of the synaptic potential elicited by stimulation of the posterior PC; both effects were antagonized by muscarinic receptor blockers (Linster et al., 1999). Other in vivo experiments utilizing optical imaging reported contrasting results suggesting that NDB stimulation reduced both mono- and disynaptic responses elicited by LOT stimulation; the mechanism and transmitter involved in these effects were not studied (Rosin et al., 1999). More recent in vivo experiments showed that topical application of muscarinic receptor antagonists did not alter spontaneous or odor-evoked activity of anterior PC cells (Wilson, 2001). However, these antagonists enhanced odor-evoked cross-habituation, perhaps indicating that ACh acts to enhance generalization between odor representations in PC.

5.6.2 Dopaminergic Inputs to PC

5.6.2.1 DA Innervation Pattern, Receptor Distribution, and Physiological Actions DA inputs to PC arise from several sources, including the substantia nigra, ventral tegmental area and, to a lesser extent, the rostrodorsal part of dorsal raphe nucleus (Datiche and Cattarelli, 1996). The dopaminergic innervation of PC exhibits a marked rostrocaudal gradient and laminar specificity (Datiche and Cattarelli, 1996). Rostrally, DA fibers are relatively sparse and primarily confined to layers II/III (➋ *Figure 6-11*). Along the rostral-to-caudal axis of PC, the density of innervation progressively increases and DA fibers invade more superficial layers of PC. A moderately dense plexus of DA fibers extends from the deep part of layer I through layer III, to the caudal limit of PC. Autoradiographic, in situ hybridization and immunocytochemical studies have revealed the presence of D1 receptors in layer II (Palacios et al., 1981a; Mansour et al., 1990a, b; Fremeau et al., 1991; Huang et al., 1992). To date, there is little definitive evidence concerning the presence of D2 receptors in PC. Relatively little is known about DA functions in PC. DA was reported to excite PC interneurons, leading to increased IPSC activity in pyramidal cells (Gellman and Aghajanian, 1993, 1994).

5.6.3 Noradrenergic Inputs to PC

5.6.3.1 Noradrenergic Innervation Pattern and Receptor Distribution Noradrenergic LC neurons project heavily to PC and represent the major source of NE input to this structure (Fallon and Loughlin, 1982; Datiche and Cattarelli, 1996). LC lesions decrease NE levels in PC by 77% (Fallon and Moore, 1978). Like the projection to the neocortex, the LC projection to PC is primarily ipsilateral with a small contralateral component. Early histofluorescence studies (Fallon and Moore, 1978) of monoaminergic innervation of PC

Figure 6-11
Neuromodulatory transmitter inputs to PC. Darkfield photomicrographs showing the distribution of dopaminergic (a), noradrenergic (b), and serotonergic (c) fibers revealed respectively with immunohistochemistry for tyrosine hydroxylase (TH), dopamine-β-hydroxylase (DBH), and serotonin (5-HT). Abbreviations: Endo, endopiriform nucleus; lot, lateral olfactory tract. Reprinted from Handbook of Chem. Neuroanat. Integrated Sys. CNS, Vol. 12, Part III, Chapter III, The Olfactory System, M. Shipley et al., pp. 469–573, 1996, with permission from Elsevier, Ltd

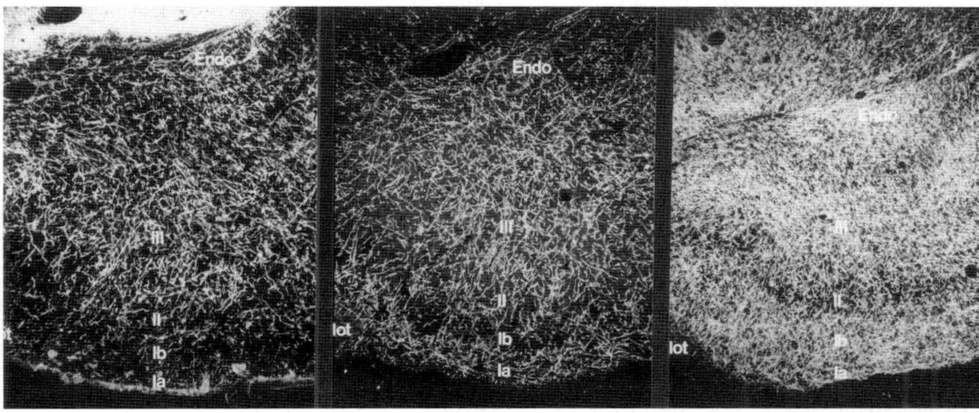

could not unambiguously distinguish among 5-HT, DA, and NE fibers. Immunohistochemistry for the NE biosynthetic enzyme dopamine-β-hydroxylase (DBH) showed that layers Ia and III of PC contained a moderate plexus of NE fibers; layer II was sparsely innervated (➜ *Figure 6-11*) (Datiche and Cattarelli, 1996). A distinctive feature of NE innervation of PC is the long fibers oriented primarily parallel to the pial surface in layer Ia. The density and laminar distribution of NE fibers are relatively uniform along the rostrocaudal axis of PC (Datiche and Cattarelli, 1996).

Both α and β receptors are present in PC (➜ *Table 6-6*). α1-receptors are highly expressed in layers II/III (Sargent-Jones et al., 1985) and layer II pyramidal cells express high levels of $\alpha1_A$, $\alpha1_B$, and low levels of $\alpha1_D$ (Pieribone et al., 1994; Day et al., 1997; Domyancic and Morilak, 1997). High levels of α2 receptor binding sites are present in PC (Young and Kuhar, 1979, 1980a; Unnerstall et al., 1984; Winzer-Serhan et al., 1997a), and $\alpha2_A$ and $\alpha2_C$ appear to be expressed by layer II pyramidal cells (Nicholas et al., 1993; Rosin et al., 1996; Talley et al., 1996; Winzer-Serhan et al., 1997a, b); $\alpha2_C$ receptors appear to be expressed by neurons in all layers of PC (Rosin et al., 1996). Weak-to-moderate expression of β receptor subtypes has also been reported (Palacios and Kuhar, 1982; Wanaka et al., 1989).

5.6.3.2 Physiological Actions of NE Early field-potential studies reported that low concentrations of NE (0.1–5 μM) enhanced, and high concentrations (20–250 μM) suppressed, LOT-evoked responses in PC (Collins et al., 1984). These effects were blocked by both α and β receptor antagonists. A subsequent study reported uniform suppression of LOT-evoked field potentials by NE (1–100 μM), although the effects were greater in layer Ib than Ia (Hasselmo et al., 1997). These effects were similar to those elicited by carbachol. The cellular actions of NE on PC neurons have received relatively little attention. NE was shown to block the slow afterhyperpolarization in guinea-pig pyramidal cells (Constanti and Sim, 1987). NE was reported to excite putative interneurons located at the layer II/III border, which in turn, increased IPSC activity in pyramidal cells in vitro (Sheldon and Aghajanian, 1990; Gellman and Aghajanian, 1993, 1994). Other studies from the same laboratory indicated that the NE-induced interneuronal excitation was mediated by $\alpha1_B$ receptors (Marek and Aghajanian, 1996a). LC activation in vivo was found to enhance odor-evoked responses in anterior and posterior PC (Bouret and Sara, 2002).

5.6.4 Serotonergic Inputs to PC

5.6.4.1 5-HT Innervation Pattern and Receptor Distribution Anterograde and retrograde tracing studies have demonstrated a rich projection from the dorsal raphe nucleus to PC (De Olmos and Heimer, 1980; Vertes, 1991; Datiche et al., 1995). Ascending serotonergic fibers from the dorsal raphe terminating in PC are believed to travel in the ventrolateral aspect of the medial forebrain bundle (Azmitia and Segal, 1978). Anterograde labeling of ascending dorsal raphe axons demonstrated that the entire PC was targeted by raphe projections (Vertes, 1991; Datiche et al., 1996). The projection was reported to be heavier to rostral than caudal PC, and heavier to the deeper than the superficial layers. The transmitter of labeled fibers could not be identified in these anterograde tracing results. The serotonergic innervation of PC is very heavy by comparison to DA and NE. 5-HT fibers are especially heavy in layers I and III and in the endopiriform nucleus (➋ *Figure 6-11*) (Datiche et al., 1996). The density of 5-HT fibers progressively decreases in the deeper parts of layer III. Layer II by contrast, is sparsely innervated.

Receptor binding and in situ hybridization studies suggest that 5-HT_{1A} receptors are located on the dendrites of pyramidal cells (layer II) and also on intrinsic cells of layer III (Pompeiano et al., 1992) (➋ *Table 6-6*). 5-HT_{2A} and 5-HT_{2C} receptors are strongly expressed by layer II pyramidal cells, as well as by interneurons in layers II and III (Pompeiano et al., 1994; Hamada et al., 1998; Cornea-Hebert et al., 1999). The 5-HT_3 subtype is present in PC and colocalizes with diverse interneuronal subtypes (Tecott et al., 1993; Morales and Bloom, 1997). 5-HT_3 colocalizes with approximately 50% of CCK-positive interneurons, and it is also expressed by interneurons containing GABA or CR, but not PV (Morales and Bloom, 1997).

5.6.4.2 Physiological Actions of 5-HT 5-HT has several actions on PC pyramidal cells as well as on putative interneurons that are mediated by distinct 5-HT receptor subtypes (Sheldon and Aghajanian, 1990, 1991; Gellman and Aghajanian, 1994). 5-HT has mixed, but predominately excitatory actions on layer II pyramidal cells. This excitation appears to be mediated by a reduction of the M-current, a noninactivating voltage-dependent outward K^+ current, as well as a decrease in a Ca^{2+}-activated K^+ current. These actions appear to be mediated via 5-HT_{1C} receptors. 5-HT also inhibits and increases the frequency of IPSPs in other pyramidal cells. This is mediated by excitation of layer III GABAergic interneurons that are directly activated (i.e., depolarized) by 5-HT acting at the 5-HT_{2A} receptor subtype (Gellman and Aghajanian, 1994; Marek and Aghajanian, 1994, 1996b). Interneurons excited by 5-HT appear to be a distinct subset distinguished by a relatively depolarized resting membrane potential, higher input resistance, and short action potential duration. The 5-HT_{2A} receptor-mediated effects on interneurons were not blocked, but instead were enhanced, by inhibitors of PKC, suggesting that PKC negatively modulates the excitation of PC interneurons (Marek and Aghajanian, 1995). In vivo, layer III putative interneurons were inhibited by 5-HT_{2A} agonists (Bloms-Funke et al., 1999). It is unclear if and how these cells correspond to those studied in vitro by Aghajanian and colleagues.

Acknowledgments

We acknowledge grant support for portions of this chapter that refer to research of the authors: PHS Grants DC03195, DC06356, DC07123, and RR020146.

References

Allen DM, Hamilton KA. 2000. Ultrastructural identification of synapses between mitral/tufted cell dendrites. Brain Res 860: 170-173.

Allison A. 1949. Quantitative observations on the olfactory system of the rabbit. Brain 72: 186-197.

Allison A. 1953. The morphology of the olfactory system in the vertebrates. Biol Rev Camb Philos Soc 28: 195-244.

Alonso JR, Arévalo R, Garcia-Ojeda E, Porteros A, Briñón JG, et al. 1995. NADPH-diaphorase active and calbindin D-28k-immunoreactive neurons and fibers in the olfactory bulb of the hedgehog (*Erinaceus europaeus*). J Comp Neurol 351: 307-327.

Alonso JR, Arévalo R, Porteros A, Briñón JG, Lara J, et al. 1993. Calbindin D-28K and NADPH-diaphorase activity are localized in different populations of periglomerular cells in the rat olfactory bulb. J Chem Neuroanat 6: 1-6.

Alonso JR, Briñón JG, Crespo C, Bravo IG, Arévalo R, et al. 1998. Chemical organization of the macaque monkey olfactory bulb: II. Calretinin, calbindin D-28k, parvalbumin, and neurocalcin immunoreactivity. J Comp Neurol 432: 389-407.

Altman J. 1969. Autoradiographic and histological studies of postnatal neurogenesis. IV. Cell proliferation and migration in the anterior forebrain, with special reference to persisting neurogenesis in the olfactory bulb. J Comp Neurol 137: 433-457.

Aroniadou-Anderjaska V, Ennis M, Shipley MT. 1997. Glomerular synaptic responses to olfactory nerve input in rat olfactory bulb slices. Neuroscience 79: 425-434.

Aroniadou-Anderjaska V, Ennis M, Shipley MT. 1999a. Current-source density analysis in the rat olfactory bulb: Laminar distribution of kainate/AMPA and NMDA receptor-mediated currents. J Neurophysiol 81: 15-28.

Aroniadou-Anderjaska V, Ennis M, Shipley MT. 1999b. Dendrodendritic recurrent excitation in mitral cells of the rat olfactory bulb. J Neurophysiol 82: 489-494.

Aroniadou-Anderjaska V, Zhou FM, Priest CA, Ennis M, Shipley MT. 2000. Tonic and synaptically evoked presynaptic inhibition of sensory input to the rat olfactory bulb via GABA (B) heteroreceptors. J Neurophysiol 84: 1194-1203.

Astic L, Saucier D. 1981. Metabolic mapping of functional activity in the olfactory projections of the rat: Ontogenetic study. Brain Res 254: 141-156.

Astic L, Saucier D. 1986. Anatomical mapping of the neuroepithelial projection to the olfactory bulb in the rat. Brain Res Bull 16: 445-454.

Astic L, Saucier D, Holley A. 1987. Topographical relationships between olfactory receptor cells and glomerular foci in the rat olfactory bulb. Brain Res 424: 144-152.

Aston-Jones G, Bloom FE. 1981. Norepinephrine-containing locus coeruleus neurons in behaving rats exhibit pronounced responses to non-noxious environmental stimuli. J Neurosci 1: 887-900.

Aungst JL, Heyward PM, Puche AC, Karnup SV, Hayar A, et al. 2003. Center-surround inhibition among olfactory bulb glomeruli. Nature 426: 623-629.

Azmitia EC, Segal M. 1978. An autoradiographic analysis of the differential ascending projections of the dorsal and median raphe nuclei in the rat. J Comp Neurol 179: 641-667.

Bailey MS, Shipley MT. 1993. Astrocyte subtypes in the rat olfactory bulb: Morphological heterogeneity and differential laminar distribution. J Comp Neurol 328: 501-526.

Baimbridge KG, Miller JJ. 1982. Immunohistochemical localization of calcium-binding protein in the cerebellum, hippocampal formation and olfactory bulb of the rat. Brain Res 245: 223-229.

Baker H. 1986. Species differences in the distribution of substance P and tyrosine hydroxylase immunoreactivity in the olfactory bulb. J Comp Neurol 252: 206-226.

Baker H. 1990. Unilateral, neonatal olfactory deprivation alters tyrosine hydroxylase expression but not aromatic amino acid decarboxylase or GABA immunoreactivity. Neuroscience 36: 761-771.

Baker H, Farbman AI. 1993. Olfactory afferent regulation of the dopamine phenotype in the fetal rat olfactory system. Neuroscience 52: 115-134.

Baker H, Kawano T, Margolis FL, Joh TH. 1983. Transneuronal regulation of tyrosine hydroxylase expression in olfactory bulb of mouse and rat. J Neurosci 3: 69-78.

Baker H, Morel K, Stone DM, Maruniak JA. 1993. Adult naris closure profoundly reduces tyrosine hydroxylase expression in mouse olfactory bulb. Brain Res 614: 109-116.

Baker H, Cummings DM, Munger SD, Margolis JW, Franzen L, et al. 1999. Targeted deletion of a cyclic nucleotide-gated channel subunit (OCNC1): Biochemical and morphological consequences in adult mice. J Neurosci 19: 9313-9321.

Baker H, Kawano T, Albert V, Joh TH, Reis DJ, et al. 1984. Olfactory bulb dopamine neurons survive deafferentation-induced loss of tyrosine hydroxylase. Neuroscience 11: 605-615.

Baker H, Towle AC, Margolis FL. 1988. Differential afferent regulation of dopaminergic and GABAergic neurons in the mouse main olfactory bulb. Brain Res 450: 69-80.

Baker H, Liu N, Chun HS, Saino S, Berlin R, et al. 2001. Phenotypic differentiation during migration of dopaminergic progenitor cells to the olfactory bulb. J Neurosci 21: 8505-8513.

Banay-Schwartz M, Lajtha A, Palkovits M. 1989a. Changes with aging in the levels of amino acids in rat CNS structural elements: I. glutamate and related amino acids. Neurochem Res 14: 555-562.

Banay-Schwartz M, Lajtha A, Palkovits M. 1989b. Changes with aging in the levels of amino acids in rat CNS structural elements: II. taurine and small neutral amino acids. Neurochem Res 14: 563-570.

Banks MI, Haberly LB, Jackson MB. 1996. Layer-specific properties of the transient K current (Ia) in piriform cortex. J Neurosci 16: 3862-3876.

Bardoni R, Magherini PC, Belluzzi O. 1996. Excitatory synapses in the glomerular triad of frog olfactory bulb in vitro. Neuroreport 7: 1851-1855.

Bassett JL, Shipley MT, Foote SL. 1992. Localization of corticotropin-releasing factor-like immunoreactivity in monkey olfactory bulb and secondary olfactory areas. J Comp Neurol 316: 348-362.

Bastianelli E, Okazaki K, Hidaka H, Pochet R. 1993. Neurocalcin immunoreactivity in rat olfactory bulb. Neurosci Lett 161: 165-168.

Bayer SA. 1983. 3H-Thymidine-radiographic studies of neurogenesis in the rat olfactory bulb. Exp Brain Res 50: 329-340.

Behan M, Haberly LB. 1999. Intrinsic and efferent connections of the endopiriform nucleus in rat. J Comp Neurol 408: 532-548.

Belluardo N, Mudò G, Trovato-Salinaro A, Le Gurun S, Charollais A, et al. 2000. Expression of connexin36 in the adult and developing rat brain. Brain Res 865: 121-138.

Belluscio L, Katz LC. 2001. Symmetry, stereotypy, and topography of odorant representations in mouse olfactory bulbs. J Neurosci 21: 2113-2122.

Belluscio L, Gold GH, Nemes A, Axel R. 1998. Mice deficient in G(olf) are anosmic. Neuron 20: 69-81.

Belluscio L, Lodovichi C, Feinstein P, Mombaerts P, Katz LC. 2002. Odorant receptors instruct functional circuitry in the mouse olfactory bulb. Nature 419: 296-300.

Belluzzi O, Benedusi M, Ackman J, Lo Turco J. 2003. Electrophysiological identification of new neurons in the olfactory bulb. J Neurosci 23: 10411-10418.

Belluzzi O, Puopolo M, Benedusi M, Kratskin I. 2004. Selective neuroinhibitory effects of taurine in slices of rat main olfactory bulb. Neuroscience 124: 929-944.

Benjamin RM, Jackson JC, Golden GT, West CH. 1982. Sources of olfactory inputs to opossum mediodorsal nucleus identified by horseradish peroxidase and autoradiographic methods. J Comp Neurol 207: 358-368.

Benson TE, Burd GD, Greer CA, Landis DM, Shepherd GM. 1985. High-resolution 2-deoxyglucose autoradiography in quick-frozen slabs of neonatal rat olfactory bulb. Brain Res 339: 67-78.

Berkowicz DA, Trombley PQ. 2000. Dopaminergic modulation at the olfactory nerve synapse. Brain Res 855: 90-99.

Biffo S, Grillo M, Margolis FL. 1990. Cellular localization of carnosine-like and anserine-like immunoreactivities in rodent and avian central nervous system. Neuroscience 35: 637-651.

Blakely RD, Ory-Lavollee L, Grzanna R, Koller KJ, Coyle JT. 1987. Selective immunocytochemical staining of mitral cells in rat olfactory bulb with affinity purified antibodies against N-acetyl-aspartyl-glutamate. Brain Res 402: 373-378.

Blakemore LJ, Trombley PQ. 2003. Kinetic variability of AMPA receptors among olfactory bulb neurons in culture. Neuroreport 14: 965-970.

Blakemore LJ, Trombley PQ. 2005. Slice electrophysiology and histological evidence of calcium-fluxing AMPA receptors in the olfactory bulb. AChemS Abstr 75.

Blanes T. 1898. Sobre algunos puntos dudosos de la estructura del bulbo olfactorio. Revista Trimestral microgafica 3: 99-127.

Bloms-Funke P, Gernert M, Ebert U, Loscher W. 1999. Extracellular single-unit recordings of piriform cortex neurons in rats: Influence of different types of anesthesia and characterization of neurons by pharmacological manipulation of serotonin receptors. J Neurosci Res 55: 608-619.

Bogan N, Brecha N, Gall C, Karten HJ. 1982. Distribution of enkephalin-like immunoreactivity in the rat main olfactory bulb. Neuroscience 7: 895-906.

Bonino M, Cantino D, Sassoe-Pognetto M. 1999. Cellular and subcellular localization of gamma-aminobutyric acidB receptors in the rat olfactory bulb. Neurosci Lett 274: 195-198.

Bonnemann C, Holland A, Meyer DK. 1989. Cholecystokinin in rat olfactory bulb is mainly contained in intrinsic neurons. Neurosci Lett 103: 281-286.

Booze RM, Crisostomo EA, Davis JN. 1989. Species differences in the localization and number of CNS beta adrenergic receptors: Rat versus guinea pig. J Pharmacol Exp Ther 249: 911-920.

Bouna S, Gysling K, Calas A, Araneda S. 1994. Some noradrenergic neurons of locus ceruleus-olfactory pathways contain neuropeptide-Y. Brain Res Bull 34: 413-417.

Bouret S, Sara SJ. 2002. Locus coeruleus activation modulates firing rate and temporal organization of odour-induced single-cell responses in rat piriform cortex. Eur J Neurosci 16: 2371-2382.

Bowery NG, Hudson AL, Price GW. 1987. GABAA and GABAB receptor site distribution in the rat central nervous system. Neuroscience 20: 365-383.

Boyd JD, Delaney KR. 2002. Tyrosine hydroxylase-immunoreactive interneurons in the olfactory bulb of the frogs Rana pipiens and Xenopus laevis. J Comp Neurol 454: 42-57.

Bradley J, Reuter D, Frings S. 2001. Facilitation of calmodulin-mediated odor adaptation by cAMP-gated channel subunits. Science 294: 2176-2178.

Breer H. 2003. Olfactory receptors: Molecular basis for recognition and discrimination of odors. Anal Bioanal Chem 377: 427-433.

Brennan P, Kaba H, Keverne EB. 1990. Olfactory recognition: A simple memory system. Science 250: 1223-1226.

Briñón JG, Alonso JR, Arévalo R, García-Ojeda E, Lara J, et al. 1992. Calbindin D-28k-positive neurons in the rat olfactory bulb. An immunohistochemical study. Cell Tissue Res 269: 289-297.

Briñón JG, Martínez-Guijarro FJ, Bravo IG, Arévalo R, Crespo C, et al. 1999. Coexpression of neurocalcin with other calcium-binding proteins in the rat main olfactory bulb. J Comp Neurol 407: 404-414.

Briñón JG, Weruaga E, Crespo C, Porteros Á, Arévalo R, et al. 2001. Calretinin-, neruocalcin-, and parvalbumin-immunoreactive elements in the olfactory bulb of the hedgehog (*Erinaceus europaeus*). J Comp Neurol 429: 554-570.

Brunet LJ, Gold GH, Ngai J. 1996. General anosmia caused by a targeted disruption of the mouse olfactory cyclic nucleotide-gated cation channel. Neuron 17: 681-693.

Brünig I, Sommer M, Hatt H, Bormann J. 1999. Dopamine receptor subtypes modulate olfactory bulb gamma-aminobutyric acid type A receptors. Proc Natl Acad Sci USA 96: 2456-2460.

Brunjes PC. 1994. Unilateral naris closure and olfactory system development. Brain Res Rev 19: 146-160.

Brunjes PC, Armstrong AM. 1996. Apoptosis in the rostral migratory stream of the developing rat. Dev Brain Res 92: 219-222.

Brunjes PC, Smith-Crafts LK, McCarty R. 1985. Unilateral odor deprivation: Effects on the development of olfactory bulb catecholamines and behavior. Dev Brain Res 22: 1-6.

Buck L, Axel R. 1991. A novel multigene family may encode odorant receptors: A molecular basis for odor recognition. Cell 65: 175-187.

Buckley NJ, Bonner TI, Brann MR. 1988. Localization of a family of muscarinic receptor mRNAs in rat brain. J Neurosci 8: 4646-4652.

Buonviso N, Revial MF, Jourdan F. 1991. The projections of mitral cells from small local regions of the olfactory bulb: An anterograde tracing study using PHA-L (*Phaseolus vulgaris leucoagglutinin*). Eur J Neurosci 3: 493-500.

Burd GD, Davis BJ, Macrides F, Grillo M, Margolis F. 1982. Carnosine in primary afferents of the olfactory system: An autoradiographic and biochemical study. J Neurosci 2: 244-255.

Cajal RSy. 1890. The origins and terminations of the olfactory, optic, and acoustic nerves in vertebrates. In studies on the cerebral cortex [Limbic structures]. Translated by Kraft LM (1955). Chicago, IL: The Year Book Publishers; pp. 1-27.

Cajal RSy. 1911a. Olfactory apparatus: Olfactory mucosa and olfactory bulb or first-order olfactory center. Histology of the nervous system, Vol. II. Translated by Swanson N, Swanson L (1995). New York: Oxford Univ. Press; pp. 532-554.

Cajal RSy. 1911b. Histologie du Systeme Nerveux de l'Hommes et des Vertebres. Maloine: Paris.

Camps M, Kelly PH, Palacios JM. 1990. Autoradiographic localization of D1 and D2 receptors in the brain of several mammalian species. J Neural Trans 80: 105-127.

Carleton A, Petreanu LT, Lansford R, Alvarez-Buylla A, Lledo P-M. 2003. Becoming a new neuron in the adult olfactory bulb. Nat Neurosci 6: 607-518.

Carlson GC, Shipley MT, Keller A. 2000. Long-lasting depolarizations in mitral cells of the rat olfactory bulb. J Neurosci 20: 2011-2021.

Carlson GC, Slawecki ML, Lancaster E, Keller A. 1997. Distribution and activation of intracellular Ca^{2+} stores in cultured olfactory bulb neurons. J Neurophysiol 78: 2176-2185.

Carson KA. 1984. Quantitative localization of neurons projecting to the mouse main olfactory bulb. Brain Res Bull 12: 629-634.

Carson KA, Burd GD. 1980. Localization of acetylcholinesterase in the main and accessory olfactory bulbs of the mouse by light and electron microscopic histochemistry. J Comp Neurol 191: 353-371.

Casabona G, Catania MV, Storto M, Ferraris N, Perroteau I, et al. 1998. Deafferentation up-regulates the expression of the mGlu1a metabotropic glutamate receptor protein in the olfactory bulb. Eur J Neurosci 10: 771-776.

Castillo PE, Carleton A, Vincent JD, Lledo PM. 1999. Multiple and opposing roles of cholinergic transmission in the main olfactory bulb. J Neurosci 19: 9180-9191.

Celio MR. 1990. Calbindin D-28k and parvalbumin in the rat nervous system. Neuroscience 35: 375-475.

Chanse NT, Kopin IJ. 1968. Stimulus-induced release of substances from olfactory bulb using the push-pull cannula. Nature 217: 466-498.

Chao TI, Kasa P, Wolff JR. 1997. Distribution of astroglia in glomeruli of the rat main olfactory bulb: Exclusion from the sensory subcompartment of neuropil. J Comp Neurol 388: 191-210.

Chen WR, Shepherd GM. 1997. Membrane and synaptic properties of mitral cells in slices of rat olfactory bulb. Brain Res 745: 189-196.

Chen WR, Xiong W, Shepherd GM. 2000. Analysis of relations between NMDA receptors and GABA release at olfactory bulb reciprocal synapses. Neuron 25: 625-633.

Chen AM, Perrin MH, Digruccio MR, Vaughan JM, Brar BK, et al. 2005. A soluble mouse brain spice variant of type 2 alpha corticotropin-releasing factory (CRF) binds ligands and modulates their activity. Proc Natl Acad Sci USA 102: 2620-2625.

Christie JM, Westbrook GL. 2003. Regulation of backpropagating action potential in mitral cell lateral dendrites by A-type potassium currents. J Neurophysiol 89: 2466-2472.

Christie JM, Schoppa NE, Westbrook GL. 2001. Tufted cell dendrodendritic inhibition in the olfactory bulb is dependent on NMDA receptor activity. J Neurophysiol 85: 169-173.

Christie JM, Bark C, Hormuzdi SG, Helbig I, Monyer H, et al. 2005. Connexin36 mediates spike synchrony in olfactory bulb glomeruli. Neuron 46: 761-772.

Cho JY, Min N, Franzen L, Baker H. 1996. Rapid down-regulation of tyrosine hydroxylase expression in the olfactory bulb of naris-occluded adult rats. J Comp Neurol 369: 264-276.

Chu DCM, Albin RL, Young AB, Penney JB. 1990. Distribution and kinetics of GABAB binding sites in rat central nervous system: A quantitative autoradiographic study. Neuroscience 34: 341-357.

Cielo MR. 1990. Calbindin D-28k and parvalbumin in the rat nervous system. Neuroscience 35: 375-475.

Cigola E, Volpe BT, Lee JW, Franzen L, Baker H. 1998. Tyrosine hydroxylase expression in primary cultures of olfactory bulb: Role of L-type calcium channels. J Neurosci 18: 7638-7649.

Cinelli AR, Hamilton KA, Kauer JS. 1995. Salamander olfactory bulb neuronal activity observed by video rate, voltage-sensitive dye imaging. III. Spatial and temporal properties of responses evoked by odorant stimulation. J Neurophysiol 73: 2053-2071.

Ciombor KJ, Ennis M, Shipley MT. 1999. Norepinephrine increases rat mitral cell excitatory responses to weak olfactory nerve input via alpha-1 receptors in vitro. Neuroscience 90: 595-606.

Clancy AN, Schoenfeld TA, Forbes WB, Macrides F. 1994. The spatial organization of the peripheral olfactory system of the hamster. Part II: Receptor surfaces and odorant passageways within the nasal cavity. Brain Res Bull 34: 211-241.

Clemett DA, Punhani T, Duxon MS, Blackburn TP, Fone KC. 2000. Immunohistochemical localisation of the 5-HT2C receptor protein in the rat CNS. Neuropharmacol 39: 123-132.

Collins GG. 1974. The rates of synthesis, uptake and disappearance of [14C]-taurine in eight areas of the rat central nervous system. Brain Res 76: 447-459.

Collins GG. 1982. Some effects of excitatory amino acid receptor antagonists on synaptic transmission in the rat olfactory cortex slice. Brain Res 244: 311-318.

Collins GG, Buckley KS. 1989. Antagonism of monosynaptic excitations in the mouse olfactory cortex slice by 6,7-dinitroquinoxaline-2,3-dione. Neuropharmacol 28: 1123-1128.

Collins GG, Probett GA. 1981a. Aspartate and not glutamate is the likely transmitter of the rat lateral olfactory tract fibres. Brain Res 209: 231-234.

Collins GG, Probett GA. 1981b. Patterns of endogenous amino acid release from slices of rat guinea-pig olfactory cortex. Brain Res 204: 103-120.

Collins GG, Richards WJ. 1990. Pharmacological evidence that protein kinase C modulates monosynaptic excitations in the olfactory cortex. Naunyn Schmiedebergs Arch Pharmacol 341: 114-122.

Collins GG, Probett GA, Anson J, McLaughlin NJ. 1984. Excitatory and inhibitory effects of noradrenaline on synaptic transmission in the rat olfactory cortex slice. Brain Res 294: 211-223.

Condorelli DF, Belluardo N, Trovato-Salimaro A, Mudò G. 2000. Expression of C×36 in mammalian neurons. Brain Res Rev 32: 72-85.

Condorelli DF, Parenti R, Spinella F, Salinaro AT, Belluardo N, et al. 1998. Cloning of a new gap junction gene (C×36) highly expressed in mammalian neurons. Eur J Neurosci 10: 1202-1208.

Conn PJ, Pin JP. 1997. Pharmacology and functions of metabotropic glutamate receptors. Ann Rev Pharmacol Toxicol 37: 205-237.

Constanti A, Galvan M. 1983. M-current in voltage-clamped olfactory cortex neurons. Neurosci Lett 39: 65-70.

Constanti A, Sim JA. 1987. Calcium-dependent potassium conductance in guinea-pig olfactory cortex neurones in vitro. J Physiol 387: 173-194.

Cooper-Kuhn CM, Winkler J, Kuhn HG. 2004. Decreased neurogenesis after cholinergic forebrain lesion in the adult rat. J Neurosci Res 77: 155-165.

Coopersmith R, Leon M. 1984. Enhanced neural response to familiar olfactory cues. Science 225: 849-851.

Cornea-Hebert V, Riad M, Wu C, Singh SK, Descarries L. 1999. Cellular and subcellular distribution of the serotonin 5-HT2A receptor in the central nervous system of adult rat. J Comp Neurol 409: 187-209.

Coronas V, Srivastava LK, Liang JJ, Jourdan F, Moyse E. 1997. Identification and localization of dopamine receptor subtypes in rat olfactory mucosa and bulb: A combined in situ hybridization and ligand binding radioautographic approach. J Chem Neuroanat 12: 243-257.

Corotto FS, Henegar JR, Maruniak JA. 1994. Odor deprivation leads to reduced neurogenesis and reduced neuronal survival in the olfactory bulb of the adult mouse. Neuroscience 61: 739-744.

Couper Leo JM, Devine AH, Brunjes PC. 2000. Focal denervation alters cellular phenotypes and survival in the developing rat olfactory bulb. J Comp Neurol 417: 325-336.

Crespo C, Blasco-Ibáñez JM, Marqués-Mari AI. 2001. Parvalbumin-containing interneurons do not innervate granule cells in the olfactory bulb. Neuroreport 12: 2553-2556.

Crespo C, Blasco-Ibáñez JM, Marqués-Marí AI, Alonso JR, Brinon JG, et al. 2002. Vasoactive intestinal polypeptide-containing elements in the olfactory bulb of the hedgehog (*Erinaceus europaeus*). J Chem Neuroanat 24: 49-63.

Cuschieri A, Bannister LH. 1975a. The development of the olfactory mucosa in the mouse: Electron microscopy. J Anat 119: 471-498.

Cuschieri A, Bannister LH. 1975b. The development of the olfactory mucosa in the mouse: Light microscopy. J Anat 119: 277-286.

Datiche F, Cattarelli M. 1996. Catecholamine innervation of the piriform cortex: A tracing and immunohistochemical study in the rat. Brain Res 671: 69-78

Datiche F, Litaudon P, Cattarelli M. 1996. Intrinsic association fiber system of the piriform cortex: a quantitative study based on a cholera toxin B subunit tracing in the rat. J Comp Neurol 376: 265-277

Datiche F, Luppi P-H, Cattarelli M. 1995. Serotonergic and non-serotonergic projections from the raphe nuclei to the piriform cortex in the rat: A cholera toxin B subunit (CTb) and 5-HT immunohistochemical study. Brain Res 671: 27-37.

Davila NG, Blakemore LJ, Trombley PQ. 2003. Dopamine modulates synaptic transmission between rat olfactory bulb neurons in culture. J Neurophysiol 90: 395-404.

Davis BJ. 1991. NADPH-diaphorase activity in the olfactory system of the hamster and rat. J Comp Neurol 314: 493-511.

Davis BJ, Macrides F. 1983. Tyrosine hydroxylase immunoreactive neurons and fibers in the olfactory system of the hamster. J Comp Neurol 214: 427-440.

Davis BJ, Burd GD, Macrides F. 1982. Localization of methionine-enkephalin, substance P, and somatostatin immunoreactivities in the main olfactory bulb of the hamster. J Comp Neurol 204: 377-383.

Davison IG, Boyd JD, Delaney KR. 2004. Dopamine inhibits mitral/tufted-granule cell synapses in the frog olfactory bulb. J Neurosci 24: 8057-8067.

Day HE, Campeau S, Watson SJ Jr, Akil H. 1997. Distribution of alpha 1a-, alpha 1b- and alpha 1d-adrenergic receptor mRNA in the rat brain and spinal cord. J Chem Neuroanat 13: 115-139.

De Lorenzo AJ. 1957. Electron microscope observations of the olfactory mucosa and olfactory nerve. J Biophys Biochem Cytol 3: 839-850.

de Olmos J, Hardy H, Heimer L. 1978. The afferent connections of the main and the accessory olfactory bulb formations in the rat: An experimental HRP-study. J Comp Neurol 181: 213-244.

De Olmos J, Heimer L. 1980. Double and triple labeling of neurons with fluorescent substances; the study of collateral pathways in the ascending raphe system. Neurosci Lett 19: 7-12.

De Saint Jan D, Westbrook GL. 2005. Detecting activity in olfactory bulb glomeruli with astrocyte recording. J Neurosci 25: 2917-2924.

De Souza EB, Insel TR, Perrin MH, Rivier J, Vale WW, et al. 1985. Corticotropin-releasing factory receptors are widely distributed within the rat central nervous system: An autoradiographic study. J Neurosci 5: 3189-3203.

Didier A, Ottersen OP, Storm-Mathisen J. 1994. Differential subcellular distribution of glutamate and taurine in primary olfactory neurons. Neuroreport 6: 145-148.

Didier A, Carleton A, Bjaalie JG, Vincent JD, Ottersen OP, et al. 2001. A dendrodendritic reciprocal synapse provides a recurrent excitatory connection in the olfactory bulb. Proc Natl Acad Sci USA 98: 6441-6446.

Domyancic AV, Morilak DA. 1997. Distribution of alpha1A adrenergic receptor mRNA in the rat brain visualized by in situ hybridization. J Comp Neurol 386: 358-378.

Doty RL, Risser JM. 1989. Influence of the D-2 dopamine receptor agonist quinpirole on the odor detection performance of rats before and after spiperone administration. Psychopharmacol 98: 310-315.

Doucette JR. 1984. The glial cells in the nerve fiber layer of the rat olfactory bulb. Anat Rec 210: 385-391.

Doucette R. 1989. Development of the nerve fiber layer in the olfactory bulb of mouse embryos. J Comp Neurol 285: 514-527.

Duchamp-Viret P, Chaput MA, Duchamp A. 1999. Odor response properties of rat olfactory receptor neurons. Science 284: 2171-2174.

Duchamp-Viret P, Delaleu J-C, Duchamp A. 2000. $GABA_B$-mediated action in the frog olfactory bulb makes odor responses more salient. Neuroscience 97: 771-777.

Duchamp-Viret P, Coronas V, Delaleu J-C, Moyse E, Duchamp A. 1997. Dopaminergic modulation of mitral cell activity in the frog olfactory bulb: A combined radioligand binding-electrophysiological study. Neuroscience 79: 203-216.

Duvoisin RM, Zhang C, Ramonell K. 1995. A novel metabotropic receptor expressed in the retina and olfactory bulb. J Neurosci 15: 3075-3083.

Ekstrand JJ, Domroese ME, Feig SL, Illig KR, Haberly LB. 2001a. Immunocytochemical analysis of basket cells in rat piriform cortex. J Comp Neurol 434: 308-328.

Ekstrand JJ, Domroese ME, Johnson DMG, Feig SL, Knodel SM, et al. 2001b. A new subdivision of anterior piriform cortex and associated deep nucleus with novel features of interest for olfaction and epilepsy. J Comp Neurol 434: 289-307.

Elaagouby A, Ravel N, Gervais R. 1991. Cholinergic modulation of excitability in the rat olfactory bulb: Effect of local application of cholinergic agents on evoked field potentials. Neuroscience 45: 653-662.

El-Etri MM, Griff ER, Ennis M, Shipley MT. 1999. Evidence for cholinergic regulation of basal norepinephrine release in the rat olfactory bulb. Neuroscience 93: 611-617.

Ennis M, Zimmer LA, Shipley MT. 1996. Olfactory nerve stimulation activates rat mitral cells via NMDA and non-NMDA receptors in vitro. Neuroreport 7: 989-992.

Ennis M, Zhou FM, Ciombor KJ, Aroniadou-Anderjaska V, Hayar A, et al. 2001. Dopamine D2 receptor-mediated presynaptic inhibition of olfactory nerve terminals. J Neurophysiol 86: 2986-2997.

Ennis M, Zhu M, Heinbockel T, Hayar A. 2006. Olfactory nerve-evoked metabotropic glutamate receptor-mediated responses in olfactory bulb mitral cells. J Neurophysiol 95: 2233-2241.

Erlich ME, Grillo M, Joh TH, Margolis FL, Baker H. 1990. Transneuronal regulation of neuronal specific gene expression in the mouse olfactory bulb. Mol Brain Res 7: 115-122.

Ezeh PI, Wellis DP, Scott JW. 1993. Organization of inhibition in the rat olfactory bulb external plexiform layer. J Neurophysiol 70: 263-274.

Fadool DA, Levitan IB. 1998. Modulation of olfactory bulb potassium current by tyrosine phosphorylation. J Neurosci 18: 6126-6137.

Fadool DA, Tucker K, Phillips JJ, Simmen JA. 2000. Brain insulin receptor causes activity-dependent current suppression in olfactory bulb through multiple phosphorylation of Kv1.3. J Neurophysiol 83: 2332-2348.

Fallon JH, Leslie FM. 1986. Distribution of dynorphin and enkephalin peptides in the rat brain. J Comp Neurol 249: 293-336.

Fallon JH, Loughlin SE. 1982. Monoamine innervation of the forebrain: Collateralization. Brain Res Bull 9: 295-307.

Fallon JH, Moore RY. 1978. Catecholamine innervation of the basal forebrain. IV. Topography of the dopamine projection to the basal forebrain and neostriatum. J Comp Neurol 180: 545-580.

Ferraris N, Perroteau I, De Marchis S, Fasolo A, Bovolin P. 1997. Glutamatergic deafferentation of olfactory bulb modulates the expression of mGluR1a mRNA. Neuroreport 8: 1949-1953.

Ferriero D, Margolis FL. 1975. Denervation in the primary olfactory pathway of mice. II. Effects on carnosine and other amine compounds. Brain Res 94: 75-86.

Ffrench-Mullen JMH, Hori N, Carpenter DO. 1986. Receptors for excitatory amino acids on neurons in rat pyriform cortex. J Neurophysiol 55: 1283-1294.

Ffrench-Mullen JM, Koller K, Zaczek R, Coyle JT, Hori N, et al. 1985. N-Acetylaspartylglutamate: Possible role as the neurotransmitter of the lateral olfactory tract. Proc Natl Acad Sci USA 82: 3897-3900.

Firestein S. 2001. How the olfactory system makes sense of scents. Nature 413: 211-218.

Fiske BK, Brunjes PC. 2001. Cell death in the developing and sensory-deprived rat olfactory bulb. J Comp Neurol 431: 311-319.

Fonseca MI, Lunt GG, Aguilar JS. 1991b. Inhibition of muscarinic cholinergic receptors by disulfide reducing agents and arsenicals. Differential effect on locust and rat. Biochem Pharmacol 41: 735-742.

Fonseca MI, Aguilar JS, Skorupa AF, Klein WL. 1991a. Cellular mapping of m2 muscarinic receptors in the rat olfactory bulb using antiserum raised against a cytoplasmic loop peptide. Brain Res 563: 163-170.

Frazier LL, Brunjes PC. 1988. Unilateral odor deprivation: Early postnatal changes in olfactory bulb cell density and number. J Comp Neurol 269: 355-370.

Frazier-Cierpial L, Brunjes PC. 1989. Early postnatal cellular proliferation and survival in the olfactory bulb and rostral migratory stream of normal and unilaterally odor-deprived rats. J Comp Neurol 289: 481-492.

Fremeau RT Jr, Duncan GE, Fornaretto MG, Dearry A, Gingrich JA, et al. 1991. Localization of D1 dopamine receptor mRNA in brain supports a role in cognitive, affective, and neuroendocrine aspects of dopaminergic neurotransmission. Proc Natl Acad Sci USA 88: 3772-3776.

Friedman D, Strowbridge BW. 2000. Functional role of NMDA autoreceptors in olfactory mitral cells. J Neurophysiol 84: 39-50.

Friedman D, Strowbridge BW. 2003. Both electrical and chemical synapses mediate fast network oscillations in the olfactory bulb. J Neurophysiol 89: 2601-2610.

Friedrich RW, Korsching SI. 1997. Combinatorial and chemotopic odorant coding in the zebrafish olfactory bulb visualized by optical imaging. Neuron 18: 737-752.

Frings S, Seifert R, Godde M, Kaupp UB. 1995. Profoundly different calcium permeation and blockage determine the specific function of distinct cyclic nucleotide-gated channels. Neuron 15: 169-179.

Fritschy JM, Benke D, Mertens S, Oertel WH, Bachi T, et al. 1992. Five subtypes of type A gamma-aminobutyric acid receptors identified in neurons by double and triple immunofluorescence staining with subunit-specific antibodies. Proc Natl Acad Sci USA 89: 6726-6730.

Frosch MP, Dichter MA. 1982. Physiology and pharmacology of olfactory bulb neurons in dissociated cell culture. Brain Res 290: 321-332.

Fuller TA, Price JL. 1988. Putative glutamatergic and/or aspartatergic cells in the main and accessory olfactory bulbs of the rat. J Comp Neurol 276: 209-218.

Gall C, Seroogy KB, Brecha N. 1986. Distribution of VIP- and NPY-like immunoreactivities in rat main olfactory bulb. Brain Res 374: 389-394.

Gall C, Sumikawa K, Lynch G. 1990. Regional distribution of mRNA for a putative kainate receptor in rat brain. Eur J Pharmacol 189: 217-221.

Gall CM, Hendry SHC, Seroogy KB, Jones EG, Haycock JW. 1987. Evidence for coexistence of GABA and dopamine in neurons of the rat olfactory bulb. J Comp Neurol 266: 307-318.

Geiling H, Schild D. 1996. Glutamate-mediated release of Ca^{2+} in mitral cells of the olfactory bulb. J Neurophysiol 76: 563-570.

Gellman RL, Aghajanian GK. 1993. Pyramidal cells in piriform cortex receive a convergence of inputs from monoamine-activated GABAergic interneurons. Brain Res 600: 63-73.

Gellman RL, Aghajanian GK. 1994. $Serotonin_2$ receptor-mediated excitation of interneurons in piriform cortex: Antagonism by atypical antipsychotic drugs. Neuroscience 58: 515-525.

Getchell ML, Getchell TV. 1992. Fine structural aspects of secretion and extrinsic innervation in the olfactory mucosa. Microsc Res Tech 23:111-127.

Getchell TV, Shepherd GM. 1975. Synaptic actions on mitral and tufted cells elicited by olfactory nerve volleys in the rabbit. J Physiol 251: 497-522.

Getchell TV, Margolis FL, Getchell ML. 1984. Perireceptor and receptor events in vertebrate olfaction. Prog Neurobiol 23: 317-345.

Gheusi G, Cremer H, McLean H, Chazal G, Vincent J-D, et al. 2000. Importance of newly generated neurons in the adult olfactory bulb for odor discrimination. Proc Natl Acad Sci USA 97: 1823-1828.

Giustetto M, Kirsch J, Fritschy JM, Cantino D, Sassoè-Pognetto M. 1998. Localization of the clustering protein gephyrin at GABAergic synapses in the main olfactory bulb of the cat. J Comp Neurol 395: 231-244.

Giustetto M, Bovolin P, Fasolo A, Bonino M, Cantino D, et al. 1997. Glutamate receptors in the olfactory bulb synaptic circuitry: Heterogeneity and synaptic localization of *N*-methyl-D-aspartate receptor subunit 1 and AMPA receptor subunit 1. Neuroscience 76: 787-798.

Golgi C. 1875. Sulla Fina Structrura del Bulbi Olfattori. Rev Sper Feniatr 1: 405-425.

Gracia-Llanes FJ, Crespo C, Blasco-Ibáñez JM, Marqués-Mari AI, Martínez-Guijarro FJ. 2003. VIP-containing deep short-axon cells of the olfactory bulb innervate interneurons different from granule cells. Eur J Neurosci 18: 1751-1763.

Graziadei PP, Graziadei GA. 1979. Neurogenesis and neuron regeneration in the olfactory system of mammals. I. Morphological aspects of differentiation and structural organization of the olfactory sensory neurons. J Neurocytol 8: 1-18.

Greer CA. 1987. Golgi analyses of dendritic organization among denervated olfactory bulb granule cells. J Comp Neurol 256: 442-452.

Greer CA, Stewart WB, Teicher MH, Shepherd GM. 1982. Functional development of the olfactory bulb and a unique glomerular complex in the neonatal rat. J Neurosci 2: 1744-1759.

Guthrie KM, Gall CM. 1995a. Functional mapping of odor-activated neurons in the olfactory bulb. Chem Senses 20: 271-282.

Guthrie KM, Gall CM. 1995b. Odors increase fos in olfactory bulb neurons including dopaminergic cells. Neuroreport 6: 2145-2149.

Guthrie KM, Pullara JM, Marshall JF, Leon M. 1991. Olfactory deprivation increases dopamine D2 receptor density in the rat olfactory bulb. Synapse 8: 61-70.

Haberly L. 1983. Structure of the piriform cortex of the opossum. I. Description of neuron types with Golgi methods. J Comp Neurol 213: 163-187.

Haberly LB, Bower JM. 1984. Analysis of association fiber system in piriform cortex with intracellular recording and staining techniques. J Neurophysiol 51: 90-112.

Haberly LB, Presto S. 1986. Ultrastructural analysis of synaptic relationships of intracellularly stained pyramidal cell axons in piriform cortex. J Comp Neurol 248: 464-474.

Haberly LB, Price JL. 1978a. Association and commissural fiber systems of the olfactory cortex of the rat. I. Systems originating in the piriform cortex and adjacent areas. J Comp Neurol 178: 711-740.

Haberly LB, Price JL. 1978b. Association and commissural fiber systems of the olfactory cortex of the rat. II. Systems originating in the olfactory peduncle. J Comp Neurol 181: 781-807.

Haberly LB, Shepherd GM. 1973. Current-density analysis of summed evoked potentials in opossum prepyriform cortex. J Neurophysiol 36: 789-802.

Haberly LB, Hansen DJ, Feig SL, Presto S. 1987. Distribution and ultrastructure of neurons in opossum piriform cortex displaying immunoreactivity to GABA and GAD and high-affinity tritiated GABA uptake. J Comp Neurol 266: 269-290.

Hack MA, Saghatelyan A, de Chevigny A, Pfeifer A, Ashery-Padan R, et al. 2005. Neuronal fate determinants of adult olfactory bulb neurogenesis. Nat Neurosci 8: 865-872.

Halabisky B, Strowbridge BW. 2003. γ-frequency excitatory input to granule cells facilitates dendrodendritic inhibition in the rat olfactory bulb. J Neurophysiol 90: 644-654.

Halabisky B, Friedman D, Radojicic M, Strowbridge BW. 2000. Calcium influx through NMDA receptors directly evokes GABA release in olfactory bulb granule cells. J Neurosci 20: 5124-5134.

Hadley WM, Dahl AR. 1982. Cytochrome P-450 dependent monooxygenase activity in rat nasal epithelial membranes. Toxicol Lett 10: 417-422.

Halász N. 1987. Autoradiographic localization of high affinity uptake sites for 3H-D-aspartate in the rat olfactory bulb. Acta Biol Hung 38: 383-392.

Halász N. 1990. The vertebrate olfactory system. Chemical neuroanatomy, function and development. Akadémiai Kiadó, Budapest, 281 pp.

Halász N, Shepherd GM. 1983. Neurochemistry of the vertebrate olfactory bulb. Neuroscience 10: 579-619.

Halász N, Ljungdahl A, Hökfelt T. 1978. Transmitter histochemistry of the rat olfactory bulb. II. Fluorescence histochemical, autoradiographic and electron microscopic localization of monoamines. Brain Res 154: 253-271.

Halász N, Hokfelt T, Norman AW, Goldstein M. 1985. Tyrosine hydroxylase and 28K-vitamin D-dependent calcium binding protein are localized in different subpopulations of periglomerular cells of the rat olfactory bulb. Neurosci Lett 61: 103-107.

Halász N, Johansson O, Hökfelt T, Ljungdahl A, Goldstein M. 1981. Immunohistochemical identification of two types of dopamine neuron in the rat olfactory bulb as seen by serial sectioning. J Neurocytol 10: 251-259.

Halász N, Ljungdahl A, Hokfelt T, Johansson O, Goldstein M, et al. 1977. Transmitter histochemistry of the rat olfactory bulb. I. Immunohistochemical localization of monoamine synthesizing enzymes. Support for intrabulbar, peri-glomerular dopamine neurons. Brain Res 126: 455-474.

Hamada S, Senzaki K, Hamaguchi-Hamada K, Tabuchi K, Yamamoto H, et al. 1998. Localization of 5-HT_{2A} receptor in rat cerebral cortex and olfactory system revealed by immunohistochemistry using two antibodies raised in rabbit and chicken. Brain Res Mol Brain Res 54: 199-211.

Hamana H, Hirono J, Kizumi M, Sato T. 2003. Sensitivity-dependent hierarchical receptor codes for odors. Chem Senses 28: 87-104.

Hamilton KA, Coppola DM. 2003. Distribution of GluR1 is altered in the olfactory bulb following neonatal naris occlusion. J Neurobiol 54: 326-336.

Hamilton KA, Heinbockel T, Ennis M, Szabó G, Erdélyi F, et al. 2005. Properties of external plexiform layer interneurons in mouse olfactory bulb slices. Neuroscience 133: 819-829.

Hara Y, Shiosaka S, Senba E, Sakanaka M, Inagaki S, et al. 1982. Ontogeny of the neurotensin-containing neuron system of the rat: Immunohistochemical analysis. I. Forebrain and diencephalon. J Comp Neurol 208: 177-195.

Hardy A, Palouzier-Paulignan B, Duchamp A, Royet JP, Duchamp-Viret P. 2005. 5-hydroxytryptamine action in the rat olfactory bulb: In vitro electrophysiological patch-clamp recordings of juxtaglomerular and mitral cells. Neuroscience 131: 717-731.

Harlan RE, Shivers BD, Romano GJ, Howells RD, Pfaff DW. 1987. Localization of preproenkephalin mRNA in the rat brain and spinal cord by in situ hybridization. J Comp Neurol 258: 159-184.

Hasselmo ME, Bower JM. 1991. Selective suppression of afferent but not intrinsic fiber synaptic transmission by 2-amino-4-phosphonobutyric acid (AP4) in piriform cortex. Brain Res 548: 248-255.

Hasselmo ME, Bower JM. 1992. Cholinergic suppression specific to intrinsic not afferent fiber synapses in rat piriform (olfactory) cortex. J Neurophysiol 67: 1222-1229.

Hasselmo ME, Anderson BP, Bower JM. 1992. Cholinergic modulation of cortical associative memory function. J Neurophysiol 67: 1230-1246.

Hasselmo ME, Linster C, Ma D, Cekic M. 1997. Noradrenergic suppression of synaptic transmission may influence cortical signal-to-noise ratio. J Neurophysiol 77: 3326-3339.

Hayar A, Shipley MT, Ennis M. 2005. Olfactory bulb external tufted cells are synchronized by multiple intraglomerular mechanisms. J Neurosci 25: 8197-8208.

Hayar A, Karnup S, Ennis M, Shipley MT. 2004b. External tufted cells: A major excitatory element that coordinates glomerular activity. J Neurosci 24: 6676-6685.

Hayar A, Karnup S, Shipley MT, Ennis M. 2004a. Olfactory bulb glomeruli: External tufted cells intrinsically burst at theta frequency and are entrained by patterned olfactory input. J Neurosci 24: 1190-1199.

Hayar A, Heyward PM, Heinbockel T, Shipley MT, Ennis M. 2001. Direct excitation of mitral cells via activation of alpha1-noradrenergic receptors in rat olfactory bulb slices. J Neurophysiol 86: 2173-2182.

Heinbockel T, Ennis M. 2003. Granule cell-mediated lateral inhibition can be driven by metabotropic glutamate receptors. AchemS Abstr XXV: 201.

Heinbockel T, Heyward P, Conquet F, Ennis M. 2004. Regulation of main olfactory bulb mitral cell excitability by metabotropic glutamate receptor mGluR1. J Neurophysiol 92: 3085-3096.

Hill JAJ, Zoli M, Bourgeois J, Changeux J. 1993. Immunocytochemical localization of a neuronal nicotinic receptor: The b2-subunit. J Neurosci 13: 1551-1568.

Hinds JW. 1968. Autoradiographic study of histogenesis in the mouse olfactory bulb. I. Time of origin of neurons and neuroglia. J Comp Neurol 134: 287-304.

Hirata Y. 1964. Some observations on the fine structure of the synapses in the olfactory bulb of the mouse, with particular reference to the atypical synaptic configurations. Arch Histol Jap 24: 293-302.

Hoffman BJ, Mezey E. 1989. Distribution of serotonin 5-HT1C receptor mRNA in adult rat brain. FEBS Lett 247: 453-462.

Hori N, Auker CR, Braitman DJ, Carpenter DO. 1982. Pharmacologic sensitivity of amino acid responses and synaptic activation of in vitro prepyriform neurons. J Neurophysiol 1289–1301.

Hormuzdi SG, Pais I, Le Bleu FEN, Towers SK, Rozoz A, et al. 2001. Impaired electrical signaling disrupts gamma frequency oscillations in connexin 36-deficient mice. Neuron 31: 487-495.

Horning MS, Blakemore LJ, Trombley PQ. 2000. Endogenous mechanisms of neuroprotection: Role of zinc, copper, and carnosine. Brain Res 852: 56-61.

Horning MS, Kwon B, Blakemore LJ, Spencer CM, Goltz M, et al. 2004. Alpha-amino-3-hydroxy-5-methyl-4-isoxazole-propionate receptor subunit expression in rat olfactory bulb. Neurosci Lett 372: 230-234.

Hsia AY, Vincent JD, Lledo PM. 1999. Dopamine depresses synaptic inputs into the olfactory bulb. J Neurophysiol 82: 1082-1085.

Huang Q, Zhou D, Chase K, Gusella JF, Aronin N, et al. 1992. Immunohistochemical localization of the D1 dopamine receptor in rat brain reveals its axonal transport, pre- and postsynaptic localization, and prevalence in the basal ganglia, limbic system, and thalamic reticular nucleus. Proc Natl Acad Sci USA 89: 11988-11992.

Hunt S, Schmidt J. 1978. Some observations on the binding patterns of alpha-bungarotoxin in the central nervous system of the rat. Brain Res 157: 213-232.

Illig KR, Haberly LB. 2003. Odor-evoked activity is spatially distributed in piriform cortex. J Comp Neurol 457: 361-373

Imaki T, Nahon JL, Sawchenko PE, Vale W. 1989. Widespread expression of corticotropin-releasing factor messenger RNA and immunoreactivity in the rat olfactory bulb. Brain Res 496: 35-44.

Inagaki S, Sakanaka M, Shiosaka S, Senba E, Takatsuki K, et al. 1982. Ontogeny of substance P-containing neuron system of the rat: Immunohistochemical analysis–I. Forebrain and upper brain stem. Neuroscience 7: 251-277.

Isaacson JS. 1999. Glutamate spillover mediates excitatory transmission in the rat olfactory bulb. Neuron 23: 377-384.

Isaacson JS, Murphy GJ. 2001. Glutamate-mediated extrasynaptic inhibition: Direct coupling of NMDA receptors to Ca^{2+}-activated K^{+} channels. Neuron 31: 1027-1034.

Isaacson JS, Strowbridge BW. 1998. Olfactory reciprocal synapses: Dendritic signaling in the CNS. Neuron 20: 749-761.

Isaacson JS. Vitten H. 2003. GABAB receptors inhibit dendrodendritic transmission in the rat olfactory bulb. J Neurosci 23: 2032-2039.

Ivic L, Pyrski MM, Margolis JW, Richards LJ, Firestein S, et al. 2000. Adenoviral vector-mediated rescue of the OMP-null phenotype in vivo. Nat Neurosci 3: 1113-1120.

Jackowski A, Parnavelas JG, Lieberman AR. 1978. The reciprocal synapse in the external plexiform layer of the mammalian olfactory bulb. Brain Res 159: 17-28.

Jacobowitz DM, Winsky L. 1991. Immunocytochemical localization of calretinin in the forebrain of the rat. J Comp Neurol 304: 198-218.

Jahr CE, Nicoll RA. 1982. Noradrenergic modulation of dendrodendritic inhibition in the olfactory bulb. Nature 297: 227-229.

Jardemark K, Nilsson M, Muyderman H, Jacobson I. 1997. Ca^{2+} ion permeability properties of (RS)α-amino-3-hydroxy-5-methyl-4-isoxazolepropionate (AMPA) receptors in isolated interneurons from the olfactory bulb of the rat. J Neurophysiol 77: 702-708.

Jin BK, Franzen L, Baker H. 1996. Regulation of c-Fos mRNA and fos protein expression in olfactory bulb from unilaterally odor-deprived mice. Int J Dev Neurosci 14: 971-982.

Jia C, Halpern M. 2004. Calbindin D28k, parvalbumin, and calretinin immunoreactivity in the main and accessory olfactory bulbs of the gray short-tailed opossum, *Monodelphis domestica*. J Morphol 259: 271-280.

Jiang MR, Griff ER, Ennis M, Zimmer LA, Shipley MT. 1996. Activation of locus coeruleus enhances the responses of olfactory bulb mitral cells to weak olfactory nerve input. J Neurosci 16: 6319-6329.

Johnson BA, Leon M. 2000a. Modular representations of odorants in the glomerular layer of the rat olfactory bulb and the effects of stimulus concentration. J Comp Neurol 422: 496-509.

Johnson BA, Leon M. 2000b. Odorant molecular length: One aspect of the olfactory code. J Comp Neurol 426: 330-338.

Johnson BA, Woo CC, Leon M. 1998. Spatial coding of odorant features in the glomerular layer of the rat olfactory bulb. J Comp Neurol 393: 457-471.

Johnson DM, Illig KR, Behan M, Haberly LB. 2000. New features of connectivity in piriform cortex visualized by intracellular injection of pyramidal cells suggest that "primary" olfactory cortex functions like "association" cortex in other sensory systems. J Neurosci 20: 6974-6982.

Jourdan F, Duveau A, Astic L, Holley A. 1980. Spatial distribution of [14C]2-deoxyglucose uptake in the olfactory bulbs of rats stimulated with two different odours. Brain Res 188: 139-154.

Kaba H, Keverne EB. 1988. The effect of microinfusions of drugs into the accessory olfactory bulb on the olfactory block to pregnancy. Neuroscience 25: 1007-1011.

Kaba H, Rosser A, Keverne B. 1989. Neural basis of olfactory memory in the context of pregnancy block. Neuroscience 32: 657-662.

Kakuta S, Oda S, Takayanagi M, Kishi K. 1998. Parvalbumin immunoreactive neurons in the main olfactory bulb of the house musk shrew, *Suncus murinus*. Brain Behav Evol 52: 285-291.

Kamisaki Y, Wada K, Nakamoto K, Itoh T. 1996. Effects of taurine on GABA release from synaptosomes of rat olfactory bulb. Amino Acids 10: 49-57.

Kanter ED, Kapur A, Haberly LB. 1996. A dendritic GABAA-mediated IPSP regulates facilitation of NMDA-mediated responses to burst stimulation of afferent fibers in piriform cortex. J Neurosci 16: 307-312.

Kaplan MS, McNelly NA, Hinds JW. 1985. Population dynamics of adult-formed granule neurons of the rat olfactory bulb. J Comp Neurol 239: 117-125.

Kapur A, Pearce RA, Lytton WW, Haberly LB. 1997. GABAA-mediated IPSCs in piriform cortex have fast and slow components with different properties and locations on pyramidal cells. J Neurophysiol 78: 2531-2545.

Kasowski HJ, Kim H, Greer CA. 1999. Compartmental organization of the olfactory bulb glomerulus. J Comp Neurol 407: 261-274.

Kato T, Yokouchi K, Kawagishi K, Fukushima N, Miwa T, et al. 2000. Fate of newly formed periglomerular cells in the olfactory bulb. Acta Otolaryngol 120: 876-879.

Kawano T, Margolis FL. 1982. Transsynaptic regulation of olfactory bulb catecholamines in mice and rats. J Neurochem 39: 342-348.

Keller A, Margolis FL. 1975. Immunological studies of the rat olfactory marker protein. J Neurochem 24: 1101-1106.

Keller A, Yagodin S, Aroniadou-Anderjaska V, Zimmer LA, Ennis M, et al. 1998. Functional organization of rat olfactory bulb glomeruli revealed by optical imaging. J Neurosci 18: 2602-2612.

Kendrick KM, Levy F, Keverne EB. 1992. Changes in the sensory processing of olfactory signals induced by birth in sheep. Science 256: 833-836.

Kerr DIB, Hagbarth KE. 1955. An investigation of the olfactory centrifugal system. J Neurophysiol 18: 362-374.

Kinoshita A, Shigemoto R, Ohishi H, van der Putten H, Mizuno N. 1998. Immunohistochemical localization of metabotropic glutamate receptors, mGluR7a and mGluR7b, in the central nervous system of the adult rat and mouse: A light and electron microscopic study. J Comp Neurol 393: 332-352.

Kinoshita A, Ohishi H, Neki A, Nomura S, Shigemoto R, et al. 1996. Presynaptic localization of metabotropic glutamate receptor, mGluR8, in the rhinencephalic areas: A light and electron microscope study in the rat. Neurosci Lett 207: 61-64.

Kinzie JM, Saugstad JA, Westbrook GL, Segerson TP. 1995. Distribution of metabotropic glutamate receptor 7 messenger RNA in the developing and adult rat brain. Neuroscience 69: 167-176.

Kinzie JM, Shinohara MM, Pol van den AN, Westbrook GL, Segerson TP. 1997. Immunolocalization of metabotropic glutamate receptor 7 in the rat olfactory bulb. J Comp Neurol 385: 372-384.

Kirschenbaum B, Doetsch F, Lois C, Alvarez-Buylla A. 1999. Adult subventricular zone neuronal precursors continue to proliferate and migrate in the absence of the olfactory bulb. J Neurosci 19: 2171-2180.

Kiyoshi K, Mori K, Ojima H. 1984. Distribution of local axon collaterals of mitral, displaced mitral, and tufted cells in the rabbit olfactory bulb. J Comp Neurol 225: 511-525.

Kiyoshi K, Mori K, Tazawa Y. 1982. Three-dimensional analysis of dendritic trees of mitral cells in the rabbit olfactory bulb. Neurosci Lett 28: 127-132.

Kleene SJ, Gesteland RC. 1991. Calcium-activated chloride conductance in frog olfactory cilia. J Neurosci 11: 3624-3629.

Kosaka T, Kosaka K. 2003. Neuronal gap junctions in the rat main olfactory bulb, with special reference to intraglomerular gap junction. Neurosci Res 45: 189-209.

Kosaka T, Kosaka K. 2004. Neuronal gap junctions between intraglomerular mitral/tufted cell dendrites in the mouse main olfactory bulb. Neurosci Res 49: 373-378.

Kosaka T, Kosaka K. 2005. Intraglomerular dendritic link connected by gap junctions and chemical synapses in the mouse main olfactory bulb: Electron microscopic serial section analyses. Neuroscience 131: 611-625.

Kosaka K, Heizmann CW, Kosaka T. 1994a. Calcium-binding protein parvalbumin-immunoreactive neurons in the rat olfactory bulb. 1. Distribution and structural features in adult rat. Exp Brain Res 99: 191-204.

Kosaka K, Heizmann CW, Kosaka T. 1994b. Calcium-binding protein parvalbumin-immunoreactive neurons in the rat olfactory bulb. 2. Postnatal development. Exp Brain Res 99: 205-213.

Kosaka K, Aika Y, Toida K, Kosaka T. 2001. Structure of intraglomerular dendritic tufts of mitral cells and their contacts with olfactory nerve terminals and calbindin-immunoreactive type 2 periglomerular neurons. J Comp Neurol 440: 219-235.

Kosaka T, Deans MR, Paul DL, Kosaka K. 2005. Neuronal gap junctions in the mouse main olfactory bulb: Morphological analyses on transgenic mice. Neuroscience 134: 757-769.

Kosaka K, Toida K, Margolis FL, Kosaka T. 1997. Chemically defined neuron groups and their subpopulations in the glomerular layer of the rat main olfactory bulb–II. Prominent differences in the intraglomerular dendritic arborization and their relationship to olfactory nerve terminals. Neuroscience 76: 775-786.

Kosaka K, Toida K, Aika Y, Kosaka T. 1998. How simple is the organization of the olfactory glomerulus?: The heterogeneity of so-called periglomerular cells. Neurosci Res 30: 101-110.

Kosaka K, Hama K, Nagatsu I, Wu JY, Kosaka T. 1988. Possible coexistence of amino acid (gamma-aminobutyric acid), amine (dopamine) and peptide (substance P); neurons containing immunoreactivities for glutamic acid decarboxylase, tyrosine hydroxylase and substance P in the hamster main olfactory bulb. Exp Brain Res 71: 633-642.

Kosaka T, Hataguchi Y, Hama K, Nagatsu I, Wu JY. 1985. Coexistence of immunoreactivities for glutamate decarboxylase and tyrosine hydroxylase in some neurons in the periglomerular region of the rat main olfactory bulb: Possible coexistence of gamma-aminobutyric acid (GABA) and dopamine. Brain Res 343: 166-171.

Kosaka T, Kosaka K, Hama K, Wu J-Y, Nagatsy I. 1987b. Differential effect of functional olfactory deprivation on the GABAergic and catecholaminergic traits in the rat main olfactory bulb. Brain Res 413: 197-203.

Kosaka K, Aika Y, Toida K, Heizmann CW, Hunziker W, et al. 1995. Chemically defined neuron groups and their subpopulations in the glomerular layer of the rat main olfactory bulb. Neurosci Res 23: 73-88.

Kosaka K, Hama K, Nagatsu I, Wu JY, Ottersen OP, et al. 1987a. Postnatal development of neurons containing both catecholaminergic and GABAergic traits in the rat main olfactory bulb. Brain Res 403: 355-360.

Kosaka T, Kosaka K, Hataguchi Y, Nagatsu I, Wu JY, et al. 1987c. Catecholaminergic neurons containing GABA-like and/or glutamic acid decarboxylase-like immunoreactivities in various brain regions of the rat. Exp Brain Res 66: 191-210.

Kosaka T, Kosaka K, Heizmann CW, Nagatsu I, Wu JY, et al. 1987d. An aspect of the organization of the GABAergic system in the rat main olfactory bulb: Laminar distribution of immunohistochemically defined subpopulations of GABAergic neurons. Brain Res 411: 373-378.

Koster NL, Norman AB, Richtand NM, Nickell WT, Puche AC, et al. 1999. Olfactory receptor neurons express D2 dopamine receptors. J Comp Neurol 411: 666-673.

Kratskin I, Belluzzi O. 2002. Immunohistochemical localization of the taurine-synthesizing enzyme in the rat olfactory bulb. Chem Senses 27: 46.

Kratskin IL, Rio JP, Kenigfest NB, Doty RL, Repérant J. 2000. A light and electron micro scopicstudy of taurine-like immunoreactivity in the main olfactory bulb of frogs. J Chem Neuroanat 18: 87-101.

Krautwurst D, Yau KW, Reed RR. 1998. Identification of ligands for olfactory receptors by functional expression of a receptor library. Cell 95: 917-926.

Kream RM, Davis BJ, Kawano T, Margolis FL, Macrides F. 1984. Substance P and catecholaminergic expression in neurons of the hamster main olfactory bulb. J Comp Neurol 222: 140-154.

Kubota Y, Jones EG. 1993. Co-localization of two calcium binding proteins in GABA cells of rat piriform cortex. Brain Res 600: 339-344.

Kunze WAA, Shafton AD, Kemm RE, McKenzie JS. 1991. Effect of stimulating the nucleus of the horizontal limb of the diagonal band on single unit activity in the olfactory bulb. Neuroscience 40: 21-27.

Kunze WAA, Shafton AD, Kemm RE, McKenzie JS. 1992. Intracellular responses of olfactory bulb granule cells to stimulating the horizontal diagonal band nucleus. Neuroscience 48: 363-369.

Laaris N, Ennis M. 2002. Distinct activity patterns evoked by activation of mitral/tufted cell and centrifugal fiber inputs to main olfactory bulb (MOB) granule cells. Soc Neurosci Abstr 561: 14.

Landis DMD, Reese TS, Raviola E. 1974. Differences in membrane structure between excitatory and inhibitory components of the reciprocal synapse in the olfactory bulb. J Comp Neurol 155: 67-92.

Laurie DJ, Seeburg PH, Wisden W. 1992. The distribution of 13 GABAA receptor subunit mRNAs in the rat brain. II. Olfactory bulb and cerebellum. J Neurosci 12: 1063-1076.

Le Jeune H, Jourdan F. 1994. Acetylcholinesterase-containing intrinsic neurons in the rat main olfactory bulb: Cytological and neurochemical features. Eur J Neurosci 6: 1432-1444.

Le Jeune H, Aubert I, Jourdan F, Quirion R. 1996. Developmental profiles of various cholinergic markers in the rat main olfactory bulb using quantitative autoradiography. J Comp Neurol 373: 433-450.

Leinders-Zufall T, Ma M, Zufall F. 1999. Impaired odor adaptation in olfactory receptor neurons after inhibition of Ca^{2+}/calmodulin kinase II. J Neurosci 19: 1-6.

Leinders-Zufall T, Greer CA, Shepherd GM, Zufall F. 1998. Imaging odor-induced calcium transients in single olfactory cilia: Specificity of activation and role in transduction. J Neurosci 18: 5630-5639.

Leinders-Zufall T, Rand MN, Shepherd GM, Greer CA, Zufall F. 1997. Calcium entry through cyclic nucleotide-gated channels in individual cilia of olfactory receptor cells: Spatiotemporal dynamics. J Neurosci 17: 4136-4148.

Levey AI, Kitt CA, Simonds WF, Price DL, Brann MR. 1991. Identification and localization of muscarinic acetylcholine receptor proteins in brain with subtype-specific antibodies. J Neurosci 11: 3218-3226.

Levey AI, Hersch SM, Rye DB, Sunahara RK, Niznik HB, et al. 1993. Localization of D1 and D2 dopamine receptors in brain with subtype-specific antibodies. Proc Natl Acad Sci USA 90: 8861-8865.

Libri V, Constanti A, Calaminci, M. Nistico G. 1994. A comparison of the muscarinic response and morphological properties of identified cells in the guinea-pig olfactory cortex in vitro. Neuroscience 59: 331-347.

Linster C, Wyble BP, Hasselmo ME. 1999. Electrical stimulation of the horizontal limb of the diagonal band of broca modulates population EPSPs in piriform cortex. J Neurophysiol 8: 2737-2742.

Liu N. 2000. Regional distribution of protein kinases in normal and odor-deprived mouse olfactory bulbs. Chem Sens 25: 401-406.

Liu G, Rao Y. 2003. Neuronal migration from the forebrain to the olfactory bulb requires a new attractant persistent in the olfactory bulb. J Neurosci 23: 6651-6659.

Liu WL, Shipley MT. 1994. Intrabulbar associational system in the rat olfactory bulb comprises cholecystokinin-containing tufted cells that synapse onto the dendrites of GABAergic granule cells. J Comp Neurol 346: 541-558.

Liu CJ, Grandes P, Matute C, Cuenod M, Streit P. 1989. Glutamate-like immunoreactivity revealed in rat olfactory bulb, hippocampus and cerebellum by monoclonal antibody and sensitive staining method. Histochemistry 90: 427-445.

Liu N, Cigola E, Tinti C, Jin BK, Conti B, et al. 1999. Unique regulation of immediate early gene and tyrosine hydroxylase expression in the odor-deprived mouse olfactory bulb. J Biol Chem 274: 3042-3047.

Lodovichi C, Belluscio L, Katz LC. 2003. Functional topography of connections linking mirror-symmetric maps in the mouse olfactory bulb. Neuron 38: 265-276.

Lois C, Alvarez-Buylla A. 1994. Long-distance neuronal migration in the adult mammalian brain. Science 264: 1145-1148.

Loo YT. 1931. The forebrain cf the opposum, *Dedelphys virginiana*. II. Histology. J Comp Neurol 52: 1-48.

López-Mascaraque L, de Carlos JA, Valverde F. 1990. Structure of the olfactory bulb of the hedgehog (*Erinaceus europaeus*): A Golgi study of the intrinsic organization of the superficial layers. J Comp Neurol 301: 243-261.

López-Mascaraque L, Villalba RM, de Carlos JA. 1989. Vasoactive intestinal polypeptide-immunoreactive neurons in the main olfactory bulb of the hedgehog (*Erinaceus europaeus*). Neurosci Lett 98: 19-21.

Lowe G. 2003. Flash photolysis reveals a diversity of ionotropic glutamate receptors on the mitral cell somatodendritic membrane. J Neurophysiol 90: 1737-1746.

Lowe G, Gold GH. 1993. Contribution of the ciliary cyclic nucleotide-gated conductance to olfactory transduction in the salamander. J Physiol 462: 175-96.

Luo M, Katz L. 2001. Response correlation maps of neurons in the mammalian olfactory bulb. Neuron 32: 1165-1179.

Luskin MB. 1993. Restricted proliferation and migration of postnatally generated neurons derived from the forebrain subventricular zone. Neuron 11: 173-189.

Luskin MB, Price JL. 1983. The topographic organization of associational fibers of the olfactory system in the rat, including centrifugal fibers to the olfactory bulb. J Comp Neurol 216: 264-291.

Lysakowski A, Wainer BH, Bruce G, Hersh LB. 1989. An atlas of the regional and laminar distribution of choline acetyltransferase immunoreactivity in rat cerebral cortex. Neuroscience 28: 291-336.

Ma J, Lowe G. 2005. Dendrites of juxtaglomerular neurons in rat olfactory bulb express Ca^{2+}-permeable AMPA receptors. AChemS Abstr 86.

Mackay-Sim A, Chuah MI. 2000. Neurotrophic factors in the primary olfactory pathway. Prog Neurobiol 62: 527-559.

MacLeod NK, Straughan DW. 1979. Responses of olfactory bulb neurones to the dipeptide carnosine. Exp Brain Res 34: 183-188.

Macrides F, Schneider SP. 1982. Laminar organization of mitral and tufted cells in the main olfactory bulb of the adult hamster. J Comp Neurol 208: 419-430.

Macrides F, Schoenfeld TA, Marchand JE, Clancy AN. 1985. Evidence for morphologically, neurochemically and functionally heterogeneous classes of mitral and tufted cells in the olfactory bulb. Chem Sens 2: 175-202.

Malioso M-L, Marqueze-Pouey B, Kuhse J, Betz H. 1991. Widespread expression of glycine receptor subunit mRNAs in the adult and developing rat brain. EMBO J 10: 2401-2409.

Malnic B, Hirono J, Sato T, Buck LB. 1999. Combinatorial receptor codes for odors. Cell 96: 713-723.

Mansour A, Meador-Woodruff JH, Bunzow JR, Civelli O, Akil H, et al. 1990a. Localization of dopamine D2 receptor mRNA and D1 and D2 receptor binding in the rat brain and pituitary: An in situ hybridization-receptor autoradiographic analysis. J Neurosci 10: 2587-2600.

Mansour A, Meador-Woodruff JH, Camp DM, Robinson TE, Bunzow J, et al. 1990b. The effects of nigrostriatal 6-hydroxydopamine lesions on dopamine D2 receptor mRNA and opioid systems. Prog Clin Biol Res 328: 227-230.

Marek GJ, Aghajanian GK. 1994. Excitation of interneurons in piriform cortex by 5-hydroxytryptamine: Blockade by MDL 100,907, a highly selective 5-HT2A receptor antagonist. Eur J Pharmacol 259: 137-141.

Marek GJ, Aghajanian GK. 1995. Protein kinase C inhibitors enhance the 5-HT2A receptor-mediated excitatory effects of serotonin on interneurons in rat piriform cortex. Synapse 21: 123-130.

Marek GJ, Aghajanian GK. 1996a. Alpha 1B-adrenoceptor-mediated excitation of piriform cortical interneurons. Eur J Pharmacol 305: 95-100.

Marek GJ, Aghajanian GK. 1996b. LSD and the phenethylamine hallucinogen DOI are potent partial agonists at 5-HT2A receptors on interneurons in rat piriform cortex. J Pharmacol Exp Ther 278: 1373-1382.

Margeta-Mitrovic M, Mitrovic I, Riley RC, Jan LY, Basbaum AI. 1999. Immunohistochemical localization of GABA(B) receptors in the rat central nervous system. J Comp Neurol 405: 299-321.

Margolis FL. 1972. A brain protein unique to the olfactory bulb. Proc Natl Acad Sci USA 69: 1221-1224.

Margolis FL. 1974. Carnosine in the primary olfactory pathway. Science 184: 909-911.

Margolis FL. 1980. Carnosine: An olfactory neuropeptide. The role of peptides in neuronal function. Barker JL, Smith TGJ, editors. New York: Marcel Dekker, Inc.; pp. 545-572.

Margolis FL, Grillo M. 1984. Carnosine, homocarnosine and anserine in vertebrate retinas. Neurochem Int'l 6: 207-209.

Margolis FL, Grillo M, Grannot-Reisfeld N, Farbman AI. 1983 Purification, characterization and immunocytochemical localization of mouse kidney carnosinase. Biochimica et Biophysica Acta 744: 237-248.

Margolis FL, Grillo M, Hempstead J, Morgan JI. 1987. Monoclonal antibodies to mammalian carnosine synthetase. J Neurochem 48: 593-600.

Margolis FL, Grillo M, Kawano T, Farbman AI. 1985. Carnosine synthesis in olfactory tissue during ontogeny: Effect of exogenous b-alanine. J Neurochem 44: 1459-1464.

Martin LJ, Blackstone CD, Huganir RL, Price DL. 1992. Cellular localization of a metabotropic glutamate receptor in rat brain. Neuron 9: 259-270.

Martin LJ, Blackstone CD, Levey AI, Huganir RL, Price DL. 1993. Cellular localizations of AMPA glutamate receptors within the basal forebrain magnocellular complex of rat and monkey. J Neurosci 13: 2249-2263.

Masu M, Tanabe Y, Tsuchida K, Shigemoto R, Nakanishi S. 1991. Sequence and expression of a metabotropic glutamate receptor. Nature 349: 760-765.

Matsutani S, Senba E, Tohyama M. 1988. Neuropeptide- and neurotransmitter-related immunoreactivities in the developing rat olfactory bulb. J Comp Neurol 272: 331-342.

McCune SK, Voigt MM, Hill JM. 1993. Expression of multiple alpha adrenergic receptor subtype messenger RNAs in the adult rat brain. Neuroscience 57: 143-151.

McLean JH, Harley CW. 2004. Olfactory learning in the rat pup: A model that may permit visualization of a mammalian memory trace. Neuroreport 15: 1691-1697.

McLean JH, Shipley MT. 1987a. Serotonergic afferents to the rat olfactory bulb: I. Origins and laminar specificity of serotonergic inputs in the adult rat. J Neurosci 7: 3016-3028.

McLean JH, Shipley MT. 1987b. Serotonergic afferents to the rat olfactory bulb: II. Changes in fiber distribution during development. J Neurosci 7: 3029-3039.

McLean JH, Shipley MT. 1988. Postmitotic, postmigrational expression of tyrosine hydroxylase in olfactory bulb dopaminergic neurons. J Neurosci 8: 3658-3669.

McLean JH, Darby-King A, Bonnell WS. 2001. Neonatal olfactory sensory deprivation decreases BDNF in the olfactory bulb of the rat. Dev Brain Res 128: 17-24.

McLean JH, Darby-King A, Hodge E. 1996. 5-HT2 receptor involvement in conditioned olfactory learning in the neonatal rat pup. Behav Neurosci 110: 1426-1434.

McLean JH, Darby-King A, Paterno GD. 1995. Localization of 5-HT2A receptor mRNA by in situ hybridization in the olfactory bulb of the postnatal rat. J Comp Neurol 353: 371-378.

McLean JH, Darby-King A, Sullivan RM, King SR. 1993. Serotonergic influence on olfactory learning in the neonate rat. Behav Neural Biol 60: 152-162.

McLean JH, Shipley MT, Nickell WT, Aston-Jones G, Reyher CK. 1989. Chemoanatomical organization of the noradrenergic input from locus coeruleus to the olfactory bulb of the adult rat. J Comp Neurol 285: 339-349.

McLennan H. 1971.The pharmacology of inhibition of mitral cells in the olfactory bulb. Brain Res 29: 177-184.

McMillian MK, Mullis SB, Wu G-C, Hudson PM, Pennypacker KR, et al. 1994. Regulation of tyrosine hydroxylase in olfactory bulb cultures: Selective inhibition of depolarization-induced increase by endogenous opioids. Brain Res 658: 105-111.

Mechawar N, Saghatelyan A, Grailhe R, Scoriels L, Gheusi G, et al. 2004. Nicotinic receptors regulate the survival of newborn neurons in the adult olfactory bulb. Proc Natl Acad Sci USA 101: 9822-9826.

Meisami E. 1989. A proposed relationship between increases in the number of olfactory receptor neurons, convergence ratio and sensitivity in the developing rat. Brain Res Dev Brain Res 46: 9-19.

Meisami E, Safari L. 1981. A quantitative study of the effects of early unilateral olfactory deprivation on the number and distribution of mitral and tufted cells and of glomeruli in the rat olfactory bulb. Brain Res 221: 81-107.

Meisami E, Sendera TJ. 1993. Morphometry of rat olfactory bulbs stained for cytochrome oxidase reveals that the entire population of glomeruli forms early in the neonatal period. Brain Res Dev Brain Res 71: 253-257.

Menco BP. 1984. Ciliated and microvillous structures of rat olfactory and nasal respiratory epithelia. A study using ultra-rapid cryo-fixation followed by freeze-substitution or freeze-etching. Cell Tissue Res 235: 225-241.

Menco BP, Farbman AI. 1985a. Genesis of cilia and microvilli of rat nasal epithelia during pre-natal development. I. Olfactory epithelium, qualitative studies. J Cell Sci 78: 283-310.

Menco BP, Farbman AI. 1985b. Genesis of cilia and microvilli of rat nasal epithelia during pre-natal development. II. Olfactory epithelium, a morphometric analysis. J Cell Sci 78: 311-336.

Mengod G, Nguyen H, Le H, Waeber C, Lubbert H, et al. 1990a. The distribution and cellular localization of the serotonin 1C receptor mRNA in the rodent brain examined by in situ hybridization histochemistry. Comparison with receptor binding distribution. Neuroscience 35: 577-591.

Mengod G, Pompeiano M, Martinez-Mir MI, Palacios JM. 1990b. Localization of the mRNA for the 5-HT2 receptor by in situ hybridization histochemistry. Correlation with the distribution of receptor sites. Brain Res 524: 139-143.

Merchenthaler I, Csernus V, Csontos C, Petrusz P, Mess B. 1988. New data on the immunocytochemical localization of thyrotropin-releasing hormone in the rat central nervous system. Am J Anat 181: 359-376.

Miller LP, Johnson AE, Gelhard RE, Insel TR. 1990. The ontogeny of excitatory amino acid receptors in the rat forebrain–II. Kainic acid receptors. Neuroscience 35: 45-51.

Miragall F, Simburger E, Dermietzel R. 1996. Mitral and tufted cells of the mouse olfactory bulb possess gap junctions and express connexin43 mRNA. Neurosci Lett 216: 199-202.

Mirich JM, Illig KR, Brunjes PC. 2004. Experience-dependent activation of extracellular signal-related kinase (ERK) in the olfactory bulb. J Comp Neurol 479: 234-241.

Molnar E, Baude A, Richmond SA, Patel PB, Somogyi P, et al. 1993. Biochemical and immunocytochemical characterization of antipeptide antibodies to a cloned GluR1 glutamate receptor subunit: Cellular and subcellular distribution in the rat forebrain. Neuroscience 53: 307-326.

Mombaerts P, Wang F, Dulac C, Chao SK, Nemes A, et al. 1996. Visualizing an olfactory sensory map. Cell 87: 675-686.

Monaghan DT, Cotman CW. 1982. The distribution of [3H] kainic acid binding sites in rat CNS as determined by autoradiography. Brain Res 252: 91-100.

Monaghan DT, Yao D, Cotman CW. 1985. L-[3H]Glutamate binds to kainate-, NMDA- and AMPA-sensitive binding sites: An autoradiographic analysis. Brain Res 340: 378-383.

Montague AA, Greer CA. 1999. Differential distribution of ionotropic glutamate receptor subunits in the rat olfactory bulb. J Comp Neurol 405: 233-246.

Monyer H, Burnashev N, Laurie DJ, Sakmann B, Seeburg PH. 1994. Developmental and regional expression in the rat brain and functional properties of four NMDA receptors. Neuron 12: 529-540.

Morales M, Bloom FE. 1997. The 5-HT3 receptor is present in different subpopulations of GABAergic neurons in the rat telencephalon. J Neurosci 17: 3157-3167.

Moran DT, Rowley JC III, Jafek BW. 1982. Electron microscopy of human olfactory epithelium reveals a new cell type: The microvillar cell. Brain Res 253: 39-46.

Mori K. 1987a. Membrane and synaptic properties of identified neurons in the olfactory bulb. Prog Neurobiol 29: 275-320.

Mori K. 1987b. Monoclonal antibodies (2C5 and 4C9) against lactoseries carbohydrates identify subsets of olfactory and vomeronasal receptor cells and their axons in the rabbit. Brain Res 408: 215-221.

Mori K, Kishi K. 1982. The morphology and physiology of the granule cells in the rabbit olfactory bulb revealed by intracellular recording and HRP injection. Brain Res 247: 129-133.

Mori K, Takagi SF. 1978. Activation and inhibition of olfactory bulb neurones by anterior commissure volleys in the rabbit. J Physiol (Lond) 279: 589-604.

Mori K, Kishi K, Ojima H. 1983. Distribution of dendrites of mitral, displaced mitral, tufted, and granule cells in the rabbit olfactory bulb. J Comp Neurol 219: 339-355.

Moriceau S, Sullivan RM. 2004. Unique neural circuitry for neonatal olfactory learning. J Neurosci 24: 1182-1189.

Morizumi T, Tsukatani H, Miwa T. 1994. Olfactory disturbance induced by deafferentation of serotonergic fibers in the olfactory bulb. Neuroscience 61: 733-738.

Morrison EE, Costanzo RM. 1990. Morphology of the human olfactory epithelium. J Comp Neurol 297: 1-13.

Mouly AM, Elaagouby A, Ravel N. 1995. A study of the effects of noradrenaline in the rat olfactory bulb using evoked field potential response. Brain Res 681: 47-57.

Mouradian LE, Scott JW. 1988. Cytochrome oxidase staining marks dendritic zones in the rat olfactory bulb external plexiform layer. J comp Neurol 271: 507-518.

Mugnaini E, Oertel WH, Wouterlood FF. 1984a. Immunocytochemical localization of GABA neurons and dopamine neurons in the rat main and accessory olfactory bulbs. Neurosci Lett 47: 221-226.

Mugnaini E, Wouterlood FG, Dahl A-L, Oertel WH. 1984b. Immunocytochemical identification of GABAergic neurons in the main olfactory bulb of the rat. Archives Italiennes de Biologie 122: 83-113.

Muller JF, Marc RE. 1984. Three distinct morphological classes of receptors in fish olfactory organs. J Comp Neurol 222: 482-495.

Munger SD, Lane AP, Zhong H, Leinders-Zufall T, Yau KW, et al. 2001. Central role of the CNGA4 channel subunit in Ca^{2+}-calmodulin-dependent odor adaptation. Science 294: 2172-2175.

Murphy GJ, Darcy DP, Isaacson JS. 2005. Intraglomerular inhibition: Signaling mechanisms of an olfactory microcircuit. Nat Neurosci 8: 354-364.

Nadi NS, Head R, Grillo M, Hempstead J, Grannot-Reisfeld N, et al. 1981. Chemical deafferentation of the olfactory bulb: Plasticity of the levels of tyrosine hydroxylase, dopamine and norepinephrine. Brain Res 213: 365-377.

Nagayama S, Takahashi YK, Yoshihara Y, Mori K. 2004. Mitral and tufted cells differ in the decoding manner of odor maps in the rat olfactory bulb. J Neurophysiol 91: 2532-2540.

Najbauer J, Leon M. 1995. Olfactory experience modulated apoptosis in the developing olfactory bulb. Brain Res 674: 245-251.

Nakajima T, Okamura M, Ogawa K, Taniguchi K. 1996. Immunohistochemical and enzyme histochemical characteristics of short axon cells in the olfactory bulb of the golden hamster. J Vet Med Sci 58: 903-908.

Nakashima M, Mori K, Takagi SF. 1978. Centrifugal influence on olfactory bulb activity in the rabbit. Brain Res 154: 301-316.

Nicholas AP, Pieribone V, Hokfelt T. 1993. Distributions of mRNAs for alpha-2 adrenergic receptor subtypes in rat brain: An in situ hybridization study. J Comp Neurol 328: 575-594.

Nickell WT, Shipley MT. 1988a. Neurophysiology of magnocellular forebrain inputs to the olfactory bulb in the rat: Frequency potentiation of field potentials and inhibition of output neurons. J Neurosci 8: 4492-4502.

Nickell WT, Shipley MT. 1988b. Two anatomically specific classes of candidate cholinoceptive neurons in the rat olfactory bulb. J Neurosci 8: 4482-4491.

Nickell WT, Shipley MT. 1993. Evidence for presynaptic inhibition of the olfactory commissural pathway by cholinergic agonists and stimulation of the nucleus of the diagonal band. J Neurosci 13: 650-659.

Nickell WT, Norman AB, Wyatt LM, Shipley MT. 1991. Olfactory bulb DA receptors may be located on terminals of the olfactory nerve. Neuroreport 2: 9-12.

Nicoll RA. 1971. Pharmacological evidence for GABA as the transmitter in granule cell inhibition in the olfactory bulb. Brain Res 35: 137-149.

Nicoll RA, Alger BE. 1980. Peptides as putative excitatory neurotransmitters: Carnosine, enkephalin, substance P and TRH. Proc R Soc Lond 210: 133-149.

Nowycky MC, Halasz N, Shepherd GM. 1983. Evoked field potential analysis of dopaminergic mechanisms in the isolated turtle olfactory bulb. Neuroscience 8: 717-722.

Nusser Z, Kay LM, Laurent G, Homanics GE, Mody I. 2001. Disruption of GABAA receptors on GABAergic interneurons leads to increased oscillatory power in the olfactory bulb network. J Neurophysiol 86: 2823-2833.

Ohishi H, Neki A, Mizuno N. 1998. Distribution of a metabotropic glutamate receptor, mGluR2, in the central nervous system of the rat and mouse: An immunohistochemical study with a monoclonal antibody. Neurosci Res 30: 65-82.

Ohishi H, Shigemoto R, Nakanishi S, Mizuno N. 1993a. Distribution of the messenger RNA for a metabotropic glutamate receptor, mGluR2, in the central nervous system of the rat. Neuroscience 53: 1009-1018.

Ohishi H, Shigemoto R, Nakanishi S, Mizuno N. 1993b. Distribution of the mRNA for a metabotropic glutamate receptor (mGluR3) in the rat brain: An in situ hybridization study. J Comp Neurol 335: 252-266.

Ohishi H, Akazawa C, Shigemoto R, Nakanishi S, Mizuno N. 1995. Distributions of mRNAs for L-2-amino-4-phoshonobutyrate-sensitive metabotropic glutamate receptors mGluR4 and mGluR7, in the rat brain. J Comp Neurol 360: 555-570.

Ohm TG, Müller H, Braak E. 1991. Calbindin-D-28k-like immunoreactive structures in the olfactory bulb and anterior olfactory nucleus of the human adult: Distribution and cell typology-partial complementarity with parvalbumin. Neuroscience 42: 823-840.

Ohm TG, Braak E, Probst A, Weindl A. 1988. Neuropeptide Y-like immunoreactive neurons in the human olfactory bulb. Brain Res 451: 295-300.

Ohm TG, Müller H, Ulfig N, Braak E. 1990. Glutamic-acid-decarboxylase- and parvalbumin-like-immunoreactive structures in the olfactory bulb of the human adult. J Comp Neurol 291: 1-8.

Ojima H, Mori K, Kishi K. 1984. The trajectory of mitral cell axons in the rabbit olfactory cortex revealed by intracellular HRP injection. J Comp Neurol 230: 77-87.

Okutani F, Kaba H, Takahashi S, Seto K. 1998. The biphasic effect of locus coeruleus noradrenergic activation on dendrodendritic inhibition in the rat olfactory bulb. Brain Res 783: 272-279.

Onoda N. 1992. Odor-induced fos-like immunoreactivity in the rat olfactory bulb. Neurosci Lett 137: 157-160.

Orona E, Rainer EC, Scott JW. 1984. Dendritic and axonal organization of mitral and tufted cells in the rat olfactory bulb. J Comp Neurol 226: 346-356.

Orona E, Scott JW, Rainer EC. 1983. Different granule cell populations innervate superficial and deep regions of the external plexiform layer in rat olfactory bulb. J Comp Neurol 217: 227-237.

Palacios JM, Kuhar MJ. 1980. Beta-adrenergic-receptor localization by light microscopic autoradiography. Science 208: 1378-1380.

Palacios JM, Kuhar MJ. 1982. Beta adrenergic receptor localization in rat brain by light microscopic autoradiography. Neurochem Int 4: 473-490.

Palacios JM, Niehoff DL, Kuhar MJ. 1981a. [3H]Spiperone binding sites in brain: Autoradiographic localization of multiple receptors. Brain Res 213: 277-289.

Palacios JM, Wamsley JK, Kuhar MJ. 1981b. High affinity GABA receptors-autoradiographic localization. Brain Res 222: 285-307.

Palouzier-Paulignan B, Duchamp-Viret P, Hardy AB, Duchamp A. 2002. GABA(B) receptor-mediated inhibition of mitral/tufted cell activity in the rat olfactory bulb: A whole-cell patch-clamp study in vitro. Neuroscience 111: 241-250.

Panzanelli P, Perazzini A-Z, Fritschy J-M, Sassoè-Pognetto M. 2005. Heterogeneity of γ-aminobutyric acid type A receptors in mitral and tufted cells of the rat main olfactory bulb. J Comp Neurol 484: 121-131.

Panzanelli P, Homanics GE, Ottersen OP, Fritschy J-M, Sassoè-Pognetto M. 2004. Pre- and postsynaptic GABAA receptors at reciprocal dendrodendritic synapses in the olfactory bulb. Eur J Neurosci 20: 2945-2952.

Paternostro MA, Reyher CKH, Brunjes PC. 1995. Intracellular injections of Lucifer Yellow into lightly fixed mitral cells reveal neuronal dye-coupling in the developing rat olfactory bulb. Dev Brain Res 84: 1-10.

Perez H, Hernandez A, Almli CR. 1987. Locus coeruleus stimulation modulates olfactory bulb evoked potentials. Brain Res Bull 18: 767-770.

Persohn E, Malherbe P, Richards JG. 1992. Comparative molecular neuroanatomy of cloned GABAA receptor subunits in the rat CNS. J Comp Neurol 326: 193-216.

Petralia RS, Wenthold RJ. 1992. Light and electron immunocytochemical localization of AMPA-selective glutamate receptors in the rat brain. J Comp Neurol 318: 329-354.

Petralia RS, Wang YX, Wenthold RJ. 1994a. Histological and ultrastructural localization of the kainate receptor subunits, KA2 and GluR6/7, in the rat nervous system using selective antipeptide antibodies. J Comp Neurol 349: 85-110.

Petralia RS, Yokotani N, Wenthold RJ. 1994b. Light and electron microscope distribution of the NMDA receptor subunit NMDAR1 in the rat nervous system using a selective anti-peptide antibody. J Neurosci 14: 667-696.

Petreanu L, Alvarez-Buylla A. 2002. Maturation and death of adult-born granule neurons: Role of olfaction. J Neurosci 22: 6106-6113.

Philpot BD, Lim JH, Brunjes PC. 1997. Activity-dependent regulation of calcium-binding proteins in the developing rat olfactory bulb. J Comp Neurol 387: 12-26.

Philpot BD, Men D, McCarty R, Brunjes. 1998. Activity-dependent regulation of dopamine content in the olfactory bulbs of naris-occluded rats. Neuroscience 85: 969-977.

Pieribone VA, Nicholas AP, Dagerlind A, Hokfelt T. 1994. Distribution of α1 adrenoreceptors in rat brain revealed by in situ hybridization experiments utilizing subtype-specific probes. J Neurosci 14: 4252-4268.

Pinching AJ, Powell TP. 1971a. The neuron types of the glomerular layer of the olfactory bulb. J Cell Sci 9: 305-345.

Pinching AJ, Powell TP. 1971b. The neuropil of the glomeruli of the olfactory bulb. J Cell Sci 9: 347-377.

Pinching AJ, Powell TP. 1971c. The neuropil of the periglomerular region of the olfactory bulb. J Cell Sci 9: 379-409.

Pinching AJ, Powell TP. 1972a. A study of terminal degeneration in the olfactory bulb of the rat. J Cell Sci 10: 585-619.

Pinching AJ, Powell TP. 1972b. Experimental studies on the axons intrinsic to the glomerular layer of the olfactory bulb. J Cell Sci 10: 637-655.

Pinching AJ, Powell TP. 1972c. The termination of centrifugal fibres in the glomerular layer of the olfactory bulb. J Cell Sci 10: 621-635.

Pissonnier D, Thiery JC, Fabre-Nys C, Poindron P, Keverne EB. 1985. The importance of olfactory bulb noradrenalin for maternal recognition in sheep. Physiol Behav 35: 361-363.

Pompeiano M, Palacios JM, Mengod G. 1992. Distribution and cellular localization of mRNA coding for 5-HT1A receptor in the rat brain: Correlation with receptor binding. J Neurosci 12: 440-453.

Pompeiano M, Palacios JM, Mengod G. 1994. Distribution of the serotonin 5-HT2 receptor family mRNAs: Comparison between 5-HT2A and 5-HT2C receptors. Brain Res Mol Brain Res 23: 163-178.

Porteros A, Brinon JG, Crespo C, Okazaki K, Hidaka H, et al. 1996. Neurocalcin immunoreactivity in the rat accessory olfactory bulb. Brain Res 729: 82-89.

Postlewaither M, Constanti A, Libri A. 1998. Muscarinic agonist-induced burst firing in immature olfactory cortex neurons in vitro. J Neurophysiol 79: 2003-2012.

Potter SM, Zheng C, Koos DS, Feinstein P, Fraser SE, et al. 2001. Structure and emergence of specific olfactory glomeruli in the mouse. J Neurosci 21: 9713-9723.

Pourcho RG, Goebel DJ, Jojich L, Hazlett JC. 1992. Immunocytochemical evidence for the involvement of glycine in sensory centers of the rat brain. Neuroscience 46: 643-656.

Pow DV, Sullivan R, Reye P, Hermanussen S. 2002. Localization of taurine transporters, taurine, and 3H taurine accumulation in the rat retina, pituitary, and brain. Glia 37: 153-168.

Pressler RT, Strowbridge BW. 2006. Blanes cells mediate persistent feedforward inhibition onto granule cells in the olfactory bulb. Neuron 49: 889-904.

Price JL, Powell TPS. 1970a. An electron-microscopic study of the termination of the afferent fibres to the olfactory bulb from the cerebral hemisphere. J Cell Sci 7: 157-187.

Price JL, Powell TPS. 1970b. The mitral and short axon cells of the olfactory bulb. J Cell Sci 7: 631-651.

Price JL, Powell TPS. 1970c. The morphology of the granule cells of the olfactory bulb. J Cell Sci 9: 91-123.

Price JL, Powell TPS. 1970d. The synaptology of the granule cells of the olfactory bulb. J Cell Sci 7: 125-155.

Price JL, Slotnick BM. 1983. Dual olfactory representation in the rat thalamus: An anatomical and electrophysiological study. J Comp Neurol 215: 63-77.

Puche AC, Shipley MT. 1999. Odor-induced, activity-dependent transneuronal gene induction in vitro: Mediation by NMDA receptors. J Neurosci 19: 1359-1370.

Puopolo M, Bean BP, Raviola E. 2005. Spontaneous activity of isolated dopaminergic periglomerular cells of the main olfactory bulb. J Neurophysiol 94: 3618-3627.

Rall W, Shepherd GM, Reese TS, Brightman MW. 1966. Dendrodendritic synaptic pathway for inhibition in the olfactory bulb. Exp Neurol 14: 44-56.

Rash JE, Staines WA, Yasmura T, Patel D, Furman DS, et al. 2000. Immunogold evidence that the neuronal gap junctions in adult rat brain and spinal cord contain connexin-36 but not connexin-32 or connexin-43. Proc Nat Acad Sci USA 97: 7573-7578.

Rash JE, Davidson KG, Kamasawa N, Yasumura T, Kamasawa M, Zhang C, Michaels R, Restrepo D, Ottersen OP, Olson CO, Nagy JI. 2005. Ultrastructural localization of connexins (Cx36, Cx43, Cx45), glutamate receptors and aquaporin-4 in rodent olfactory mucosa, olfactory nerve and olfactory bulb. J Neurocytol 34: 307-341.

Rawson NE, Eberwine J, Dotson R, Jackson J, Ulrich P, et al. 2000. Expression of mRNAs encoding for two different olfactory receptors in a subset of olfactory receptor neurons. J Neurochem 75: 185-195.

Reed RR. 1992. Signaling pathways in odorant detection. Neuron 8: 205-209.

Ressler KJ, Sullivan SL, Buck LB. 1993. A zonal organization of odorant receptor gene expression in the olfactory epithelium. Cell 73: 597-609.

Ressler KJ, Sullivan SL, Buck LB. 1994. Information coding in the olfactory system: Evidence for a stereotyped and highly organized epitope map in the olfactory bulb. Cell 79: 1245-1255.

Reyher CK, Lubke J, Larsen WJ, Hendrix GM, Shipley MT, et al. 1991. Olfactory bulb granule cell aggregates: Morphological evidence for interperikaryal electrotonic coupling via gap junctions. J Neurosci 11: 1485-1495.

Ribak CE, Vaughn JE, Saito K, Barber R, Roberts E. 1977. Glutamate decarboxylase localization in neurons of the olfactory bulb. Brain Res 126: 1-18.

Richards JG, Schoch P, Haring P, Takacs B, Mohler H. 1987. Resolving GABAA/benzodiazepine receptors: Cellular and subcellular localization in the CNS with monoclonal antibodies. J Neurosci 7: 1866-1886.

Roberts GW, Woodhams PL, Polak JM, Crow TJ. 1982. Distribution of neuropeptides in the limbic system of the rat: The amygdaloid complex. Neuroscience 7: 99-131.

Rochefort C, Gheusi G, Vincent J-D, Lledo P-M. 2002. Enriched odor exposure increases the number of newborn neurons in the adult olfactory bulb and improves odor memory. J Neurosci 22: 2679-2689.

Rochel S, Margolis FL. 1982. Carnosine release from olfactory bulb synaptosomes is calcium-dependent and depolarization-stimulated. J Neurochem 38: 505-1514.

Rodriguez R, Haberly LB. 1989. Analysis of synaptic events in the opossum piriform cortex with improved current source-density techniques. J Neurophysiol 61: 702-718.

Rogers JH. 1992. Immunohistochemical markers in rat brain: Colocalization of calretinin and calbindin-D28k with tyrosine hydroxylase. Brain Res 587: 203-210.

Romano C, Sesma MA, McDonald CT, O'Malley K, van den Pol AN, et al. 1995. Distribution of metabotropic glutamate receptor mGluR5 immunoreactivity in rat brain. J Comp Neurol 355: 455-469.

Rosin DL, Talley EM, Lee A, Stornetta RL, Gaylinn BD, et al. 1996. Distribution of alpha 2C-adrenergic receptor-like immunoreactivity in the rat central nervous system. J Comp Neurol 372: 135-165.

Rosin J-F, Datiche F, Cattarelli M. 1999. Modulation of the piriform cortex activity by the basal forebrain: An optical recording study in the rat. Brain Res 820: 105-111.

Ross CD, Godfrey DA, Parli JA. 1995. Amino acid concentrations and selected enzyme activities in rat auditory, olfactory, and visual systems. Neurochem Res 20: 1483-1490.

Rosser AE, Keverne EB. 1985. The importance of central noradrenergic neurones in the formation of an olfactory memory in the prevention of pregnancy block. Neuroscience 15: 1141-1147.

Rotter A, Birdsall NJ, Field PM, Raisman G. 1979. Muscarinic receptors in the central nervous system of the rat. II. Distribution of binding of [3H]propylbenzilylcholine mustard in the midbrain and hindbrain. Brain Res 180: 167-183.

Rowley JC III, Moran DT, Jafek BW. 1989. Peroxidase backfills suggest the mammalian olfactory epithelium contains a second morphologically distinct class of bipolar sensory neuron: The microvillar cell. Brain Res 502: 387-400.

Royet J, Distel H, Hudson R, Gervais R. 1998. A re-estimation of the number of glomeruli and mitral cells in the olfactory bulb of rabbit. Brain Res 788: 35-42.

Royet JP, Souchier C, Jourdan F, Ploye H. 1988. Morphometric study of the glomerular population in the mouse olfactory bulb: Numerical density and size distribution along the rostrocaudal axis. J Comp Neurol 270: 559-568.

Rozental R, Srinivas M, Gökhan S, Urban M, Dermietzel R, et al. 2000. Temporal expression of neuronal connexins during hippocampal ontogeny. Brain Res Rev 32: 57-71.

Rubin BD, Katz LC. 1999. Optical imaging of odorant representations in the mammalian olfactory bulb. Neuron 23: 499-511.

Rubin BD, Katz LC. 2001. Spatial coding of enantiomers in the rat olfactory bulb. Nat Neurosci 4: 355-356.

Saghatelyan A, de Chevigny A, Schachner M, Lledo P-M. 2004. Tenascin-R mediates activity-dependent recruitment of neuroblasts in the adult mouse forebrain. Nat Neurosci 7: 347-356.

Sahara Y, Kubota T, Ichikawa M. 2001. Cellular localization of metabotropic glutamate receptors mGluR1, 2/3, 5 and 7 in the main and accessory olfactory bulb of the rat. Neurosci Lett 312: 59-62.

Sahin M, Bowen WD, Donoghue JP. 1992. Location of nicotinic and muscarinic cholinergic and mu-opiate receptors in rat cerebral neocortex: Evidence from thalamic and cortical lesions. Brain Res 579: 135-147.

Saito N, Kikkawa U, Nishizuka Y, Tanaka C. 1988. Distribution of protein kinase C-like immunoreactive neurons in rat brain. J Neurosci 8: 369-382.

Salin PA, Lledo PM. Vincent JD, Charpak S. 2001. Dendritic glutamate autoreceptors modulate signal processing in rat mitral cells. J Neurophysiol 85: 1275-1282.

Sallaz M, Jourdan F. 1992. Apomorphine disrupts odour-induced patterns of glomerular activation in the olfactory bulb. Neuroreport 3: 833-836.

Salmoiraghi GC, Bloom FE, Costa E. 1964. Adrenergic mechanisms in rabbit olfactory bulb. Am J Physiol 207: 1417-1424.

Sanides-Kohlrausch C, Wahle P. 1990a. Morphology of neuropeptide Y-immunoreactive neurons in the cat olfactory bulb and olfactory peduncle: Postnatal development and species comparison. J Comp Neurol 291: 468-489.

Sanides-Kohlrausch C, Wahle P. 1990b. VIP- and PHI-immunoreactivity in olfactory centers of the adult cat. J Comp Neurol 294: 325-339.

Sargent-Jones L, Gauger LL, Davis JN. 1985. Anatomy of brain alpha1-adrenergic receptors: In vitro autoradiography with [125I]-heat. J Comp Neurol 231: 190-208.

Sashihara S, Waxman SG, Greer CA. 1997. Downregulation of Na^+ channel mRNA in olfactory tufted cells following deafferentation. Neuroreport 8: 1289-1293.

Sassoè-Pognetto M, Ottersen OP. 2000. Organization of ionotropic glutamate receptors at dendrodendritic synapses in the rat olfactory bulb. J Neurosci 20: 2192-2201.

Sassoè-Pognetto M, Cantino D, Panzanelli P. 1993. Presynaptic co-localization of carnosine and glutamate in olfactory neurons. Neuroreport 5: 7-10.

Sassoè-Pognetto M, Panzanelli P, Siegert W, Fritschy J-M. 2000. Colocalization of multiple GABAA receptor subtypes with gephyrin at postsynaptic sites. J Comp Neurol 420: 481-498.

Sassoè-Pognetto M, Utvik JK, Camoletto P, Watanabe M, Stephenson FA, et al. 2003. Organization of postsynaptic density proteins and glutamate receptors in axodendritic dendrodendritic synapses of the rat olfactory bulb. J Comp Neurol 463: 237-248.

Satou M, Mori K, Tazawa Y, Takagi SF. 1983a. Interneurons mediating fast postsynaptic inhibition in pyriform cortex of the rabbit. J Neurophysiol 50: 89-101.

Satou M, Mori K, Tazawa Y, Takagi SF. 1983b. Monosynaptic and disynaptic activation of pyriform cortex neurons by synchronous lateral olfactory tract volleys in the rabbit. Exp Neurol 81: 571-585.

Satou M, Mori K, Tazawa Y, Takagi SF. 1983c. Neuronal pathways for activation of inhibitory interneurons in pyriform cortex of the rabbit. J Neurophysiol 50: 74-88.

Saucier D, Astic L. 1986. Analysis of the topographical organization of olfactory epithelium projections in the rat. Brain Res Bull 16: 455-462.

Saugstad JA, Kinzie JM, Shinohara MM, Segerson TP, Westbrook GL. 1997. Cloning and expression of rat metabotropic glutamate receptor 8 reveals a distinct pharmacological profile. Mol Pharmacol 51: 119-125.

Schild D, Restrepo D. 1998. Transduction mechanisms in vertebrate olfactory receptor cells. Physiol Rev 78: 429-466.

Schneider SP, Macrides F. 1978. Laminar distributions of interneurons in the main olfactory bulb of the adult hamster. Brain Res Bull 3: 73-82.

Schneider SP, Scott JW. 1983. Orthodromic response properties of rat olfactory mitral and tufted cells correlate with their projection patterns. J Neurophysiol 50: 358-378.

Schoenfeld TA, Macrides F. 1984. Topographic organization of connections between the main olfactory bulb and pars externa of the anterior olfactory nucleus in the hamster. J Comp Neurol 227: 121-135.

Schoenfeld TA, Marchand JE, Macrides F. 1985. Topographic organization of tufted cell axonal projections in the hamster main olfactory bulb: An intrabulbar associational system. J Comp Neurol 235: 503-518.

Schoenfeld TA, Clancy AN, Forbes WB, Macrides F. 1994. The spatial organization of the peripheral olfactory system of the hamster. Part I: Receptor neuron projections to the main olfactory bulb. Brain Res Bull 34: 183-210.

Schoppa NE, Westbrook GL. 1997. Modulation of mEPSCs in olfactory bulb mitral cells by metabotropic glutamate receptors. J Neurophysiol 78: 1468-1475.

Schoppa NE, Westbrook GL. 1999. Regulation of synaptic timing in the olfactory bulb by an A-type potassium current. Nat Neurosci 2: 1106-13.

Schoppa NE, Westbrook GL. 2001. Glomerulus-specific synchronization of mitral cells in the olfactory bulb. Neuron 31: 639-651.

Schoppa NE, Westbrook GL. 2002. AMPA receptors drive correlated spiking in olfactory bulb glomeruli. Nat Neurosci 5: 1194-1202.

Schoppa NE, Kinzie JM, Sahara Y, Segerson TP, Westbrook GL. 1998. Dendrodendritic inhibition in the olfactory bulb is driven by NMDA receptors. J Neurosci 18: 6790-6802.

Scott JW. 1986. The olfactory bulb and central pathways. Experientia 42: 223-232.

Scott JW, Harrison TA. 1987. The olfactory bulb: Anatomy and physiology. Neurobiology of taste and smell. Finger TE, Silver WF, editors. NY: Wiley; pp. 151-178.

Scott JW, McDonald JK, Pemberton JL. 1987. Short axon cells of the rat olfactory bulb display NADPH-diaphorase activity, neuropeptide Y-like immunoreactivity, and somatostatin-like immunoreactivity. J Comp Neurol 260: 378-391.

Seguela P, Datiche J, Donnelley-Miller E, Danni JA, Patrick JW. 1993. Molecular cloning, functional properties, and distribution of rat brain alpha 7: a nicotinic cation channel highly permeable to calcium. J Neurosci 13: 596-604.

Seroogy KB, Brecha N, Gall C. 1985. Distribution of cholecystokinin-like immunoreactivity in the rat main olfactory bulb. J Comp Neurol 239: 373-383.

Seroogy K, Hokfelt T, Buchan A, Brown JC, Terenius L, et al. 1989. Somatostatin-like immunoreactivity in rat main olfactory bulb: Extent of coexistence with neuropeptide Y-, tyrosine hydroxylase- and vitamin D-dependent calcium binding protein-like immunoreactivities. Brain Res 496: 389-396.

Sheldon PW, Aghajanian GK. 1990. Serotonin (5-HT) induces IPSPs in pyramidal layer cells of rat piriform cortex: Evidence for the involvement of a 5-HT2-activated interneuron. Brain Res 506: 62-69.

Sheldon PW, Aghajanian GK. 1991. Excitatory responses to serotonin (5-HT) in neurons of the rat piriform cortex: Evidence for mediation by 5-HT_{1C} receptors in pyramidal cells and 5-HT2 receptors in interneurons. Synapse 9: 208-218.

Shepherd GM. 1972. Synaptic organization of the mammalian olfactory bulb. Physiol Rev 52: 864-917.

Shepherd GM, Greer CA. 1998. The olfactory bulb. The synaptic organization of the brain. Shepherd GM, editor. New York: Oxford UP; pp. 159-203.

Shepherd GM, Chen WR, Greer CA. 2004. Olfactory bulb. The synaptic organization of the brain, 5th Edn. Shepherd GM, editor. New York: Oxford Univ. Press; pp. 165-216.

Shigemoto R, Nakanishi S, Mizuno N. 1992. Distribution of the mRNA for a metabotropic glutamate receptor (mGluR1) in the central nervous system: An in situ hybridization study in adult and developing rat. J Comp Neurol 322: 121-135.

Shigemoto R, Nomura S, Ohishi H, Sugihara H, Nakanishi S, et al. 1993. Immunohistochemical localization of a metabotropic glutamate receptor, mGluR5, in the rat brain. Neurosci Lett 163: 53-57.

Shipley MT, Adamek GD. 1984. The connections of the mouse olfactory bulb: A study using orthograde and retrograde transport of wheat germ agglutinin conjugated to horseradish peroxidase. Brain Res Bull 12: 669-688.

Shipley MT, Ennis M, Puche A. 2004. The olfactory system. The rat nervous system. Paxinos G, editor. New York: Academic Press; pp. 923-964.

Shipley MT, Halloran FJ, de la Torre J. 1985. Surprisingly rich projection from locus coeruleus to the olfactory bulb in the rat. Brain Res 329: 294-299.

Shipley MT, McLean JH, Zimmer LA, Ennis M. 1996. The olfactory system. Integrated systems of the CNS, part III. Swanson LW, Bjorklund A, Hokfelt T, editors. New York: Elsevier; pp. 469-575.

Shults CW, Quirion R, Chronwall B, Chase TN, O'Donohue TL. 1984. A comparison of the anatomical distribution of substance P and substance P receptors in the rat central nervous system. Peptides 5: 1097-1128.

Siklos L, Rickmann M, Joo F, Freeman WJ, Wolff JR. 1995. Chloride is preferentially accumulated in a subpopulation of dendrites and periglomerular cells of the main olfactory bulb in adult rats. Neuroscience 64: 165-172.

Smith C, Luskin MB. 1998. Cell cycle length of olfactory bulb neuronal progenitors in the rostral migratory stream. Dev Dyn 213: 220-227.

Smith TC, Jahr CE. 2002. Self-inhibition of olfactory bulb neurons. Nat Neurosci 5: 760-766.

Smith RL, Baker H, Kolstad K, Spencer DD, Greer CA. 1991. Localization of tyrosine hydroxylase and olfactory marker protein immunoreactivities in the human and cacaque olfactory bulb. Brain Res 548: 140-148.

Spencer DG Jr, Horvath E, Traber J. 1986. Direct autoradiographic determination of M1 and M2 muscarinic acetylcholine receptor distribution in the rat brain: Relation to cholinergic nuclei and projections. Brain Res 380: 59-68.

Stenman J, Toresson H, Campbell K. 2003. Identification of two distinct progenitor populations in the lateral ganglionic eminence: Implications for striatal and olfactory bulb neurogenesis. J Neurosci 23: 167-174.

Stone DM, Wessel T, Joh TH, Baker H. 1990. Decrease in tyrosine hydroxylase, but not aromatic L-amino acid decarboxylase, messenger RNA in rat olfactory bulb following neonatal, unilateral odor deprivation. Brain Res Mol Brain Res 8: 291-300.

Strotmann J, Wanner I, Helfrich T, Beck A, Breer H. 1994. Rostro-caudal patterning of receptor-expressing olfactory neurones in the rat nasal cavity. Cell Tissue Res 278: 11-20.

Sullivan SL, Ressler KJ, Buck LB. 1995. Spatial patterning and information coding in the olfactory system. Curr Opin Genet Dev 5: 516-523.

Sullivan RM, Wilson DA, Leon M. 1989. Norepinephrine and learning-induced plasticity in infant rat olfactory system. J Neurosci 9: 3998-4006.

Sullivan RM, Zyzak DR, Skierkowski P, Wilson DA. 1992. The role of olfactory bulb norepinephrine in early olfactory learning. Brain Res Dev 70: 279-282.

Sullivan RM, Stackenwalt G, Nasr F, Lemon C, Wilson DA. 2000. Association of an odor with activation of olfactory bulb noradrenergic beta-receptors or locus coeruleus stimulation is sufficient to produce learned approach responses to that odor in neonatal rats. Behav Neurosci 114: 957-962.

Sun L, Shipley MT, Lidow MS. 2000. Expression of NR1, NR2A-D, and NR3 subunits of the NMDA receptor in the cerebral cortex and olfactory bulb of adult rat. Synapse 35: 212-221.

Talley EM, Rosin DL, Lee A, Guyenet PG, Lynch KR. 1996. Distribution of α2a-adrenergic receptor like immunoreactivity in the rat central nervous system. J Comp Neurol 372: 111-134.

Tanabe Y, Masu M, Ishii T, Shigemoto R, Nakanishi S. 1992. A family of metabotropic glutamate receptors. Neuron 8: 169-179.

Tang AC, Hasselmo ME. 1994. Selective suppression of intrinsic but not afferent fiber synaptic transmission by baclofen in the piriform (olfactory) cortex. Brain Res 659: 75-81.

Tecott LH, Maricq AV, Julius D. 1993. Nervous system distribution of the serotonin 5-HT3 receptor mRNA. Proc Natl Acad Sci USA 90: 1430-1434.

Teicher MH, Stewart WB, Kauer JS, Shepherd GM. 1980. Suckling pheromone stimulation of a modified glomerular region in the developing rat olfactory bulb revealed by the 2-deoxyglucose method. Brain Res 194: 530-535.

Teubner B, Degen J, Söhl G, Güldenagel M, Bukauskas FF, et al. 2000. Functional expression of the murine connexin 36 gene coding for a neuron-specific gap junctional protein. J Membrane Biol 176: 249-263.

Theis M, Söhl G, Speidel D, Willecke K. 2003. Connexin43 is not expressed in principal cells of mouse cortex and hippocampus. Eur J Neurosci 18: 267-274.

Toida K, Kosaka K, Aika Y, Kosaka T. 2000. Chemically defined neuron groups and their subpopulations in the glomerular layer of the rat main olfactory bulb-IV. Intraglomerular synapses of tyrosine hydroxylase-immunoreactive neurons. Neuroscience 101: 11-17.

Toida K, Kosaka K, Heizmann CW, Kosaka T. 1994. Synaptic contacts between mitral/tufted cells and GABAergic neurons containing calcium-binding protein parvalbumin in the rat olfactory blob, with special reference to reciprocal synapses between them. Brain Res 650: 347-352.

Toida K, Kosaka K, Heizmann CW, Kosaka T. 1996. Electron microscopic serial-sectioning/reconstruction study of parvalbumin-containing neurons in the external plexiform layer of the rat olfactory bulb. Neuroscience 72: 449-466.

Toida K, Kosaka K, Heizmann CW, Kosaka T. 1998. Chemically defined neuron groups and their subpopulations in the glomerular layer of the rat main olfactory bulb: III. Structural features of calbindin D28K-immunoreactive neurons. J Comp Neurol 392: 179-198.

Treloar HB, Feinstein P, Mombaerts P, Greer CA. 2002. Specificity of glomerular targeting by olfactory sensory axons. J Neurosci 22: 2469-2477.

Trombley PQ. 1992. Norepinephrine inhibits calcium currents and EPSPs via a G-protein-coupled mechanism in olfactory bulb neurons. J Neurosci 12: 3992-3998.

Trombley PQ, Shepherd GM. 1992. Noradrenergic inhibition of synaptic transmission between mitral and granule cells in mammalian olfactory bulb cultures. J Neurosci 12: 3985-3991.

Trombley PQ, Shepherd GM. 1994. Glycine exerts potent inhibitory actions on mammalian olfactory bulb neurons. J Neurophysiol 71: 761-767.

Trombley PQ, Shepherd GM. 1996. Differential modulation zinc and copper of amino acid receptors from rat olfactory bulb neurons. J Neurophysiol 76: 2536-2546.

Trombley PQ, Westbrook GL. 1992. L-AP4 inhibits calcium currents and synaptic transmission via a G-protein-coupled glutamate receptor. J Neurosci 12: 2043-2050.

Trombley PQ, Horning MS, Blakemore LJ. 1998. Carnosine modulates zinc and copper effects on amino acid receptors and synaptic transmission. Neuroreport 9: 3503-3507.

Trombley PQ, Hill BJ, Horning MS. 1999. Interactions between GABA and glycine at inhibitory amino acid receptors on rat olfactory bulb neurons. J Neurophysiol 82: 3417-3422.

Tseng G-F, Haberly LB. 1988. Characterization of synaptically mediated fast and slow inhibitory processes in piriform cortex in an in vitro slice preparation. J Neurophysiol 59: 1352-1376.

Tseng G-F, Haberly LB. 1989. Deep neurons in piriform cortex. II. Membrane properties that underlie unusual synaptic responses. J Neurophysiol 62: 386-400.

Tsuruo Y, Hokfelt T, Visser TJ. 1988. Thyrotropin-releasing hormone (TRH)-immunoreactive neuron populations in the rat olfactory bulb. Brain Res 447: 183-187.

Tucker K, Fadool DA. 2002. Neurotrophin modulation of voltage-gated potassium channels in rat through TrkB receptors is time and sensory experience dependent. J Physiol (Lond) 542.2: 413-429.

Uchida N, Takahashi YK, Tanifuji M, Mori K. 2000. Odor maps in the mammalian olfactory bulb: Domain organization and odorant structural features. Nat Neurosci 3: 1035-1043.

Unnerstall JR, Kopajtic TA, Kuhar MJ. 1984. Distribution of alpha 2 agonist binding sites in the rat and human central nervous system: Analysis of some functional, anatomic correlates of the pharmacologic effects of clonidine and related adrenergic agents. Brain Res 319: 69-101.

Urban NN, Sakmann B. 2002. Reciprocal intraglomerular excitation and intra- and interglomerular lateral inhibition between mouse olfactory bulb mitral cells. J Physiol (Lond) 542: 355-367.

Valverde F. 1965. Studies on the piriform lobe. Cambridge, MA: Harvard Univ Press.

van den Pol AN. 1995. Presynaptic metabotropic glutamate receptors in adult and developing neurons: Autoexcitation in the olfactory bulb. J Comp Neurol 359: 253-271.

van den Pol AN, Gorcs T. 1988. Glycine and glycine receptor immunoreactivity in brain and spinal cord. J Neurosci 8: 472-492.

Van Gehuchten LE, Martin A. 1891. Le bulbe olfactif chez quelques mammiferes. Cellule 205–237.

Van Pett K, Viau V, Bittencourt JC, Chan RKW, Li H-Y, et al. 2000. Distribution of mRNAs encoding CRF receptors in the brain and pituitary of rat and mouse. J Comp Neurol 428: 191-212.

Vassar R, Ngai J, Axel R. 1993. Spatial segregation of odorant receptor expression in the mammalian olfactory epithelium. Cell 74: 309-318.

Vassar R, Chao SK, Sitcheran R, Nunez JM, Vosshall LB, et al. 1994. Topographic organization of sensory projections to the olfactory bulb. Cell 79: 981-991.

Vertes RP. 1991. A PHA-L analysis of ascending projections of the dorsal raphe nucleus in the rat. J Comp Neurol 313: 643-668.

Villalba RM, Rodrigo J, Alvarez FJ, Achaval M, Martinez-Murillo R. 1989. Localization of C-PON immunoreactivity in the rat main olfactory bulb. Demonstration that the population of neurons containing endogenous C-PON display NADPH-diaphorase activity. Neuroscience 33: 373-382.

von Baumgarten R, Green JD, Mancia M. 1962. Slow waves in the olfactory bulb and their relation to unitary discharges. Electroenceph Clin Neurophysiol 14: 621-634.

Wachowiak M, Cohen LB. 1999. Presynaptic inhibition of primary olfactory afferents mediated by different mechanisms in lobster and turtle. J Neurosci 19: 8808-8817.

Wachowiak M, Cohen LB. 2001. Representation of odorants by receptor neuron input to the mouse olfactory bulb. Neuron 32: 723-735.

Wachowiak M, Zochowski M, Cohen LB, Falk CX. 2000. The spatial representation of odors by olfactory receptor neuron input to the olfactory bulb is concentration invariant. Biol Bull 199: 162-163.

Wada E, Shigemoto R, Kinoshita A, Ohishi H, Mizuno N. 1998. Metabotropic glutamate receptor subtypes in axon terminals of projection fibers from the main and accessory olfactory bulbs: A light and electron microscopic immunohistochemical study in the rat. J Comp Neurol 393: 493-504.

Wahle P, Sanides-Kohlrausch S, Meyer G, Lubke J. 1990. Substance P- and opioid immunoreactive structures in olfactory centers of the cat: Adult pattern and postnatal development. J Comp Neurol 302: 349-369.

Walsh RR. 1959. Olfactory bulb potentials evoked by electrical stimulation of the contralateral bulb. Am J Physiol 196: 327-329.

Wanaka A, Kiyama H, Murakami T, Matsumoto M, Kamada T, et al. 1989. Immunocytochemical localization of beta-adrenergic receptors in the rat brain. Brain Res 485: 125-140.

Wang F, Nemes A, Mendelsohn M, Axel R. 1998. Odorant receptors govern the formation of a precise topographic map. Cell 93: 47-60.

Warden MK, Young WS. 1988. Distribution of cells containing mRNAs encoding substance P and neurokinin B in the rat central nervous system. J Comp Neurol 272: 90-113.

Watanabe K, Kawana E. 1984. Selective retrograde transport of tritiated D-aspartate from the olfactory bulb to the anterior olfactory nucleus, pyriform cortex and nucleus of the lateral olfactory tract in the rat. Brain Res 296: 148-151.

Watanabe M, Inoue Y, Sakimura K, Mishina M. 1993. Distinct distributions of five N-methyl-D-aspartate receptor channel subunit mRNAs in the forebrain. J Comp Neurol 338: 377-390.

Wellis DP, Kauer JS. 1994. GABAergic and glutamatergic synaptic input to identified granule cells in salamander olfactory bulb. J Physiol (Lond) 475.

Wellis DP, Scott JW. 1990. Intracellular responses of identified rat olfactory bulb interneurons to electrical and odor stimulation. J Neurophysiol 64: 932-947.

Wellis DP, Scott JW, Harrison TA. 1989. Discrimination among odorants by single neurons of the rat olfactory bulb. J Neurophysiol 61: 1161-1177.

Westenbroek RE, Westrum LE, Hendrickson AE, Wu JY. 1987. Immunocytochemical localization of cholecystokinin and glutamic acid decarboxylase during normal development in the prepyriform cortex of rats. Brain Res 431: 191-206.

Whitaker-Azmitia PM, Clarke C, Azmitia EC. 1993. Localization of 5-HT1A receptors to astroglial cells in adult rats: Implications for neuronal-glial interactions and psychoactive drug mechanism of action. Synapse 14: 201-205.

White EL. 1972. Synaptic organization in the olfactory glomerulus of the mouse. Brain Res 37: 69-80.

Whittemore ER, Koerner JF. 1989. An explanation for the purported excitation of piriform cortical neurons by N-acetyl-L-aspartyl-L-glutamic acid (NAAG). Proc Natl Acad Sci USA 86: 9602-9605.

Wichterle H, Turnbull DH, Nery S, Fishell G, Alvareq-Buylla A. 2001. In utero fate mapping reveals distinct migratory pathways and fates of neurons born in the mammalian basal forebrain. Development 128: 3759-3771.

Wilson DA. 1995. NMDA receptors mediate expression of one form of functional plasticity induced by olfactory deprivation. Brain Res 677: 238-242.

Wilson DA. 2001. Scopolamine enhances generalization between odor representations in rat olfactory cortex. Learn Mem 8: 279-85.

Wilson M, Bower JM. 1992. Cortical oscillations and temporal interactions in a computer simulation of piriform cortex. J Neurophysiol 67: 981-995.

Wilson DA, Leon M. 1988. Noradrenergic modulation of olfactory bulb excitability in the postnatal rat. Dev Brain Res 42: 69-75.

Wilson DA, Sullivan RM. 1991. Olfactory associative conditioning in infant rats with brain stimulation as reward: II. Norepinephrine mediates a specific component of the bulb response to reward. Behav Neurosci 105: 843-849.

Wilson DA, Sullivan RM. 1995. The D2 antagonist spiperone mimics the effects of olfactory deprivation on mitral/tufted cell odor response patterns. J Neurosci 15: 5574-5581.

Wilson DA, Wood JG. 1992. Functional consequences of unilateral olfactory deprivation: Time-course and age sensitivity. Neuroscience 49: 183-192.

Winzer-Serhan UH, Raymon HK, Broide RS, Chen Y, Leslie FM. 1997a. Expression of $\alpha 2$ adrenoreceptors during rat brain development-I. α_{2A} messenger RNA expression. Neuroscience 76: 241-260.

Winzer-Serhan UH, Raymon HK, Broide RS, Chen Y, Leslie FM. 1997b. Expression of $\alpha 2$ adrenoreceptors during rat brain development-II. α_{2C} messenger RNA expression and [^{3}H]rauwolscine binding. Neuroscience 76: 261-272.

Wisden W, Seeburg PH. 1993. A complex mosaic of high-affinity kainate receptors in rat brain. J Neurosci 13: 3582-3598.

Wong ST, Trinh K, Hacker B, Chan GC, Lowe G, et al. 2000. Disruption of the type III adenylyl cyclase gene leads to peripheral and behavioral anosmia in transgenic mice. Neuron 27: 487-497.

Woo CC, Leon M. 1995. Distribution and development of beta-adrenergic receptors in the rat olfactory bulb. J Comp Neurol 352: 1-10.

Woolf TB, Shepherd GM, Greer CA. 1991. Serial reconstruction of granule cell spines in the mammalian olfactory bulb. Synapse 7: 181-191.

Wright DE, Seroogy KB, Lundgren KH, Davis BM, Jennes L. 1995. Comparative localization of serotonin1A, 1C, and 2 receptor subtype mRNAs in rat brain. J Comp Neurol 351: 357-373.

Yamamoto C, Yamamoto T, Iwama K. 1963. The inhibitory systems in the olfactory bulb studied by intracellular recording. J Neurophysiol 26: 403-415.

Yang X, Renken R, Hyder F, Siddeek M, Greer CA, et al. 1998. Dynamic mapping at the laminar level of odor-elicited responses in rat olfactory bulb by functional MRI. Proc Natl Acad Sci USA 95: 7715-7720.

Young WS, Kuhar MJ. 1979. Noradrenergic a1 and a2 receptors: Autoradiographic visualization. Eur J Pharmacol 59: 317-319.

Young WS, Kuhar MJ. 1980a. Noradrenergic alpha 1 and alpha 2 receptors: Light microscopic autoradiographic localization. Proc Natl Acad Sci USA 77: 1696-1700.

Young WS, Kuhar MJ. 1980b. Radiohistochemical localization of benzodiazepine receptors in rat brain. J Pharm Exp Ther 212: 337-346.

Youngentob SL, Pyrski MM, Margolis FL. 2004. Adenoviral vector-mediated rescue of the OMP-null behavioral phenotype: Enhancement of odorant threshold sensitivity. Behav Neurosci 118: 636-42.

Yuan Q, Harley CW, McLean JH. 2003. Mitral cell $\alpha 1$ and 5-HT2A receptor colocalization and cAMP coregulation: A new model of norepinephrine-induced learning in the olfactory bulb. Learn Mem 10: 5-15.

Zaborszky L, Carlsen J, Brashear HR, Heimer L. 1986. Cholinergic and GABAergic afferents to the olfactory bulb in the rat with special emphasis on the projection neurons in the nucleus of the horizontal limb of the diagonal band. J Comp Neurol 243: 488-509.

Zhang C, Restrepo D. 2002. Expression of connexin 45 in the olfactory system. Brain Res 929: 37-47.

Zhang C, Restrepo D. 2003. Heterogeneous expression of connexin 36 in the olfactory epithelium and glomerular layer of the olfactory bulb. J Comp Neurol 459: 426-439.

Zhang JH, Sato M, Tohyama M. 1991. Different postnatal development profiles of neurons containing distinct GABAA receptor beta subunit mRNAs (beta 1, beta 2, and beta 3) in the rat forebrain. J Comp Neurol 308: 586-613.

Zhao H, Ivic L, Otaki JM, Hashimoto M, Mikoshiba K, et al. 1998. Functional expression of a mammalian odorant receptor. Science 279: 237-242.

Zheng LM, Caldani M, Jourdan F. 1988. Immunocytochemical identification of luteinizing hormone-releasing hormone-positive fibres and terminals in the olfactory system of the rat. Neurosci 24: 567-578.

Zimmer LA, Ennis M, Shipley MT. 1999. Diagonal band stimulation increases piriform cortex pyramidal cell excitability in vivo. Neuroreport 10: 2101-2105.

Zou Z, Horowitz LF, Montmayeur JP, Snapper S, Buck LB. 2001. Genetic tracing reveals a stereotyped sensory map in the olfactory cortex. Nature 414: 173-179.

Zufall F, Firestein S. 1993. Divalent cations block the cyclic nucleotide-gated channel of olfactory receptor neurons. J Neurophysiol 69: 1758-1768.

Zufall F, Munger SD. 2001. From odor and pheromone transduction to the organization of the sense of smell. Trends Neurosci 24: 191-193.

Zufall F, Firestein S, Shepherd GM. 1991. Analysis of single cyclic nucleotide-gated channels in olfactory receptor cells. J Neurosci 11: 3573-3580.

7 Glucose Sensing Neurons

V. H Routh · J. J. McArdle · N. M. Sanders · Z. Song · R. Wang

Abstract: It is clear that the brain plays a key role in the maintenance of glucose homeostasis. The exact mechanism(s) by which this occurs remains a mystery. However, glucose sensing neurons stand out as prime candidates which enable the brain to sense and respond to changing glucose levels. These neurons are located in key brain regions involved in the regulation of glucose and energy homeostasis. They are also located in the periphery. Glucose sensing neurons are exquisitely sensitive to small changes in extracellular glucose within the physiological range. Their glucose sensitivity becomes impaired under conditions where central glucose sensing mechanisms become dysfunctional. This review discusses the locations of central and peripheral glucose sensing neurons and the mechanisms by which they sense glucose. Putative physiological roles of both central and peripheral glucose sensors are described. Finally, the relationship between glucose and other nutrient signals to the brain is discussed.

List of Abbreviations: Ach, acetylcholine; AMPK, AMP activated protein kinase; ARC, arcuate nucleus of the hypothalamus; CRR, counterregulatory response to hypoglycemia; CSF, cerebrospinal fluid; GABA, γ-aminobutyric acid; GE neurons, glucose-excited neurons; GI neurons, glucose-inhibited neurons; GK, glucokinase; GLUT, glucose transporter; HGE, high glucose excited; HGI, high glucose inhibited; HK, hexokinase; KATP channel, ATP-sensitive K^+channel; LDH, lactate dehydrogenase; LTD, long term potentiation; MCT, Monocarboxylate transporter; αMSH, alpha melanocyte stimulating hormone; nNOS, neuronal nitric oxide synthase; NO, nitric oxide; NPY, neuropeptide Y; NTS, nucleus of the solitary tract; OA, oleic acid; PED, presynaptically excited by decreased glucose; PER, presynaptically excited by raised glucose; PIR, presynaptically inhibited by raised glucose; POMC, proopiomelanocortin; PVN, paraventricular nucleus of the hypothalamus; 5TG, 5 thioglucose; T2DM, type 2 diabetes mellitus; VSAC, volume sensitive anion channel; VMH, Ventromedial hypothalamus; VMN, ventromedial hypothalamic nucleus

1 Introduction

The prevalence of obesity and Type 2 diabetes mellitus (T2DM) in the developed world has risen at an alarming rate over the past decade. In fact, obesity and its associated comorbidities (e.g., heart disease, hypertension, stroke, and cancer) are the second leading cause of death in the U.S. Although obesity and T2DM clearly pose a serious growing health threat, the underlying cause of these related conditions remains uncertain. What is certain is that obesity and T2DM are linked by an underlying dysfunction in the regulation of energy balance that occurs before the clinical manifestation of obesity and frank diabetes. Peripheral insulin resistance combined with impaired insulin secretion in response to glucose is one of the earliest indications of disease. Growing evidence indicates that central dysfunction drives these disorders. There is no longer any doubt that the brain plays a major role in the regulation of energy balance (Levin and Routh, 1996). However, the mechanism whereby the brain actually monitors and responds to the metabolic status of the body remains a mystery.

Mayer's glucostatic hypothesis (Mayer and Thomas, 1967) implies that plasma glucose levels are sensed by the brain and that an increase in glucose utilization triggers meal initiation. Although increased glucose utilization may serve as a signal for meal initiation (Campfield and Smith, 1986), it is more likely to be a critical signal for integrating autonomic activity (Levin et al., 1980) and whole body metabolism (Steffens et al., 1972). The regulation of glucose is so important that the brain has glucose sensors located on neurons in key areas such as the hypothalamus (Kow and Pfaff, 1989; Ashford et al., 1990a; Shiraishi, 1991; Song et al., 2001), nucleus solitarius (NTS) (Mizuno and Oomura, 1984), and amygdala (Nakano and Oomura, 1986), which monitor glucose status in the body and initiate appropriate sympathoadrenal and neurohumoral responses to maintain homeostasis. Interestingly, although glucose serves as the metabolic substrate for all neurons, only 10–40% of the neurons in these brain regions actually alter their action potential frequency in response to changes in extracellular glucose levels (Nakano and Oomura, 1986; Ashford et al., 1990a; Song et al., 2001). This review discusses the location of the glucose sensors, the mechanisms by which glucose sensing occurs and the putative physiological roles of peripheral and central glucose sensors. A major tenet of this review is that glucose sensing neurons in the hypothalamus are critical to the maintenance of energy homeostasis. It is also possible that these neurons link neuronal activity with nutrient availability throughout the brain. As such, glucose sensing neurons would be important not only to obesity and diabetes but also in stroke and other neurodegenerative disorders where neuronal energy supply is disrupted.

2 Levels of Glucose Being Sensed

2.1 Periphery

Before discussing the glucose sensors themselves, it is necessary to define the physiological and pathological parameters, both peripheral and central, under which glucose sensing occurs. In plasma, euglycemia is considered to be 80–100 mg/dL or 4.4–5.5 mM. Prediabetes is defined as plasma glucose levels above 100 and below 120 mg/dL (5.5–6.7 mM). Though, plasma levels above 120 mg/dL define the diabetic state, plasma glucose levels may be as high as 400 mg/dL (22 mM) in severe uncompensated diabetes (American Diabetes Association). On the other hand, when glucose levels fall below 60–70 mg/dL (~3.6 mM), powerful hormonal and neuroendocrine responses are generated to restore glucose to safe levels [i.e., counterregulatory response to hypoglycemia (CRR)] (Cryer, 2001). Thus, for a peripheral glucose sensor to play a role in the regulation of glucose homeostasis, one would expect its glucose sensitivity to be in the range of 3–20 mM glucose.

2.2 Central Nervous System

Plasma glucose levels have served as the standard for studying neurons. In fact, the early studies suggesting the existence of central glucose sensors used extracellular glucose levels from 0 to 10 or 20 mM (Oomura et al., 1964, 1969; Oomura, 1983; Mizuno and Oomura, 1984; Nakano and Oomura, 1986; Ashford et al., 1990a, b; Spanswick et al., 1997, 2000). Obviously, a cerebrospinal fluid (CSF) level of 0 mM glucose would be incompatible with life. Moreover, 10 mM glucose is considered the standard concentration for the solutions that bathe neurons in the majority of neurophysiological studies. This assumes that the extracellular fluid levels of glucose in the brain equal those of the plasma. It is noteworthy that even if this were true, 10 mM plasma glucose is not euglycemia (~5 mM) but rather hyperglycemia. Furthermore, it has become clear that extracellular brain glucose levels are much lower than in plasma.

The elegant studies of Silver and Erecinska (1994, 1998) were the first to simultaneously measure extracellular brain and plasma glucose levels. In their studies, Silver and Erecinska clamped peripheral glucose and measured extracellular brain glucose with a glucose oxidase electrode. Using this technique, they measured extracellular glucose levels in the brain of ~2.5 mM in a fed, anesthetized rat when plasma glucose levels were 7.6 mM. This plasma level is on the high side of euglycemia. Thus, according to their studies plasma levels in the euglycemic range of 5–7 mM corresponded to brain levels of ~1–2 mM. When plasma levels fell to 2–3 mM during insulin-induced hypoglycemia, brain levels fell to ~0.2 mM. Interestingly, even when plasma glucose levels were elevated to 20 mM, brain glucose levels never exceeded 4.5 mM—significantly lower than the experimental standard of 10 mM. Although these studies were performed in anesthetized animals in response to pharmacological manipulation of plasma glucose, they are consistent with recent studies in conscious animals. Using the zero net flux method for microdialysis, DeVries and colleagues (2003) show that glucose levels in the ventromedial hypothalamus nucleus (VMH) were ~1.5 mM in fed unanesthetized rats and 0.7 mM after an overnight fast. Interestingly, using similar techniques, McNay and colleagues have shown that brain glucose levels vary with rat strain, brain region, and neuronal activity (McNay and Gold, 1999; McNay et al., 2000, 2001). However, all these studies consistently indicate that physiological levels of glucose within the brain vary within a fairly tight range from 0.7 to 2.5 mM. On the other hand, extracellular brain glucose levels below 0.7 mM and from 2.5 to 5 mM are associated pathological hypo- and hyperglycemia, respectively.

Although we discuss specific central glucose sensors that respond to changes in extracellular glucose from 5 to 20 mM (Fioramonti et al., 2004), it is not clear whether brain glucose levels ever exceed 5 mM in vivo. One caveat has been suggested regarding the arcuate nucleus (ARC), a region very important in the regulation of energy balance. That is a "leaky" blood–brain barrier in the median eminence may expose ARC neurons to higher levels of glucose (Peruzzo et al., 2000). It is unlikely that glucose diffuses from the plasma or CSF to the ARC due to the specialized glial cells, tanycytes, which line the third ventricle (Navarro et al., 1996; Cheunsuang and Morris, 2005). It is possible, however, that neurons adjacent to the median eminence and other circumventricular organs have projections into the CSF and thus may be exposed to higher glucose levels.

One final consideration relating to the physiological relevance of a particular range of glucose sensitivity as measured in vitro is the interaction between glucose and other components of the extracellular milieu. That is, in vivo neurons are exposed to a variety of compounds that could modify their glucose sensitivity. These compounds include nutrients (e.g., lactate and fatty acids) as well as hormones (e.g., leptin, insulin, corticosterone). Our recent results indicate that the effects of insulin (Wang et al., 2004) and fatty acids (Wang et al., 2005) are dependent on the extracellular glucose level. It is likely that the converse is also true. Thus, when these same neurons are isolated and exposed to glucose in the absence of these compounds, it is likely that the observed in vitro glucose sensitivity differs significantly from that occurring in vivo.

3 Location of Glucose Sensors

3.1 Periphery

3.1.1 Pancreas

The function of pancreatic glucose sensors is to directly sense and regulate plasma glucose levels. The pancreas contains two types of glucose sensing cells, the glucagon secreting α-cells and the insulin secreting β-cells in the pancreatic islets of Langerhans (Heimberg et al., 1996; Matschinsky and Collins, 1997). The β-cells are the best described of the glucose sensors in the body. Interestingly, a recent paper shows that insulin-secreting cells exist within the nervous system of *Drosophila* (Iuras et al., 2005). This suggests that the maintenance of glucose homeostasis, and presumably, glucose sensors may have evolved first in the brain and migrated to the pancreas.

3.1.2 Carotid Body

The carotid bodies are typically associated with blood pressure and chemosensing. However, they also respond to changes in extracellular glucose (Alvarez-Buylla and de Alvarez-Buylla, 1998). The carotid body glucose sensor may be involved in generating the CRR (Pardal and Lopez-Barneo, 2002). Dogs with carotid body resections had blunted glucagon and cortisol secretion as well as decreased endogenous glucose production in response to insulin-induced hypoglycemia. Epinephrine and norepinephrine release, however, did not differ between sham and carotid-resected dogs (Koyama et al., 2000). The carotid body also appears to coordinate ventilation with energy status. Insulin-induced hypoglycemia increases spontaneous ventilation in control rats. Carotid resection abolishes this effect (Bin-Jaliah et al., 2004).

3.1.3 Portal Vein

A hepatic portal vein glucose sensor was identified in the early 1980s (Shimizu et al., 1983). The physiological role of this glucose sensor has been the focus of considerable study in recent years (Donovan et al., 1991, 1994; Hevener et al., 1997, 2000, 2001).The portal vein glucose sensors communicate with the brain via sympathetic afferents and are involved in the initiation of the CRR (Hevener et al., 2000). The argument has been made that it is the peripheral glucose sensors, which are solely responsible for detecting hypoglycemia and initiating the CRR. Insulin-induced hypoglycemia produced in the hepatic-portal region during central euglycemia in dogs is sufficient to elicit the CRR, suggesting the existence of a liver glucose sensor (Donovan et al., 1991). In contrast, Biggers and group (1989) showed that glucagon release, as well as hepatic glucose production was significantly attenuated during insulin-induced hypoglycemia with cerebral euglycemia. In this study, selective hypoglycemia in either the carotid or the vertebrobasilar arteries during systemic euglycemia produced the full CRR in dogs. However, selective euglycemia in either location during peripheral hypoglycemia only slightly inhibited the CRR (Frizzell et al., 1993). The authors conclude that multiple brain regions and redundant central pathways are important for the CRR. Finally, Donovan and colleagues (1994) have shown that when hypoglycemia develops slowly over 2–3 h, the contribution of the portal vein to the CRR is enhanced. Thus, the rate of hypoglycemia development also plays a role in the relative activation of a particular glucose sensor.

3.1.4 Myenteric Plexus

Glucose sensing neurons exist in the myenteric and submucosal plexus within the ileum. Inhibition of these neurons decreases colonic motility (Liu et al., 1999). Glucose infusion increases c-fos in the calbindin and calrexin expressing cells of the submucosal plexus, as well as in neuronal nitric oxide synthase (nNOS)-expressing cells of the myenteric plexus (Sayegh et al., 2004).Thus, enteric glucose sensing neurons may modulate glucose-induced reflexes in the gut.

3.2 Central Nervous System

3.2.1 Hypothalamus

Although glucose sensing neurons exist in many brain regions, those in the hypothalamus are well characterized. The hypothalamus, particularly the VMH, monitors glucose status and initiates a sympathoadrenal response to deficits in glucose availability (Borg et al., 1994, 1995). VMH glucose sensing neurons increase their action potential frequency in response to local application of glucose (Oomura et al., 1964) or peripheral glucose infusion (Silver and Erecinska, 1994, 1998). Electrical stimulation of the VMH activates the sympathoadrenal system in a manner similar to that observed during the CRR (Stoddard-Apter et al., 1986). Conversely, VMH lesions decrease sympathetic and increase parasympathetic tone (Yoshimatsu et al., 1984). This increases basal and glucose-stimulated insulin release as well as blood flow to the pancreas (Penicaud and Ferre, 1988). Thus, the VMH is an integral regulator of autonomic activity which directly affects glucose metabolism (Yoshimatsu et al., 1984; Steffens et al., 1988). The VMH consists of two distinct nuclei: the ventromedial hypothalamic nucleus (VMN) and the ARC. The ARC neuropeptide Y (NPY) neurons that project to the paraventricular hypothalamic nucleus (PVN) (Bai et al., 1985) stand out as prime candidates as a final common pathway in the integration of signals relating to daily energy balance. This pathway favors anabolic processes. NPY injected into the PVN potently stimulates food intake (Stanley et al., 1989). Independent of its effects on food intake, NPY also decreases sympathetic activity (van Dijk et al., 1994) and increases carbohydrate oxidation at the expense of fat oxidation, promoting lipid deposition in adipose stores (White, 1993). ARC NPY neurons receive input from central and peripheral metabolic signals regulating food intake and energy balance (e.g., monoamines, insulin, leptin) (Li and Pelletier, 1986; Billington and Levine, 1992; Schwartz et al., 1992, 1996). They also receive direct neural afferents from the brainstem and from other hypothalamic nuclei involved in energy homeostasis (e.g., VMN) (Chronwall et al., 1985; Magoul et al., 1993). The ARC also possesses the proopiomelanocortin (POMC) neurons that mediate catabolic responses. The POMC cleavage product, α-melanocyte stimulating hormone (αMSH), suppresses food intake and increases energy expenditure (Boston et al., 1997). Moreover, POMC neurons are altered in genetic and dietary obesity with T2DM (Thornton et al., 1997). We have shown that ARC glucose sensing neurons undergo a graded change in activity as glucose levels vary throughout the entire physiologic range (from 0.1 to 5 mM glucose) (Wang et al., 2004). Thus, ARC neurons integrate and regulate systems involved in the central control of energy homeostasis.

On the other hand, VMN glucose sensing neurons selectively detect hypoglycemia. We have shown that VMN glucose sensing neurons only respond to changes in glucose below 2.5 mM; their greatest response occurs as glucose levels decrease below 0.7 mM. Thus, the VMN appears to be critical for the detection of hypoglycemia and generation of the CRR. However, it is also apparent that multiple hypothalamic sites contribute to the full expression of hypoglycemia-induced CRR. For example, temporary inactivation of the PVN impairs catecholamine and adrenocorticotropin hormone secretion in response to hypoglycemia (Evans et al., 2003). On the other hand, inactivation of the dorsomedial nucleus selectively inhibits hypothalamic-pituitary adrenal axis activation (Evans et al., 2004). Glucoprivation also increases the expression of specific hypothalamic neuropeptides, such as NPY (Li and Ritter, 2004) and agouti gene-related peptide (Andrew and Ritter, 2003). Recruitment of these powerful orexigenic neuropeptides contributes, in part, to hypoglycemia-induced food intake (Fraley and Ritter, 2003; Sindelar et al., 2004). Finally, it is not clear whether the initial detection of hypoglycemia occurs in the hypothalamus (or even in the brain itself). Glucose sensing neurons are widely distributed throughout the brain, as well as in the periphery. Extrahypothalamic brain regions possessing glucose sensing neurons include the hippocampus, substantia nigra, septal nucleus, and the hindbrain (Shoji, 1992; During et al., 1995; Ritter et al., 1998; Levin, 2000; Ritter et al., 2000; Sanders and Ritter, 2000; Zawar and Neumcke, 2000).

3.2.2 Hindbrain

There is a long history of anatomical, physiological, and pharmacological evidence supporting the role of the hindbrain in glucose sensing mechanisms (DiRocco and Grill, 1979; Ritter et al., 1981, 2000). Early on it was demonstrated that lateral ventricular delivery of the glucoprivic agent, 5-thio-glucose (5TG) stimulates feeding and the CRR. These responses, however, were blocked if the cerebral aqueduct, connecting the third and fourth ventricles, was temporarily obstructed (Ritter et al., 1981). Additionally, chronically decerebrate rats initiate a sympathoadrenal response to glucoprivation. Thus, these initial studies demonstrated the existence of hindbrain glucoreceptors and a hindbrain efferent pathway controlling sympathoadrenal secretion in response to glucoprivation.

More recently, the discrete hindbrain sites controlling glucoprivic feeding and adrenal medullary secretion have been identified (Ritter et al., 2000). Localized injection of 5TG into hindbrain, but not hypothalamic sites, potently stimulates food intake, sympathoadrenal activation (Ritter et al., 2000) corticosterone and glucagon secretion (Andrew and Ritter, 2003). One of these hindbrain sites, the caudal dorsal medial medulla, overlaps with serotonin neurons in the raphe pallidus, raphe obscurus, and raphe magnus. These serotonin neurons coexpress the low-affinity glycolytic enzyme, glucokinase (GK) (Maekawa et al., 2000), an important regulator of glucose sensing in the pancreatic β-cell (Matschinsky et al., 1998) and the hypothesized glucose sensor in neurons (Dunn-Meynell et al., 2002). It is not known if serotonin neurons are glucose sensing neurons. However, they do exhibit diverse and widespread projections to effector nuclei capable of modulating glucose homeostasis and behavior (Ad, 1981; Tache et al., 1995; Bago et al., 2002; Thor and Helke, 2002). Although it is clear that the hindbrain is capable of glucose sensing, it is also evident that the full expression of neuroendocrine and behavioral CRR requires coordinated input from both hindbrain and hypothalamic sites (Ritter et al., 2001; Sanders et al., 2004).

Elegant anatomical studies have revealed critical neural pathways that link hindbrain glucoprivic signals to effector nuclei mediating CRR. Glucoprivation activates specific subpopulations of hindbrain catecholamine neurons (Ritter et al., 1998). These neurons reside near hindbrain glucoreceptive sites (Ritter et al., 2000) and exhibit extensive projections to both the hypothalamus and spinal cord (Sawchenko and Swanson, 1982; Byrum and Guyenet, 1987; Tucker et al., 1987). Using a selective neurotoxin, Ritter and colleagues demonstrated that the destruction of the rostral-projecting hindbrain catecholamine neurons permanently eliminates the feeding but not sympathoadrenal response to glucoprivation. In addition, increases in hypothalamic fos-ir (Ritter et al., 2001) and feeding-relevant peptides (Fraley et al., 2001; Fraley and Ritter, 2003) are abolished. Alternatively, if the spinal-projecting catecholamine neurons are destroyed, rats continue to exhibit a robust feeding response to glucoprivation but adrenal medullary secretion is blocked (Ritter et al., 2001). These studies demonstrate that feeding and sympathoadrenal responses to glucoprivation require distinct subpopulations of rostral and caudal projecting hindbrain catecholamine neurons.

3.2.3 Other Brain Regions

Glucose also regulates the activity of neurons in other brain regions including the amygdala (Nakano and Oomura, 1986), hippocampus (Zawar and Neumcke, 2000), lateral septal nuclei (Shoji, 1992), and the substantia nigra (During et al., 1995; Levin, 2000; Roeper and Ashcroft, 2002). The role of these glucose sensing neurons is not known. It is possible that the presence of glucose sensing neurons in different brain areas allows for comparison of glucose levels between multiple brain regions as well as peripheral sites. Another possibility is that there is a division in the central nervous system between those sensors involved in restoring blood glucose to normal levels, those involved in daily (meal to meal) glucose balance, and those which are primarily neuroprotective. This latter group could be further subdivided into self-preservation and preservation of function within a particular brain region (e.g., memory or cognitive function).

4 Mechanisms of Glucose Sensing

4.1 Peripheral Sensors

4.1.1 Pancreatic β-cell

The pancreatic β-cell is the most well characterized of all glucose sensors. The β-cell possesses a unique hexokinase (HK IV or GK) whose K_m is in the physiological range for plasma glucose. In addition, the β-cell also possesses a low-affinity glucose transporter (GLUT2). These enzymes transduce changes in plasma glucose into a change in intracellular metabolism. Thus, increased extracellular glucose increases the ATP/ ADP ratio. This inhibits the ATP-sensitive K^+ (KATP) channel on the plasma membrane and depolarizes the cell. Depolarization results in opening of voltage-sensitive calcium channels, calcium influx, and an increase in the exocytosis of insulin-containing vesicles. KATP channel independent mechanisms for glucose-induced insulin secretion have also been described although they are less clearly characterized. Possible mechanisms include an increase in reactive oxygen species as a result of increased flux through the electron transport chain or direct activation of calcium channels (Best and McLaughlin, 2004; Goldstein et al., 2005). Another interesting hypothesis involves the volume-sensitive anion channel (VSAC) in conjunction with the $Na^+/K^+/Cl^-$ cotransporter. This transporter maintains the chloride concentration above that of the Nernst potential for chloride (E_{Cl}). Thus, the activation of a chloride channel would result in chloride efflux and depolarization of the cell. Since glucose activates the VSAC, this mechanism could explain the KATP independent mechanism for insulin release (Best et al., 1997; Best and McLaughlin, 2004). The focus of this review is neuronal glucose sensors, thus if the reader has further interest in β-cell glucose sensing, we refer them to a number of excellent reviews on this topic (Matschinsky and Collins, 1997; Matschinsky et al., 1998; Ashcroft and Gribble, 1999; Bataille et al., 1999; Schuit et al., 2001; Best and McLaughlin, 2004).

4.1.2 Pancreatic α-Cell

Pancreatic α-cells secrete the hormone glucagon in response to a decrease in plasma glucose. The mechanism by which they are inhibited by increased glucose is much less clearer than the KATP channel mediated excitation by glucose in the β-cell. One possible hypothesis is again via the VSAC, now in conjunction with the presence of the K^+/Cl^- cotransporter, the dominant Cl^- cotransporter in the α-cells (Davies et al., 2004).

In contrast to the situation discussed earlier, this transporter maintains the chloride concentration below E_{Cl}. Thus, in this case, activation of the VSAC by glucose would result in chloride influx and inhibition of the α-cell (Best and McLaughlin, 2004; Davies et al., 2004).

4.1.3 Carotid Body

Insulin-induced hypoglycemia significantly increases chemoafferent discharge from the carotid body glomus cells (Bin-Jaliah et al., 2004). Furthermore, increased plasma glucose appears to inhibit a voltage-gated K^+ channel (Pardal and Lopez-Barneo, 2002; Lopez-Barneo, 2003). However, whether this is a direct effect of glucose or whether it is mediated through some other blood-borne metabolic factor is controversial. Lopez-Borneo and colleagues found that the effects of glucose on the carotid body could be reproduced on thin tissue slices in vitro (Lopez-Barneo, 2003). In contrast, Bin-Jaliah and colleagues saw no change in chemoreceptor discharge frequency in carotid slices when glucose was lowered by sucrose substitution (Bin-Jaliah et al., 2004).

4.2 Central Sensors

Glucose sensing neurons were first identified in the hypothalamus by the laboratories of Oomura and Anand in the 1960s (Anand et al., 1964; Oomura et al., 1964). It was 30 years later that Ashford and colleagues demonstrated that VMH neurons whose action potential frequency was increased by glucose (glucose-excited or GE neurons) utilized the KATP channel to sense glucose (Ashford et al., 1988, 1990b). Neurons whose activity was inhibited by glucose (glucose-inhibited or GI neurons) were also observed, but their glucose sensing mechanism was not certain (Ashford et al., 1988, 1990a, b). In these early studies, glucose sensing neurons were characterized in response to large and nonphysiological changes in extracellular glucose (e.g., 0–10 or 20 mM). However, as mentioned earlier, it is now understood that brain glucose levels are actually much lower. When glucose sensing neurons are evaluated throughout the physiological range for central glucose concentration, their characterization is more complex. We have found at least five subtypes of glucose sensing neurons in the VMN and ARC which respond to changes in extracellular glucose from 0.1 to 5 mM (Song et al., 2001; Wang et al., 2004). Of these, two subtypes directly sense glucose and are similar to the GE and GI neurons described above. An additional three subtypes are presynaptically modulated by glucose. Fioramonti and group have recently described another two subtypes of glucose sensing neurons, which directly sense increases in extracellular glucose above 5 mM (Fioramonti et al., 2004). Finally, there is also evidence that glial cells may be glucose sensors (Garcia et al., 2003; Sanders et al., 2004).

4.2.1 Direct effects of glucose: 0.1–5 mM

Two subtypes of VMN and ARC glucose sensing neurons respond directly to changes in extracellular glucose. GE neurons increase and GI neurons decrease their action potential frequency directly as extracellular glucose increases (Song et al., 2001). The opening of a chloride channel appears to mediate the glucose-induced inhibition of GI neurons. The identity of this chloride channel is not yet known; however, we hypothesize that it may belong to the cystic fibrosis transmembrane regulator family. This family of chloride channels is inhibited by sulfonylureas and activated by ATP (Sheppard and Welsh, 1992; Winter et al., 1994). It is also possible that a combination of the VSAC and K^+/Cl^- cotransporter also exist on GI neurons as described above for the pancreatic α-cells (Best and McLaughlin, 2004). The glucose concentration–response relationship for VMN GI neurons is linear between 0.1 and 1 mM glucose, with an apparent plateau at 2.5 mM glucose (Song and Routh, 2005b).

GE neurons show a dose-dependent inhibition of a KATP channel in response to increased glucose (Song et al., 2001; Wang et al., 2004; Song and Routh, 2005b). Interestingly, the glucose sensitivity of VMN and ARC GE neurons differed. The glucose concentration–response relationship for VMN GE neurons was fit by an equation for a rectangular hyperbole; with the steepest slope below 0.5 mM and reaching a plateau at ~2.5 mM (Parkinson et al., 1997). Input resistance (an index of the degree of ion channel closure) in 2.5 and 5 mM glucose was not significantly different, indicating that these VMN GE neurons do not directly sense increases in extracellular glucose above 2.5 mM. The concentration-response for ARC GE neurons also followed a rectangular hyperbole. However, the steepest slope occurred between 0.5 and 2 mM, with the response reaching a plateau at 5 mM. In contrast to VMN GE neurons, ARC GE neurons showed a significant increase in action potential frequency and input resistance from 2.5 to 5 mM glucose (Wang et al., 2004). The decreased input resistance and increased KATP channel current show closely related reciprocal relationships suggesting that regulation of the KATP channel is the dominant effect of glucose in these neurons. Thus, there is a striking similarity between the range of brain glucose levels and the glucose sensitivity of ARC and VMN GE and GI neurons. ARC and VMN GE neurons are capable of sensing daily (meal-induced) fluctuations in extracellular glucose levels (Silver and Erecinska, 1998; De Vries et al., 2003). Moreover, in contrast to VMN GE neurons, ARC GE neurons are capable of altering their activity in response to increases in glucose above 2.5 mM. However, since their greatest glucose sensitivity occurred below 2.5 mM glucose, these data suggest that the ARC GE neurons may play a role in sensing glucose deficits within the normal daily range. This is consistent with evidence suggesting that the ARC plays a role in the regulation of food intake and energy balance. On the other hand, VMN GE neurons appear to be more finely tuned to sense dips in glucose below 0.5 mM. This suggests that they may play a role in sensing profound decreases in glucose, which would initiate the CRR. Support for a role of VMN GE neurons in the CRR comes from the work of Miki and group (2001) showing that the CRR is impaired in mice lacking functional KATP channels. Furthermore, both the CRR and the glucose sensitivity of VMN GI neurons are severely impaired following recurrent bouts of insulin-induced hypoglycemia (Heller and Cryer, 1991; Cryer, 2001; Song and Routh, 2005a).

Another important issue concerning the physiologic relevance of GE and GI neurons involves the mechanism by which alterations in extracellular glucose levels are transduced into an intracellular signal (e.g., ATP, ADP) which regulates action potential frequency. The most likely candidate for this mechanism is the rate-limiting glycolytic enzyme, hexokinase (HK). Most neurons contain HK I which is saturated at physiologic glucose levels (Liu et al., 1991; Matschinsky et al., 1998). In these neurons, altered extracellular glucose within the physiologic range would not result in changes in intracellular metabolism. However, like pancreatic β-cells, GE and GI neurons possess GK, whose K_m for glucose is in the physiologic range for extracellular glucose (Matschinsky et al., 1998). GK is located in brain regions that contain glucose sensing neurons (Jetton et al., 1994; Navarro et al., 1996; Romero-Navarro et al., 1999; Yang et al., 1999; Lynch et al., 2000). Moreover, single VMN GE and GI neurons express GK mRNA, and GK inhibitors prevent glucose responses in these neurons (Dunn-Meynell et al., 2002; Kang et al., 2004). Thus, GE and GI neurons within the VMN and ARC have the necessary biochemical machinery to sense glucose within the physiological range.

4.2.2 Direct Effects of Glucose: 5–20 mM

Several investigators have identified VMH neurons that are either excited or inhibited in response to changes in extracellular glucose between 5 and 20 mM (Yang et al., 1999; Ibrahim et al., 2003; Fioramonti et al., 2004). Fioramonti and group (2004) have recently used patch clamp techniques in the ARC to characterize these neurons. They have defined two subtypes of glucose sensing neurons: high-glucose-excited (HGE) and high-glucose-inhibited (HGI) neurons. Their data suggest that at glucose concentrations above 5 mM, a KATP-independent mechanism mediates the glucose-induced excitation of the HE neurons. Instead, a nonselective cationic conductance appears to be responsible since the glucose effect reversed near –23 mV. The HGI neurons were encountered too rarely in the ARC to determine the mechanism of glucose inhibition.

4.2.3 Presynaptic Effects of Glucose

Three subtypes of VMN sense glucose via presynaptic mechanisms (Song et al., 2001). One subtype is presynaptically excited as extracellular glucose levels decrease below 2.5 mM (PED neurons). The other two subtypes of presynaptically modulated glucose sensing neurons respond to increasing extracellular glucose from 2.5 to 5 mM. PER neurons are excited when extracellular glucose is raised to 5 mM, while PIR neurons are inhibited. The synaptic mechanisms mediating PED and PER neuronal responses to glucose are unclear. However, PIR neurons receive glucose controlled inhibitory γ-aminobutyric acid (GABA)-ergic input. The antidiabetic sulfonylurea drug, tolbutamide, which closes the KATP channel, mimics the effect of increased glucose on PIR neurons. This suggests KATP channel involvement. We hypothesize that this KATP channel is located on the terminal of the GABAergic neuron. Such presynaptic channels are present in the substantia nigra where glucose has been shown to regulate GABA release (During et al., 1995; Levin, 2000).

The interactions between pre- and postsynaptic influences are complex. For example, some GE neurons exhibit biphasic responses to changes in extracellular glucose. These GE neurons respond directly with increased action potential frequency as glucose increases from 0.1 to 2.5 mM, but are inhibited via presynaptic inputs at glucose levels above 2.5 mM. Biphasic responses have also been observed for GI and PED neurons (Song et al., 2001). In these cases, the neurons are quiescent at 2.5 mM extracellular glucose and excited when glucose levels increase or decrease. Thus, neurons in the VMN which directly sense glucose are also modulated by glucose sensing neurons originating elsewhere in the brain. Furthermore, VMN neurons lacking intrinsic glucose sensing ability themselves receive input from glucose sensing neurons. Interestingly, while there were no direct responses to glucose above 2.5 mM, VMN neurons were presynatically regulated by an increase in extracellular glucose to 5 mM. The adjacent ARC might be the origin of these presynaptic inputs since ARC GE and HE neurons respond to glucose levels above 2.5 mM (Fioramonti et al., 2004; Wang et al., 2004).

4.2.4 Specialized Glial cells: Tanycytes

Tanycytes are specialized glial cells that line the dorsal lateral, ventral lateral, and floor of the third cerebral ventricle. Anatomically and functionally, tanycytes are poised to sense glucose and communicate with hypothalamic glucose sensing sites. For example, the tanycytes lining the dorsal and ventral lateral third ventricle preferentially express the glucose transporter, GLUT2, in the proximal portion of the cell body that is in direct contact with CSF (Garcia et al., 2003). In addition, dorsal and ventral lateral tanycytes express GK (Roncero et al., 2000) and the KATP channel (Garcia et al., 2003). Thus, the same proteins that determine the glucose sensing capacity of the pancreatic β-cell and glucose sensing neurons are expressed in tanycytes. Anatomically, dorsal lateral tanycytes send long processes that terminate within the VMN whereas the ventral lateral tanycytes terminate in both the ARC and the median eminence, where they contact portal blood vessels (Garcia et al., 2003).

In vivo findings also support a role for third ventricular tanycytes in both glucose sensing and the CRR to glucoprivation. We previously reported that alloxan, a GK inhibitor, delivered into the third ventricle temporarily impaired glucoprivic CRR (Sanders et al., 2004). Impaired CRR were associated with the apparent loss or retraction of the third ventricular tanycytes that also express GK. In addition, third ventricular alloxan blocked the sympathoadrenal response normally elicited by hindbrain glucoprivic stimulation, suggesting that substrates in and around the third ventricle are required for the expression of this CRR elicited by hindbrain stimulation. However, two weeks after alloxan injection, third ventricle tanycyte cell bodies and processes were restored along with the recovery of glucoprivic CRR. Since alloxan is preferentially transported into cells by GLUT2, it is likely that third ventricular tanycytes, which express GLUT2 (Garcia et al., 2003), were the primary target of alloxan. Additional lines of research also suggest that tanycytes may be directly sensitive to changes in metabolic status, given that their cell bodies are in contact with CSF and their end feet, in some cases, with blood vessels. Uncontrolled diabetes causes retraction and shrinkage of tanycytic processes away from hypothalamic termination sites (Bestetti and Rossi, 1980, 1982). Thus, it is possible that tanycytes participate in glucose sensing via rapid and dynamic morphological changes in their end feet, a process well-characterized for the β2 tanycytes lining the floor of the third ventricle which engulf and regulate neuroendocrine terminals such as gonadotropin-releasing hormone (Prevot et al., 1999, 2004).

4.2.5 Glucose-Regulated Nitric Oxide Signaling

NO is an exciting and novel candidate as a potential mediator of insulin, leptin, and glucose signaling. Neuronal NO synthase (nNOS) produces NO from L-arginine. Many of the effects of NO are mediated by soluble guanylyl cyclase (sGC), which exists both in the soma and presynaptic terminals. sGC produces cyclic guanosine monophosphate (cGMP) which, in turn, activates several downstream targets including protein kinase G (PKG) (Boehning and Snyder, 2003). As a gas, NO readily diffuses from its site of synthesis, allowing it to potentially alter presynaptic input. Thus, NO is a favored candidate for the retrograde signal which may mediate long-term potentiation (LTP) and long-term depression (LTD) (Boehning and Snyder, 2003). However, this has never been shown directly. A number of studies suggest that NO mediates the effects of insulin and leptin. Insulin injected into the NTS stimulates nNOS (Tseng et al., 2003). Inhibition of nNOS in the NTS prevents insulin's ability to lower blood pressure (Tseng et al., 2003). In contrast, leptin decreases nNOS activity in the brain (Das, 2001). NO may also play a role in glucose sensing. NO released from neurons during glucose–oxygen deprivation increases glycolysis and decreases oxidative phosphorylation in astrocytes. This leads to an increase in astrocytic lactate production, presumably to fuel neurons during energy deficit (Almeida et al., 2001a, 2002; Gegg et al., 2003). Furthermore, the signal of energy deficit, AMPK, phosphorylates and activates nNOS in skeletal muscle.(Li et al., 2004) Additionally, Zucker *fa/fa* rats show impaired NO activation of KATP channels (Erdos et al., 2003). Finally, NO is involved in the regulation of food intake. Inhibition of NOS decreases daily food intake (Morely et al., 2003). Furthermore, nNOS knockout mice are refractory to NPY- and orexin-induced feeding (Morely et al., 2003). Central inhibition of NOS reduces food in take in obese but not lean Zucker rats (Bai et al., 1985). Finally, we have recently shown that glucose inhibits NO production in VMH GI neurons via activation of the putative cellular fuel sensor, AMP-activated protein kinase (AMPK) (Cinco et al., 2005).

5 Beyond Glucose

5.1 Lactate

Glucose is not the only fuel sensed by GE and GI neurons. In fact, lactate may be the primary fuel source of the brain (Ames, 2000). We found that lactate did indeed reverse the inhibitory effect of low glucose on action potential frequency and input resistance in GE neurons. The reversal potential for the lactate response in the presence of 0.1 or 0.5 mM glucose was near E_K. Furthermore, the KATP channel opener, diazoxide, reversed the effects of lactate. These data suggest that lactate closes the KATP channel on VMN GE neurons. Lactate, in the presence of 2.5 mM glucose, causes further excitation and increases input resistance even though these neurons do not respond to increases in glucose above 2.5 mM. However sulfonylureas did increase neuronal activity in the presence of 2.5 mM glucose. These data suggest that the KATP channel in VMN GE neurons is more sensitive to ATP generated from lactate than glucose. Interestingly, although GI neurons are inhibited by glucose, lactate further excited GI neurons in both 0.1 and 0.5 mM glucose. Lactate also excited VMN GI neurons in 2.5 mM glucose (when they are normally inhibited). Interestingly, the reversal potential for the excitatory response of lactate on GI neurons was significantly different in 0.1 or 0.5 mM versus 2.5 mM glucose. The former is between E_{Cl} and E_K whereas the latter is near E_K. This suggests that the effect of lactate may be mediated *via* both Cl^- and K^+ channels. The fact that this effect is reversed by diazoxide suggests that the KATP channel may be involved. In fact, we have shown in both electrophysiological and calcium imaging studies that GI neurons are excited by sulfonylureas, indicating that they do possess KATP channels. Non glucose sensing neurons did not respond to lactate in either low or high glucose. Finally, sc RT-PCR studies (Kang et al., 2004) show that over 50% of

VMN neurons, regardless of glucose sensing ability express mRNA for the monocarboxylic acid transporters (MCT1 & 2) which transport lactate across the cell membrane (McKenna et al., 1998; Halestrap and Price, 1999). Furthermore, more than 70% of all neurons expressed mRNA for lactate dehydrogenase (LDH) which converts lactate to pyruvate for entry into the Krebs cycle (Bittar et al., 1996; Laughton et al., 2000). Thus, even though the majority of VMH neurons express MCT and LDH, only glucose sensing neurons use lactate as a signal to regulate neuronal activity. Presumably, non glucose sensing neurons possess MCT and LDH to enable them to use lactate as a fuel source (Ames, 2000). These data show that glucose sensing neurons do respond to other fuel substrates, although not necessarily in the same manner as they do to glucose. They also suggest that the KATP channel mediates the lactate response on both GE and GI neurons. Interestingly, VMH lactate infusion during insulin-induced hypoglycemia blunts the CRR, suggesting that increased lactate levels during hypoglycemia decrease the glucose sensitivity of glucose sensing neurons (Borg et al., 2003). This is consistent with our recent data where exogenous lactate impairs the glucose sensitivity of VMN GI neurons in brain slices (Song and Routh, 2005a, Song and Routh, 2006).

Extracellular lactate also regulates the activity of VMH HGE and HGI neurons. In these studies using extracellular recording techniques, lactate stimulates HGE, and inhibits HGI neurons. However, pyruvate did not alter their activity. The authors suggest that NADH may be the metabolic signal which regulates the activity of the HE and HGI neurons (Yang et al., 1999, 2004).

5.2 Ketones

The effects of ketones have been investigated in GE neurons, which were identified by a change in glucose from 0 to 20 mM. In these neurons, β-hydroxybutyrate caused an excitation and increase in input resistance similar to that seen with glucose (Minami et al., 1990). Moreover, ketone infusion during hypoglycemia attenuates the CRR suggesting that ketones can substitute for glucose in glucose sensing neurons, at least under conditions of energy deficit (Amiel et al., 1991). There is even speculation that the anorexic effect of the currently popular ketogenic diet may be mediated through central KATP channels on glucose sensing neurons (Vamecq et al., 2005).

5.3 Long-Chain Fatty Acids

Over the past 5 years, it has become apparent that hypothalamic fatty acid signaling plays a role in the regulation of glucose and energy homeostasis (Loftus et al., 2000; Clement et al., 2002; Thupari et al., 2002; Obici et al., 2003; Cruciani-Guglielmacci et al., 2004; Kim et al., 2004; Miller et al., 2004; Lam et al., 2005; Tu et al., 2005). We have shown that the long-chain fatty acid, oleic acid (OA), regulates the activity of distinct populations of ARC neurons. Like glucose sensing neurons, there are both OA-excited and OA-inhibited neurons. Interestingly, the ability of ARC neurons to sense OA is dependent on extracellular glucose levels. However, neurons that responded to OA only rarely also responded to glucose (Wang et al., 2005).

5.4 Indices of Peripheral Energy Stores

5.4.1 Insulin and Leptin

Leptin, in 10 mM glucose, opens the KATP channel and inhibits VMH GE neurons (Spanswick et al., 1997). However, we found that the same concentration of leptin (10 nM) used here opened the KATP channel and inhibited VMN GE neurons in 2.5 mM glucose (Wang et al., 2004). However, in the presence of 2.5 or 0.1 mM glucose, leptin did not alter the action potential frequency of ARC GE neurons. Although leptin had no effect on ARC GE neurons, insulin did modulate their activity in a glucose-dependent fashion (Wang et al., 2004). We previously showed that insulin in the presence of

10 mM glucose opened the KATP channel and inhibited VMH GE neurons (Spanswick et al., 2000). We recently found that insulin has no effect on ARC GE neurons in 2.5 mM glucose. However, in the presence of 0.1 mM glucose, insulin actually increases the action potential frequency of ARC GE neurons (Wang et al., 2004). We hypothesize that there are multiple mechanisms (i.e., excitatory and inhibitory) by which insulin modulates the activity of ARC GE neurons. In high glucose, the inhibitory effect on the KATP channel may predominate. However, as glucose levels decrease, the KATP channel moves progressively toward the open state. As this occurs, the effect of opening the KATP channel has less of an overall effect on neuronal activity. In 2.5 mM glucose, these opposing effects may be balanced, with the excitatory effect becoming dominant as glucose levels decrease to 0.1 mM. The excitatory effect of insulin may be due to increased glucose transport into GE neurons since a subpopulation of VMN GE neurons express mRNA coding for both the insulin receptor and the insulin-dependent glucose transporter (GLUT4) (Kang et al., 2004). An interaction between glucose and insulin is not surprising since the brain is never exposed to either insulin or glucose in isolation.

5.4.2 NPY/aMSH

NPY directly inhibits ARC GE neurons. On the other hand, αMSH excites GE neurons by both pre- and postsynaptic mechanisms (Wang et al., 2004). Whether NPY and POMC neurons are also glucose sensing is a matter of some controversy. Measurement of intracellular calcium suggests that NPY neurons are GI neurons (Muroya et al., 1999). However, patch clamp studies reveal that this may be true for only ~25% of the NPY neurons (Fioramonti et al., 2005). Ibrahim and group showed that POMC neurons were excited by increases in extracellular glucose from 5 to 10 mM suggesting that they are HGE neurons (Ibrahim et al., 2003). However, this is in contrast to the recent work of Fioramonti and group who showed that POMC neurons were insensitive to changes in glucose levels between 0.5 and 20 mM (Fioramonti et al., 2005). Furthermore, we have shown that ARC GE neurons were not POMC immunoreactive (Wang et al., 2004). Whether extracellular glucose levels modulate the activity of NPY and POMC neurons is uncertain. However, due to the changes in their peptide gene expression as a result of feeding state, one might hypothesize this to be the case (White, 1993; Mizuno et al., 1999; Bi et al., 2001; Hansen et al., 2004).

6 Physiological Roles for Glucose Sensors

6.1 Peripheral Glucose Sensors

The role of the pancreatic glucose sensors is clearly the direct regulation of plasma glucose levels through the secretion of insulin and glucagon. Those in the myenteric plexus regulate gut motility and are thus positioned to link nutrient absorption with nutrient status. The carotid body and portal vein sensors communicate with the brain via sympathetic afferents (Hevener et al., 2000; Bin-Jaliah et al., 2004). As discussed earlier, they appear to play a role in the detection of hypoglycemia and the generation of the CRR (Hevener et al., 1997; Pardal and Lopez-Barneo, 2002; Lopez-Barneo, 2003). Carotid body glucose sensors also mediate the increased ventilatory response to insulin-induced hypoglycemia (Sayegh et al., 2004). Thus, overall, the actions of the peripheral glucose sensors appear to be reflexive in nature. Each sensor has a defined role with respect to glucose sensing that is consistent with the overall function of the organ system involved.

6.2 The Neuromuscular Junction

Activity of voluntary skeletal muscle normally fulfills the nutrient needs signaled by the glucose sensors discussed earlier; that is, food seeking and ingestive behavior. At the same time, voluntary skeletal muscle utilizes glucose as its major energy source. Delivery of glucose to its metabolic machinery is so critical that skeletal muscle utilizes insulin-dependent as well as insulin-independent mechanisms to take up glucose (Balon and Nadler, 1997; Bradley et al., 1999; Roberts et al., 1999; Higaki et al., 2001; Kingwell et al., 2002). For these reasons, activity of voluntary skeletal muscle is a key determinant of glucose homeostasis.

The neuromuscular junction is the final synapse linking CNS pathways with skeletal muscle activity. The link is highly reliable. That is, the neuromuscular junction evolved to initiate muscle contraction in response to each motor nerve action potential. Decades of research have resulted in detailed understanding of the plasticity and function of the proteins, which make possible the unique reliability of transmission across the neuromuscular junction. However, understanding of the energy cost of operating and maintaining the neuromuscular junction is lacking. Such understanding is important since the function and morphology of the neuromuscular junction changes early after the onset of chronic hyperglycemia (Constantini et al.1987; Schiller and Rahamimoff, 1989; Fahim et al., 1999, 2000; Marques and Santo Neto, 2005). These changes contribute to the decline of muscle activity during chronic hyperglycemia (Fahim et al., 1998; Lesniewski et al., 2003). Thus, hyperglycemia induces a vicious cycle of pathology in which changes at the neuromuscular junction reduce muscle activity and so exacerbates the overall hyperglycemic state.

Understanding the processes contributing to neuromuscular junction changes will improve management of hyperglycemic states. Furthermore, similar process may initiate changes of CNS synapses more intimately involved in glucose sensing. Accessibility to experimental study makes the neuromuscular junction an ideal model in which to explore the energetics of synapse function and plasticity. Of immediate interest is the influence of nerve terminal metabolism on the process(es) of transmitter release.

Functional neuromuscular transmission depends on the synchronized release of 50–100 synaptic vesicles. Each vesicle contains 5,000–10,000 molecules of acetylcholine (Ach) as well as ATP (Van der Kloot, 2003). Though the function of Ach in neuromuscular transmission is well understood, that of ATP remains a mystery. Available data suggest that adenosine resulting from the degradation of coreleased ATP interacts with presynaptic receptors to modulate the probability of vesicle release. This modulation may involve regulation of voltage-activated calcium channels (Silinsky, 2004). These channels allow the influx of extracellular calcium, which couples the axon's action potential to the terminal's secretory process. The magnitude and duration of the increase of nerve terminal calcium concentration determines the number of vesicles released (Katz and Miledi, 1965, 1967a, b). Thus, ATP released from motor nerve endings may modulate stimulus-evoked synchronous release of synaptic vesicles.

In addition to functionally important stimulus-evoked release, motor nerve terminals spontaneously release synaptic vesicles. These vesicles are regarded as the "quanta" underlying transmission across the neuromuscular junction (Katz, 1971). Katz and his colleagues (Fatt and Katz, 1950, 1952; Del Castillo and Katz, 1954) demonstrated that, unlike stimulus-evoked release, spontaneous vesicle release does not require extracellular calcium. In 1975, Alnaes and Rahamimoff showed that inhibitors of the electron transport chain as well as oxidative phosphorylation increased spontaneous vesicle release from nerve muscle preparations bathed in a calcium-free extracellular solution. Their observations led Alnaes and Rahamimoff to conclude that "...mitochondria play a role in transmitter release by participating in the regulation of the intracellular free Ca." David and colleagues (David et al., 1998; David, 1999; David and Barrett, 2000, 2003) convincingly show that mitochondrial regulation of intracellular calcium is an important regulator of stimulus-evoked as well as spontaneous vesicle release.

Three studies suggest additional differences of spontaneous and stimulus-evoked transmitter release. First, these two forms may use distinct vesicle release mechanisms (Geppert et al., 1994). Second, spontaneous and stimulus-evoked transmitter release may utilize distinct vesicle populations (Sara et al., 2005). Third, inhibition of mitochondrial $F_oF_1H^+$ATP synthase enhances spontaneous but not evoked transmitter release (McArdle et al., 2005). These differences raise further questions regarding the physiologic significance of spontaneous vesicle release.

The information discussed earlier suggests a working hypothesis. That is, distinct populations of vesicles and/or release sites respond to an intracellular pool of calcium, which is under mitochondrial control. Thus, the frequency of spontaneous transmitter release responds to the metabolic state of the nerve terminal. Discharge of spontaneously released vesicles activates postsynaptic Ach receptors, which mediate a highly localized calcium influx (Villarroel and Sakmann, 1996). Because calcium-activated NOS colocalizes with the muscle endplate Ach receptor (Luck et al., 2000; Blottner and Luck, 2001), each spontaneously released vesicle stimulates production of NO. NO acts on the motor nerve terminal in a retrograde fashion to modulate transmitter release (Lindgren and Laird, 1994; Ambiel and Alves Do Prado, 1997; Thomas and Robitaille, 2001; Etherington and Everett, 2004; Graves et al., 2004; Zhu et al., 2006) and/or neuromuscular junction stability (Christova et al., 1997; Descarries et al., 1998; Grozdanovic and Baumgarten, 1999; Tews, 2001; Godfrey and Schwarte, 2003; Schwarte and Godfrey, 2004; Mulvey et al., 2005; Puttmann et al., 2005).

This hypothesis guides our current studies of the neuromuscular junction. The hope is that these studies will provide useful insights into how metabolic activity controls transmission and storage of information within the peripheral and CNS. Such knowledge is likely to improve understanding and treatment of diseases like diabetes and obesity.

6.3 Central Nervous System

There are several possible functions for central glucose sensing neurons. First, regional detection and regulation of glucose and/or nutrient availability are necessary to meet the demands of local synaptic activity. Synaptic activity is an energy consuming process. Therefore, it is imperative that the brain maintain local energy reserves to support synaptic function. Glucose is the primary fuel of the brain, although recent studies suggest an important role for glial-derived lactate (Magistretti and Pellerin, 1996; Tsacopoulos and Magistretti, 1996). Thus, we hypothesize that glucose sensing neurons evolved primarily to maintain local energy need to couple nutrient availability with synaptic function. This is likely to be regulated by a complex interaction between neurons and glial cells involving NO signaling, as described later.

A separate function relates to whole body energy and glucose homeostasis. The maintenance of energy balance is clearly a critical function of the brain. Therefore, it is reasonable that glucose sensing neurons also sense and respond to both nutrient (i.e., glucose, lactate, fatty acids) and hormonal (i.e., insulin, leptin) changes in the extracellular milieu of the brain. In regions of the brain associated with the regulation of energy balance, glucose sensing neurons are capable of detecting and integrating changes in whole body energy balance. As a result of their location, they are then poised to either directly or indirectly signal the sympathoadrenal and neuroendocrine systems to make the appropriate adjustments in peripheral energy balance to restore homeostasis. When the ability of glucose sensing neurons to sense changes in local nutrients becomes compromised, it can lead to diseases such as diabetes and obesity. Conversely, the alterations in peripheral glucose and energy balance in diabetes and obesity cause further dysfunction of glucose sensing neurons.

These functions are not mutually exclusive. We hypothesize that the dysfunctional glucose and energy homeostasis observed in diabetes and obesity results, in part, from changes in the function of glucose sensing neurons. These changes are initially protective in terms of maintaining an ideal extracellular environment for synaptic function. However, impairment of glucose sensing neurons is ultimately maladaptive since it disrupts the ability of the brain to sense and maintain normal energy balance. This is described in ❯ *Figure 7-1.* We hypothesize that glucose sensing neurons or at least GI neurons, evolved primarily to ensure that local energy needs for synaptic function are met. Thus, they are probably located throughout the brain, as they are peripherally. When local glucose levels fall, even slightly, AMPK activity increases in GI neurons, increasing NO production (Minokoshi et al., 2004; Cinco et al., 2005; Lee et al., 2005). Three sequelae of elevated NO production could increase nutrient availability to the surrounding region. First, NO causes dilation of cerebral vasculature, which would increase nutrient delivery (Santizo et al., 2001). Second, NO stimulates glial cells to produce lactate (Almeida and Bolanos, 2001; Almeida et al., 2001, 2002, 2003, 2004, 2005; Bolanos et al., 2001; Bolanos et al., 2004). Increased glial lactate production presumably increases lactate availability to neurons for fuel (Tsacopoulos and Magistretti, 1996). Third, glucose and lactate transporter activity, as well as glycolytic enzyme activity are enhanced during hypoglycemia (Simpson et al., 1999; Dunn-Meynell et al., 2002; Vannucci and Hagberg, 2004). In support of this, the MCT inhibitor, 4-CIN, prevents septal nuclei neurons from adapting to recurrent hypoglycemia (Shoji, 1992; Hasuo and Akasu, 2003). These effects may be mediated by NO (Tsuura et al., 1998; Bolanos et al., 2001; Cidad et al., 2001; Rizzo and Piston, 2003; Almeida et al., 2004; Li et al., 2004; Garcia et al., 2005). Since the majority of VMN neurons express sGC (Cinco et al., 2005), NO diffusing from

Figure 7-1
The nitric oxide (NO) hypothesis. Decreases in extracellular glucose activates AMP-activated protein kinase (AMPK), which in turn phosphorylates and activates NO synthase (NOS), increasing NO production in glucose-inhibited (GI) neurons. NO acts in at least three ways to increase the availability and/or utilization of neuronal fuels. First, NO stimulates glycolysis and inhibits oxidative phosphorylation in glial cells, increasing lactate production. This presumably increases lactate efflux and uptake into neurons for fuel. NO also causes local vasodilation, increasing glucose delivery. Finally, NO increases the expression of glycolytic enzymes and glucose transporters. Since NO is a diffusible messenger, it can influence presynaptic inputs to the GI neuron. NO can also influence adjacent neurons (e.g., GE neurons)

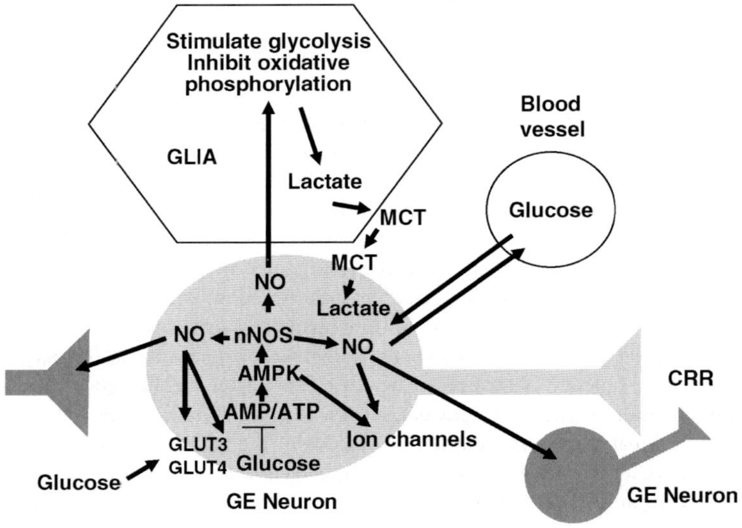

VMN GI neurons could modulate the activity of adjacent neurons involved in glucose homeostasis. In fact, this might explain why we found that GE neurons also decrease their glucose sensitivity after recurrent hypoglycemia even though they do not produce NO (Song and Routh, 2005a). Moreover, GI neurons may project to both inter- and intrahypothalamic neurons involved in the regulation of glucose and energy homeostasis. These mechanisms would be very beneficial during small transient decreases in glucose or following a single hypoglycemic or ischemic episode. However, the brain did not evolve to see repetitive hypoglycemic insults, such as those occurring during intensive insulin therapy (Cryer, 2001). These severe and repetitive insults may result in sustained upregulation of nutrient availability/utilization to protect the brain via the mechanisms described earlier. Thus, if subsequent hypoglycemic incidents were to occur within a "refractory period" when local nutrient availability/utilization is upregulated, then GI and/or GE neurons would not perceive the intensity of the glucose deficit and signal for an appropriate CRR.

7 Conclusions

In conclusion, glucose sensors exist throughout the body, both centrally and peripherally. We hypothesize that they form a hierarchical reflex network whose teleological function is to maintain appropriate energy levels for local and whole body function. The peripheral glucose sensors serve as the front line regulators of plasma glucose levels (α- and β-cells), oxygen availability (via ventilation), nutrient absorption (myenteric plexus), and motor function at the neuromuscular junction. Carotid and portal vein glucose sensors also relay this information to higher centers via the hindbrain. Central glucose sensors are poised to serve multiple functions. First, they may play a role in the maintenance of local energy needs for synaptic function. Secondly, they are capable of communicating information regarding nutrient status to surrounding projection neurons.

This would then link the function of discrete brain regions to whole body energy status. That is, glucose sensors located in the hindbrain and/or hypothalamus are well situated to regulate the neuroendocrine response to changes in glucose levels. The hypothalamus appears to be critical for mediating alterations in feeding behavior (Billington and Levine, 1992; Ritter et al., 2003; Fraley and Ritter, 2003). Glucose sensors located in higher centers may further refine behavioral and cognitive aspects of whole body nutrient regulation and food intake. Finally, we would like to propose that diabetes and obesity may result from disruption of the brain's ability to link synaptic function and nutrient availability. This disruption in synaptic function further impairs central glucose sensing mechanisms. This vicious cycle may be initiated and/or perpetuated by genetic, hormonal, or dietary factors.

Acknowledgments

This work was supported, in part, by DK55619 (VHR), DK64566 (VHR), American Diabetes Association Junior Faculty Award (NMS), NS045979 (JJM), and the Kirby Foundation (JJM).

References

Ad L. 1981. Raphe pallidus and raphe obscurus projections to the intermediolateral cell column in the rat. Brain Res 222: 129-133.

Almeida A, Bolanos JP. 2001. A transient inhibition of mitochondrial ATP synthesis by nitric oxide synthase activation triggered apoptosis in primary cortical neurons. J Neurochem 77: 676-690.

Almeida A, Cidad P, Bolanos JP. 2002. Nitric oxide accounts for an increased glycolytic rate in activated astrocytes through a glycogenolysis-independent mechanism. Brain Res 945: 131-134.

Almeida A, Moncada S, Bolanos JP. 2003. Nitric oxide switches on glycolysis through the AMP protein kinase and 6-phosphofructo-2-kinase pathway. Nat Cell Biol 6: 45-52.

Almeida A, Moncada S, Bolanos JP. 2004. Nitric oxide switches on glycolysis through the AMP protein kinase and 6-phosphofructo-2-kinase pathway. Nat Cell Biol 6: 45-51.

Almeida A, Almeida J, Bolanos JP, Moncada S. 2001. Different responses of astrocytes and neurons to nitric oxide: The role of glycolytically generated ATP in astrocyte protection. Proc Natl Acad Sci 98: 15294-15299.

Almeida A, Cidad P, Delgado-Esteban M, Fernandez E, Garcia-Nogales P, et al. 2005. Inhibition of mitochondrial respiration by nitric oxide: Its role in glucose metabolism and neuroprotection. J Neurosci Res 79: 166-171.

Alnaes E, Rahamimoff R. 1975. On the role of mitochondria in transmitter release from motor nerve terminals. J Physiol 248: 285-306.

Alvarez-Buylla R, de Alvarez-Buylla ER. 1998. Carotid sinus receptors participate in glucose homeostasis. Respir Physiol 72: 347-360.

Ambiel CR, Alves-Do-Prado W. 1997. Neuromuscular facilitation and blockade induced by L-arginine and nitric oxide in the rat isolated diaphragm. Gen Pharmacol 28: 789-794.

Ames AI. 2000. CNS energy metabolism as related to function. Brain Res Rev 34: 42-68.

Amiel SA, Archibald HR, Chusney G, Williams AJ, Gale EA. 1991. Ketone infusion lowers hormonal responses to hypoglycaemia: Evidence for acute cerebral utilization of a non-glucose fuel. Clin Sci (Lond) 81: 189-194.

Anand BK, China GS, Sharma KN, Dua S, Singh B. 1964. Activity of single neurons in the hypothalamus feeding centers: Effect of glucose. Am J Physiol 2207: 1146-1154.

Andrew S, Ritter S. 2003. Localized glucoprivation of hindbrain but not hypothalamic sites stimulates CORT and glucagon secretion. Appetite 4: 315.

Ashcroft FM, Gribble FM. 1999. ATP-sensitive K^+ channels and insulin secretion: Their role in health and disease. Diabetologia 42: 903-919.

Ashford MLJ, Boden PR, Treherne JM. 1990a. Glucose-induced excitation of hypothalamic neurones is mediated by ATP-sensitive K^+ channels. Pfugers Arch Eur J Physiol 415: 479-483.

Ashford MLJ, Boden PR, Treherne JM. 1990b. Tolbutamide excites rat glucoreceptive ventromedial hypothalamic neurones by indirect inhibition of ATP-sensitive K^+ channels. Br J Pharmacol 101: 531-540.

Ashford MLJ, Sturgess NC, Trout NJ, Gardner NJ, Hales CN. 1988. Adenosine-5′-triphosphate-sensitive ion channels in neonatal rat cultured central neurons. Pfugers Arch Eur J Physiol 412: 297-304.

Bago M, Marson L, Dean C. 2002. Serotonergic projections to the rostroventrolateral medulla from midbrain and raphe nuclei. Brain Res 945: 249-258.

Bai FL, Yamano M, Shiotani Y, Emson PC, Smith AD, et al. 1985. An arcuato-paraventricular and -dorsomedial hypothalamic neuropeptide Y-containing system which lacks noradrenaline in the rat. Brain Res 331: 172-175.

Balon TW, Nadler JL. 1997. Evidence that nitric oxide increases glucose transport in skeletal muscle. J Appl Physiol 82: 359-363.

Bataille D, Heron L, Virsolvy A, Peyrollier K, LeCam A, et al. 1999. alpha-Endosulfine, a new entity in the control of insulin secretion. Cell Mol Life Sci 56: 78-84.

Best L, McLaughlin J. 2004. Nutrients as regulators of endocrine and neuroendocrine secretion. Topics in Current Genetics. Winderickx, Taylor P, editors. Berlin: Springer-Verlag; pp. 70-111.

Best L, Brown PD, Tomlinson S. 1997. Anion fluxes, volume regulation and electrical activity in the mammalian pancreatic beta-cell. Exp Physiol 82: 957-966.

Bestetti G, Rossi GL. 1980. Hypothalamic lesions in rats with longterm streptozotocin-induced diabetes mellitus. A semiquantitative light and electron-microscopic study. Acta Neuropathol 52: 119-127.

Bestetti G, Rossi GL. 1982. Hypothalamic changes in diabetic Chinese hamsters: A semiquantitative, light and electron microscopic study. Lab Invest 47: 516-522.

Bi S, Ladenheim EE, Schwartz GJ, Moran TH. 2001. A role for NPY overexpression in the dorsomedial hypothalamus in hyperphagia and obesity of OLETF rats. Am J Physiol 281: R254-R260.

Biggers DW, Myers SR, Neal D, Stinson R, Cooper NB, et al. 1989. Role of brain in counterregulation of insulin-induced hypoglycemia in dogs. Diabetes 38: 7-16.

Billington CJ, Levine AS. 1992. Hypothalamic neuropeptide Y regulation of feeding and energy metabolism. Curr Opin Neurobiol 2: 847-851.

Bin-Jaliah I, Maskell PD, Kumar P. 2004. Indirect sensing of insulin-induced hypoglycaemia by the carotid body in the rat. J Physiol Online 556: 255-266.

Bittar PG, Charnay Y, Pellerin L, Bouras C, Magistretti PJ. 1996. Selective distribution of lactate dehydrogenase isoenzymes in neurons and astrocytes of human brain. J Cereb Blood Flow Metab 16: 1079-1089.

Blottner D, Luck G. 2001. Just in time and place: NOS/NO system assembly in neuromuscular junction formation. Micros Res Tech 55: 171-180.

Boehning D, Snyder SH. 2003. Novel neural modulators. Ann Rev Neurosci 26: 105-131.

Bolanos JP, Cidad P, Garcia-Nogales P, Delgado-Esteban M, Fernandez E, et al. 2004. Regulation of glucose metabolism by nitrosative stress in neural cells. Mol Aspects Med 25(1–2): 61-73.

Bolanos JP, Garcia-Nogales P, Vega-Agapito V, Delgado-Esteban M, Cidad P, et al. 2001. Nitric oxide-mediated mitochondrial impairment in neural cells: A role for glucose metabolism in neuroprotection. Prog Brain Res 132: 441-454.

Borg MA, Tamborlane WV, Shulman GI, Sherwin RS. 2003. Local lactate perfusion of the ventromedial hypothalamus suppresses hypoglycemic counterregulation. Diabetes 52: 663-666.

Borg WP, During MJ, Sherwin RS, Borg MA, Brines ML, et al. 1994. Ventromedial hypothalamic lesions in rats suppress counterregulatory responses to hypoglycemia. J Clin Invest 93: 1677-1682.

Borg WP, Sherwin RS, During MJ, Borg MA, Shulman GI. 1995. Local ventromedial hypothalamus glucopenia triggers counterregulatory hormone release. Diabetes 44: 180-184.

Boston BA, Blaydon KM, Varnerin J, Cone RD. 1997. Independent and additive effects of central POMC and leptin pathways on murine obesity. Science 278: 1641-1644.

Bradley SJ, Kingwell BA, McConell GK. 1999. Nitric oxide synthase inhibition reduces leg glucose uptake but not blood flow during dynamic exercise in humans. Diabetes 48: 1815-1821.

Byrum DC, Guyenet PG. 1987. Afferent and efferent connections of the A5 noradrenergic cell group in the rat. J Comp Neurol 261: 529-542.

Campfield LA, Smith FJ. 1986. Functional coupling between transient declines in blood glucose and feeding behavior: Temporal relationships. Brain Res Bull 17: 427-549.

Cheunsuang O, Morris R. 2005. Astrocytes in the arcuate nucleus and median eminence that take up a fluorescent dye from the circulation express leptin receptors and neuropeptide Y Y1 receptors. Glia 52: 228-233.

Christova T, Grozdanovic Z, Gossrau R. 1997. Nitric oxide synthase (NOS) I during postnatal development in rat and mouse skeletal muscle. Acta Histochem 99(3): 311–324.

Chronwall BM, DiMaggio DA, Massari VJ, Pickel VM, Ruggiero DA, et al. 1985. The anatomy of neuropeptide-Y-containing neurons in rat brain. Neuroscience 15: 1159-1181.

Cidad P, Garcia-Nogales P, Almeida A, Bolanos JP. 2001. Expression of glucose transporter GLUT3 by endotoxin in cultured rat astrocytes: The role of nitric oxide. J Neurochem 79: 17-24.

Cinco DD, Potian JG, Beuve A, McArdle JJ, Routh VH. 2005. Nitric oxide integrates leptin, insulin and glucose signaling in ventromedial hypothalamic (VMH) neurons. 2005 Abstract Viewer/Itinerary Planner. Washington, DC: Society for Neuroscience, 2005. Online. 533.8.

Clement L, Cruciani-Guglielmacci C, Magnan C, Vincent M, Douared L, et al. 2002. Intracerebroventricular infusion of a triglyceride emulsion leads to both altered insulin secretion and hepatic glucose production in rats. Pflugers Arch 445: 375-380.

Constantini S, Schiller Y, Cohen AM, Rahamimoff R. 1987. Pathophysiology of the neuromuscular junction in diabetic rats. Isr J Med Sci 23: 101-106.

Cruciani-Guglielmacci C, Hervalet A, Douared L, Sanders NM, Levin BE, et al. 2004. Beta oxidation in the brain is required for the effects of non-esterified fatty acids on glucose-induced insulin secretion in rats. Diabetologia 47: 2032-2038.

Cryer PE. 2001. Hypoglycemia-associated autonomic failure in diabetes. Am J Physiol 281: E1115-E1121.

Das UN. 2001. Is obesity an inflammatory condition? Nutrition 17: 953-966.

David G. 1999. Mitochondrial clearance of cytosolic $Ca^{(2+)}$ in stimulated lizard motor nerve terminals proceeds without progressive elevation of mitochondrial matrix [$Ca^{(2+)}$]. J Neurosci 19: 7495-7506.

David G, Barrett EF. 2000. Stimulation-evoked increases in cytosolic [$Ca^{(2+)}$] in mouse motor nerve terminals are limited by mitochondrial uptake and are temperature-dependent. J Neurosci 20: 7290-7296.

David G, Barrett EF. 2003. Mitochondrial Ca^{2+} uptake prevents desynchronization of quantal release and minimizes depletion during repetitive stimulation of mouse motor nerve terminals. The J Physiol 548: 425-438.

David G, Barrett JN, Barrett EF. 1998. Evidence that mitochondria buffer physiological Ca^{2+} loads in lizard motor nerve terminals. J Physiol 509: 59-65.

Davies SL, Roussa E, Le Rouzic P, Thevenod F, Alper SL, et al. 2004. Expression of K+-Cl- cotransporters in the [alpha]-cells of rat endocrine pancreas. Biochim Biophys Acta (BBA)-Biomembranes 1667: 7-14.

Del Castillo J, Katz B. 1954. Statistical factors involved in neuromuscular facilitation and depression. J Physiol 124: 574-585.

Descarries LM, Cai S, Robitaille R, Josephson EM, Morest DK. 1998. Localization and characterization of nitric oxide synthase at the frog neuromuscular junction. J Neurocytol 27: 829-840.

De Vries MG, Arseneau LM, Lawson ME, Beverly JL. 2003. Extracellular glucose in rat ventromedial hypothalamus during acute and recurrent hypoglycemia. Diabetes 52: 2767-2773.

DiRocco RJ, Grill HJ. 1979. The forebrain is not essential for sympathoadrenal hyperglycemic response to glucoprivation. Science 204: 1112-1114.

Donovan CM, Cane P, Bergman RN. 1991. Search for the hypoglycemia receptor using the local irrigation approach. Adv Exp Med Biol 291: 185-196.

Donovan CM, Hamilton-Wessler M, Halter JB, Bergman RN. 1994. Primacy of liver glucosensors in the sympathetic response to progressive hypoglycemia. Proc Natl Acad Sci 91: 2863-2867.

Dunn-Meynell AA, Routh VH, Kang L, Gaspers L, Levin BE. 2002. Glucokinase is the likely mediator of glucosensing in both glucose-excited and glucose-inhibited central neurons. Diabetes 51: 2056-2065.

During MJ, Leone P, Davis KE, Kerr D, Sherwin RS. 1995. Glucose modulates rat substantia nigra GABA release in vivo via ATP-sensitive potassium channels. J Clin Invest 95: 2403-2408.

Erdos B, Miller AM, Busija DW. 2003. Cerebrovascular dysfunction in insulin Zucker obese rats. Soc. Neurosci. Abstr. Viewer/Itinerary Planner 7.12.

Etherington SJ, Everett AW. 2004. Postsynaptic production of nitric oxide implicated in long-term depression at the mature amphibian (*Bufo marinus*) neuromuscular junction. J Physiol 559: 507-517.

Evans SB, Wilkinson CW, Gronbeck P, Bennett JL, Taborsky GJ Jr, et al. 2003. Inactivation of the PVN during hypoglycemia partially simulates hypoglycemia-associated autonomic failure. Am J Physiol 284: R57-R65.

Evans SB, Wilkinson CW, Gronbeck P, Bennett JL, Zavosh A, et al. 2004. Inactivation of the DMH selectively inhibits the ACTH and corticosterone responses to hypoglycemia. Am J Physiol Regul Integr Comp Physiol 286: R123-R128.

Fahim MA, Al Shuaib W, Davidson N. 1999. Depolarization affects neuromuscular junction of streptozotocin-diabetic mice. Cell Mol Biol (Noisy-le-grand) 45: 259-263.

Fahim MA, el-Sabban F, Davidson N. 1998. Muscle contractility decrement and correlated morphology during the pathogenesis of streptozotocin-diabetic mice. Anat Rec 251: 240-244.

Fahim MA, Hasan MY, Alshuaib WB. 2000. Early morphological remodeling of neuromuscular junction in a murine model of diabetes. J Appl Physiol 89: 2235-2240.

Fatt P, Katz B. 1950. Some observations on biological noise. Nature 166: 597-598.

Fatt P, Katz B. 1952. Spontaneous subthreshold activity at motor nerve endings. J Physiol 117: 109-128.

Fioramonti X, Lorsignol A, Song Z, Routh VH, Penicaud L. 2005. Characterization of different glucose sensing neuron subpopulations in rat and mouse arcuate nucleus. 2005 Abstract Viewer/Itinerary Planner. Washington, DC: Society for Neuroscience, 2005. Online. 812.6.

Fioramonti X, Lorsignol A, Taupignon A, Penicaud L. 2004. A New ATP-Sensitive K+ Channel-Independent Mechanism Is Involved in Glucose-Excited Neurons of Mouse Arcuate Nucleus. Diabetes 53: 2767-2775.

Fraley GS, Ritter S. 2003. Immunolesion of Norepinephrine and Epinephrine Afferents to Medial Hypothalamus Alters Basal and 2-Deoxy-D-Glucose-Induced Neuropeptide Y and Agouti Gene-Related Protein Messenger Ribonucleic Acid Expression in the Arcuate Nucleus. Endocrinology 144: 75-83.

Fraley GS, Dinh TT, Ritter S. 2001. Immunotoxin lesion of catecholaminergic neurons innervating the medial hypothalamus elevates basal expression of and attenuates glucoprivation-induced increases in agouti gene related protein (AGRP). Abstr Soc Neurosci 31, 947.

Frizzell RT, Jones EM, Davis SN, Biggers DW, Myers SR, et al. 1993. Counterregulation during hypoglycemia is directed by widespread brain regions. Diabetes 42: 1253-1261.

Garcia M, Millan C, Balmaceda-Aguilera C, Castro T, Pastor P, et al. 2003. Hypothalamic ependymal-glial cells express the glucose transporter GLUT2, a protein involved in glucose sensing. J Neurochem 86: 709-724.

Garcia O, Almeida A, Massieu L, Bolanos JP. 2005. Increased mitochondrial respiration maintains the mitochondrial membrane potential and promotes survival of cerebellar neurons in an endogenous model of glutamate receptor activation. J Neurochem 92: 183-190.

Gegg ME, Beltran B, Salas-Pino S, Bolanos JP, Clark JB, et al. 2003. Differential effect of nitric oxide on glutathione metabolism and mitochondrial function in astrocytes and neurones: Implications for neuroprotection/neurodegeneration? J Neurochem 86: 228-237.

Geppert M, Goda Y, Hammer RE, Li C, Rosahl TW, et al. 1994. Synaptotagmin I: A major Ca^{2+} sensor for transmitter release at a central synapse. Cell 79: 717-727.

Godfrey EW, Schwarte RC. 2003. The role of nitric oxide signaling in the formation of the neuromuscular junction. J Neurocytol 32: 591-602.

Goldstein BJ, Mahadev K, Wu X. 2005. Redox paradox: Insulin action is facilitated by insulin-stimulated reactive oxygen species with multiple potential signaling targets. Diabetes 54: 311-321.

Graves AR, Lewin KA, Lindgren C. 2004. Nitric oxide, cAMP and the biphasic muscarinic modulation of ACh release at the lizard neuromuscular junction. J Physiol 559: 423-432.

Grozdanovic Z, Baumgarten HG. 1999. Nitric oxide synthase in skeletal muscle fibers: A signaling component of the dystrophin-glycoprotein complex. Histol Hisotpathol 14: 243-256.

Halestrap AP, Price NT. 1999. The proton-linked monocarboxylate transporter (MCT) family: Structure, function and regulation. Biochem J 343(2): 281-299.

Hansen MJ, Jovanovska V, Morris MJ. 2004. Adaptive responses in hypothalamic neuropeptide Y in the face of prolonged high-fat feeding in the rat. J Neurochem 88: 909-916.

Hasuo H, Akasu T. 2003. Monocarboxylate transporters contribute to the adaptation of neuronal activity to repeated glucose deprivation in the rat lateral septal nucleus. Synapse 49: 97-105.

Heimberg H, De Vos A, Moens K, Quartier E, Bouwens L, et al. 1996. The glucose sensor protein glucokinase is expressed in glucagon-producing α-cells. Proc Natl Acad Sci 93: 7036-7041.

Heller SR, Cryer PE. 1991. Reduced neuroendocrine and symptomatic responses to subsequent hypoglycemia after 1 episode of hypoglycemia in nondiabetic humans. Diabetes 40: 223-226.

Hevener AL, Bergman RN, Donovan CM. 1997. Novel glucosensor for hypoglycemic detection localized to the portal vein. Diabetes 46: 1521-1525.

Hevener AL, Bergman RN, Donovan CM. 2000. Portal afferents are critical for the sympathoadrenal response to hypoglycemia. Diabetes 49: 8-12.

Hevener AL, Bergman RN, Donovan CM. 2001. Hypoglycemic detection does not occur in the hepatic artery or liver: Findings consistent with a portal vein glucosensor locus. Diabetes 50: 399-403.

Higaki Y, Hirshman MF, Fujii N, Goodyear LJ. 2001. Nitric oxide increases glucose uptake through a mechanism that is distinct from the insulin and contraction pathways in rat skeletal muscle. Diabetes 50: 241-247.

Ibrahim N, Bosch MA, Smart JL, Qiu J, Rubinstein M, et al. 2003. Hypothalamic proopiomelanocortin neurons are glucose responsive and express K(ATP) channels. Endocrinology 144: 1331-1340.

Iuras A, Telles MM, Bertoncini CRA, Ko GM, de Andrade IS, et al. 2005. Central administration of a nitric oxide precursor abolishes both the hypothalamic serotonin release and the hypophagia induced by interleukin-1[beta] in obese Zucker rats. Regul Pept 124: 145-150.

Jetton TL, Liang Y, Pettepher CC, Zimmerman EC, Cox FG, et al. 1994. Analysis of upstream glucokinase promoter activity in transgenic mice and identification of glucokinase in rare neuroendocrine cells in the brain and gut. J Biol Chem 269: 3641-3654.

Kang L, Routh VH, Kuzhikandathil EV, Gaspers L, Levin BE. 2004. Physiological and molecular properties of rat hypothalamic ventromedial nucleus glucosensing neurons. Diabetes 53: 559

Katz B. 1971. Quantal mechanism of neural transmitter release. Science 173: 123-126.

Katz B, Miledi R. 1965. The effect of calcium on acetylcholine release from motor nerve terminals. Proc R Soc Lond B Biol Sci 161: 496-503.

Katz B, Miledi R. 1967a. The release of acetylcholine from nerve endings by graded electric pulses. Proc R Soc Lond B Biol Sci 23–38.

Katz B, Miledi R. 1967b. The timing of calcium action during neuromuscular transmission. J Physiol 189: 535-544.

Kim EK, Miller I, Aja S, Landree LE, Pinn M, et al. 2004. C75, a Fatty Acid Synthase Inhibitor, Reduces Food Intake via Hypothalamic AMP-activated Protein Kinase. J Biol Chem 279: 19970-19976.

Kingwell BA, Formosa M, Muhlmann M, Bradley SJ, McConell GK. 2002. Nitric oxide synthase inhibition reduces glucose uptake during exercise in individuals with type 2 diabetes more than in control subjects. Diabetes 51: 2572-2589.

Kow LM, Pfaff DW. 1989. Responses of hypothalamic paraventricular neurons in vitro to norepinephrine and other feeding-relevant agents. Physiol Behav 46: 265-271.

Koyama Y, Coker RH, Stone EE, Lacy DB, Jabbour K, et al. 2000. Evidence that carotid bodies play an important role in glucoregulation in vivo. Diabetes 49: 1434-1442.

Lam TKT, Pocai A, Gutierrez-Juarez R, Obici S, Bryan J, et al. 2005. Hypothalamic sensing of circulating fatty acids in required for glucose homeostasis. Nat Med 11: 320-327.

Laughton JD, Charnay Y, Belloir B, Pellerin L, Magistretti PJ, et al. 2000. Differential messenger RNA distribution of lactate dehydrogenase LDH-1 and LDH-5 isoforms in the rat brain. Neuroscience 96: 619-625.

Lee K, Li B, Xi X, Suh Y, Martin RJ. 2005. Role of Neuronal Energy Status in the Regulation of Adenosine 5′-Monophosphate-Activated Protein Kinase, Orexigenic Neuropeptides Expression, and Feeding Behavior. Endocrinology 146: 3-10.

Lesniewski LA, Miller TA, Armstrong RB. 2003. Mechanisms of force loss in diabetic mouse skeletal muscle. Muscle Nerve 28: 493-500.

Levin BE. 2000. Glucose-regulated dopamine release from substantia nigra neurons. Brain Res 874: 158-164.

Levin BE, Routh VH. 1996. Role of the brain in energy balance and obesity. Am J Physiol 271: R491-R500.

Levin BE, Triscari J, Sullivan AC. 1980. Abnormal sympathoadrenal function and plasma catecholamines in obese Zucker rats. Pharmacol Biochem Behav 13: 107-113.

Li AJ, Ritter S. 2004. Glucoprivation increases expression of neuropeptide Y mRNA in hindbrain neurons that innervate the hypothalamus. Eur J Neurosci 19: 2147-2154.

Li J, Hu X, Selvakumar P, Russell RR III, Cushman SW, et al. 2004. Role of the nitric oxide pathway in AMPK-mediated glucose uptake and GLUT4 translocation in heart muscle. Am J Physiol 287: E834-E841.

Li S, Pelletier G. 1986. The role of dopamine in the control of neuropeptide Y neurons in the rat arcuate nucleus. Neurosci Lett 69: 74-77.

Lindgren CA, Laird MV. 1994. Nitroprusside inhibits neurotransmitter release at the frog neuromuscular junction. Neuroreport 5: 2205-2208.

Liu F, Dong Q, Myers AM, Fromm HJ. 1991. Expression of human brain hexokinase in Escherichia coli: Purification and characterization of the expressed enzyme. Biochem Biophys Res Commun 177: 305-311.

Liu M, Seino S, Kirchgessner AL. 1999. Identification and characterization of glucoresponsive neurons in the enteric nervous system. J Neurosci 19: 10305-10317.

Loftus TM, Jaworsky DE, Frehywot GL, Townsend CA, Ronnett GV, et al. 2000. Reduced food intake and body weight in mice treated with fatty acid synthase inhibitors. Science 288: 2379-2381.

Lopez-Barneo J. 2003. Oxygen and glucose sensing by carotid body glomus cells. Curr Opin Neurobiol 13: 493-499.

Luck G, Hoch W, Hopf C, Blottner D. 2000. Nitric oxide synthase (NOS-1) coclustered with agrin-induced AChR-specializations on cultured skeletal myotubes. Mol Cell Neurosci 16: 268-281.

Lynch RM, Tompkins LS, Brooks HL, Dunn-Meynell AA, Levin BE. 2000. Localization of glucokinase gene expression in the rat brain. Diabetes 49: 693-700.

Maekawa F, Toyoda Y, Torii N, Miwa I, Thompson RC, et al. 2000. Localization of glucokinase-like immunoreactivity in the rat lower brain stem: For possible location of brain glucose-sensing mechanisms. Endocrinology 141: 375-384.

Magistretti PJ, Pellerin L. 1996. Cellular mechanisms of brain energy metabolism. Relevance to functional brain imaging and to neurodegenerative disorders. Ann NY Acad Sci 777: 380-387.

Magoul R, Onteniente B, Benjelloun W, Tramu G. 1993. Tachykinergic afferents to the rat arcuate nucleus. A combined immunohistochemical and retrograde tracing study. Peptides 14: 275-286.

Marques MJ, Santo Neto H. 2005. Acetylcholine receptors and nerve terminal distribution at the neuromuscular junction of non-obese diabetic mice. Anat Rec 267: 112-119.

Matschinsky FM, Collins HW. 1997. Essential biochemical design features of the fuel-sensing system in pancreatic beta-cells. Chem Biol 4: 249-257.

Matschinsky FM, Glaser B, Magnuson MA. 1998. Pancreatic beta-cell glucokinase: Closing the gap between theoretical concepts and experimental realities. Diabetes 47: 307-315.

Mayer J, Thomas DW. 1967. Regulation of food intake and obesity. Science 156: 328-337.

McArdle JJ, Sellin LC, Coakley KM, Potian JG, Hognason K. 2006. Mefloquine selectively increases asynchronous acetylcholine release from motor nerve terminals. Neuropharmacology 50(3): 345-353.

McKenna MC, Tildon JT, Stevenson JH, Hopkins IB, Huang X, et al. 1998. Lactate transport by cortical synaptosomes from adult rat brain: Characterization of kinetics and inhibitor specificity. Dev Neurosci 20: 300-309.

McNay EC, Gold PE. 1999. Extracellular glucose concentrations in the rat hippocampus measured by zero-net-flux: Effects of microdialysis flow rate, strain, and age. J Neurochem 72: 785-790.

McNay EC, Fries TM, Gold PE. 2000. Decreases in rat extracellular hippocampal glucose concentration associated with cognitive demand during a spatial task. Proc Nat Acad Sci 97: 2881-2885.

McNay EC, McCarty RC, Gold PE. 2001. Fluctuations in brain glucose concentration during behavioral testing: Dissociations between brain areas and between brain and blood. Neurobiol Learn Mem 75: 325-337.

Miki T, Liss B, Minami K, Shiuchi T, Saraya A, et al. 2001. ATP-sensitive potassium channels in hypothalamic neurons play an essential role in the maintenance of glucose homeostasis by controlling glucagon release and food intake. Nat Neurosci 5: 507-512.

Miller I, Ronnett GV, Moran TH, Aja S. 2004. Anorexigenic C75 alters c-Fos in mouse hypothalamic and hindbrain subnuclei. Neuroreport 15: 925-929.

Minami T, Shimizu N, Duan S, Oomura Y. 1990. Hypothalamic neuronal activity responses to 3-hydroxybutyric acid, an endogenous organic acid. Brain Res 509: 351-354.

Minokoshi Y, Alquier T, Furukawa N, Kim YB, Lee A, et al. 2004. AMP-kinase regulates food intake by responding to hormonal and nutrient signals in the hypothalamus. Nature 428: 569-574.

Mizuno TM, Makimura H, Silverstein J, Roberts JL, Lopingco T, et al. 1999. Fasting regulates hypothalamic neuropeptide Y, agouti-related peptide, and proopiomelanocortin in diabetic mice independent of changes in leptin or insulin. Endocrinology 140: 4551-4557.

Mizuno Y, Oomura Y. 1984. Glucose responding neurons in the nucleus tractus solitarius of the rat, in vitro study. Brain Res 307: 109-116.

Morely JE, Farr SA, Gaskin FS, Kumar VB, Banks WA. 2003. The orexigenic peptides NPY and orexin - A have NO effect on behavior in neuronal-NOS knockout mice. Soc Neurosci Abstr Viewer/Itinerary Planner 615.14.

Mulvey C, Harno E, Keenan A, Ohlendieck K. 2005. Expression of the skeletal muscle dystrophin-dystroglycan complex and syntrophin-nitric oxide synthase complex is severely affected in the type 2 diabetic Goto-Kakizaki rat. Eur J Cell Biol 84: 867-883.

Muroya S, Yada T, Shioda S, Takigawa M. 1999. Glucose-sensitive neurons in the rat arcuate nucleus contain neuropeptide Y. Neurosci Lett 264: 113-116.

Nakano Y, Oomura Y. 1986. Feeding-related activity of glucose- and morphine-sensitive neurons in the monkey amygdala. Brain Res 399: 167-172.

Navarro M, Rodriquez dF, Alvarez E, Chowen JA, Zueco JA, et al. 1996. Colocalization of glucagon-like peptide-1 (GLP-1) receptors, glucose transporter GLUT-2, and glucokinase mRNAs in rat hypothalamic cells: Evidence for a role of GLP-1 receptor agonists as an inhibitory signal for food and water intake. J Neurochem 67: 1982-1991.

Obici S, Feng Z, Arduini A, Conti R, Rossetti L. 2003. Inhibition of hypothalamic carnitine palmitoyltransferase-1 decreases food intake and glucose production. Nat Med 9: 756-761.

Oomura Y. 1983. Glucose as a regulator of neuronal activity. Adv Metab Disord 10: 31-65.

Oomura Y, Ono H, Ooyama H, Wayner MJ. 1969. Glucose and osmosensitive neurons of the rat hypothalamus. Nature 222: 282-284.

Oomura Y, Kimura K, Ooyama H, Maeo T, Iki M, et al. 1964. Reciprocal activities of the ventromedial and lateral hypothalamic area of cats. Science 143: 484-485.

Pardal R, Lopez-Barneo J. 2002. Low glucose-sensing cells in the carotid body. Nat Neurosci 5: 197-198.

Parkinson JF, Mitrovic B, Merrill JE. 1997. The role of nitric oxide in multiple sclerosis. J Mol Med 75: 174-186.

Penicaud L, Ferre P. 1988. Development of insulin resistance during the course of obesity: Lessons from animal models. J Obesity Weight Reg 7: 91-109.

Peruzzo B, Pastor FE, Blazquez JL, Schobitz K, Pelaez B, et al. 2000. A second look at the barriers of the medial basal hypothalamus. Exp Brain Res 132: 10-26.

Prevot V, DeSeranno S, Estrella C. 2004. Glial-neuronal-endothelial interactions and the neuroendocrine control of GnRH secretion. In: Non-Neuronal Cells of the Nervous System: Function and Dysfunction, Elsevier, Amsterdam pp. 99-214.

Prevot V, Croix D, Bouret S, Dutoit S, Tramu G, et al. 1999. Definitive evidence for the existence of morphological plasticity in the external zone of the median eminence during the rat estrous cycle: Implication of neuro-glio-endothelial interactions in gonadotropin-releasing hormone release. Neuroscience 94: 809-819.

Puttmann B, Gerlach EM, Kruger M, Blottner D. 2005. Neuromuscular contacts induce nitric oxide signals in skeletal myotubes in vitro. Neurosignals 14: 85-95.

Ritter R, Slusser P, Stone S. 1981. Glucoreceptors controlling feeding and blood glucose: Location in the hindbrain. Science 213: 451-452.

Ritter S, Bugarith K, Dinh TT. 2001. Immunotoxic destruction of distinct catecholamine subgroups produces selective impairment of glucoregulatory responses and neuronal activation. J Comp Neurol 432: 197-216.

Ritter S, Dinh TT, Zhang Y. 2000. Localization of hindbrain glucoreceptive sites controlling food intake and blood glucose. Brain Res 856: 37-47.

Ritter S, Llewellyn-Smith I, Dinh TT. 1998. Subgroups of hindbrain catecholamine neurons are selectively activated by 2-deoxy-D-glucose induced metabolic challenge. Brain Res 805: 41-54.

Ritter S, Watts AG, Dinh TT, Sanchez-Watts G, Pedrow C. 2003. Immunotoxin lesion of hypothalamically projecting norepinephrine and epinephrine neurons differentially affects circadian and stressor-stimulated corticosterone secretion. Endocrinology 144: 1357-1367.

Rizzo MA, Piston DW. 2003. Regulation of beta cell glucokinase by S-nitrosylation and association with nitric oxide synthase. J Cell Biol 161: 243-248.

Roberts CK, Barnard RJ, Scheck SH, Balon TW. 1999. Exercise-stimulated glucose transport in skeletal muscle is nitric oxide dependent. Am J Physiol 273: E220-E225.

Roeper J, Ashcroft FM. 2002. Metabolic inhibition and low internal ATP activate K-ATP channels in rat dopaminergic substantia nigra neurones. Pflugers Arch 405: 305-309.

Romero-Navarro G, Cabrera-Valladares G, German MS, Matschinsky FM, Velazquez A, et al. 1999. Biotin regulation of pancreatic glucokinase and insulin in primary cultured rat islets and in biotin-deficient rats. Endocrinology 140: 4595-4600.

Roncero I, Alvarez E, Vazquez P, Blazquez E. 2000. Functional glucokinase isoforms are expressed in rat brain. J Neurochem 74: 1848-1857.

Sanders NM, Ritter S. 2000. Repeated 2-deoxy-D-glucose-induced glucoprivation attenuates Fos expression and glucoregulatory responses during subsequent glucoprivation. Diabetes 49: 1865-1874.

Sanders NM, Dunn-Meynell AA, Levin BE. 2004. Third Ventricular Alloxan Reversibly Impairs Glucose Counterregulatory Responses. Diabetes 53: 1230-1236.

Santizo RA, Koenig HM, Pelligrino DA. 2001. {beta}-Adrenoceptor and nNOS-derived NO interactions modulate hypoglycemic pial arteriolar dilation in rats. Am J Physiol 280: H562-H568.

Sara Y, Virmani T, Deák F, Liu X, Kavalali ET. 2005. An isolated pool of vesicles recycles at rest and drives spontaneous neurotransmission. Neuron 45: 563-573.

Sawchenko PE, Swanson RA. 1982. The organization of noradrenergic pathways from the brainstem to the paraventricular and supraoptic nuclei in the rat. Brain Res 257: 275-325.

Sayegh AI, Covasa M, Ritter RC. 2004. Intestinal infusions of oleate and glucose activate distinct enteric neurons in the rat. Auton Neurosci 115: 54-63.

Schiller Y, Rahamimoff R. 1989. Neuromuscular transmission in diabetes: Response to high-frequency activation. J Neurosci 9: 3709-3719.

Schuit FC, Huypens P, Heimberg H, Pipeleers DG. 2001. Glucose sensing in pancreatic β-cells: A model for the study of other glucose-regulated cells in gut, pancreas, and hypothalamus. Diabetes 50: 1-11.

Schwartz MW, Baskin DG, Bukowski TR, Kuijper JL, Foster D, et al. 1996. Specificity of leptin action on elevated blood glucose levels and hypothalamic neuropeptide Y gene expression in ob/ob mice. Diabetes 45: 531-535.

Schwarte RC, Godfrey EW. 2004. Nitric oxide synthase activity is required for postsynaptic differentiation of the embryonic neuromuscular junction. Dev Biol 273: 276-284.

Schwartz MW, Sipols AJ, Marks JL, Sanacora G, White JD, et al. 1992. Inhibition of hypothalamic neuropeptide Y gene expression by insulin. Endocrinology 130: 3608-3616.

Sheppard DN, Welsh MJ. 1992. Effect of ATP-sensitive K^+ channel regulators on cystic fibrosis transmembrane conductance regulator chloride currents. J Gen Physiol 100: 573-591.

Shimizu N, Oomura Y, Novin D, Grijalva CV, Cooper PH. 1983. Functional correlations between lateral hypothalamic glucose-sensitive neurons and hepatic portal glucose-sensitive units in rat. Brain Res 265: 49-54.

Shiraishi T. 1991. Noradrenergic neurons modulate lateral hypothalamic chemical and electrical stimulation induced feeding. Brain Res Bull 27: 347-351.

Shoji S. 1992. Glucose regulation of synaptic transmission in the dorsolateral septal nucleus of the rat. Synapse 12: 322-332.

Silinsky EM. 2004. Adenosine decreases both presynaptic calcium currents and neurotransmitter release at the mouse neuromuscular junction. J Physiol 558: 389-401

Silver IA, Erecinska M. 1994. Extracellular glucose concentration in mammalian brain: Continuous monitoring of changes during increased neuronal activity and upon limitation in oxygen supply in normo-, hypo, and hyperglycemic animals. J Neurosci 14: 5068-5076.

Silver IA, Erecinska M. 1998. Glucose-induced intracellular ion changes in sugar-sensitive hypothalamic neurons. J Neurophysiol 79: 1733-1745.

Simpson IA, Appel NM, Hokari M, Oki J, Holman GD, et al. 1999. Blood-brain barrier glucose transporter: Effects of hypo- and hyperglycemia revisited. J Neurochem 72: 238-247.

Sindelar DK, Marie L, Miura GI, Palmiter RD, McMinn JE, et al. 2004. Neuropeptide Y Is Required for Hyperphagic Feeding in Response to Neuroglucopenia. Endocrinology 145: 3363-3368.

Song Z, Routh VH. 2005a. Recurrent hypoglycemia reduces the glucose sensitivity of glucose sensing neurons in the ventromedial hypothalamic nucleus (VMN). 2005 Abstract Viewer/Itinerary Planner. Washington, DC: Society for Neuroscience, 2005. Online. 532.9.

Song Z, Routh VH. 2005b. Differential effects of glucose and lactate on glucosensing neurons in the ventromedial hypothalamic nucleus. Diabetes 54: 15-22.

Song Z, Routh VH. 2006. Recurrent hypoglycemia decreases the glucose sensitivity of glucose-inhibited (G1) neurons in the ventromedial hypothalamic nucleus (VMN). Am J Physiol Regul Integr Comp Physiol Jun 22 (epub ahead of print) doi: 10.1152/ajpregu 00148.2006.

Song Z, Levin BE, McArdle JJ, Bakhos N, Routh VH. 2001. Convergence of pre- and postsynaptic influences on glucosensing neurons in the ventromedial hypothalamic nucleus. Diabetes 50: 2673-2681.

Spanswick D, Smith MA, Groppi VE, Logan SD, Ashford MLJ. 1997. Leptin inhibits hypothalamic neurons by activation of ATP-sensitive potassium channels. Nature 390: 521-525.

Spanswick D, Smith MA, Mirshamsi S, Routh VH, Ashford MLJ. 2000. Insulin activates ATP-sensitive K^+ channels in hypothalamic neurons of lean, but not obese rats. Nat Neurosci 3: 757-758.

Stanley BG, Anderson KC, Grayson MH, Leibowitz SF. 1989. Repeated hypothalamic stimulation with neuropeptide Y increases daily carbohydrate and fat intake and body weight gain in female rats. Physiol Behav 46: 173-177.

Steffens AB, Mogenson GJ, Stevenson JA. 1972. Blood glucose, insulin, and free fatty acids after stimulation and lesions of the hypothalamus. Am J Physiol 222: 1446-1452.

Steffens AB, Scheurink AJ, Luiten PG, Bohus B. 1988. Hypothalamic food intake regulating areas are involved in the homeostasis of blood glucose and plasma FFA levels. Physiol Behav 44: 581-589.

Stoddard-Apter SL, Bergdall VK, Townsend DW, Levin BE. 1986. Plasma catecholamines associated with hypothalamically-elicited defense behavior. Physiol Behav 36: 867-873.

Tache Y, Yang H, Kaneko H. 1995. Caudal raphe-dorsal vagal complex peptidergic projections: Role in gastric vagal control. Peptides 16: 431-435.

Tews DS. 2001. Role of nitric oxide and nitric oxide synthases in experimental models of denervation and reinnervation. Microsc Res Tech 55: 181-186.

Thomas S, Robitaille R. 2001. Differential Frequency-Dependent Regulation of Transmitter Release by Endogenous Nitric Oxide at the Amphibian Neuromuscular Synapse. J Neurosci 21: 1087-1095.

Thor K, Helke CJ. 2002. Serotonin and substance P colocalization in medullary projections to the nucleus tractus solitarius: Dual-colour immunohistochemistry combined with retrograde tracing. J Chem Neuroanat 2: 139-148.

Thornton JE, Cheung CC, Clifton DK, Steiner RA. 1997. Regulation of hypothalamic proopiomelanocortin mRNA by leptin in ob/ob mice. Endocrinology 138: 5063-5066.

Thupari JN, Landree LE, Ronnett GV, Kuhajda FP. 2002. C75 increases peripheral energy utilization and fatty acid oxidation in diet-induced obesity. Proc Natl Acad Sci 99: 9498-9502.

Tsacopoulos M, Magistretti PJ. 1996. Metabolic coupling between glia and neurons. J Neurosci 16: 877-885.

Tseng C, Huang S, Lo W, Lin C, Hsiao M. 2003. Insulin molecular signaling is defective in NTS neurons of adult spontaneous hypertensive rat. Society for Neuroscience Abstract Viewer/Itinerary Planner 922.1.

Tsuura Y, Ishida H, Shinomura T, Nishimura M, Seino Y. 1998. Endogenous nitric oxide inhibits glucose-induced insulin secretion by suppression of phosphofructokinase activity in pancreatic islets. Biochem Biophys Res Commun 252: 34-38.

Tu Y, Thupari JN, Kim EK, Pinn ML, Moran TH, et al. 2005. C75 Alters Central and Peripheral Gene Expression to Reduce Food Intake and Increase Energy Expenditure. Endocrinology 146: 486-493.

Tucker DC, Saper CB, Ruggiero DA, Reis DJ. 1987. Organization of central adrenergic pathways: I. Relationships of ventrolateral medullary projections to the hypothalamus and spinal cord. J Comp Neurol 259: 591-603.

Vamecq J, Vallee L, Lesage F, Gressens P, Stables JP. 2005. Antiepileptic popular ketogenic diet: Emerging twists in an ancient story. Prog Neurobiol 75: 1-28.

Van der Kloot W. 2003. Loading and recycling of synaptic vesicles in the Torpedo electric organ and the vertebrate neuromuscular junction. Prog Neurobiol 71: 269-303.

van Dijk G, Bottone AE, Strubbe JH, Steffens AB. 1994. Hormonal and metabolic effects of paraventricular hypothalamic administration of neuropeptide Y during rest and feeding. Brain Res 660: 96-103.

Vannucci SJ, Hagberg H. 2004. Hypoxia-ischemia in the immature brain. J Exp Biol 207: 3149-3154.

Villarroel A, Sakmann B. 1996. Calcium permeability increase of endplate channels in rat muscle during postnatal development. J Physiol 496: 331-338.

Wang R, Liu X, Hentges ST, Dunn-Meynell AA, Levin BE, et al. 2004. The regulation of glucose-excited neurons in the hypothalamic arcuate nucleus by glucose and feeding-relevant peptides. Diabetes 53: 1959-1965.

Wang R, Cruciani-Guglielmacci C, Migrenne S, Magnan C, Cotero VE, et al. 2006. The effects of oleic-acid (OA) on distinct populations of neurons in the hypothalamic arcuate nucleus (ARC) are dependent on extracellular glucose levels. J Neurophysiol 95(3): 1491-1498.

White JD. 1993. Neuropeptide Y: A central regulator of energy homeostasis. Regul Pept 49: 93-107.

Winter MC, Sheppard DN, Carson MR, Welsh MJ. 1994. Effect of ATP concentration on CFTR Cl- channels: A kinetic analysis of channel regulation. Biophys J 66: 1398-1403.

Yang XJ, Kow LM, Funabashi T, Mobbs CV. 1999. Hypothalamic glucose sensor: Similarities to and differences from pancreatic beta-cell mechanisms. Diabetes 48: 1763-1772.

Yang XJ, Kow LM, Pfaff DW, Mobbs CV. 2004. Metabolic Pathways That Mediate Inhibition of Hypothalamic Neurons by Glucose. Diabetes 53: 67-73.

Yoshimatsu H, Niijima A, Oomura Y, Yawabe K, Katafuchi T. 1984. Effects of hypothalamic lesion on pancreatic autonomic nerve activity in the rat. Brain Res 303: 147-152.

Zawar C, Neumcke B. 2000. Differential activation of ATP-sensitive potassium channels during energy depletion in CA1 pyramidal cells and interneurones of rat hippocampus. Pfugers Arch 439: 256-262.

Zhu X, Heunks LM, Ennen L, Machiels HA, Van Der Heijden HF et al. 2006. Nitric oxide modulates neuromuscular transmission during hypoxia in rat diaphragm. Muscle Nerve 33: 104-112.

8 CO_2/H^+ Homeostasis: Role of Central and Peripheral Chemoreceptors in Adult Mammals

S. Lahiri · S. M. Baby · C. D. Giulio · A. Roy

Abstract: CO_2 and H^+ are metabolic end-products, which are produced continuously and excreted steadily to maintain steady-state concentrations of CO_2/H^+ in the body, primarily by the chemoreceptors. To help maintain an adequate speed of reaction compatible with life, these reactions are enhanced by carbonic anhydrase (CA) present in the chemoreceptor cells. The role of chemoreceptors in H^+ homeostasis is the focus of this chapter. The peripheral chemoreceptors very readily sense CO_2/H^+ and stimulate ventilation in order to enhance CO_2 exhalation. Central chemoreceptors are stimulated similarly but slowly, also resulting in increased ventilation. Altogether, this phase, assisted by respiration alone, can be defined as the acute response. However, the H^+ left behind is excreted by the renal system, rather slowly (chronic phase) without any direct intervention by the chemoreceptors. Thus, respiratory and renal systems are integrated in the long run to maintain CO_2/H^+ homeostasis.

A consensus model of CO_2/H^+ stimulatory response is as follows:

CO_2/H^+stimulus → Cellular K^+current suppression → Cell depolarization → Voltage-dependent Ca^{2+}gate opens → Ca^{2+}influx andstore-operated Ca^{2+}release → $[Ca^{2+}]_i$ rise → Neurotransmission → Instantaneous chemosensory discharge (acute effects like hypoxia).

At the same time, ATP comes into play as a neurotransmitter, but experimental data suggest a complex mechanism of interactions. Spyer et al. (2004) proposed ATP as the key neurotransmitter, but that the CO_2 response is not affected by PPADS (ATP antagonist) in in vitro preparation. Additionally, recent data from P_2X_2 knockout mice showed a normal response to CO_2, suggesting that purinergic mechanisms are not involved in central chemoreception. However, this leaves out a role of CO_2/H^+ sensing in the peripheral chemoreceptors.

Turning to constancy of $[H^+]$ in the fluid matrix of the body at the body temperature, it remains a fundamental condition of life, a legacy of Claude Bernard.

List of Abbreviations: CA, carbonic anhydrase; Hx, hypoxia; IbTX, iberiotoxin; PGN, Petrosal ganglion neurons; RTN, retrotrapezoid; VLM, ventrolateral medullary area

1 Introduction

1.1 CO_2/H^+ Sensing

CO_2 and H^+ are metabolic end-products, which are produced continuously and excreted steadily to maintain a steady-state concentration in the body. The role of chemoreceptors in maintaining $[H^+]$ homeostasis in the body is the focus of this chapter. As Walter Cannon (1929) elaborated, following Claude Bernard (1878), the stability of the internal environment, the body fluid matrix (blood and lymph) independent of the external environment, is the goal of the homeostatic mechanisms. This self-regulation of bodily $[H^+]$ is space and time dependent. CO_2 in gaseous phase is rapidly eliminated from the lungs by ventilation (space). $[H^+]$ that remains in body fluids is excreted slowly by the kidneys. Thus, $[H^+]$ is regulated in an integrative fashion by the respiratory and the renal systems.

After watching abortive attempts by previous scientists investigating self-regulation in animals and also from his many experiments and insights, Claude Bernard (1978) finally enunciated the fixity of the milieu interieur—cells strive to maintain the condition of the life. Acid–base concepts were not fully developed until some years later (e.g., Henderson, 1928), and the new era of constant pH regulation was fully born. Fifty years later, several crucial studies appeared. Among these was the hypothesis of Walter Cannon (1929, 1932), who formulated after a decade of research and exploration, that all the systems of an organism tend to achieve equilibrium somewhat similar to that described by Le Chatelier (1888) for physical systems. However, physiological systems achieve instead, a semiequilibrium with a physiological twist. Cannon (1929) termed this condition as homeostasis. He described such homeostasis for each systems of the body in his book—*The Wisdom of the Body,* first published in 1932. Since then, this term has been universally accepted and increasingly used by the physiologists, biochemists, and the molecular biologists.

$[H^+]$ in the body fluid matrix is stabilized at around pH 7.4 at a body temperature of about 37°C. This is always dependent on temperature, as the physical chemistry demands, and this is the first line of defense. The intracellular pH, being more acidic, runs parallel to it (*Figure 8-1*).

Figure 8-1
Blood pH (extracellular) and intracellular pH (muscle) as a function of body temperature. With decreased temperature, the pH becomes alkaline (From Rahn, 1974)

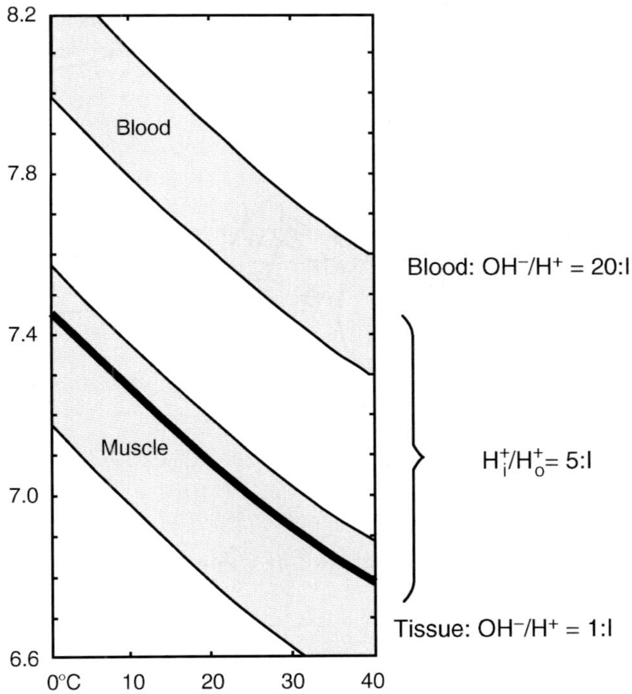

The consensus model of this cellular $[H^+]$ homeostasis at a given temperature is as follows:

Hypercapnia $\rightarrow$ CO_2/H^+stimulus $\rightarrow$ cellular $\rightarrow$ K^+current suppression $\rightarrow$ cell deploarization
$\rightarrow$ voltage-dependent Ca^{2+}gate opens $\rightarrow$ Ca^{2+}influx and store-operated Ca^{2+}release
$\rightarrow$ $[Ca^{2+}]_i$ rise $\rightarrow$ neurotransmission $\rightarrow$ instantaneous chemosensory discharge
(acute effects like hypoxia) and ventilation in consequence. See Buckler and Vaughan-Jones, 1994.

1.2 Sites of Chemoreception

There are two sets of sensors: peripheral and central chemoreceptors. The peripheral receptors reside in the glomus cells of the tiny organ of carotid and aortic bodies, which are located in the carotid bifurcations and in the aortic arch, respectively, and they respond instantaneously. This response is transmitted to the cardio-respiratory centers, which reflexively respond by increasing the ventilation, for example, which will blow off

Figure 8-2
(a) Effect of ATP on central chemoreceptors. (b) Glomus cell releases ATP upon stimulus. This ATP works on the carotid sinus nerve endings (From Spyer et al., 2004)

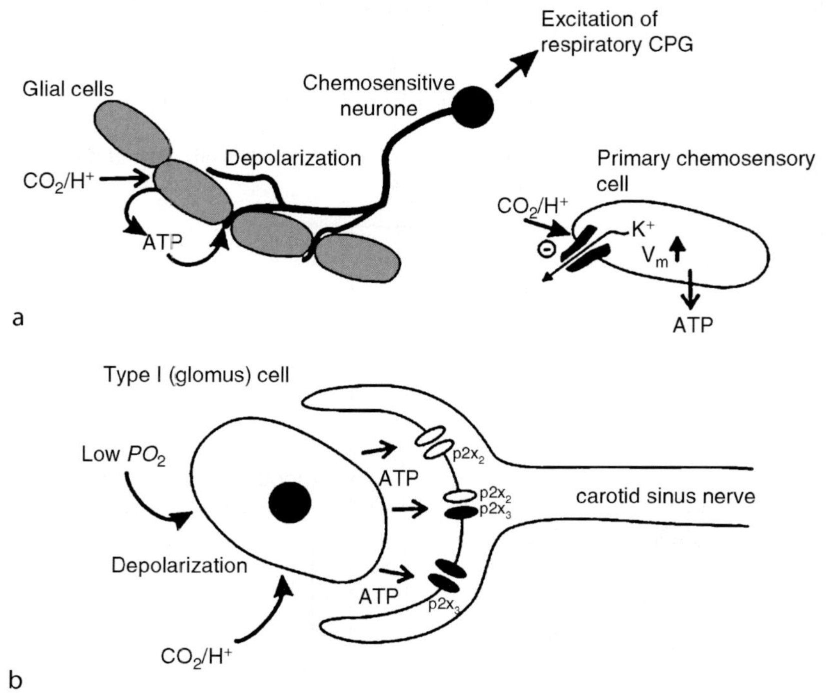

CO_2. This is the acute phase, which will be followed by a slow renal H^+ excretion (chronic phase) to maintain $[H^+]$ homeostasis.

1.3 General Principle

The general principle involving cellular CO_2 reactions is mediated by carbonic anhydrase in the sensors:

$$CO_2 + H_2O \xrightarrow{CA} H_2CO_3 \rightarrow H^+ + HCO_3^-.$$

At chemical equilibrium, the whole reaction can be expressed as

$$K^1 = \frac{[H^+][HCO_3^-]}{[CO_2]}.$$

Taking logarithm to the base 10 and rearranging we obtain the familiar Henderson–Hasselbalch equation,

$$pH = pK^- + \log 10 \frac{[H^+][HCO_3^-]}{[CO_2]},$$

where pH is the negative logarithm of hydrogen ion concentration $[H^+]$ and pK^- is the negative logarithm of K^+, under a given condition equals to 6.07 at 37°C.

In addition to the obvious elimination of CO_2 from the body, CA accelerates what otherwise would be a rate-limiting step. A classical example is the so-called Jacob–Stewart cycle (Jacobs and Stewart, 1942), which carries $[H^+]$ across cell membranes.

CO_2 is highly permeable through the lipid membrane. HCO_3^- is carried across by a transport protein in exchange for another anion. $[H^+]$ is impermeable, but rather uses specific transport mechanisms, such as Na^+–H^+ exchange. An increase in intracellular $[H^+]$ reacts with HCO_3^- to form CO_2, which diffuses across and reacts to form H^+ and HCO_3^- again in the cell. The net effect is to transfer H^+ across a lipid membrane, which is impermeable.

In addition, central chemoreceptor will also respond by increasing ventilation, which will be followed by corrective renal response. These general considerations are given in detail elsewhere (Lahiri and Forster, 2003).

2 Central CO_2/H^+ Chemosensing

2.1 CO_2 Sensing

Molecular mechanisms for CO_2 sensing are rather general in the brain stem areas, including the respiration-related neurons. The sensing molecule detects changes in pH rather than molecular CO_2. Evident changes in ventilation occur with a rise in PCO_2 as low as 1 Torr (Feldman et al., 2003), as shown some 100 years ago by Haldane and Priestley (1905).

2.2 $[H^+]$ Sensing

The putative CO_2/H^+-sensing molecules are expressed in the brain stem, which are connected to the network controlling ventilation. Protonation of these residues may lead to a change in protein conformation that is coupled to a network of channel activity. Depending on the location of the protonation sites, a membrane protein can detect extracellular and/or intracellular pH (Nattie et al., 2001; Guyenet et al., 2005).

2.3 Respiratory Neurons

It is now known that multiple sites in the brain stem have CO_2/H^+ chemosensitivity:

1. the ventral chemosensitive groups are located in the ventrolateral medullary area (VLM);
2. the dorsal chemosensitive group;
3. the middle chemosensitive group;
4. the chemosensitive group made of middle raphe nuclei; and
5. the glial cells.

These networks of respiratory neurons are responsible for generation of respiratory rhythm. There is evidence suggesting that at least some VLM neurons are intrinsically sensitive to the change in $[H^+]$ that follows changes in arterial PCO_2 (Kawai et al., 1996). The multiple chemosensitive sites may be involved in the control of breathing and cardiovascular regulation (Jiang et al., 2005).

Some of these neurons encode $[H^+]$ binding and discharge tonically. In vivo, the retrotrapezoid nucleus (RTN) receives respiratory inhibitory inputs that may serve as feedback for respiratory modulation (Guyenet et al., 2005).

2.4 ATP is a Key for Central Chemosensory Function

ATP is released extracellularly in the ventrolateral medulla during hypercapnia because of activation of central chemoreceptors. The action of hypercapnia is on P_2 receptors localized in close proximity to the VLM inspiratory neurons. But the cellular sources of ATP released are yet to be investigated (Spyer and Thomas, 2000; Spyer et al., 2004, Gourine, 2005).

Purinoreceptor P_2X_2 subunit deficiency results in an attenuation of ventilatory responses to CO_2 (Rong et al., 2003). The lack of function in this subunit deficiency is particularly useful. P_2X_2 subunits containing

ATP receptors appear to contribute to central chemosensitivity in vivo (Spyer and Thomas, 2000; Gourine and Spyer, 2003). However, recent data from P_2X_2 knockout mice (Rong et al., 2003) indicate that they are not essential. The effects of P_2 receptor agonists/antagonists were further examined (Lorier et al., 2004) and they found that ATP potentiates respiratory frequency and PPADS-sensitive neurons (P_2 receptor antagonist) did not interfere with respiratory frequency by increased PCO_2. Thus, these studies conclude that P_2 receptors do not contribute to central chemosensitivity in vitro, although these receptors modulate respiratory rhythm. But see Gourine, 2005.

2.5 Controversy

Since the CO_2 response is not affected by PPADS in in vitro preparations from P_2X_2 knockout mice (Lorier et al., 2004), in which CO_2 sensitivity is normal (Rong et al., 2003), it is strongly suggested that P_2X_2 receptors do not contribute to the central CO_2 chemosensitivity. Obviously, more work is needed for understanding this lack of response. But see Gourine, 2005.

3 Peripheral CO_2/H^+ Sensing

3.1 Rapid Release of ATP from the Peripheral Chemoreceptors

ATP has been implicated as a neurotransmitter in the peripheral chemoreceptors as a result of hypoxic (Prasad et al., 2001; Buttigieg and Nurse, 2004) and CO_2/pH stimulation (Zhang and Nurse, 2004). ATP was presumed to be costored with other classical transmitters (e.g., ACh, dopamine, 5-HT, and GABA in the synaptic vesicles in the chemoreceptor cells (type I cells) of the carotid body) (Gonzalez et al., 1994; Zhang et al., 2000). More recently, a direct stimulus-evoked ATP release from the type-I cells has been demonstrated (Buttigieg and Nurse, 2004), and this release was dependent on extracellular Ca^{2+} via mainly L-type Ca^{2+} channels, blocked by nifedipine (50 μm) and cadmium (50 μm).

3.2 K^+ Channel Blockers and ATP Release

Hypoxia inhibits voltage-gated and Ca^{2+}-dependent maxi K^+ or BK channels (Williams et al., 2004), in addition to background or leak K^+ channels (Buckler et al., 2004) in rat glomus cells. Blockade of BK channels with iberiotoxin (IbTX) or TEA, leads to catecholamine secretion. IbTX (100 nm) also stimulates ATP release (Buttigieg and Nurse, 2004). But when applied under hypoxic conditions ($PO_2 \simeq 15 - 20$), their combined effect on ATP release was largely blocked. Also, by closing K^+ channels that are open at rest, both hypoxia and K^+ channel blockers depolarize the glomus cells, leading to Ca^{2+} entry and ATP release (Peers, 1990).

ATP coreleased with ACh mediates fast excitation by acting on postsynaptic ionotropic P_2X_2–P_2X_3 purinergic receptors on afferent nerve terminal, and consequently a ventilatory response to hypoxia (Spyer et al., 2004).

3.3 Coculture

Using the same coculture techniques, glomus cells with petrosal ganglion cells (PGN) and recording from the PGN neurons, it was shown that PGN cells and their terminals responded to hypoxia and CO_2/pH and not the PGN cells alone (Zhong et al., 1997; Prasad et al., 2001; Zhang and Nurse, 2004). This means that neurotransmitters released from the glomus cells excited the PGN cells. These transmitters consisted of ATP and ACh. Similar results were obtained by Varas et al. (2003). Recorded intracellularly from identified PGN functionally connected with the carotid body (CB) in vitro, and which responded to CB stimulation by stop flow or by acidosis, these neurons responded largely to application of ATP and ACh. That is, the PG neuron terminals in the CB were responsive. They concluded that ATP and ACh mediated the excitatory transmission in carotid body–identified chemoreceptor units in vitro. Also, more recently Roy, Mitchell, and Lahiri found that high PCO administration in normoxia immediately resulted in ATP release (in the dark) just as the chemosensory discharge is elicited (unpublished observation).

3.4 ATP Interaction in the Afferents: Hypoxic and Hypercapnic Stimulation

ATP release from the CB induced by hypoxia (Buttigieg and Nurse, 2004) and hypercapnia (Prasad et al., 2001) has been demonstrated, but the release of ATP due to combined stimulation has not been shown. Such ATP interaction is expected, but its demonstration in the CB and its effects on the afferent terminals would serve as an unifier of all functions.

4 An Unified Hypothesis

These observations provide apparently convincing evidence in favor of ATP-mediated purinergic signaling, having a pivotal role in chemosensory control carotid body function. Given the available evidence, it has been naturally proposed that ATP plays a common signaling function in respiratory control (➋ *Figure 8-3*; Spyer et al., 2004). From a global point of view of energy metabolism, ATP can be recognized as a common "coin" but how unified, remains to be seen (see Lahiri et al., 2006).

5 CO_2/H^+ Chemoreception is Augmented in the Absence of Hypoxic Chemoreception

The following examples show dominance of CO_2/H^+ in life.

5.1 Blunted Ventilatory Response to Hypoxia in High-Altitude Natives

There are changes in ventilatory response to $PaCO_2$ at different levels of PaO_2. As the PaO_2 decreases, the response becomes steeper (➋ *Figure 8-4*).

When the effect of PaO_2 is lost, although the $PaCO_2$ effect is normal, it is due primarily to central chemoreceptor stimulation. The indications are that the stimulation due to carotid chemoreceptor is minimal (Milledge and Lahiri, 1967).

5.2 Blunted Response to Hypoxia Due to Exposure of Cats to Normobaric Oxygenation

After exposure of cats to normobaric hyperoxia for over 36 h, their carotid chemoreceptors did not respond to hypoxia but responded to hypercapnia with an overshoot and enhanced activity. The overshoot was abolished after inhibition of carbonic anhydrase with acetazolamide, showing that carotid body chemoreception responded to H^+ (Lahiri et al., 1987).

5.3 Hypoxic Response of Carotid Chemosensory Discharge was Attenuated with Substance P Antagonist (Prabhakar et al., 1987)

SP receptor antagonists in nanomolar concentration provert the excitation effect of SP, and abolish the sensory response to hypoxia while leaving the sensitory response of hypercapnia unaffected in cats (Prabhakar et al., 1987).

Figure 8-3
Human ventilatory responses to CO_2 at different levels of PO_2, decreasing levels of PO_2 increased the response to CO_2 (*upper panel*) but not in the lower panel. The response to CO_2 remains in the face of blunted response to hypoxia (From Milledge and Lahiri, 1967)

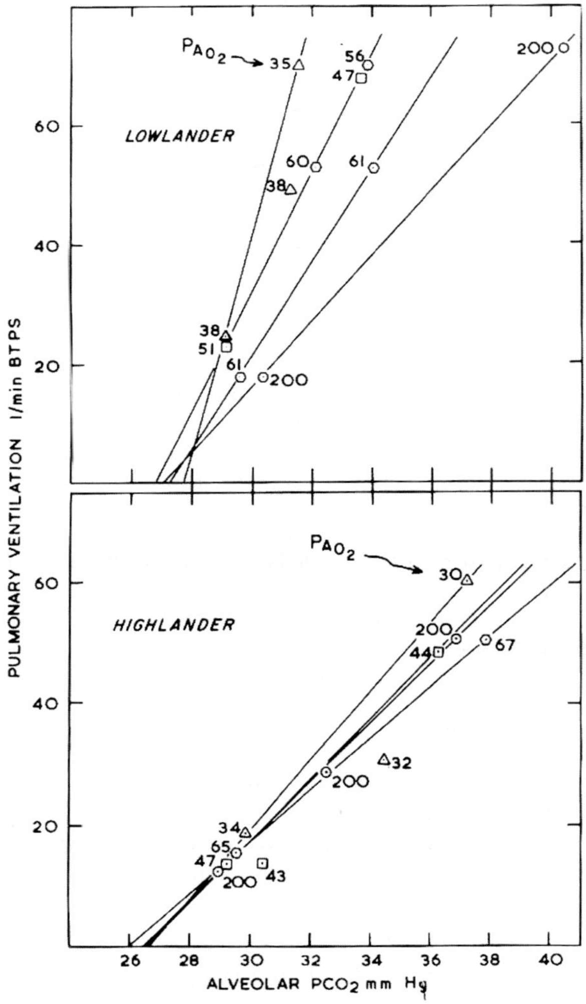

5.4 IP_3 Antagonist, Abolished the Hypoxic Chemosensory Responses but not the Hypercapnic Responses see McKashi et al. (2003)

This paper describes the reduction of hypoxic but not hypercapnic chemosensory discharge of not carotid body key cell permanent IP_3 inhibitors, 2-aminoethoxydiphenylgorene (2-APB, 50–200 μM) in a dose-dependent way.

Figure 8-4

Rat carotid sinus nerve activity is diminished by 2-APB during hypoxia (From Roy et al., 2006)

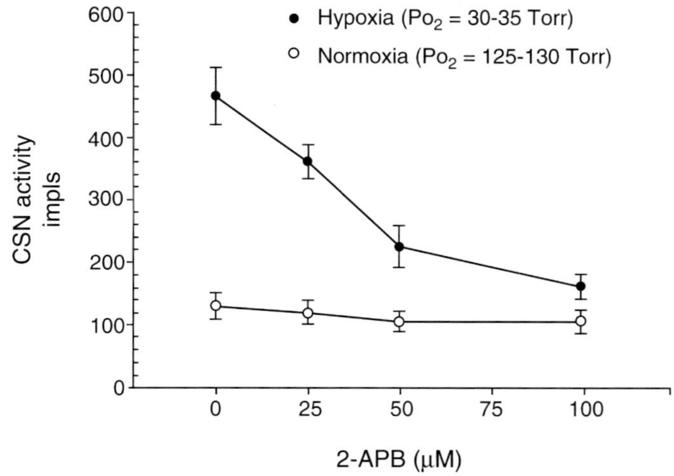

5.5 Oxidative Phosphorylation Antagonists, Oligomycin, and Antimycin Abolished the Hypoxic Chemosensory Response, but not the Hypercapnic Chemosensory Response of the Carotid Chemoreceptor (Mulligan and Lahiri, 1982)

This paper describes the response of the same chemoreceptor afferent to hypoxic were abolished by the drugs but to those hypercapnic remains infact in the cats.

Figure 8-5

Glomus cell responses (a) Oligomycin abolished the calcium response to hypoxia (Hx), but exaggerated the same to hypercapnia. (b) 2-APB (IP_3 antagonist) drastically reduced (From Lahiri, 2004)

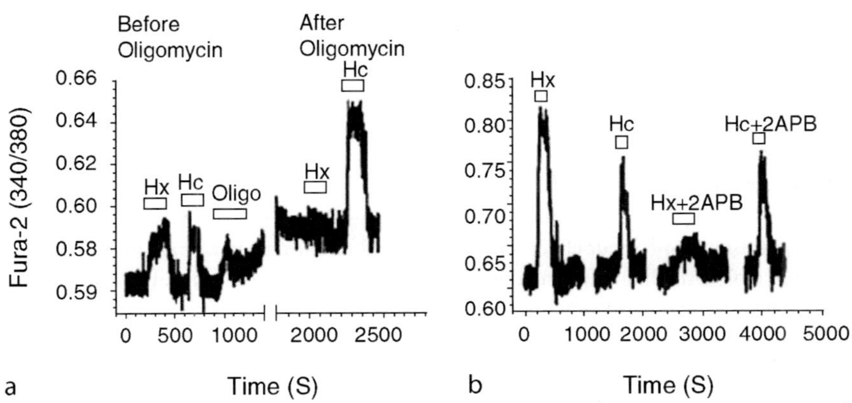

5.6 Oligomycim and Antimycin A Abolished the Calcium Response of the Glomus Cells but Enhances the Hypercapnic Response

Aortic chemoreceptors which show feeble response to CO_2 were enhanced after treating the aortic body with oligomycin (Erhan et al., 1981). The overshoot of chemosensory response was abolished by the carbonic anhydrase inhibitor, indicating that the hypercapnic response was due to H^+ being responsible. This shows (➋ *Figure 8-6*) dominance of CO_2/H^+ response.

Figure 8-6

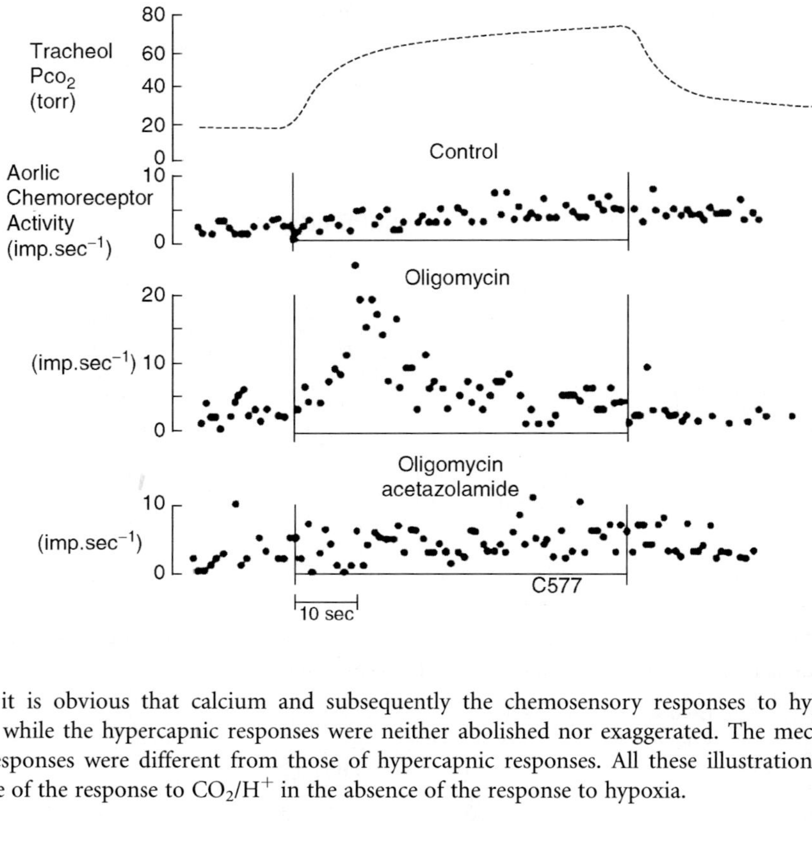

Thus, it is obvious that calcium and subsequently the chemosensory responses to hypoxia were abolished, while the hypercapnic responses were neither abolished nor exaggerated. The mechanisms of hypoxic responses were different from those of hypercapnic responses. All these illustrations show the dominance of the response to CO_2/H^+ in the absence of the response to hypoxia.

6 Pasteur Effect

In terms of mechanisms, the Pasteur effect came into play when the decreased oxidative phosphoynglasin in hypoxic response is abolished; that is, when the aerobic metabolism is abolished, the anaerobic metabolism takes over. Perhaps, the same phenomenon is working here, at least in part.

6.1 Hypoxia-Inducible Factor-1α is a Necessary Mediator of the Pasteur Effect

The Pasteur effect includes decreased oxidative phosphorylation and an increase in anaerobic fermentation. Because fermentation produces far less ATP than oxidative phosphorylation per molecule of glucose, increased activity of the glycolytic pathway is necessary to maintain the ATP levels in the hypoxic cells.

These results dramatically lower free ATP levels in HIF-1α–deficient hypoxic cells (Seagraves et al., 2001). It is interesting to see that this switch of energy metabolism is regulated by one transcription factor HIF-1α, the primary regulator of oxygen homeostasis, because HIF-1α deficiency reduces energy metabolism.

7 Perspectives

H^+ response remains the fundamental conditions of life in the internal environment even when the oxygen supply is diminished. This was the legacy of Claude Bernard. This legacy has been repeated many times in the past, for example, by Haldane and Priestley (1905), Henderson (1928), Cannon (1932), and Rahn (1974), to name a few. Constancy of "milieu interieur" in $[H^+]$ remains the last word in this treatise.

Acknowledgement

This work was supported by the NIH grant, HL43413-15.

References

Bernard C. 1878. Lecons Sur Les Phenomenes De La Vie Communs Aux Animaux Et Aux Vegetaux. Paris: Bailliers; pp. 113-121.

Buckler KJ, Vaughan-Jones RD. 1994. Role of intracellular pH and $[Ca^{2+}]$ in chemoreceptor cells of carotid body. In Arterial Chemoreceptors: Cell to System. pp. 41-55.

Buckler KJ, Williams BA, Honore E. 2004. An oxygen-, acid- and anaesthetic-sensitive TASK-like background potassium channel in rat arterial chemoreceptor cells. J Physiol 525: 135-142.

Buttigieg J, Nurse CA. 2004. Detection of hypoxia-evoked ATP release from chemoreceptor cells of the rat carotid body. Biochem Biophys Res Commun 322: 82-87.

Cannon WB. 1929. Organization and physiological homeostasis. Physiol Rev 9: 399-431.

Cannon WB. 1932. The Wisdom of the Body. New York: Norton and Company.

Erhan B, Mulligan S, Lahiri S. 1981. Metabolic regulation of aortic chemoreceptor response to CO2. Neurosci Lett 24: 143-147.

Feldman JL, Mitchell GS, Nattie EE. 2003. Breathing: Rhythmicity, plasticity, chemosensitivity. Annu Rev Neurosci 26: 239-266.

Gonzalez C, Almaraz L, Obeso A, Rigual R. 1994. Carotid body chemoreceptors: From natural stimuli to sensory discharges. Physiol Rev 74: 829-898.

Gourine AV, Spyer KM. 2003. Chemosensitivity of medullary respiratory neurones. A role for ionotropic P2X and GABA (A) receptors. Adv Exp Med Biol 536: 375-387.

Gourine AV. 2005. On the peripheral and central chemoreception and control of breathing: An emerging role of ATP. J Physiol 568: 715-724.

Guyenet PG, Mulkey DK, Stornetta RL, Bayliss DA. 2005. Regulation of ventral surface chemoreceptors by the central respiratory pattern generator. J Neurosci 25: 8938-8947.

Haldane JS, Priestley JG. 1905. The regulation of the lung ventilation. J Physiol 32: 225-266.

Henderson LJ. 1928. Blood: A study in General Physiology. New Haven, Connecticut: Yale University Press.

Jacobs MH, Stewart DR. 1942. The role of carbonic anhydrase in central ionic exchange involving the erythrocyte. J Gen Physiol 25: 539-552.

Jiang C, Rojas A, Wang R, Wang X. 2005. CO_2 central chemosensitivity: Why are there so many sensing molecules? Respir Physiol Neurobiol 145: 115-126.

Kawai A, Ballantyne D, Muckenhoff K, Scheid P. 1996. Chemosensitive medullary neurones in the brainstem–spinal cord preparation of the neonatal rat. J Physiol 492: 277-292.

Lahiri S. 2004. Highlighted topic: Oxygen sensing in health and disease. J Appl Physiol 96: 1170-1172.

Lahiri S, Forster RE IInd. 2003. CO_2/H^+ sensing: Peripheral and central chemoreception. Int J Biochem Cell Biol 35: 1413-1435.

Lahiri S, Mulligan E, Andronikon S, Shriharta M, Mokashi A. 1987. Carotid body chemosensory functions in prolonged normobaric hyperoxia in the cat. J Appl Physiol 62: 1924-1931.

Lahiri S, Roy A, Li J, Baby SM, Mokashi A, et al. 2004. Role of Fe^{2+} in oxygen sensing in the carotid body. Adv Exp Med Biol 551: 59-64.

Lahiri S, Roy A, Baby SM, Hoshi T, Semenza GL, et al. 2006. Oxygen sensing in the body. Prog Biophys Mol Biol 91: 249-286.

Le Chatelier H. 1888. Recherches Experimentales et Theoretiques sur les Equilibres Chimiques. Paris: Dunod.

Lorier AR, Peebles K, Brosenitsch T, Robinson DM, Housley GD, et al. 2004. P_2 receptors modulate respiratory rhythm but do not contribute to central CO_2 sensitivity in vitro. Respir Physiol Neurobiol 142: 27-42.

Milledge JS, Lahiri S. 1967. Respiratory control in lowlanders and Sherpa highlanders at altitude. Respir Physiol 2: 310-322.

Mokashi A, Roy A, Baby SM, Daudu PA, Lahiri S 2003. Hypoxic and not hypercapnic chemoreception is mediated by IP_3 receptors in adult rat carotid body. FASEB J 17, A15.

Mulligan E, Lahiri S. 1982. Separation of carotid body chemoreceptor responses to O_2 and CO_2 by oligomycin and by antimycin A. Am J Physiol 242: C200-C206.

Mulligan E, Lahiri S, Storey BT. 1981. Carotid body O2 chemoreception and mitochondrial oxidative phosphorylation. J Appl Physiol 51: 438-446.

Nattie E, Shi J, Li A. 2001. Bicuculline dialysis in the retrotrapezoid nucleus (RTN) region stimulates breathing in the awake rat. Respir Physiol 124: 179-193.

Peers C. 1990. Hypoxic suppression of K^+ currents in type I carotid body cells: Selective effect on the Ca^{2+}-activated K^+ current. Neurosci Lett 119: 253-256.

Prabhakar NR, Mitra J, Cherniack NS. 1987. Role of substance P in hypercapnic excitation of carotid chemoreceptors. J Appl Physiol 63: 2418-2425.

Prasad M, Fearon IM, Zhang M, Laing M, Vollmer C, et al. 2001. Expression of P_2X_2 and P_2X_3 receptor subunits in rat carotid body afferent neurones: Role in chemosensory signalling. J Physiol 537: 667-677.

Rahn H. 1974. Body temperature and acid-base regulation (Review article). Pneumonologie 151: 87-94.

Rahn H, Reeves RB, Howell BJ. 1975. Hydrogen ion regulation, temperature, and evolution. Am Rev Respir Dis 112: 165-172.

Rong W, Gourine AV, Cockayne DA, Xiang Z, Ford AP, et al. 2003. Pivotal role of nucleotide P2X2 receptor subunit of the ATP-gated ion channel mediating ventilatory responses to hypoxia. J Neurosci 23: 11315-11321.

Roy A, Mokashi A, Baby SM, DiGiulio C, Lahiri S. 2006. 2-APB and curcumin inhibit the hypoxic but not the hypercapnic carotid chemosensory discharge. (communicated).

Seagraves TN, Ryna HE, Lu H, Wouters BG, Knapp M, et al. 2001. Transcription factor HIF-1 is a necessary mediator of the Pasteur effect in mammalian cells. Mol Cell Biol 21: 3436-3444.

Spyer KM, Thomas T. 2000. A role for adenosine in modulating cardio-respiratory responses: A mini-review. Brain Res Bull 53: 121-124.

Spyer KM, Dale N, Gourine AV. 2004. ATP is a key mediator of central and peripheral chemosensory transduction. Exp Physiol 89: 53-59.

Varas R, Alcayaga J, Iturriaga R. 2003. Carotid chemosensory neurons in the petrosal ganglia are excited by ACh and ATP. Adv Exp Med Biol 536: 321-326.

Williams SE, Wootton P, Mason HS, Bould J, Iles DE, et al. 2004. Hemoxygenase-2 is an oxygen sensor for a calcium-sensitive potassium channel. Science 306: 2093-2097.

Zhang M, Nurse CA. 2004. CO_2/pH chemosensory signaling in co-cultures of rat carotid body receptors and petrosal neurons: role of ATP and ACh. J Neurophysiol 92: 3433-3445.

Zhang M, Zhong H, Vollmer C, Nurse CA. 2000. Co-release of ATP and ACh mediates hypoxic signalling at rat carotid body chemoreceptors. J Physiol 525: 143-158.

Zhong H, Zhang M, Nurse CA. 1997. Synapse formation and hypoxic signalling in co-cultures of rat petrosal neurones and carotid body type 1 cells. J Physiol 503: 599-612.

9 Sensory Neuropathies

M. Auer-Grumbach · P. Auer-Grumbach · T. Pieber

Abstract: Sensory neuropathies comprise a heterogeneous group of diseases which are caused by damage of the sensory nerves and lead to sensory disturbances mainly affecting the feet and legs, and less frequently the hands and arms. This article defines sensory neuropathies and introduces clinical features and the underlying etiology as well as risk factors and possible complications. Physical as well as laboratory findings are presented and the diagnostic role of nerve conduction studies, nerve biopsy and skin biopsy studies are shortly discussed. Important clinical entities like diabetic or hereditary neuropathies amongst others are described in more detail. Finally, an overview addressing patient management and particularly, management of foot complications is given.

List of Abbreviations: HSAN, Hereditary sensory and autonomic neuropathies; HSN, Hereditary sensory neuropathies; NCS, Nerve conduction studies; QST, Quantitative somatosensory testing

1 Introduction

Peripheral neuropathies are common neurological disorders that result from damage to the peripheral nerves (Pascuzzi, 2003). The peripheral nerves represent the nerves outside the brain and spinal cord and consist of motor, sensory, and autonomic nerve fibers. Depending on the fiber classes affected, neuropathies are subdivided into motor, motor and sensory, and pure sensory forms. In addition, autonomic disturbances may be present leading to the additional classification into autonomic or autonomic and sensory neuropathies.

Sensory neuropathies affect the sensory nerves, which are the nerves responsible for sensation throughout the body. Most commonly sensory neuropathies affect the feet and legs, and less frequently the hands and arms. Other parts of the body are rarely affected. Large fiber (responsible for touch, vibration, and joint position sense) and small fiber (responsible for pain and temperature sensation) sensory modalities can be distinguished. The diagnosis of sensory neuropathy requires a history compatible with predominantly sensory dysfunction and a physical examination notable for abnormal sensory findings at least in the feet, with reduced or absent ankle jerks. The prevalence of sensory neuropathies might be underestimated due to the fact that dysfunction of nerve fiber loss may remain clinically silent for a long time, even until irreversible nerve damage.

This chapter focuses on sensory neuropathies and discusses the causes, pathogenesis, clinical features, and management of these disorders. Some common and distinct forms of sensory neuropathies are described in more detail.

2 Etiology and Risk Factors

Sensory neuropathies may be associated with a large number of diseases, but they can also be the result of systemic illness. ➋ *Table 9-1* summarizes the most important known causes of sensory neuropathies and provides a list of diseases that should be considered in patients affected with sensory neuropathies. The most common forms are diabetic and alcoholic sensory neuropathies. Frequently, the cause of the disease remains unknown throughout the patient's life.

3 Clinical Features

As the name implies, sensory disturbances are the predominant symptoms of sensory neuropathies. Initially, patients complain of tingling paresthesia, pain, dysesthesia, and loss of feeling distally in the extremities (Thomas and Tomlinson, 1999). Early in the course of the disease, dysesthesia is confined to the toes or soles only. Later on, dysesthesia tends to ascend symmetrically up to the ankles and knees. Similar symptoms may be present with progression of the disease in the fingertips extending to the hands in a glove-and-stocking pattern of sensory loss. Patients often feel as though they are "walking on a bed of

Table 9-1
Causes of sensory neuropathies

Diabetes mellitus
Alcoholism
Other neurotoxic substances (isoniazid, metronidazole, dapsone, vincristine, pyridoxine, stavidine, zalcitabine, and others)
HIV infection (AIDS)
Inherited sensory neuropathies caused by mutations in different genes (HSN I–V)
Endocrinological diseases (thyroid dysfunction)
Vitamin deficiencies
Vascular and collagen disorders (systemic lupus erythematosus, scleroderma, rheumatoid arthritis, and others)
Paraneoplastic syndrome (associated with lung, prostate, breast cancer, or hematological diseases)
Renal and liver diseases
Amyloidosis
Nutritional deficiencies

coals" or that they are wearing an invisible sock or glove. Sensory disturbances exhibit a length-related pattern and often begin gradually. Patients complain of sensations such as burning, tingling, shooting and lancinating pain, numbness, throbbing, and aching. In some instances pain limits walking.

4 Physical Findings

On examination, there may be an extreme sensitivity to touch. This contact hypersensitivity may be evoked by lightly stroking the plantar and dorsal surfaces of the feet. Some patients develop hyperalgesia and can hardly bear socks and tight-fitting shoes. Occasionally, the skin may become so sensitive that the slightest touch is agonizing. There may also be numbness or complete lack of feeling in the feet.

In most patients, pain and vibratory sensation are abnormal. Pain sensation is assessed by lightly applying a pin on the skin and asking the patient if it feels sharp. The pin is applied to the toes, the dorsum of the foot, and then to the ankle and along the calf to the knee. Then the patient is asked if the pin feels equally sharp throughout. Most patients with sensory neuropathy have some abnormality of pin sensation in the toes and over the dorsum of the foot. Vibration sense is best assessed by applying a tuning fork over the great toe. Patients with sensory neuropathy will generally feel the "buzzing" for only a few seconds or it will even be absent.

Patients with sensory neuropathies often exhibit dry and fissured skin pointing to additional autonomic disturbances. Sometimes the skin is moist and of red-bluish color (glossy skin) because of the wide opening of small vessels owing to paresis of autonomic nerves.

The vast majority of patients with neuropathies have either absent or reduced ankle jerks (Thomas and Tomlinson, 1999).

If the motor nerves are damaged there may be mild distal muscle weakness, but this is not a regular and distinct feature. However, intrinsic foot muscle weakness may lead to deformation of feet and hammertoes.

5 Ancillary Studies

5.1 Electrophysiological Examinations: Nerve Conduction Studies

Electrophysiological exploration is a standard method of investigating the dysfunction of myelinated larger fibers. Nerve conduction studies (NCS) are a helpful and noninvasive tool in the assessment of nerve damage. They are used to define the severity of a neuropathy, and can also be used to detect subclinical neuropathies (Thomas and Tomlinson, 1999). Surface electrodes are placed on a distal muscle or along a sensory nerve, and electric stimuli are given to the corresponding nerve to receive a motor or sensory nerve action potential. Then, conduction velocities and amplitudes can be calculated. The changes involve a reduction of velocity together with diminished amplitudes and increased temporal dispersion of the sensory action potentials. In severe cases, sensory nerve action potentials are often absent.

5.2 Nerve Biopsy Studies

Nerve biopsy studies are very rarely indicated and are usually reserved for those patients in whom diagnosis is uncertain, for example, where vasculitis or amyloidosis is a concern.

5.3 Skin Biopsy Studies

In the past years, skin biopsy studies have become a widely used tool to investigate small caliber sensory nerves including somatic unmyelinated intraepidermal nerve fibers, dermal myelinated nerve fibers, and autonomic nerve fibers in peripheral neuropathies and other conditions. The techniques used are described elsewhere (Cornblath et al., 2005). Skin biopsy measures the intraepidermal nerve fiber density and morphology, and appears to be a sensitive measure of nerve fiber integrity. Training in established cutaneous nerve laboratories is recommended before using skin biopsy as a diagnostic tool in peripheral neuropathies (Cornblath et al., 2005).

5.4 Quantitative Somatosensory Testing

Quantitative somatosensory testing (QST) uses calibrated tools to assess the function of all the sensory modalities. The smaller caliber nerves are evaluated by measuring pain and temperature (hot and cold) thresholds, and larger caliber nerves are evaluated by measuring the thresholds for perception of vibration, joint position, and touch. This is done by touching the patients' skin with stimuli of defined characteristics, such as a computer-controlled probe that can heat or cool to specific temperatures. The effectiveness of QST is limited because it requires subject cooperation and is inherently subjective, as it relies on the reported interpretation of sensory stimulation from the subject. This also influences reproducibility of this sensory test (Fink and Oaklander, 2006).

6 Diabetic Neuropathy

Diabetes is by far the most frequent cause of peripheral sensory neuropathy. Typically, sensory abnormalities are present in a symmetrical distal distribution and patients also present with autonomic features (Thomas and Tomlinson, 1999). These can be recognized by skin abnormalities ranging from hyperhidrosis to hypohidrosis or anhidrosis (dry or wet skin). Autonomic dysfunction can also affect the cardiovascular, gastrointestinal, and urogenital systems. The process by which the peripheral nerves are damaged is not entirely clear but it is probably due to high blood-glucose changes that affect the metabolism of cells. All sensory qualities may be affected, but loss of pain and temperature sensation is often a leading feature pointing to a predominant small fiber neuropathy (Thomas and Tomlinson, 1999).

Diabetic neuropathy has become a serious health problem as it can be associated with severe complications such as the diabetic foot syndrome (Ramsey et al., 1999; Mueller et al., 2003). Patients with diabetic neuropathies often develop chronic foot ulcerations. The ulcers are sometimes located on the balls of the feet over the metatarsal heads but are more commonly encountered on the toes and sometimes over the heels or malleoli. They are primarily related to tissue damage, of which the patient is unaware because of loss of pain and temperature sensation. Other contributory factors are autonomic skin changes with fissuring as well as foot deformity and hammertoes that promote pressure points (Thomas and Tomlinson, 1999). Foot ulcerations are frequently followed by bone infections and necrosis and may necessitate amputations (*Figure 9-1*).

Figure 9-1
Chronic foot ulceration of the great toe in a patient with diabetic neuropathy

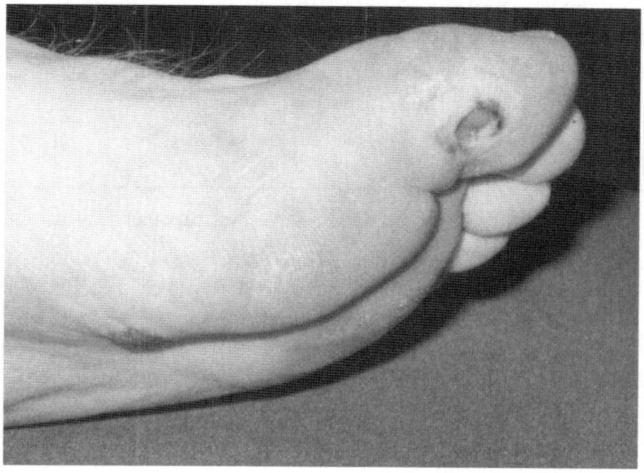

7 Alcoholic Neuropathies

Chronic use of alcohol may lead to nerve damage. The clinical picture is similar to that seen in other neuropathies; but in patients with alcoholic neuropathy, tendon reflexes are often preserved.

8 HIV Neuropathy

Peripheral neuropathy may be the most frequent neurologic disorder associated with HIV infection (Cherry et al., 2005). Its symptoms cause substantial morbidity and discomfort to patients with AIDS. A 30%–35% prevalence of peripheral neuropathy has been documented in patients with HIV infection, but autopsy-based studies have found it in nearly 100% of patients who died of AIDS. The most common peripheral neuropathy associated with HIV occurs in the later stages of HIV disease, usually after the patient has had other AIDS defining illnesses. Symptoms of HIV-associated sensory neuropathies are almost identical to those of other sensory neuropathies. Both distal sensory neuropathy due to HIV infection (seen mainly in late disease) and antiretroviral toxic neuropathy occur, or sensory neuropathy is caused by a combination of both (Cherry et al., 2005).

9 Hereditary Sensory Neuropathies

The hereditary sensory (and autonomic) neuropathies (HSN/HSAN) are a clinically and genetically heterogeneous group of disorders, which are characterized by prominent distal sensory loss and acromutilating complications (Auer-Grumbach et al., 2006; Verhoeven et al., 2006). Their classification is complicated and has led to controversies in terminology. Originally, Dyck and Ohta proposed a clinically descriptive classification of five different subtypes of HSN (HSN I–V) (Dyck, 1993). With the exception of HSN type I, which is transmitted in an autosomal dominant way, the other HSNs are transmitted as autosomal recessive disorders and most of them are congenital forms. In each type, distinct populations of nerve fibers are affected, providing additional diagnostic criteria when nerve biopsies are available (Dyck, 1993). Molecular genetic research has now shown that at least eight loci and six genes are associated with HSN (Verhoeven et al., 2006). The phenotypes that are associated with mutations in the SPTLC1, RAB7, and HSN2 genes show several similarities to diabetic neuropathies and are also often associated with severe acromutilating complications (Dawkins et al., 2001; Verhoeven et al., 2003; Lafreniere et al., 2004). They thus often mimic a pseudodiabetic foot syndrome (Auer-Grumbach et al., 2006). It is important to assess a patient's family history in order to separate inherited forms from acquired neuropathies, which is important for genetic counseling.

10 Management

Management of sensory neuropathies is targeted to relieve painful neuropathic symptoms and to teach patients how to avoid foot complications.

Patients with mild pain that is not functionally limiting may tolerate their symptoms without any medication. However, symptomatic treatment with pharmacologic agents is often requested by patients, but no specific characteristics of the pain can predict which agents will be beneficial. The choice of medication is based on the severity of the patients' symptoms and the side-effect profile of the medication. In some patients with mild neuropathic pain, the use of nonsteroidal antiinflammatory agents such as ibuprofen might be successful.

When pain or other dysesthetic symptoms begin to limit functional ability, tricyclic antidepressants may be useful such as oral amitriptyline or related agents, which can be given at bedtime and improves nocturnal pain (Dworkin et al., 2003). The anticholinergic side effects of these agents (dry mouth, urinary retention, orthostatic hypotension, and sedation, among others) can be minimized by beginning with low doses and gradually titrating to the minimal effecting dose. It may then take up to two or three weeks for a beneficial effect to occur. Other agents that have been shown to be successful in the treatment of neuropathic pain are anticonvulsants such as carbamazepine, phenytoin, gabapentin, and pregabalin (Gidal, 2006; Gilron and Flatters, 2006). In particular, controlled studies of gabapentin and pregabalin have recently shown great benefit in the treatment of patients with painful peripheral neuropathies. A topical agent is capsaicin but the possible transient increase in pain has shown that this substance is often poorly tolerated by the patients.

For patients with very severe neuropathic pain, the use of opiates and narcotic agents may be necessary. Like the tricyclics, these agents must be titrated to the minimum effective dose required for improvement. Patients should be evaluated frequently for side effects and for ineffective symptom control.

Vitamin combinations can be used and some people have reported improvement and additional relief of symptoms. Caution is urged with vitamin B_6 (pyridoxine) since it will cause peripheral neuropathy if high doses are used.

Research is in progress on other agents including nerve growth factor to reverse nerve fiber destruction, and the first clinical trials have been carried out (Wellmer et al., 2001). However, no therapies exist currently that halt or reverse nerve fiber destruction.

10.1 Management of Foot Complications

As causal therapy is not available for many forms of foot complications, continuous care and patient education are mandatory. This is best developed for the prevention of the diabetic foot syndrome. In many countries, training programs as well as printed and electronic information are available. The key problem is that patients must compensate their diminished or lost sensory qualities by daily inspection of their feet for changes of color, blisters, fissures, or any other skin changes; and also, daily inspection of shoes for foreign bodies is necessary. However, this might sometimes be problematic, particularly in patients who have additional eye problems or reduced mobility.

Patients must also be trained to be careful with hot water as they could burn their feet. They must be informed that they should always wear shoes that fit well. Tight-fitting shoes and socks must be avoided. They might exacerbate pain and tingling and can also promote the development of foot ulcerations. Shoes should also offer good air circulation and must not be excessively loose. Special shoes for diabetic patients already exist with no sutures but soft material inside to avoid pressure points. Patients who do already have deformed feet should receive individualized orthopedic shoes, which have to be continuously adapted. All this helps to reduce amputations and high costs of late-stage disease, which have been documented (Ramsey et al., 1999). Therefore, a foot deformation or even an ulcer is not a minor problem but a possible life threatening event. To achieve this goal, close caretaking over years by family members and a variety of health care workers like nurses, podiatrists, physiotherapists, and doctors in different fields is necessary.

References

Auer-Grumbach M, Mauko B, Auer-Grumbach P, Pieber TR. 2006. Molecular genetics in sensory neuropathies. Neuromolecular Med 8: 147-158.

Cherry CL, Wesselingh SL, Lal L, McArthur JC. 2005. Evaluation of a clinical screening tool for HIV-associated sensory neuropathies. Neurology 65: 1778-1781.

Cornblath LG, Johansson O, McArthur JC, Mellgren SI, Nolano M, et al. 2005. European federation of neurological societies. EFNS guidelines on the use of skin biopsies in the diagnosis of peripheral neuropathy. Eur J Neurol 10: 747-758.

Dawkins JL, Hulme DJ, Brahmbhatt SB, Auer-Grumbach M, Nicholson GA. 2001. Mutations in SPTLC1, encoding serine palmitoyltransferase, long chain base subunit-1, cause hereditary sensory neuropathy type I. Nat Genet 27: 309-312.

Dworkin RH, Backonja M, Rowbotham MC 2003. Advances in neuropathic pain. Arch Neurol 60: 1524-1534.

Dyck PJ. 1993. Neuronal atrophy and degeneration predominantly affecting peripheral sensory and autonomic neurons. Peripheral Neuropathy. 3rd ed. Dyck PJ, Thomas PK, Griffin JW, Low PA, Poduslo JF, editors. Philadelphia: W.B. Saunders; pp. 1065-1093.

Fink E, Oaklander AL. 2006. Small-fiber neuropathy: Answering the burning questions. Sci Aging Knowl Environ 6: pe7.

Gidal BE. 2006. New and emerging treatment options for neuropathic pain. Am J Manag Care 12: 269-278.

Gilron I, Flatters SJ. 2006. Gabapentin and pregabalin for the treatment of neuropathic pain: A review of laboratory and clinical evidence. Pain Res Manag Summer 11 (Suppl. A): 16A-29A.

Lafreniere RG, Mac Donald ML, Dube MP, Mac Farlane J, O'Driscoll M, et al. 2004. Identification of a novel gene (HSN2) causing hereditary sensory and autonomic neuropathy type II through the study of Canadian genetic isolates. Am J Hum Genet 74: 1064-1073.

Mueller MJ, Hastings M, Commean PK, Smith KE, Pilgram TK, et al. 2003. Forefoot structural predictors of plantar pressures during walking in people with diabetes and peripheral neuropathy. J Biomech 36: 1009-1017.

Pascuzzi RM. 2003. Peripheral neuropathies in clinical practice. Med Clin North Am 87: 697-724.

Ramsey SD, Nemton K, Blough D, et al. 1999. Incidence, outcomes, and cost of foot ulcers in patients with diabetes. Diabetes care 22: 382-387.

Thomas PK, Tomlinson DR. 1999. Diabetic and hypoglycemic neuropathy. Diabetic Neuropathy. Dyck PJ, Thomas PK, editors. 2nd ed, Elsevier, pp. 1219–1250.

Verhoeven K, De Jonghe P, Coen K, Verpoorten N, Auer-Grumbach M, et al. 2003. Mutations in the small GTP-ase late endosomal protein RAB7 cause Charcot-Marie-Tooth type 2B neuropathy. Am J Hum Genet 72: 722-727.

Verhoeven K, Timmerman V, Mauko B, Pieber TR, De Jonghe P, et al. 2006. Recent advances in hereditary sensory and autonomic neuropathies. Curr Opin Neurol 124: 271-276.

Wellmer A, Misra VP, Sharief MK, Kopelman PG, Anand P. 2001. A double-blind placebo-controlled clinical trial of recombinant human brain-derived neurotrophic factor (rhBDNF) in diabetic polyneuropathy. J Peripher Nerv Syst 6: 204-210.

10 Amino Acid-Sensing Mechanisms: Biochemistry and Behavior

D. W. Gietzen · S. Hao · T. G. Anthony

Abstract: Sensors are crucial components in homeostatic control systems. Amino acids (AAs) are the building blocks of protein; all 20 standard AAs that are used in protein synthesis must be available simultaneously for an organism to maintain homeostasis and survive. Therefore, a sensory system for AAs is essential for protein synthesis and survival. In living organisms, biochemical and behavioral observations demonstrate AA sensing.

Taste receptors that respond to both L- and D-AAs, as well as umami taste, exist in animals and humans; they may signal high dietary protein. The gastrointestinal (GI) tract may taste ingested nutrients as well. Postabsorptively, the conserved mammalian target of rapamycin (mTOR) is activated by increases in nutrients, particularly leucine. These three: taste, gut sensors, and mTOR, may serve as "adequate diet" sensors. In addition, mTOR signals tissue accretion and satiety. Transporters for AAs respond to changes in substrate availability; sodium-dependent transporters can be electrogenic and activate neural feeding systems. Also conserved, the general AA control non-derepressing system (GCN) is activated by AA deprivation via uncharged tRNA, leading to AA biosynthesis in yeast or to diet rejection and increased foraging in animals. Thus, two postabsorptive systems, conserved from yeast to mammals, have recently been reported; the results suggest that there are separate biochemical systems for sensing high and low levels of AAs. These systems demonstrate conservation of AA sensory mechanisms across eukaryotic species.

List of Abbreviations: AAs, amino acids; APC, anterior piriform cortex; ATF4, activating transcription factor 4; CaMKII, calcium/calmodulin-dependent protein kinase II; CCK, cholecystokinin; DMH, dorsomedial hypothalamus; ERK, extracellular signal related protein kinase; GI, gastrointestinal; IAA, indispensable amino acid; LTP, long-term potentiation; MAPK, mitogen-activated protein kinase; MeAIB, 2-methyl-aminoisobutryic acid; mGluR, metabotropic glutamate receptor; MSG, monosodium glutamate; mTORC1, mTOR complex 1; mTOR, mammalian target of rapamycin; PepT1, protein-coupled oligopeptide transporter

1 Introduction

The ability to synthesize protein is crucial for survival. All organisms must maintain a full complement of the amino acids (AAs) and/or nitrogen sources, which are the protein precursors, or they will become catabolic and eventually weaken and die. There are wide differences in how organisms respond to differing levels of AAs. For example, in deficiency, yeast activate their biosynthetic enzymes to replenish their AAs, whereas animals, lacking the ability to synthesize all AAs, reject the deficient food and begin foraging for a better diet. In contrast, upon encountering a rich nutrient source, organisms switch on their anabolic systems for tissue accretion and growth. Thus, the responses to increased AAs differ from those to AA deficiency. Rather than a single bidirectional signal, it appears that one set of kinases responds to deficiency and another to excess or repletion. In this chapter, we approach AA sensing from a systems/behavioral perspective and discuss mechanistic observations that have informed our current understanding of these systems.

Nearly half of the AAs present in protein cannot be synthesized or stored in metazoans; the genes for their synthesis were lost early in evolution (reviewed by Gietzen and Rogers, 2006). These are the essential or dietary "indispensable" amino acids (IAA) and are listed in, ➋ *Table 10-1*. The behavioral responses to differing IAA proportions in the diet have been well studied (Harper et al., 1970; Rogers and Leung, 1973, 1977; Harper, 1976; Anderson, 1977; Harper and Peters, 1983; Gietzen et al., 1986; Gietzen, 1993, 2000). Here, we address the responses to varying proportions of dietary IAA:

1. Adequate protein/IAA allows normal feeding and growth. Feeding proceeds to satiety.
2. Diets containing very high or very low protein/IAA are rejected. AA in excess can be toxic (Harper et al., 1970; Garlick, 2001). Although adaptation can occur to relatively high levels of protein or balanced IAA, severe deficiencies are lethal.
3. Diets that contain imbalances of IAA, or that are devoid of a single IAA, are rejected (Harper et al., 1970; Rogers and Leung, 1973, 1977). Adaptation can occur over time to an imbalanced diet, but as with severe protein restriction, prolonged ingestion of a diet devoid of even one IAA is incompatible with life.

Table 10-1
Standard dietary amino acids (AA) incorporated into protein

Indispensable	Dispensable
L-Histidine[1]	L-Alanine
L-Isoleucine[2]	L-Arginine
L-Leucine[2]	L-Aspartic acid[7]
L-Lysine[6]	L-Asparagine[6]
L-Methionine[3]	L-Cystine[3]
L-Threonine	L-Glutamic acid[7]
L-Tryptophan[4]	L-Glutamine[6]
L-Tyrosine[5]	Glycine[8]
L-Valine[2]	L-Phenylalanine
	L-Proline
	L-Serine

1. Conditionally essential for adults, essential for infants, precursor for histamine;
2. Branched-chain AA;
3. Sulfur containing AA;
4. Precursor for serotonin;
5. Precursor for catecholamines, epinephrine, norepinephrine and dopamine;
6. Basic AA;
7. Acidic AA, also serve as excitatory neurotransmitters;
8. Has no stereoisomer

2 Adequate Protein and Balanced AA Profile—Selection Within Normal Limits

Simpson and Raubenheimer (1999) provided an elegant mathematical model accounting for many variables that impact nutrient selection and concluded that organisms tend to use integrated systems to maintain growth and health. Their model assumes nutrient sensing as part of a controlled system, and seems to fit the normal situation, albeit with excursions. How well animals control their intake of protein or IAA has been controversial (reviewed in Berthoud and Seeley, 2000 (see Galef therein); Koehnle and Gietzen, 2005; Gietzen and Rogers, 2006). Yet, although the limits within which animals select the protein and thus the IAA fraction of their diet are relatively broad if viewed in well-nourished subjects over time (Harper and Peters, 1989; Makarios-Lahham et al., 2004), even these studies show limits of acceptability.

2.1 The Taste Hypothesis: Amino Acids Are Sensed by Taste Receptors

Taste may promote selection and ingestion of a diet containing an adequate level of protein (*Figure 10-1*). An AA taste receptor has been thoroughly investigated; it is a protein dimer, termed T1R1/T1R3 (Li et al., 2002; Nelson et al., 2002; Taylor-Burds et al., 2004; Scott, 2005; Smith and Boughter, 2007). Glutamate and glutamine, both nonessential AAs, i.e., dispensable in the dietary sense (*Table 10-1*), have been suggested to signal high quality protein preabsorptively. Glutamate is sensed by plant roots as a nitrogen source (Walch-Liu et al., 2006), yeast have a membrane-associated AA sensor that responds to glutamine (Powers et al., 2004), and the catfish is a well-studied model for AA taste as it can detect nanomoles of AAs in water. The catfish is sensitive to a variety of L-AAs and to different proportions of IAA (reviewed by Brand et al., 1991).

In animals the specific taste "umami," associated with savory foods, relies on glutamate, monosodium glutamate (MSG), and a few other non-AA compounds, termed umami taste substances (Brand et al., 1991; Kurihara and Kashiwayanagi, 2000). Taste buds that respond to umami have been found in the tongues of many mammals, including humans. The umami taste receptor may be a dimer of T1RN and T2RN (Scott, 2005).

Figure 10-1
Neural connections project information from taste receptors in the taste buds of the mouth, via cranial nerves VII and IX, and from nutrient sensors in the gastrointestinal (GI) tract, via cranial nerve X. The nutrient sensors of the GI tract may be activated by cholecystokinin (CCK) or serotonin (5-HT). The afferent fibers end at the nucleus of the solitary tract in the brainstem. From there, after integration with other signals, projections to the forebrain carry the sensory information to higher structures

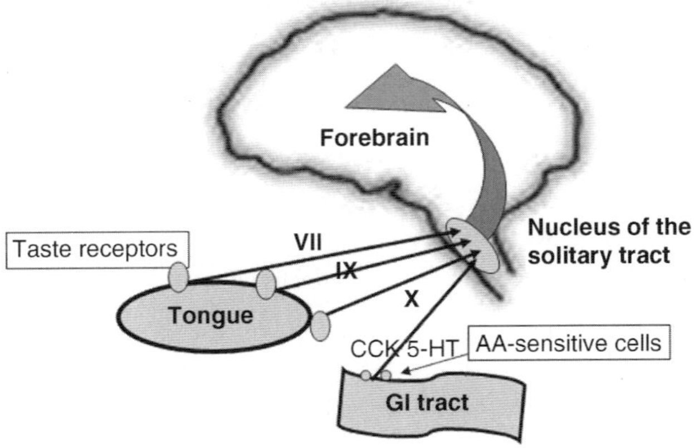

While somewhat controversial, the receptor types and species differences make umami taste a most interesting and well-studied topic (reviewed by Scott, 2005; Smith and Boughter, 2007). There may be more umami taste receptors to be described, or differing combinations of T[N]R[N] proteins in the dimers (Maruyama et al., 2006), and there appears also to be a role for the metabotropic glutamate receptor (mGluR) in umami taste (Chaudhari et al., 2000). The situation is complicated by an interaction with sweet taste and the T1R2 dimer (Damak et al., 2003; Xu et al., 2004), although sweet and AA taste have been separable in studies by Zhao and coworkers (2003) and Eylam and Spector (2004).

Taste preferences for umami are seen in rats when dietary protein is within the normal range; in a protein-deficient state, rats prefer glycine or NaCl (Mori et al., 1991). Thus, taste preferences for umami may indicate the protein-nutritional state of the animal as does dietary choice.

It should be noted that animals do not avoid an unpleasant taste that is associated with correcting a deficiency of the limiting IAA. When they are deficient for that IAA, they will replete their deficiency by drinking a bitter-tasting fluid that contains the limiting IAA rather than avoid the unpleasant taste (Rogers and Leung, 1977). Remarkably, if given a choice between a deficient diet containing the sweet tastant, saccharin, and a balanced diet containing the bitter tastant, quinine, rats show a preference for the balanced diet containing quinine. Apparently, the usual taste preferences are not strong enough to overcome the metabolic consequences of the IAA deficiency (Leung and Rogers, 1987). This supports the notion that taste does not serve as the primary mechanism for sensing IAA deficiency (reviewed in Harper et al., 1970; Rogers and Leung, 1977; Gietzen, 1993, 2000). Rather, AA taste receptors may provide the ability to sense that protein is being eaten.

2.2 The GI/Vagal Hypothesis: Amino Acid Sensing in the Gastrointestinal Tract

The sensory functions of the gastrointestinal (GI) tract have been studied over many years (reviewed in Raybould, 1998; Berthoud et al., 2004; Raybould et al., 2006). The lumen of the GI tract has the first access to ingested foods after the taste receptors in the mouth. Good candidates for GI taste cells include one for protein/oligopeptide, the cholecystokinin (CCK) endocrine cell (Raybould et al., 2006). Another is the serotoninergic enterochromaffin cell (Nilsson et al., 1987) that may respond to free AAs as well as peptides and protein.

There is a plethora of information about nutrient sensing in the GI tract. Chemoreceptors in the gut respond to AAs along with glucose and fatty acids (Powley and Berthoud, 1986). There is good evidence that the protein-coupled oligopeptide transporter (PepT1) stimulates release of CCK from intestinal endocrine cells in response to short di- or tripeptides (Darcel et al., 2005b; Raybould, 2006). Such sensing is thought to be transmitted to the brain via afferents of the vagus nerve (cranial nerve X) (➋ *Figure 10-1*). The vagus has been associated with satiety induced by maltose, oleic acid, and L-phenylalanine (Yox et al., 1991), or both IAA and nonessential AAs (Jeanningros, 1982), although duodenal peptone is more effective than glucose in stimulating vagal afferent activity (Schwartz and Moran, 1998). Hepatic vagal afferents carry AA information (IAA and non-IAA similarly) from the liver (Tanaka et al., 1990; Niijima and Meguid, 1995), and lysine deficiency upregulates hepatic vagal afferents, tested after five days eating a lysine-deficient diet (Torii and Niijima, 2001). Yet, total hepatic denervation failed to alter the responses to IAA-imbalanced diets on the first day (Bellinger et al., 1993, 1996).

The peptide products of protein digestion activate both CCK and the oligo- (di- and tri-) peptide transporter, PepT1 (Darcel et al., 2005b). It has long been appreciated that the peripheral effects of CCK are blocked by vagotomy (Smith et al., 1981), showing that peripheral effects of CCK are dependent on an intact vagus. Vagal afferent terminals have type 3 receptors for serotonin (5-HT_3) (Round and Wallis, 1986). Recent evidence for AA sensing in the GI tract includes responses to glutamate in the stomach, transmitted via 5-HT_3 receptors on vagal afferents (Uneyama et al., 2006). Thus, both CCK and 5-HT-related signals and interactions (Aja et al., 1999) may relay information on ingested protein and AAs in general from the ingested food to the brain. Vagal afferent signaling to the nucleus of the solitary tract is activated in rats adapted to high protein diets (Darcel et al., 2005a). These GI nutrient-sensing mechanisms do not appear to recognize IAA deficiency or protein quality in the short term (less than 3 h) (Washburn et al., 1994; Dixon et al., 2000; reviewed in Gietzen, 1993, 2000). However, they may signal the normal or high levels of protein or AAs in the stomach or intestinal lumen that arrive after a recent meal.

3 Selection with High Levels of AAs and Protein

Herbivores and omnivores select the diet containing the higher level of protein or AA mixtures when the choices contain balanced IAA profiles (Peters and Harper, 1984). Indeed, rats can select levels of protein that are considerably higher than their requirement (Makarios-Lahham et al., 2004). Although the mechanism for selection of high levels of dietary protein is not known, it appears that it occurs in the absence of the learned aversion that accompanies avoidance of an IAA-deficient diet (Tomé, 2004).

On the other hand, if the levels of IAA or protein exceed tolerable levels a decrease in food intake is seen (Harper et al., 1970). High levels of leucine (like excesses of most IAA) decrease food intake (Rogers and Leung, 1973), as do leucine injections into the third ventricle (Cota et al., 2006). The notion of protein as the most satiating of the three macronutrients is generally accepted, although somewhat controversial (Bensaid et al., 2002; Westerterp-Plantenga, 2003; Anderson and Moore, 2004; Tomé, 2004). Because high levels of protein in the diet have been associated with satiety and a decrease in food intake, high-protein diets are used in weight-loss regimens (reviewed by Westerterp-Plantenga, 2003; Last and Wilson, 2006).

3.1 The Mammalian Target of Rapamycin Hypothesis

A good candidate for an intracellular sensory system for rich nutrient sources is the mammalian target of rapamycin (mTOR) (reviewed by Jacinto and Hall, 2003; Kimball and Jefferson, 2006). TOR was discovered in yeast (Heitman et al., 1991) and has been associated with its responses to AAs (Powers et al., 2004). TOR appears to recognize the dispensable AA glutamine, which is a primary nitrogen source for yeast (Powers et al., 2004) as it is for many organisms that have AA biosynthetic systems (reviewed by Gietzen and Rogers, 2006).

The mammalian version, mTOR, may recognize normal amounts of protein in the diet and may also sense AAs at high levels, particularly the IAA, L-leucine, which increases the activity of this threonine–serine kinase (Lynch et al., 2000), albeit with tissue and cell culture differences (Pham et al. 2000) (➋ *Figure 10-2*).

◘ Figure 10-2

The hypothalamus, here shown to contain mammalian target of rapamycin complex 1 (mTORC1), receives nutrients from the bloodstream via the cerebrospinal fluid. After activation by increased nutrients such as leucine, mTOR activation leads to pathways affecting motor systems for feeding in the forebrain and in the brainstem autonomic centers, such as the dorsal motor complex of the vagus. The result is a decrease in food intake

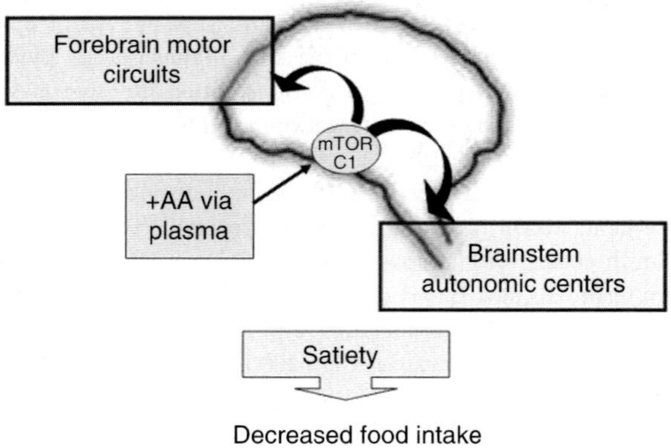

Leucine is found in high concentrations in many protein-containing foods, and the stimulation of mTOR seems to be selective for leucine (reviewed by Kimball and Jefferson, 2006). Mammalian TOR has been studied in vivo, in situ, and in cultured muscle (Anthony et al., 2000; Bolster et al., 2004; Kimball and Jefferson, 2006), adipocytes (Lynch et al., 2000), and liver (Kimball et al., 1991; Anthony et al., 2001a, b, 2004). mTOR is clearly implicated in the anabolic responses to refeeding after starvation in adipocytes (Lynch et al., 2000); it is associated with hypertrophy, particularly in muscle and adipose tissue, and synaptic plasticity in neurons. The details of the signaling pathways, both upstream and downstream of mTOR, are currently under intense investigation. A recently described subset of two mTOR-associated protein complexes has led to the designation mTOR complex 1 (mTORC1), which is sensitive to rapamycin, and mTORC2, which appears to be rapamycin insensitive (Sarbassov et al., 2004; Jacinto et al., 2004; Inoki and Guan, 2006; reviewed in Wullschleger et al., 2006) (➋ *Figure 10-3*). This discovery presents a challenge. The observation that TORC2 may not be rapamycin sensitive calls into question previously "rapamycin insensitive" data, which were interpreted to rule out a role for mTOR. These issues are currently under study.

Intra-third-ventricular injections of leucine (but not isoleucine or valine, the other two branched chain AAs) reduce food intake within 4 h, apparently via mTOR, in a signaling cascade that is rapamycin sensitive and appears to include the adipose-derived satiety hormone, leptin (Cota et al., 2006). This suggests that mTORC1 signaling in brain is important for satiety during ingestion of rich nutrients, as signaled by leucine. It is not known if physiological levels of AAs regulate food intake via this pathway (Flier, 2006). If they do, such a mechanism, when fully described, could help to explain how high-protein diets provide satiety.

Although Cota and coworkers (2006) focused primarily on the arcuate nucleus in the hypothalamus, they did note that mTOR was phosphorylated on serine 2448, showing activation of the kinase in other brain areas and importantly in the hippocampus. The mTOR system has a role in learning; it is important in hippocampal long-term potentiation (LTP), a well-studied in vitro electrophysiological model of learning that requires local protein synthesis in dendritic spines (Tang et al., 2002; Cammalleri et al., 2003; Jaworski et al., 2005; Lee et al., 2005; Tsokas et al., 2005). It should be noted that analogous LTP is also seen in the IAA-sensitive anterior piriform cortex (APC) (reviewed by Larson et al., 2005).

The ability to recall the location of a satisfactory food would be adaptive. There is ample documentation of conditioned responses associated with food. The learned aversion to a flavor associated with malaise, i.e., conditioned taste aversion, is also known as "bait shyness," or the Garcia effect (Garcia et al., 1955).

Figure 10-3

The mammalian target of rapamycin complex 1 (mTORC1) system as it is affected by increased nutrients (↑AA). The increased AA may interact with the tuberous sclerosis complex 1 or 2 (TSC1 or TSC2), which inhibit Ras homologue enriched in brain (Rheb). The increased AA may also act on Rheb or mTOR directly. Rapamycin inhibits mTOR in this complex, which includes Raptor and the G protein, GβL, which binds to the kinase domain of mTOR; both facilitate mTOR signaling. Output from the Raptor protein in mTORC1 includes the activation of the ribosomal subunit, S6K, and inhibition of the 4E-binding protein (4E-BP), both of which lead to protein synthesis and cellular hypertrophy. mTOR directly stimulates ribosome synthesis and inhibits autophagy. *Arrows* indicate stimulatory effects, *T-bars* indicate inhibition

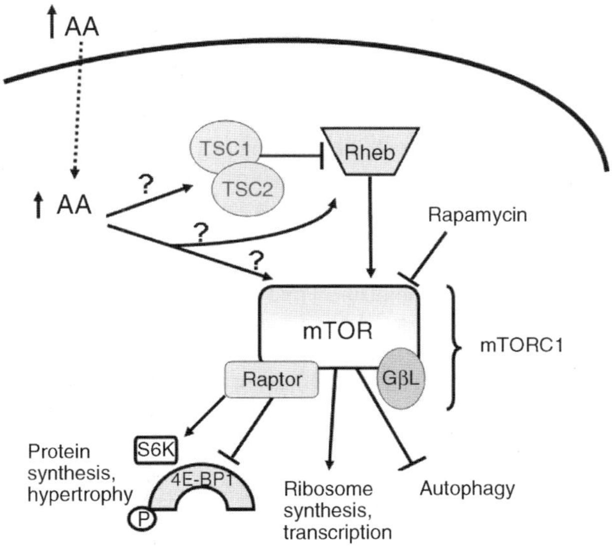

Conditioned aversions and preferences have been associated with deficient and replete foods, respectively (Simson and Booth, 1973; Rogers and Leung, 1977; Gietzen et al., 1992; Naito-Hoopes et al., 1993; Fromentin et al., 1997, 1998; Markison et al., 1999; Feurté et al., 2000). Threonine-deficient rats show a preference for the place where they were given threonine intragastrically, demonstrating a conditioned place preference associated with IAA repletion (Fromentin et al., 1998). Memory for a place is well known to be associated with hippocampal function. In this context, it is interesting that mTOR is associated with LTP, as discussed above. Because mTOR is activated by other nutrients as well as the IAA leucine (Dennis et al., 2001; Yeshao et al., 2005), it is tempting to suggest that activation of mTOR in neurons could provide the molecular basis for remembering where rich sources of nutrients were found, in addition to satiety discussed above. This clearly would aid appropriate food selection.

4 Sensing IAA Deficiency with Diets Either Imbalanced or Devoid of an IAA

Throughout evolution, challenges to nutrient supply have been more prevalent than surpluses. Many prokaryotes and some primitive eukaryotes have the necessary control systems for sensing AA deprivation and are able to initiate AA biosynthesis when necessary. Strategies for repletion across evolution, from single-celled organisms to humans, have been detailed (reviewed by Gietzen and Rogers, 2006). For herbivores and omnivores, diet selection must provide a full supply of IAA in a timely fashion or general protein synthesis is halted and degradation exceeds synthesis (Kadowaki and Kanazawa, 2003). The inability of incomplete proteins (i.e., those missing or having inadequate IAA in the food) to support human health was appreciated as early as the 1800s (reviewed in Carpenter, 2003). When animals are given an IAA-deficient or IAA-imbalanced diet, they do not grow. Rejection of the diet was shown in the 1930s (McCoy et al., 1935) to explain this growth failure (reviewed in Harper et al., 1970; Rogers and Leung, 1973, 1977; Gietzen, 1993, 2000; Gietzen and Rogers, 2006). Rejection of a test diet and dietary choice in laboratory animals remain useful tools for evaluating protein quality and IAA balance.

4.1 Diet Selection—Behavioral Practices Demonstrate IAA Sensing

The most efficient means of maintaining IAA homeostasis is to select a diet that includes animal source proteins (Milton, 2003). Where animal source foods are not available, or there are cultural histories of famine or vegetarian food preference, complimentary nonanimal proteins, such as rice and beans, are used routinely. These dietary practices in humans predate discovery of IAA (reviewed in Gietzen and Rogers, 2006). If rats are given even a limited choice, as with near-basal levels of lysine, the more adequate diet is chosen at lysine levels that do not decrease food intake (Hrupka et al., 1999). Therefore, given a choice, rats will switch diets rather than stop eating; this indicates that the meal termination seen with IAA-deficient diets is not due either to toxicity or to satiety. The threshold for sensing IAA using a variety of limiting IAA is in the range of 90–120 parts per million (Hrupka et al., 1997, 1999), showing exquisite sensitivity to IAA depletion.

The behavioral strategies for dealing with limiting amounts of IAA include altered food choice, meal termination, foraging for foods that will complement or correct the deficiency, the development of a learned aversion to a deficient or imbalanced food in order to avoid that food in the future, and memory for the place associated with replenishing food (Fromentin et al., 1997, 1998; Gietzen, 2000; Koehnle et al., 2003, 2004a, b; Koehnle and Gietzen, 2005). Any of the first three of these strategies will help the animal obtain a complete meal only if it occurs as a consequence of sensing the IAA deficiency within that meal.

4.2 The Time Course for Postabsorptive AA Sensing

Using computerized on-line meal pattern analysis and a threonine-devoid diet, the determination of the time course for detecting IAA depletion was made on the basis of time to meal termination; 90% of the rats eating an IAA-deficient meal terminated that meal in 20 min, whereas 50% of the control rats continued eating their balanced meal (Koehnle et al., 2003, 2004a, b). In a bar-pressing paradigm, after prefeeding a threonine-limiting "basal" diet, the fastest rat stopped bar pressing for a threonine-devoid pellet in 28 min. These meal terminations are not due to satiety, as shown using observations of postmeal behaviors, including the absence of the "satiety sequence" in rats given an IAA-deficient diet (Feurté et al., 2000). In a choice situation, the better diet was selected between 15 and 30 min (Gietzen et al., 1986). When exposed to a lysine-deficient diet, within five days groups of rats learned to find the bottle containing lysine HCl (which has a bitter taste) from amongst 15 bottles containing various solutions. As long as the rats remained lysine deficient, they continued to ingest enough lysine HCl solution to replete their lysine stores. Very shortly (within 30 min) (Torii, K., personal communication) after the rats are given a diet containing adequate lysine, they switch their fluid intake to MSG, which rats apparently prefer (Mori et al., 1991). Also, rats resume a rapid rate of feeding within 30 min when switched from an IAA-deficient or imbalanced diet to a complete diet (Rogers and Leung, 1973). Thus, the time to sensing IAA repletion is less than 30 min. In the absence of foraging opportunities or dietary choices leading to repletion, the animals simply stop eating the deficient diet by $\sim$20 min (Gietzen, 2000; Koehnle et al., 2003, 2004a, b).

4.3 Postabsorptive Sensing of IAA Depletion. Hypothesis: The Brain Senses AA Deficiency

As reviewed above, AA sensing either by AA taste receptors in the oral cavity, or in the gut, or the intracellular mTOR system appears to sense increases in AA, rather than deficiencies. The involvement of the brain in sensing IAA depletion was supported by the finding that the rate of decrease of the limiting IAA in brain tissue is as rapid as that in plasma (Peng et al., 1972). Also, the limiting IAA reverses the feeding response to an IAA-imbalanced diet at a much lower concentration when infused into the carotid artery than into the jugular vein (Leung and Rogers, 1969). Beginning in the 1960s, using classical brain lesioning techniques, Leung and Rogers ablated a series of sites in the feeding and food selection circuitry (reviewed in Leung and Rogers, 1987; Gietzen, 1993). After a student misplaced the ear bars of the stereotaxic apparatus too far forward in the head, these rats failed to reject an IAA imbalanced diet. With careful reconstruction of the lesion sites and appropriate replication of the study, the chemosensor for IAA depletion was shown to be in the very rostral brain area, the APC (Leung and Rogers, 1971) (➋ *Figure 10-4*).

Figure 10-4

Effects of IAA depletion are sensed in the anterior piriform cortex (APC). The initial signals include a decrease in the limiting IAA, increases in uncharged transfer ribonucleic acid (tRNA), and activation of the well-conserved general amino acid control (GC) system in the pyramidal cells of layer II in the APC. Glutamate output projections lead to increased foraging via motor circuits and aversive conditioning via the amygdala, hippocampus, and parabrachial nucleus (PBN). These lead to long-term rejection of cues associated with the deficient diet

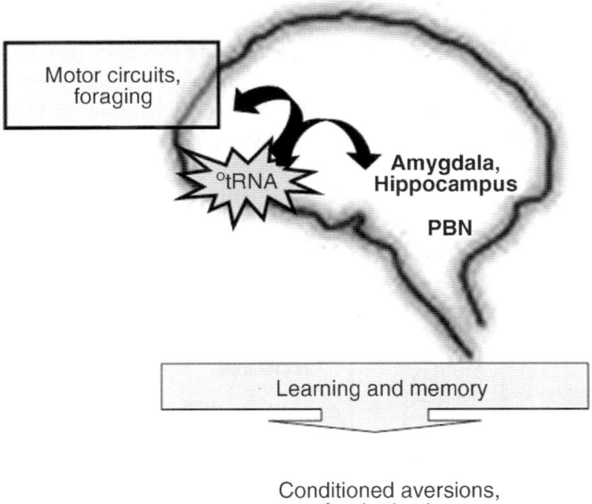

Subsequently, this finding was replicated in rats (Noda and Chikamori, 1976) and birds (Firman and Kuenzel, 1988). Later, microinjection and histochemical studies have confirmed this finding. Russell and coworkers (2003) confirmed and extended the injection studies of Beverly and coworkers (1990) and Monda and coworkers (1997) using injections into the APC. Animals injected with saline into the APC terminate a threonine-devoid meal within 20 min, but those rats injected with 2 nmol of threonine into the APC continue to eat, as do injected animals given a complete control diet. This study also confirms the 20 min time to meal termination with IAA deficiency (Koehnle et al., 2003, 2004a, b).

4.4 The Mechanism for Sensing IAA Deficiency in the Brain

We have reported a biochemical pathway in the APC that serves as an IAA detection system. It is conserved across eukaryotic species (Hao et al., 2005a), and it is activated within the 20 min time frame for identification of the IAA sensor. The initial event, after ingestion of an IAA-imbalanced or -deficient diet, is a decrease in the concentration of the limiting IAA in the APC; the limiting IAA is decreased in APC tissue by 56% at 21 min (Koehnle et al., 2004a).

In the earliest steps leading to the initiation of mRNA translation, AAs are acylated to tRNA by their cognate aminoacyl-tRNA synthetase enzymes, a process termed "tRNA charging." Even if dispensable AAs are continuously available by definition, a decrease in IAA below a given threshold will trigger tRNA deacylation. The existence of deacylated tRNA in vivo is controversial. The K_m for the amino acyl synthetases are very low (Shenoy and Rogers, 1978). Experimental differences such as timing (Magrum et al., 2002) or tissue source (Dennis et al., 2001) may affect the results, as the initial signal occurred by 20 min after introduction of an IAA-deficient diet, and the responsive cells are in a very restricted area of the APC (Sharp et al., 2006). Still, deacylated tRNA is known to initiate signaling for AA deficiency in vivo in single-cell systems, such as yeast (Hinnebusch and Natarajan, 2002). A role for uncharged tRNA in sensing IAA deficiency by animals was shown directly by injecting tRNA synthetase inhibitors into the rat APC and observing the behavioral response (Hao et al., 2005a) (➋ *Figure 10-5*).

◘ Figure 10-5

Effects of uncharged tRNA include dimerization and phosphorylation of GCN2 (GCN2-P), which causes an increase of the phosphorylation of eukaryotic initiation factor 2α (eIF2) and binding to the ternary initiation complex (left), for each round in the initiation of translation. The ternary initiation complex includes methionyl-tRNA, eIF2α-P, GTP, and the 40S ribosome. The phosphorylated eIF2α must be dephosphorylated in order to begin another round of translation. When it is bound to eIF2B, the necessary guanine exchange does not take place (middle of the figure). When eIF2 is bound to GDP, it becomes inhibitory to new translation, as is the inactive complex of eIF2α-P with eIF2B and GDP. The result of increased phosphorylation of eIF2α is inhibition of global protein synthesis, but increased translation of a few key proteins for rescue, via open reading frames, leading to signaling

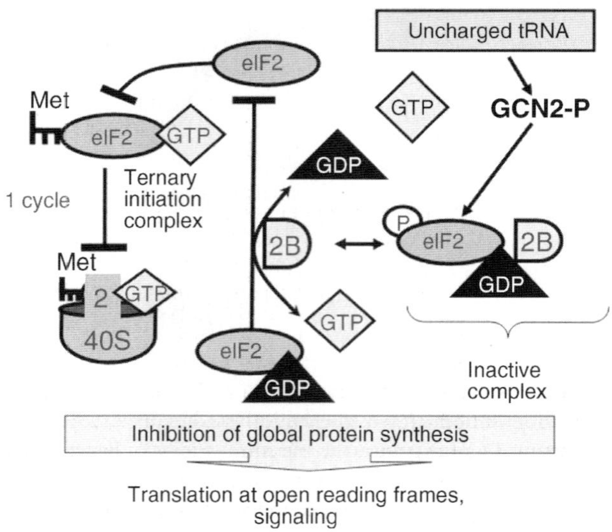

Amino-alcohols inhibit their respective aminoacyl-tRNA synthetases, thereby increasing the concentration of deacylated tRNA (Hansen et al., 1972). Injection of L-threoninol or L-leucinol, but not D-threoninol or a dispensable AA, into the APC decreases food intake at 20 min and increases other downstream events the same as eating a threonine-devoid diet. The effects of L-amino-alcohols are stereospecific, competitive, selective for their respective AAs, and mimic the effects of IAA depletion (Hao et al., 2005a).

The next step in the yeast pathway is the activation of the GCN2 kinase (Hinnebusch, 2000), which dimerizes and autophosphorylates when uncharged tRNA binds to its nonselective histidine tRNA site

(HisRS) (Wek et al., 1995; Padayana et al., 2005). When a threonine-free diet is fed to naive mice deleted for the gene encoding the GCN2 kinase (*GCN2*$^{-/-}$ mice), they fail to reject the IAA-depleting diet (Hao et al., 2005a) but the naive intact mice do so. This was confirmed when Maurin and coworkers (2005) reported that brain-specific *GCN2*$^{-/-}$ mice also fail to reject an IAA-deficient diet, and reportedly do not develop the usual learned aversion to the IAA-depleting diet. Interestingly, in *GCN2*$^{-/-}$ mice, long-term synaptic plasticity and memory consolidation (both dependent on *GCN2*) are impaired under strong training protocols (Costa-Mattioli et al., 2005). Thus, conditioned aversion to an IAA-depleting diet may be affected in these knockout animals. In accord with this idea, hippocampal lesions interfere with learning in rats fed IAA-imbalanced diets; the expected conditioned taste aversion is delayed (Leung and Rogers, 1979).

The activated kinase, GCN2-P, phosphorylates the α subunit of eukaryotic initiation factor 2 (eIF2α) (reviewed by Dever, 2002; Wek et al., 2006), a pivotal factor in the control of the initiation of translation in protein synthesis (Anthony et al., 2001a; Wek et al., 2006). Intact mice (Hao et al., 2005a, Maurin et al., 2005) as well as rats (Gietzen et al., 2004) not only reject a threonine-devoid diet but also respond by increasing the phosphorylation of eIF2α in APC neurons. The level of eIF2α-P is in proportion to the amount of the devoid diet eaten before the detection of the deficiency in the control wild-type mice ($r^2 = 0.99$, $P < 0.001$). The *GCN2*$^{-/-}$ mouse neither rejects the diet nor phosphorylates eIF2α, and there is no such correlation in the null mice ($r^2 = 0.02$, $P =$ NS) (Hao et al., 2005a).

GCN2 has interactions with TOR and mTOR in several models (Iiboshi et al. 1999; Hinnebusch and Natarajan, 2002; Cherkasova and Hinnebusch, 2003; Kubota et al., 2003 Anthony et al., 2004). Not all cell types regulate mTOR through charging levels of tRNA, as some are resistant to activation by amino-alcohols (Pham et al., 2000). To our knowledge, mTOR, despite its role in sensing lysine and other nutrients as described in detail above, has not been shown to have a direct role in the detection of IAA deficiency. Still, IAA depletion, by limiting its agonists, could reduce its activation via a number of different pathways (reviewed by Wullschleger et al., 2006). To address this question, Hao and coworkers (Nemanic et al., 2004; Hao et al., 2006) injected rapamycin into the APC and saw no effect on intake either of a control (basal) or IAA-deficient (threonine devoid) diet from 20 min to 21 h after diet introduction. Therefore, the feeding response to IAA deficiency in the rat is not sensitive to rapamycin. In the absence of rapamycin sensitivity, mTORC2 could be involved in the APC's responses to IAA deficiency, as it is associated with the cytoskeleton. Of interest in this context, activation of GCN2 can be inhibited by an actin-binding protein, IMPACT (Pereira et al., 2005).

5 Activating the APC After AA Sensing: The Transporter Hypothesis

The various AA receptors and transporters are attractive candidates to provide information to the organism about the availability of AAs in the medium/plasma/extracellular spaces. AA limitation has been known to affect transporters since the early 1970s (Gazzola et al., 1972) and has been reviewed (Christensen, 1990; Kilberg et al., 1993, 2005; Palacin et al., 1998; Hyde et al., 2003; Mackenzie and Erickson, 2004). A role for AA sensing also has been proposed for a calcium receptor (Conigrave et al., 2000), but this receptor responds stereospecifically to most of the 20 standard L-AAs used in protein synthesis; so selectivity for any particular IAA cannot be assumed. Yeast have an AA-sensitive permease regulator, Ssy1p, that responds to changes in the concentrations of AAs in the media (Gagiano et al., 2002), but it appears to require phosphorylation either for activation or for translocation to the membrane; so, upstream signaling must be involved.

In addition, cationic AA transporters are inducible by limitation for any IAA, and not by dispensable AA, but there is a delay of 2 h for induction and dependence on phosphorylation of eIF2α, again suggesting that synthesis of new transporters is involved (Hatzoglou et al., 2004). Such delays indicate that these transporters are downstream from IAA sensing, although they help remediate the IAA deficiency in response to other signals by importing limiting IAA into the cell.

The classical system A amino acid transporter (ATA[N]) family is sodium dependent and was defined using 2-methyl-aminoisobutryic acid (MeAIB) (reviewed by Palacin et al., 1998; Mackenzie and Erickson, 2004). MeAIB-blockable transport of labeled threonine is activated within 10 min after exposing neuron-rich cultures from the IAA-sensitive APC to a threonine-devoid medium (Blais et al., 2003). Such early activation is likely due to recruitment rather than gene expression (Ling et al., 2001). Recent elegant molecular studies have cloned and renamed the system A transporters the *s*odium-coupled *n*eutral *a*mino acid *t*ransporters SNAT1, SNAT2, and SNAT4; both SNAT1 (old ATA1) and SNAT2 (old ATA2) are found in the brain (Mackenzie and Erickson, 2004). The IAA-sensitive cells in the APC are the glutamatergic pyramidal cells (Jung et al., 1990), which could include them into either of the new SNAT1 or SNAT2 categories. It is not known whether the neurons of the APC use SNAT1 and/or SNAT2 to import IAA. MeAIB, at 0.1mM, was used as the inhibitor to identify system A transport. The probes used, with threonine as the limiting IAA, were interpreted at that time to suggest that the AA transporter in APC neurons was ATA1/SNAT1 (Blais et al., 2003). Although threonine is not the preferred substrate for either SNAT1 or SNAT2, it is carried by system A as shown in Albers' studies (Albers et al., 2001) using voltage clamp recordings with rat ATA1 (SNAT1) transfected into *Xenopus* oocytes.

The MeAIB-sensitive uptake of threonine in threonine-deficient APC cells (activated by 10 min) is dependent on phosphorylation for activation and movement to the membrane; so these transporters, like the others mentioned above, are likely to be downstream of the sensor for IAA depletion (Blais et al., 2003). However, the MeAIB transporter activation in threonine-deficient APC neurons is likely upstream of gene expression as we saw increases in eIF2α-P, but not the activating transcription factor 4 (ATF4) protein at 20 min (Gietzen et al., 2004). ATF4 is a mammalian ortholog of GCN4, the putative master regulator of gene expression in AA starvation in yeast (Hinnebusch and Natarajan, 2002). It is downstream from GCN2-activated eIF2α. Increased translation of mRNA for SNAT2 in mouse embryonic fibroblasts deprived of IAA for 1 h depends on phosphorylation of eIF2α and translation at an internal ribosome entry site (Gaccioli et al., 2006). Increased eIF2α phosphorylation is also seen in rat and mouse neurons from the IAA-sensitive region of the APC, taken after 20 min eating a threonine-devoid diet (Gietzen et al., 2004; Hao et al., 2005a; Maurin et al., 2005; Sharp et al., 2006). Of particular interest in terms of neuronal signaling, the increased intracellular sodium from cotransport by system A (either ATA1 or ATA2 (SNAT1 or SNAT2)) is electrogenic (Yao et al., 2000; Albers et al., 2001); and should also be electrogenic in the neurons of the APC. Although not shown directly, as far as we know, this could link phosphorylation of eIF2α to sodium-coupled AA transport, and thus to the signaling of IAA status in the glutamatergic output cells of the APC (➋ *Figure 10-6*, left column).

5.1 Activating the APC After AA Sensing: The Glutamate Hypothesis

Glutamate is the transmitter of the primary output cells of the APC (reviewed by Jung et al., 1990; Doherty et al., 2000; Dybdal and Gale, 2000; Blevins et al., 2004). If the sensor is indeed in the APC's glutamatergic output cells and they are activated in response to IAA deficiency, then these cells should signal appropriate brain circuitry and initiate IAA replenishing strategies. Releasable glutamate could be increased in the cells by transamination (Hutson et al., 2001) of the IAA that are in relative excess with the relative imbalance, and because they are not used for protein synthesis owing to the inhibition of global translation by eIF2α-P (Anthony et al., 2001b, 2004; Wek et al., 2006). Several studies suggest the importance of glutamate AMPA receptors in the behavioral and biochemical responses to IAA deficiency, secondary to the initial signal in the APC (Gietzen and Magrum, 2001; Sharp et al., 2004; Hao et al., 2005b) (➋ *Figure 10-6*, top right). There is good evidence that glutamate, at several different receptor subtypes, is associated with the synaptic plasticity of LTP in the hippocampus (Hou and Klann, 2004; Klann and Dever, 2004; Gong et al., 2006; Page et al., 2006).

5.2 Activating the APC After IAA Sensing: The Calcium Hypothesis

In addition to sodium and glutamate, both intra- and extracellular sources of calcium are involved in the acute responses of the APC to IAA depletion. We have reported changes in intracellular calcium in the APC with changes in IAA but not dispensable AAs (Magrum et al., 1999), as well as increased phosphorylation of

Figure 10-6

Activation of the anterior piriform cortex (APC) may occur by at least two mechanisms. On the left, uncharged tRNA activates GCN2 for phosphorylating eIF2α (top cell on left is a neuron showing labeling for eIF2α-P in the cytoplasm). ERK1 and ERK2 are phosphorylated in layer II cells (lower neuron shows ERK1/2-P labeling, also in the cytoplasm). System A, the electrogenic sodium-dependent neutral AA transporter (SNAT1/2) is activated in these neurons, leading to the suggestion that this is the mechanism for activation of the APC by IAA depletion. The *star* labeled "1" indicates activation of the primary sensory cell. These cells are glutamatergic and activate AMPA receptors on neighboring cells in layer II (*upward arrow*) to *star* labeled "2."An alternative mechanism could be that increased IAA, relative to the depleted one, could be transaminated to glutamate, which would be available for signaling at the glutamate AMPA receptor in this scheme. Activation of the AMPA receptor leads to increased intracellular calcium and activation of calcium/calmodulin-dependent protein kinase II (CaMKII-P), which also serves as an indicator of increased intracellular calcium (neuron on the right is labeled for CaMKII-P). Such activity would lead to further signaling and synaptic plasticity

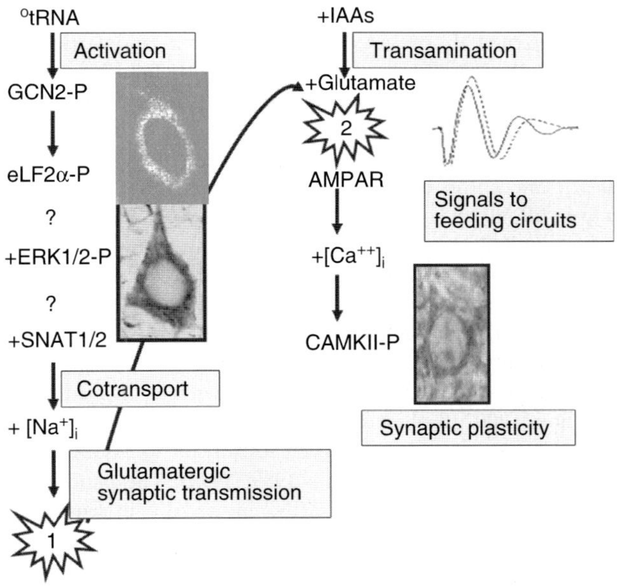

calcium/calmodulin-dependent protein kinase II (CaMKII) in the cell-body layer that houses the primary output neurons of the APC (Sharp et al., 2004). CaMKII-P is elevated within the 20 min window for the behavioral rejection of an IAA-depleting diet. CaMKII, among other functions, activates the glutamate AMPA receptor, GluR1, by phosphorylation in the postsynaptic density (Vinade and Dosemeci, 2000). Both CaMKII-P and the glutamate AMPA receptor subunit (GluR1), which are colocalized within GluR1-P in APC neurons, are increased by 20 min after eating an IAA-deficient diet. This provides evidence for both calcium and glutamate in signaling IAA deficiency in the APC (Sharp et al., 2004). CaMKII-P appears in far more APC cells than does eIF2α-P, and its increase depends on extracellular calcium (Gietzen et al., unpublished results); so this phosphorylated kinase is likely to be present in the postsynaptic (secondary) cells of the APC (*Figure 10-6*, bottom right), which are activated in the positive feedback loop that characterizes the piriform cortex (reviewed by Larson et al., 2005).

5.3 Activating the APC After IAA Sensing: The ERK Hypothesis

Alternatively, the neurons could be activated by increased intracellular sodium via the sodium-dependent AA cotransporters, discussed above (Blais et al., 2003; Gaccioli et al., 2006), which require activation by eIF2α-P (Gaccioli et al., 2006) and mitogen-activated protein kinase (MAPK/ERK1/2) (Franchi-Gazzola et al., 1999). MAPK-P is associated with the cells positive for eIF2α-P in the APC (Sharp et al., 2002, 2006) and with LTP in the hippocampus (Page et al., 2006). Colocalization of MAPK-P and eIF2α-P occurs in cells that are limited to in a narrow (<1 mm) rostrocaudal segment of the cell-body layer of the pyramidal output neurons of the APC (Sharp et al., 2006), and, in contrast to the many cells staining positively for CaMKII-P (Sharp et al., 2004), the for-fewer MAPK-P/eIF2α-P-positive neurons may be the primary sensory cells that recognize IAA deficiency in the APC.

6 Neural Circuitry and Involvement of Other Brain Areas

The APC is located in the anterior ventrolateral forebrain, and the neural circuitry associated with the feeding responses to IAA depletion has been reviewed (Gietzen et al., 1998; Gietzen, 2000; Koehnle and Gietzen, 2005). The APC has long been known to project to areas of the brain important for the control of food intake (Haberly and Price, 1978; Gietzen et al., 1998; Aja, 1999). Yet, several laboratories have indicated that other areas of the brain, particularly the hypothalamus (Mori et al., 1991; Tabuchi et al., 1991; Torii et al., 1996; Blevins et al. 2004; Tomé, 2004), are involved in the IAA response. The lateral hypothalamus is involved in the hyperphagia to 10% (moderately low) protein (White et al., 2003). The vagus projects AA-related information to the lateral hypothalamus (Jeanningros, 1984). However, roles for the ventromedial hypothalamus and lateral hypothalamus as primary sensors of IAA deficiency were ruled out in the 1970s (reviewed in Rogers and Leung, 1973). Data from Monda and coworkers (1997) and Blevins and coworkers (2004) suggest that the lateral hypothalamus acts secondarily, receiving signals generated in the APC.

Sattlegger, Castilho, and coworkers (Pereira et al., 2005) have described an actin-binding protein "IMPACT" that is preferentially expressed in brain tissue, and which inhibits the activation of GCN2. Interestingly, the level of IMPACT is inversely correlated with the phosphorylation of eIF2α and is much higher in hypothalamic areas than in the APC (Pereira et al., 2005). This could explain why the hypothalamus, which has classically been thought to house important nuclei in feeding circuits (see references above and reviews by Berthoud, 2002; Broberger, 2005) and where mTOR responds to intra-third-ventricular injections of leucine (Cota et al., 2006), has been so difficult to associate with the sensing of IAA deficiency (Rogers and Leung, 1973; Gietzen et al., 1989, 1998). The paucity of the GCN2 inhibitor, IMPACT, in the APC is consistent with a role for GCN2 in the APC in sensing IAA deficiency (Pereira et al., 2005).

After c-fos expression was seen in the dorsomedial hypothalamus (DMH) when animals had eaten an IAA-imbalanced diet (Wang et al., 1996), this hypothalamic region was studied. Cutting fibers running anteriorly from the DMH increases intake of an IAA-imbalanced diet for the entire first day, and the nucleus itself may be involved in the first 3 h of the responses (Bellinger et al., 1998, 1999); these intriguing findings deserve further study. It will be interesting to learn if the GCN2 inhibitor, IMPACT, is expressed in the neurons of the DMH.

7 Concluding Remarks

The discoveries of roles for the GCN pathway in the APC for the sensing of IAA deficiency, and the mTOR system in the hypothalamus for sensing AA sufficiency, underscore the importance of AA homeostasis as a crucial biological mechanism, and suggest that the responses to high and low levels of AAs are mediated by distinct brain systems. The recent demonstrations of the sensory limbs of the postabsorptive responses to IAA limitation (Hao et al., 2005a) and to rich supplies of the IAA, leucine, and general energy substrates (Cota et al., 2006) make significant contributions to our understanding of AA sensing. Future studies will be required to elucidate the transduction mechanisms by which these sensory systems activate their respective neural substrates and the neural circuits supporting the behavioral responses.

Acknowledgments

The authors appreciate the helpful comments of Drs. Beatriz Castilho, Evelyn Sattlegger, and Mauro Costa-Mattioli about the manuscript. The work was supported by NIH grants NS 043210 and NS 33347 to D.W.G. Portions of this material, reviewed also in trends in neurosciences 2006, 29: 91-99 and Ann Rev Natl, 2007, in press, are used here with permission.

References

Aja SM, Barrett JA, Gietzen DW. 1999. CCK_A and $5\text{-}HT_3$ receptors cooperate in anorectic responses to amino acid imbalanced diets. Pharmacol Biochem Behav 62: 487-491.

Aja SM. 1999. Efferent projections from the region of the anterior piriform cortex involved in aminoprivic feeding. Neural Transmitters and Neural Circuits Supporting Aminoprivic Feeding. Ph.D. dissertation. University of California, Davis, pp. 323–343.

Albers A, Bröer A, Wagner CA, Setiawan I, Lang PA, et al. 2001. Na^+ transport by the neural glutamine transporter ATA1. Pflügers Arch 443: 92-101. (Provides an electrophysiological demonstration of electrogenic activity in the System A amino acid transporter)

Anderson GH. 1977. Regulation of protein intake by plasma amino acids. Advances in Nutritional Research. Vol. 1. Draper HH, editor. New York: Plenum Press; pp. 135-166.

Anderson GH, Moore SE. 2004. Dietary proteins in the regulation of food intake and body weight in humans. J Nutr 134: 974S-979S.

Anthony JC, Yoshizawa F, Anthony TG, Vary TC, Jefferson LS, et al. 2000. Leucine stimulates translation initiation in skeletal muscle of postabsorptive rats via a rapamycin-sensitive pathway. J Nutr 130: 2413-2419. (The first report of the mechanism underlying the stimulation of protein synthesis by leucine in the whole animal)

Anthony TG, Anthony JC, Yoshizawa F, Kimball SR, Jefferson LS. 2001a. Oral administration of leucine stimulates ribosomal protein mRNA translation but not global rates of protein synthesis in the liver of rats. J Nutr 131: 1171-1176.

Anthony TG, Reiter AK, Anthony JC, Kimball SR, Jefferson LS. 2001b. Deficiency of dietary EAA preferentially inhibits mRNA translation of ribosomal proteins in liver of meal-fed rats. Am J Physiol 281: E430-E439.

Anthony TG, McDaniel BJ, Byerley RL, McGrath BC, Cavener DR, et al. 2004. Preservation of liver protein synthesis during dietary leucine deprivation occurs at the expense of skeletal muscle mass in mice deleted for eIF2 kinase GCN2. J Biol Chem 279: 36553-36561.

Bellinger LL, Evans JL, Gietzen DW. 1998. Dorsomedial hypothalamic lesions alter intake of an imbalanced amino acid diet in rats. J Nutr 128: 1213-1217.

Bellinger LL, Evans JF, Tillberg CM, Gietzen DW. 1999. Effects of dorsomedial hypothalamic nuclei lesions on intake of an imbalanced amino acid diet. Am J Physiol 277: R250-R262.

Bellinger LL, Gietzen DW, Williams FE. 1993. Liver denervation, $5\text{-}HT_3$ receptor antagonist, and intake of imbalanced amino acid diet. Brain Res Bull 32: 549-554.

Bellinger LL, Williams FE, Rogers QR, Gietzen DW. 1996. Liver denervation attenuates the hypophagia produced by an imbalanced amino acid diet. Physiol Behav 59: 925-929.

Bensaid A, Tomé D, Gietzen DW, Even P, Morens C, et al. 2002. Protein is more potent than carbohydrate for reducing appetite in rats. Physiol Behav 75: 577-582.

Berthoud HR. 2002. Multiple neural systems controlling food intake and body weight. Neurosci Biobehav Rev 26: 393-428. (A detailed review of the neural systems controlling food intake in response to energy sources and demands. Includes extrahypothalamic structures and circuitry)

Berthoud HR, Seeley RJ. 2000. Neural and Metabolic Control of Macronutrient Intake. Boca Raton, FL: CRC Press.

Berthoud HR, Blackshaw LA, Brookes SJ, Grundy D. 2004. Neuroanatomy of extrinsic afferents supplying the gastrointestinal tract. Neurogastroenterol Motil 16(Suppl. 1): 28-33.

Beverly JL, Gietzen DW, Rogers QR. 1990. Effect of dietary limiting amino acid in prepyriform cortex on food intake. Am J Physiol 259: R709-R715. (First demonstration that nanomoles of the limiting IAA injected into the APC restore feeding of IAA imbalanced diets)

Blais A, Huneau JF, Magrum LJ, Koehnle TJ, Sharp JW, et al. 2003. Threonine deprivation rapidly activates the system A amino acid transporter in primary cultures of rat neurons from the essential amino acid sensor in the anterior piriform cortex. J Nutr 133: 2156-2164.

Blevins JE, Truong BG, Gietzen DW. 2004. NMDA receptor function within the anterior piriform cortex and lateral hypothalamus in rats on the control of intake of amino acid-deficient diets. Brain Res 1019: 124-144.

Bolster DR, Jefferson LS, Kimball SR. 2004. Regulation of protein synthesis associated with skeletal muscle hypertrophy by insulin-, amino acid-, and exercise-induced signalling. Proc Nutr Soc 63: 351-356.

Brand JG, Teeter JH, Kumazawa T, Huque T, Bayley DL. 1991. Transduction mechanisms for the taste of amino acids. Physiol Behav 49: 899-904. (Covers taste responses for AA in catfish, preliminary studies of similar mechanisms in mouse)

Broberger C. 2005. Brain regulation of food intake and appetite: Molecules and networks. J Intern Med 258: 301-327. (Presents neurobiology of feeding from a clinical and basic neuroscience perspective)

Cammalleri M, Lütjens R, Berton F, King AR, Simpson C, et al. 2003. Time-restricted role for dendritic activation of the mTOR-p70^{S6K} pathway in the induction of late-phase long-term potentiation in the CA1. Proc Natl Acad Sci USA 100: 14368-14373.

Carpenter KJ. 2003. A short history of nutritional science: Part 1 (1785–1885). J Nutr 133: 638-645.

Chaudhari N, Landin AM, Roper SD. 2000. A metabotropic glutamate receptor variant functions as a taste receptor. Nat Neurosci 3: 113-119.

Cherkasova VA, Hinnebusch AG. 2003. Translational control by TOR and TAP42 through dephosphorylation of eIF2α kinase GCN2. Genes Dev 17: 859-872.

Christensen HN. 1990. Role of amino acid transport and countertransport in nutrition and metabolism. Physiol Rev 70: 43-77.

Conigrave AD, Quinn SJ, Brown EM. 2000. L-amino acid sensing by the extracellular Ca^{2+}-sensing receptor. Proc Natl Acad Sci USA 97: 4814-4819.

Costa-Mattioli M, Gobert D, Harding H, Herdy B, Azzi M, et al. 2005. Translational control of hippocampal synaptic plasticity and memory by the eIF2α kinase GCN2. Nature 436: 1166-1170.

Cota D, Proulx K, Blake Smith KA, Kozma SC, Thomas G, et al. 2006. Hypothalamic mTOR signaling regulates food intake. Science 312: 927-930. (Demonstrates that rapamycin-sensitive activation of mTOR in the hypothalamus affects food intake)

Damak S, Rong M, Yasumatsu K, Kokrashvili Z, Varadarajan V, et al. 2003. Detection of sweet and umami taste in the absence of taste receptor T1r3. Science 301: 850-853.

Darcel N, Fromentin G, Raybould HE, Gougis S, Gietzen DW, et al. 2005a. Fos-positive neurons are increased in the nucleus of the solitary tract and decreased in the ventromedial hypothalamus and amygdala by a high-protein diet in rats. J Nutr 135: 1486-1490.

Darcel NP, Liou AP, Tomé D, Raybould HE. 2005b. Activation of vagal afferents in the rat duodenum by protein digests requires PepT1. J Nutr 135: 1491-1495.

Dennis PB, Jaeschke A, Saitoh M, Fowler B, Kozma SC, et al. 2001. Mammalian TOR: A homeostatic ATP sensor. Science 294: 1102-1105.

Dever TE. 2002. Gene-specific regulation by general translation factors. Cell 108: 545-556.

Dixon KD, Williams FE, Wiggins RL, Pavelka J, Lucente J, et al. 2000. Differential effects of selective vagotomy and tropisetron in aminoprivic feeding. Am J Physiol 279: R997-R1009.

Doherty J, Gale K, Eagles DA. 2000. Evoked epileptiform discharges in the rat anterior piriform cortex: Generation and local propagation. Brain Res 861: 77-87.

Dybdal D, Gale K. 2000. Postural and anticonvulsant effects of inhibition of the rat subthalamic nucleus. J Neurosci 20: 6728-6733.

Eylam S, Spector AC. 2004. Stimulus processing of glycine is dissociable from that of sucrose and glucose based on behaviorally measured taste signal detection in Sac 'taster' and 'nontaster' mice. Chem Senses 29: 639-649.

Feurté S, Nicolaïdis S, Berridge KC. 2000. Conditioned taste aversion for a threonine-deficient diet in the rat: Demonstration by the taste reactivity test. Physiol Behav 68: 423-429.

Firman JD, Kuenzel WJ. 1988. Neuroanatomical regions of the chick brain involved in monitoring amino acid deficient diets. Brain Res Bull 21: 637-642.

Flier JS. 2006. Regulating energy balance: The substrate strikes back. Science 312: 861-864.

Franchi-Gazzola R, Visigalli R, Dall'Asta V, Gazzola GC. 1999. Adaptive increase of amino acid transport system A requires ERK1/2 activation. J Biol Chem 274: 28922-28928. (Demonstrates a role for MAPK/ERK1/2 in activation of the System A amino acid transporter)

Fromentin G, Feurté S, Nicolaïdis S. 1998. Spatial cues are relevant for learned preference/aversion shifts due to amino-acid deficiencies. Appetite 30: 223-234. (Demonstrates that animals can make an association between the place where they sensed repletion of IAA deficiency and the relief of the deficiency: not single trial learning, but place preference.

Fromentin G, Gietzen DW, Nicolaïdis S. 1997. Aversion-preference patterns in amino acid- or protein-deficient rats: A comparison with previously reported responses to thiamin-deficient diets. Br J Nutr 77: 299-314.

Gaccioli F, Huang CC, Wang C, Bevilacqua E, Franchi-Gazzola R, et al. 2006. Amino acid starvation induces the SNAT2 neutral amino acid transporter by a mechanism that involves eukaryotic initiation factor 2α phosphorylation and cap-independent translation. J Biol Chem 281: 17929-17940.

Gagiano M, Bauer FF, Pretorius IS. 2002. The sensing of nutritional status and the relationship to filamentous growth in *Saccharomyces cerevisiae*. FEMS Yeast Res 2: 433-470.

Garcia J, Kimeldorf DJ, Koelling RA. 1955. Conditioned aversion to saccharin resulting from exposure to γ radiation. Science 122: 157-158. (This was the first demonstration of single trial learning in animals. Prior to this work, it was believed that all conditioning required repetitive trials. The phenomenon of conditioned taste aversion became known as the Garcia effect.)

Garlick PJ. 2001. Assessment of the safety of glutamine and other amino acids. J Nutr 131: 2556S-2561S.

Gazzola GC, Franchi R, Saibene V, Ronchi P, Guidotti GG. 1972. Regulation of amino acid transport in chick embryo heart cells. I. Adaptive system of mediation for neutral amino acids. Biochim Biophys Acta 266: 407-421.

Gietzen DW. 1993. Neural mechanisms in the responses to amino acid deficiency. J Nutr 123: 610-625.

Gietzen DW. 2000. Amino acid recognition in the central nervous system. Neural and Metabolic Control of Macronutrient Intake, Chapter 23. Berthoud HR, Seeley RJ, editors. New York: CRC Press; pp. 339-357. (Includes pharmacology of neurotransmitter signaling in the APC downstream of AA sensing)

Gietzen DW, Magrum LJ. 2001. Molecular mechanisms involved in the anorexia of branched chain amino acid deficiency. J Nutr 131: 851S-855S.

Gietzen DW, Rogers QR. 2006. Nutritional homeostasis and indispensable amino acid sensing: A new solution to an old puzzle. Trends Neurosci 29: 91-99. (Brief review of the sensing of IAA deficiency from an evolutionary perspective)

Gietzen DW, Erecius LF, Rogers QR. 1998. Neurochemical changes after imbalanced diets suggest a brain circuit mediating anorectic responses to amino acid deficiency. J Nutr 128: 771-781.

Gietzen DW, Leung PMB, Castonguay TW, Hartman WJ, Rogers QR. 1986. Time course of food intake and plasma and brain amino acid concentrations in rats fed amino acid-imbalanced or-deficient diets. Interactions of the Chemical Senses with Nutrition. Kare MR, Brand JG, editors. Philadelphia: Academic Press; pp. 415-459.

Gietzen DW, Leung PMB, Rogers QR. 1989. Dietary amino acid imbalance and neurochemical changes in three hypothalamic areas. Physiol Behav 46: 503-511.

Gietzen DW, McArthur LH, Theisen JC, Rogers QR. 1992. Learned preference for the limiting amino acid in rats fed a threonine-deficient diet. Physiol Behav 51: 909-914.

Gietzen DW, Ross CM, Hao S, Sharp JW. 2004. Phosphorylation of eIF2α is involved in the signaling of indispensable amino acid deficiency in the anterior piriform cortex of the brain in rats. J Nutr 134: 717-723. (First demonstration of the phosphorylation of eIF2α in the APC in response to IAA deficiency)

Gong R, Park CS, Abbassi NR, Tang SJ. 2006. Roles of glutamate receptors and the mammalian target of rapamycin (mTOR) signaling pathway in activity-dependent dendritic protein synthesis in hippocampal neurons. J Biol Chem 281: 18802-18815.

Haberly LB, Price JL. 1978. Association and commissural fiber systems of the olfactory cortex of the rat: 1. Systems originating in the piriform cortex and adjacent areas. J Comp Neurol 178: 711-740. (Shows broad divergence of output from the piriform cortex)

Hansen BS, Vaughan MH, Wang LJ. 1972. Reversible inhibition by histidinol of protein synthesis in human cells at the activation of histidine. J Biol Chem 247: 3854-3857. (Mechanistic study validating the use of amino-alcohols as AA-synthetase inhibitors)

Hao S, Ross-Inta CM, Nemanic S, Gietzen DW. 2006. Rapamycin independent responses to amino acid deficiency in the anterior piriform cortex.

Hao S, Sharp JW, Ross-Inta CM, McDaniel BJ, Anthony TG, et al. 2005a. Uncharged tRNA and sensing of amino acid deficiency in mammalian piriform cortex. Science 307: 1776-1778. (First report of the full biochemical pathway from a measured decrease in the limiting IAA to phosphorylation of eIF2α in the brain area that serves as the sensor for IAA depletion)

Hao S, Ross-Inta CM, Rudell JB, Sharp JW, Gietzen DW. 2005b. The role of glutamate AMPA receptors in the recognition of amino acid deficiency in the anterior piriform cortex of rats. FASEB J 20: Abstract 970.6.

Harper AE. 1976. Protein and amino acids in the regulation of food intake. Hunger: Basic Mechanisms and Clinical Implications. Novin D, Wyrwicka W, Bray G, editors. New York: Raven Press; pp. 103-113.

Harper AE, Peters JC. 1983. Amino acid signals and food intake and preference: Relation to body protein metabolism. Experientia 44(Suppl.): 107-134.

Harper AE, Peters JC. 1989. Protein intake, brain amino acid and serotonin concentrations and protein self-selection. J Nutr 119: 677-689.

Harper AE, Benevenga NJ, Wohlhueter RM. 1970. Effects of ingestion of disproportionate amounts of amino acids. Physiol Rev 50: 428-558. (A very thorough, detailed review of the work covering the effects of AA on feeding prior to 1970)

Hatzoglou M, Fernandez J, Yaman I. 2004. Regulation of cationic amino acid transport: The story of the CAT-1 transporter. Ann Rev Nutr 24: 377-399.

Heitman J, Movva NR, Hall MN. 1991. Targets for cell cycle arrest by the immunosuppressant rapamycin in yeast. Science 253: 905-909.

Hinnebusch AG. 2000. Mechanism and regulation of initiator methionyl-tRNA binding to ribosomes. Translational Control of Gene Expression. Sonenberg N, Hershey JWB, Matthews MB, editors. New York: Cold Spring Harbor

Press, Cold Spring Harbor; pp. 185-243. (Review of the role of GCN2 in response to IAA deprivation in yeast)

Hinnebusch AG, Natarajan K. 2002. Gcn4p, a master regulator of gene expression, is controlled at multiple levels by diverse signals of starvation and stress. Eukaryot Cell 1: 22-32.

Hou L, Klann E. 2004. Activation of the phosphoinositide 3-kinase-akt-mammalian target of rapamycin signaling pathway is required for metabotropic glutamate receptor-dependent long-term depression. J Neurosci 24: 6352-6361.

Hrupka BJ, Lin Y, Gietzen DW, Rogers QR. 1997. Small changes in essential amino acid concentrations dictate diet selection in amino acid-deficient states. J Nutr 127: 777-784. (Shows threshold for discrimination of IAA in the parts-per-million range after basal diet prefeeding)

Hrupka BJ, Lin Y, Gietzen DW, Rogers QR. 1999. Lysine deficiency alters diet selection without depressing food intake in rats. J Nutr 129: 424-430. (Shows that one sees dietary choice for the limiting IAA before food intake depression)

Hutson SM, Lieth E, La Noue KF. 2001. Function of leucine in excitatory neurotransmitter metabolism in the central nervous system. J Nutr 131: 846S-850S.

Hyde R, Taylor PM, Hundal HS. 2003. Amino acid transporters: Roles in amino acid sensing and signalling in animal cells. Biochem J 373: 1-18.

Iiboshi Y, Papst PJ, Kawasome H, Hosoi H, Abraham RT, et al. 1999. Amino acid-dependent control of $p70^{s6k}$: Involvement of tRNA aminoacylation in the regulation. J Biol Chem 274: 1092-1099.

Inoki K, Guan KL. 2006. Complexity of the TOR signaling network. Trends Cell Biol 18: 206-212.

Jacinto E, Hall MN. 2003. TOR signaling in bugs, brain and brawn. Nat Rev Mol Cell Biol 4: 117-126. (Comparative review of (yeast) TOR, (drosophila) dTOR and mTOR)

Jacinto E, Loewith R, Schmidt A, Lin S, Rüegg MA, et al. 2004. Mammalian TOR complex 2 controls the actin cytoskeleton and is rapamycin insensitive. Nature Cell Biol (Nov.) 6: 1122-1128.

Jaworski J, Spangler S, Seeburg DP, Hoogenraad CC, Sheng M. 2005. Control of dendritic arborization by the phosphoinositide-3′-kinase-akt-mammalian target of rapamycin pathway. J Neurosci 25: 11300-11312.

Jeanningros R. 1982. Vagal unitary responses to intestinal amino acid infusions in the anesthetized cat: A putative signal for protein induced satiety. Physiol Behav 28: 9-21.

Jeanningros R. 1984. Lateral hypothalamic responses to preabsorptive and post-absorptive signals related to amino acid ingestion. J Auton Nerv Syst 10: 261-268.

Jung MW, Larson J, Lynch G. 1990. Role of NMDA and non-NMDA receptors in synaptic transmission in rat piriform cortex. Exp Brain Res 82: 451-455.

Kadowaki M, Kanazawa T. 2003. Amino acids as regulators of proteolysis. J Nutr 133: 2052S-2056S.

Kilberg MS, Pan YX, Chen H. 2005. Nutritional control of gene expression: How mammalian cells respond to amino acid limitation. Ann Rev Nutr 25: 59-85. (Excellent review of the responses to IAA limitation in terms of the effects on protein synthesis that follow AA sensing)

Kilberg MS, Stevens BR, Novak C. 1993. Recent advances in mammalian amino acid transport. Ann Rev Nutr 13: 137-165.

Kimball SR, Jefferson LS. 2006. New functions for amino acids: Effects on gene transcription and translation. Am J Clin Nutr 83: 500S-507S.

Kimball SR, Antonetti DA, Brawley RM, Jefferson LS. 1991. Mechanism of inhibition of peptide chain initiation by amino acid deprivation in perfused rat liver. Regulation involving inhibition of eukaryotic initiation factor 2α phosphatase activity. J Biol Chem 266: 1969-1976.

Klann E, Dever TE. 2004. Biochemical mechanisms for translational regulation in synaptic plasticity. Nat Rev Neurosci 5: 931-942.

Koehnle TJ, Gietzen DW. 2005. Modulation of feeding behavior by amino acid deficient diets; present findings and future directions. Nutritional Neuroscience, Chapter 10. Lieberman HR, Kanarek RB, Prasad C, editors. Boca Raton, FL: CRC Press; pp. 147-161.

Koehnle TJ, Russell MC, Gietzen DW. 2003. Rats rapidly reject diets deficient in essential amino acids. J Nutr 133: 2331-2335.

Koehnle TJ, Russell MC, Morin AS, Erecius L, Gietzen DW. 2004a. Diets deficient in indispensable amino acids rapidly decrease the concentration of the limiting amino acid in the anterior piriform cortex of rats. J Nutr 134: 2365-2371.

Koehnle TJ, Stephens AL, Gietzen DW. 2004b. Threonine imbalanced diet alters first meal microstructure in rats. Physiol Behav 81: 15-21.

Kubota H, Obata T, Ota K, Sasaki T, Ito T. 2003. Rapamycin-induced translational derepression of *GCN4* mRNA involves a novel mechanism for activation of the eIF2α kinase GCN2. J Biol Chem 278: 20457-20460.

Kurihara K, Kashiwayanagi M. 2000. Physiological studies on umami taste. J Nutr 130: 931S-934S.

Larson J, Jessen RE, Kim D, Fine AK, du Hoffmann J. 2005. Age-dependent and selective impairment of long-term potentiation in the anterior piriform cortex of mice lacking the fragile X mental retardation protein. J Neurosci 25: 9460-9469. (Recent study using the LTP protocol in the APC slice)

Last AR, Wilson SA. 2006. Low-carbohydrate diets. Am Fam Physician 73: 1942-1948.

Lee CC, Huang CC, Wu MY, Hsu KS. 2005. Insulin stimulates postsynaptic density-95 protein translation via the phosphoinositide 3-kinase-akt-mammalian target of rapamycin signaling pathway. J Biol Chem 280: 18543-18550.

Leung PM, Rogers QR. 1979. Effects of hippocampal lesions on adaptive intake of diets with disproportionate amounts of amino acids. Physiol Behav 23: 129-136.

Leung PMB, Rogers QR. 1969. Food intake: Regulation by plasma amino acid pattern. Life Sci 8: 1-9. (First indication that IAA-deficiency sensing occurs in the brain)

Leung PMB, Rogers QR. 1971. Importance of prepyriform cortex in food-intake response of rats to amino acids. Am J Physiol 221: 929-935. (First report that anterior piriform cortex functions as a sensor for IAA deficiency)

Leung PMB, Rogers QR. 1987. The effect of amino acids and protein on dietary choice. Umami: A Basic Taste. Kawamura Y, Kare MR, editors. New York: Marcel Dekker; pp. 565-610.

Li X, Staszewski L, Durick K, Zoller M, Adler E. 2002. Human receptors for sweet and umami taste. Proc Natl Acad Sci USA 99: 4692-4696.

Ling R, Bridges CC, Sugawara M, Fujita T, Leibach FH, et al. 2001. Involvement of transporter recruitment as well as gene expression in the substrate-induced adaptive regulation of amino acid transport system A. Biochim Biophys Acta 1512: 15-21.

Lynch CJ, Fox HL, Vary TC, Jefferson LS, Kimball SR. 2000. Regulation of amino acid-sensitive TOR signaling by leucine analogues in adipocytes. J Cell Biochem 77: 234-251.

Mackenzie B, Erickson JD. 2004. Sodium-coupled neutral amino acid (system N/A) transporters of the SLC38 gene family. Pflügers Arch 447: 784-795.

Magrum LJ, Hickman MA, Gietzen DW. 1999. Increased intracellular calcium in rat anterior piriform cortex in response to threonine after threonine deprivation. J Neurophysiol 81: 1147-1149.

Magrum LJ, Teh PS, Kreiter MR, Hickman MA, Gietzen DW. 2002. Transfer ribonucleic acid charging in rat brain after consumption of amino acid-imbalanced diets. Nutr Neurosci 5: 125-130.

Makarios-Lahham L, Roseau SM, Fromentin G, Tomé D, Even PC. 2004. Rats free to select between pure protein and a fat-carbohydrate mix ingest high-protein mixed meals during the dark period and protein meals during the light period. J Nutr 134: 618-624.

Markison S, Gietzen DW, Spector AC. (1999) Essential amino acid deficiency enhances long-term intake but not short-term licking of the required nutrient. J Nutr 129: 1604-1612.

Maruyama Y, Pereira E, Margolskee RJ, Chaudhari N, Roper SD. 2006. Umami responses in mouse taste cells indicate more than one receptor. J Neurosci 26: 2227-2234.

Maurin AC, Jousse C, Averous J, Parry L, Bruhat A, et al. 2005. The GCN2 kinase biases feeding behavior to maintain amino acid homeostasis in omnivores. Cell Metab 1: 273-277. (The first use of brain-specific *GCN2* knockout mice in the study of IAA-deficiency sensing)

McCoy RH, Meyer CE, Rose WC. 1935. Feeding experiments with mixtures of highly purified amino acids. VIII. Isolation and identification of a new essential amino acid. J Biol Chem 112: 283-302. (Reports the discovery of the final IAA, threonine)

Milton K. 2003. The critical role played by animal source foods in human (Homo) evolution. J Nutr 133: 3886S-3892S.

Monda M, Sullo A, De Luca V, Pellicano MP, Viggiano A. 1997. L-Threonine injection into PPC modifies food intake, lateral hypothalamic activity, and sympathetic discharge. Am J Physiol 273: R554-R559. (Shows signaling from APC to the lateral hypothalamus with IAA repletion in APC and direct electrophysiological recordings in the lateral hypothalamus)

Mori M, Kawada T, Ono T, Torii K. 1991. Taste preference and protein nutrition and L-amino acid homeostasis in male Sprague-Dawley rats. Physiol Behav 49: 987-995.

Naito-Hoopes M, McArthur LH, Gietzen DW, Rogers QR. 1993. Learned preference and aversion for complete and isoleucine-devoid diets in the rat. Physiol Behav 53: 485-494.

Nelson G, Chandrashekar J, Hoon MA, Feng L, Zhao G, et al. 2002. An amino-acid taste receptor. Nature 41: 199-202. (Shows that the taste receptor subunits, T1R1 and T1R3, but not their D-isomers, combine to act as sensors for the 20 L-AA that are incorporated into protein)

Nemanic S, Ross CM, Hao S, Rudell JB, Gietzen DW. 2004. The effects of rapamycin injected into the anterior piriform cortex on food intake in rats. FASEB J 19: Abstract 125.3.

Niijima A, Meguid MM. 1995. An electrophysiological study on amino acid sensors in the hepatoportal system in the rat. Obes Res 3: 741S-745S.

Nilsson O, Ahlman H, Geffard M, Dahlstrom A, Ericson LE. 1987. Bipolarity of duodenal enterochromaffin cells in the rat. Cell Tiss Res 248: 49-54.

Noda K, Chikamori K. 1976. Effect of ammonia via prepyriform cortex on regulation of food intake in the rat. Am J Physiol 231: 1263-1266.

Padyana AK, Qiu H, Roll-Mecak A, Hinnebusch AG, Burley SK. 2005. Structural basis for autoinhibition and mutational activation of eukaryotic initiation factor 2α protein kinase GCN2. J Biol Chem 280: 29289-29299.

Page G, Khidir FAL, Pain S, Barrier L, Fauconneau B, et al. 2006. Group 1 metabotropic glutamate receptors activate the p70S6 kinase via both mammalian target of rapamycin (mTOR) and extracellular signal-related kinase (ERK 1/2) signaling pathways in rat striatal and hippocampal synaptoneurosomes. Neurochem Int 49: 413-421.

Palacin M, Estevez R, Bertran J, Zorzano A. 1998. Molecular biology of mammalian plasma membrane amino acid transporters. Physiol Rev 78: 969-1054. (Thorough review of AA transport studies up to 1998)

Peng Y, Tews JK, Harper AE. 1972. Amino acid imbalance, protein intake, and changes in rat brain and plasma amino acids. Am J Physiol 222: 314-321.

Pereira CM, Sattlegger E, Jiang HY, Longo BM, Jaqueta CB, et al. 2005. IMPACT, a protein preferentially expressed in the mouse brain, binds GCN1 and inhibits GCN2 activation. J Biol Chem 280: 28316-28323. (First report of IMPACT, an inhibitor of GCN2 function in brain)

Peters JC, Harper AE. 1984. Influence of dietary protein level on protein self-selection and plasma and brain amino acid concentrations. Physiol Behav 33: 783-790.

Pham PTT, Heydrick SJ, Fox HL, Kimball SR, Jefferson LS Jr, et al. 2000. Assessment of cell-signaling pathways in the regulation of mammalian target of rapamycin (mTOR) by amino acids in rat adipocytes. J Cell Biochem 79: 427-441.

Powers T, Dilova I, Chen CY, Wedaman K. 2004. Yeast TOR signaling: A mechanism for metabolic regulation. Curr Top Microbiol Immunol 279: 39-51. (Review of TOR signaling in yeast)

Powley TL, Berthoud HR. 1986. Participation of the vagus and other autonomic nerves in the control of food intake. Feeding Behavior: Neural and Humoral Controls. Ritter RC, Ritter S, Barnes CD, editors. New York: Academic Press; pp. 67-101.

Raybould HE. 1998. Does your gut taste? Sensory transduction in the gastrointestinal tract. News Physiol Sci 13: 275-280.

Raybould HE, Glatzle J, Freeman SL, Whited K, Darcel N, et al. 2006. Detection of macronutrients in the intestinal wall. Auton Neurosci 125: 28-33. (Review of nutrient sensing in the GI tract)

Rogers QR, Leung PMB. 1973. The influence of amino acids on the neuroregulation of food intake. Fed Proc 32: 1709-1719.

Rogers QR, Leung PMB. 1977. The control of food intake: When and how are amino acids involved? The Chemical Senses and Nutrition. Kare MR, Maller O, editors. New York: Academic Press; pp. 213-249.

Round A, Wallis DI. 1986. The depolarizing action of 5-hydroxytryptamine on rabbit vagal afferent and sympathetic neurons in vitro and its selective blockade by ICS 205–930. Br J Pharmacol 88: 485-494.

Russell MC, Koehnle TJ, Barrett JA, Blevins JE, Gietzen DW. 2003. The rapid anorectic response to a threonine imbalanced diet is decreased by injection of threonine into the anterior piriform cortex of rats. Nutr Neurosci 6: 247-251.

Sarbassov DD, Ali SM, Kim DH, Guertin DA, Latek RR, et al. 2004. Rictor, a novel binding partner of mTOR, defines a rapamycin-insensitive and raptor-independent pathway that regulates the cytoskeleton. Curr Biol (Jul.) 14: 1296-1302. (First report of rapamycin-insensitive mTOR complex)

Schwartz GJ, Moran TH. 1998. Duodenal nutrient exposure elicits nutrient-specific gut motility and vagal afferent signals in rat. Am J Physiol 274: R1236-R1242.

Scott K. 2005. Taste recognition: Food for thought. Neuron 48: 455-464. (Review of taste mechanisms)

Sharp JW, Magrum LJ, Gietzen DW. 2002. Role of MAP kinase in signaling indispensable amino acid deficiency in the brain. Brain Res Mol Brain Res 105: 11-18.

Sharp JW, Ross-Inta CM, Hao S, Rudell JB, Gietzen DW. 2006. Co-localization of phosphorylated extracellular signal-related protein kinases 1/2 (ERK1/2) and phosphorylated eukaryotic initiation factor 2α (eIF2α) in response to a threonine-devoid diet. J Comp Neurol 494: 485-494. (Gives A-P dimension and location of APC cells that may be the primary sensory cells for IAA deficiency)

Sharp JW, Ross CM, Koehnle TJ, Gietzen DW. 2004. Phosphorylation of Ca^{2+}/calmodulin dependent protein kinase type II and the α-amino-3-hydroxy-5-methyl-4-isoxazole propionate (AMPA) receptor in response to a threonine-devoid diet. Neuroscience 126: 1053-1062.

Shenoy ST, Rogers QR. 1978. Effect of dietary amino acids on transfer ribonucleic acid charging levels in rat liver. J Nutr 108: 1412-1421.

Simpson SJ, Raubenheimer D. 1999. Assuaging nutritional complexity: A geometrical approach. Proc Nutr Soc 58: 779-789.

Simson PC, Booth DA. 1973. Effect of CS-US interval on the conditioning of odour preferences by amino acid loads. Physiol Behav 11: 801-808.

Smith DV, Boughter JD. 2007. Neurochemistry of the gustatory system, Chapter 5. Neurochemistry and Molecular Neurobiology of the Sensory Systems. D. Johnson, editor. Springer-Verlag: New York.

Smith GP, Jerome C, Cushin BJ, Eterno R, Simansky KJ. 1981. Abdominal vagotomy blocks the satiety effect of cholecystokinin in the rat. Science 213: 1036-1037.

Tabuchi E, Ono T, Nishijo H, Torii K. 1991. Amino acid and NaCl appetite, LHA neuron responses of lysine-deficient rat. Physiol Behav 49: 951-964.

Tanaka K, Inoue S, Nagase H, Takamura Y, Niijima A. 1990. Amino acid sensors sensitive to alanine and leucine exist in the hepato-portal system in the rat. J Auton Nerv Syst 31: 41-46.

Tang SJ, Reis G, Kang H, Gingras AC, Sonenberg N, et al. 2002. A rapamycin-sensitive signaling pathway contributes to long-term synaptic plasticity in the hippocampus. Proc Natl Acad Sci USA 99: 467-472.

Taylor-Burds CC, Westburg AM, Wilfall TC, Delay ER. 2004. Behavioral comparisons of the tastes of L-alanine and monosodium glutamate in rats. Chem Senses 29: 807-814.

Tomé D. 2004. Protein, amino acids and the control of food intake. Br J Nutr 92: S27-S30.

Torii K, Niijima A. 2001. Effect of lysine on afferent activity of the hepatic branch of the vagus nerve in normal and L-lysine-deficient rats. Physiol Behav 72: 685-690.

Torii K, Yokawa T, Tabuchi E, Hawkins RL, Mori M, et al. 1996. Recognition of deficient nutrient intake in the brain of rat with L-lysine deficiency monitored by functional magnetic resonance imaging, electrophysiologically and behaviorally. Amino Acids 10: 73-81.

Tsokas P, Grace EA, Chan PM, Ma T, Sealfon SC, et al. 2005. Local protein synthesis mediates a rapid increase in dendritic elongation factor 1A after induction of late long-term potentiation. J Neurosci 25: 5833-5843.

Uneyama H, Niijima A, San Gabriel A, Torii K. 2006. Luminal amino acid sensing in the rat gastric mucosa. Am J Physiol 291: G1163-1170. (This paper reports a glutamate sensing system in the stomach, acting via the serotonin 3 receptor on vagal afferents.)

Vinade L, Dosemeci A. 2000. Regulation of the phosphorylation state of the AMPA receptor GluR1 subunit in the postsynaptic density. Cell Mol Neurobiol 20: 451-463.

Walch-Liu P, Ivanov. II, Filleur S, Gan Y, Remans T, et al. 2006. Nitrogen regulation of root branching. Ann Botany 97: 875-881.

Wang Y, Cummings SL, Gietzen DW. 1996. Temporal-spatial pattern of c-Fos expression in the rat brain in response to amino acid deficiency I. The initial recognition phase. Mol Brain Res 40: 27-34.

Washburn BS, Jiang JC, Cummings SL, Dixon K, Gietzen DW. 1994. Anorectic responses to dietary amino acid imbalance: Effects of vagotomy and tropisetron. Am J Physiol 266: R1922-R1927. (Shows that the vagally-mediated responses to IAA deficiency occur too late to be involved in AA sensing)

Wek RC, Jiang HY, Anthony TG. 2006. Coping with stress: eIF2 kinases and translational control. Biochem Soc Trans 34: 7-11.

Wek SA, Zhu S, Wek RC. 1995. The histidyl-tRNA synthetase-related sequence in the eIF-2α protein kinase GCN2 interacts with tRNA and is required for activation in response to starvation for different amino acids. Mol Cell Biol 15: 4497-4506. (Demonstrates the mechanism of transduction of the uncharged tRNA signal by the eIF2α kinase, GCN2)

Westerterp-Plantenga MS. 2003. The significance of protein in food intake and body weight regulation. Curr Opin Clin Nutr Metab Care 6: 635-638.

White BD, Du F, Higgenbotham DA. 2003. Low dietary protein is associated with an increase in food intake and a decrease in the in vitro release of radiolabeled glutamate and GABA from the lateral hypothalamus. Nutr Neurosci 6: 361-367.

Wullschleger S, Loewith R, Hall MN. 2006. TOR signaling in growth and metabolism. Cell 124: 471-484. (Good review of TOR, including the recently discovered mammalian TOR complexes, mTORC1, and the rapamycin-insensitive mTORC2)

Xu H, Staszewski L, Tang H, Adler E, Zoller M, et al. 2004. Different functional roles of T1R subunits in the heteromeric taste receptors. Proc Natl Acad Sci USA 101: 13972-13973.

Yao D, Mackenzie B, Ming H, Varoqui H, Zhu H, et al. 2000. A novel system A isoform mediating Na^+/neutral amino acid cotransport. J Biol Chem 275: 22790-22797.

Yeshao W, Gu J, Peng X, Nairn AC, Nadler JL. 2005. Elevated glucose activates protein synthesis in cultured cardiac myocytes. Metabolism 54: 1453-1460. (Shows that mTOR can be activated by glucose)

Yox DP, Stokesberry H, Ritter RC. 1991. Vagotomy attenuates suppression of sham feeding induced by intestinal nutrients. Am J Physiol 260: R503-R508.

Zhao GQ, Zhang Y, Hoon MA, Chandrashekar J, Erlenbach I, et al. 2003. The receptors for mammalian sweet and umami taste. Cell 115: 255-266.

11 Oxygen Sensors of the Peripheral and Central Nervous Systems

G. M. C. Renshaw · M. Nikinmaa

Abstract: Neural systems exposed to diminished oxygen availability have a compromised metabolism that leads to pathophysiological changes or neuronal death, depending on the severity and duration of oxygen deprivation. A distributed network of oxygen sensors responds to protect cells by slowing or ameliorating pathophysiological changes and forestalling neuronal death via short-term or long-term changes involving gene expression and the modification of sensors and effectors. In mammalian systems such protective changes are not sufficient to prevent damage under extreme conditions, unlike some hypoxia- and anoxia-tolerant vertebrates which demonstrate oxygen-dependent, reversible reprogramming to protect vital organs such as the brain and heart.

This chapter examines (1) the nature of the signal for oxygen sensors; (2) the molecules used to sense oxygen; (3) how the primary signal is generated, converted, and used in an oxygen-dependent manner; (4) how effector systems function in different cell types; and (5) how oxygen-sensing pathways are interconnected to more general protective stress responses which confer cross-protection for a number of physiological stressors.

While future therapies may focus on the activation of hypoxia-inducible factor (HIF) and its downstream gene products, selected gene products could be administered to reduce neuronal loss and improve recovery after acute insults due to ischemic events and degenerative diseases of the brain and retina. Activation of neuroprotective pathways by oxygen sensors and other physiological stressors could be used as pretreatment to minimize neurotrauma associated with neurosurgical procedures and as an ancillary treatment during early stages of rehabilitation.

List of Abbreviations: HIF, hypoxia-inducible factor; HREs, hypoxia response elements; Hsps, heat-shock proteins; iNOS, inducible nitric oxide synthase; NE, noradrenaline; ODG, oxygen-dependent genes; ROS, reactive oxygen species; TH, tyrosine hydroxylase

1 Introduction

Diminished oxygen availability is a physiological stressor that compromises metabolism by slowing or halting aerobic ATP production and oxygen-dependent enzymatic reactions. In any neural system this can result in pathophysiological changes or neuronal death, depending on the severity and duration of oxygen deprivation. Oxygen-dependent responses mediated by oxygen sensors via their effectors can elicit neuroprotection in the face of significantly decreased oxygen levels, such that pathophysiological changes are ameliorated and neuronal death forestalled. Disruptions to oxygen availability in humans can range from a severe acute insult occurring as a result of heart attack, stroke, and traumatic brain injury to a chronic mild insult occurring as a result of sleep apnea and early stage cardiovascular disease. The prognosis for short-term survival of these events depends upon how well and how quickly the neural systems can be protected or how fast they can recover. There are clearly several functionally related questions that pertain to oxygen-dependent phenomena. (1) What is the nature of the signal sensed by oxygen sensors? (2) Which molecules and molecular pathways are used to sense oxygen? (3) How is the primary signal generated, converted, and used in an oxygen-dependent manner? (4) How do effector systems function in different cell types, particularly in neural systems? (5) How are oxygen-sensing pathways connected to more general protective stress responses which confer cross-protection for a number of physiological stressors?

For the purpose of this chapter, we define oxygen sensors as molecules and proteins that respond to changes in tissue oxygen levels with a direct change in their structure that results in changes in the tertiary structure and function of existing proteins and the upregulation of oxygen-dependent genes (ODG) often as a result of transcription factor function. Thus, we consider the whole sequence of events from primary oxygen-sensing to oxygen-dependent effects and their significance to equate to oxygen sensors. Consequently, we cover heme-containing proteins (globins and cytochromes), membrane-bound ion channels, mitochondria and their subtypes which appear to be multifunctional (sensing oxygen-dependent signals, generating signals to be sensed by other oxygen sensors and as an oxygen-dependent effector site), cellular systems, and specialized organs such as the carotid body. It is also apparent that oxygen sensors are located in a number of cellular compartments ranging from extracellular sensors (including erythrocytes as sensing cells) to cytoplasmic compartments (neuroglobin) and even intranuclear domains as exemplified by cytoglobin distribution in the mammalian brain. Also, evidence for interactions between oxygen-dependent and other stress pathways are discussed.

Since molecular oxygen is rarely sensed as a pure signal this review will discuss which additional signals can be sensed when oxygen level approaches a critical threshold, and how changes in the level of some molecules (metabolites) such as reactive oxygen species (ROS) (including hydroxyl radicals) and nitric oxide (NO) as well as changes in NADPH, ATP, pH, and general redox status can modulate the activity of the oxygen sensors and their downstream effectors. We will also examine the extent of cross-talk between oxygen-dependent pathways and pathways originating from other sensors detecting physiological status because it is difficult to separate them entirely; for example, energy-dependent pathways are profoundly altered when oxygen levels fall below a critical threshold.

There are two important points to note throughout. First, while single-cell organisms such as bacteria and yeast contain a single oxygen-sensitive molecular pair to regulate the expression of oxygen-sensitive genes (reviewed by Bunn and Poyton, 1996) no universal oxygen sensor has been identified in vertebrates (reviewed by Wenger, 2000; Lopez-Barneo et al., 2001; Lutz and Prentice, 2002; Cummins and Taylor, 2005; Lahiri et al., 2006). Instead, there appear to be several different primary oxygen sensors which are linked to a number of different effector systems that synergistically mediate oxygen-dependent responses in which neurotransmitters and modulators may or may not be involved.

Secondly most studies on oxygen sensors have used hypoxia as a stressor to elicit oxygen-dependent changes. Usually the physiologically effective degree of hypoxia has not been characterized in detail because the actual oxygen level experienced at the cellular level was not measured. The oxygen tension experienced by any cell in an organism depends on its location, especially its distance from an arterial blood vessel, and on its oxygen consumption. Consequently, there are marked differences of cellular oxygen tensions between cell types even in a normoxic body. For example, at normal ambient oxygen levels, arterial P_{O_2} is approximately 100 mmHg while the P_{O_2} in muscle interstitium in vivo has been measured at 3.3–24.2 mmHg and the P_{O_2} at the muscle mitochondria was 4–20 mmHg (Richmond et al., 1997). The mean cerebral P_{O_2} is higher than that of muscle, at around 20 mmHg and the P_{O_2} at the renal medulla is close to 50 mmHg (Johannes et al., 2006). These P_{O_2} values can be considered to reflect normoxia for the respective sites because this is the $[O_2]$ at which the animal displays its routine metabolic rate. Thus, it is clear that normoxia is very different depending on whether the frame of reference is a particular tissue or the mitochondria within that tissue (see also review by Lutz and Prentice, 2002). Furthermore, the oxygen tensions experienced by lung epithelial cells and hepatocytes are markedly different. Since vertebrates characteristically display a heterogeneous oxygen map, perhaps one should define hypoxia as an oxygen tension that is below the tension at which aerobic metabolism becomes limited, i.e., the critical P_{O_2}. In this paradigm, muscle hypoxia would be below 2.4–2.9 mmHg, its critical P_{O_2} (Richmond et al., 1997) and hypoxia for the mitochondria would be below its critical P_{O_2} of 1 mmHg (Rosenthal et al., 1976). Similarly, in vitro work on established cell lines should take into account the origin of the cell type used in the definition of hypoxia. In terms of oxygen-dependent responses, it is also important to differentiate between decreased oxygen level (hypoxia) and anoxia, i.e., total lack of oxygen (Wenger and Gassmann, 1996).

In defining hypoxia as the $[O_2]$ at which aerobic metabolism becomes limited, one also needs to consider how limited the aerobic metabolism needs to become, before oxygen sensors initiate signal cascades which result in the upregulation of oxygen-dependent genes and whether general physiological stress per se can turn on what are currently thought of as oxygen-dependent genes. We discuss the possibility that whenever anoxic conditions prevail, a response attributed to the involvement of oxygen sensors could be a more general stress response, which has little to do with changes in oxygen tension.

2 The Nature of the Signal Sensed by Oxygen Sensors

Among the systems that are usually classified as oxygen sensitive, only ones that include heme-containing molecule (Lopez-Barneo et al., 2001) or prolyl/asparagine hydroxylase use molecular oxygen as a substrate (Berra et al., 2006). In addition, the level of ROS may be the primary signal that is affecting the activity of oxygen-dependent systems. While earlier on it was thought that ROS are mainly conferring oxygen toxicity, recent findings indicate that at low levels they are important signaling molecules (Finkel, 1998; Weir et al., 2002; Werner, 2004; Wolin et al., 2005). The effects of ROS are often considered to be mediated via effects on conserved cysteine residues (Michiels et al., 2002). Especially hydrogen peroxide and hydroxyl radicals appear to be important in signaling (Gloire et al., 2006). Hydrogen peroxide may be important, since it is quite stable, membrane permeant (Lesser, 2006), and can affect the activity of tyrosine phosphatases by oxidizing cysteines in the catalytic center (Gloire et al., 2006). Hydroxyl radicals can also carry out the oxidation, and their use in cellular signaling would introduce spatial resolution in the system, because the short life time of the molecule (10^{-7} s) restricts the diffusion distance to 4–5 nm (Lesser, 2006). Notably, it is very difficult to separate hydrogen peroxide and hydroxyl radicals, since hydrogen peroxide is converted to hydroxyl radicals in the Fenton reaction, if adequate iron (or copper) ion stores are available (Bogdanova and Nikinmaa, 2001; Lesser, 2006). While NO has been recognized as an important signaling molecule, less emphasis has been paid to the fact that NO can affect the oxygen affinity of mitochondrial function (Koivisto et al., 1997). Also, NO can react with superoxide anion, and the formed peroxynitrite anion is a powerful membrane-permeant oxidant (Fridovich, 1986; Marla et al., 1997) with a lifetime near 0.1 s. Consequently, NO-dependent pathways may play a role in any effect that is considered oxygen sensitive. Two other gaseous molecules may play a role in oxygen-dependent signaling, namely carbon monoxide (CO) and hydrogen sulfide (H_2S). Heme oxidase enzyme, which has CO as one end product, is regulated by hypoxia (Lee et al., 1997). CO may control the neural discharge from rat carotid body (Lahiri and Acker, 1999). H_2S is involved in the oxygen-dependent regulation in vascular tone (Olson, 2005). Any marked decrease in oxygen availability leads to a decrease in cellular ATP concentration (Lutz and Nilsson, 2004), which appears to be the first step in induction of neural death (Lipton, 1999) and does not occur in the very anoxia-tolerant crucian carp (Lutz et al., 2003) or the tropical epaulette shark (Renshaw and Dyson, 1999). Thus, any mechanism detecting disturbances in the energy balance with oxygen depletion would be highly useful for maintaining cellular function in general and neural function in particular. From an energetic point of view the ADP/ATP ratio is the primary regulated function, but since the AMP/ATP ratio varies as a square of the ADP/ATP ratio, it enables a more sensitive regulation of the energy balance, with the end product that the energy-producing/consuming systems are adjusted (Hardie, 2003). AMP kinase senses changes in AMP and, ultimately, the AMP/ATP ratio, (Hardie, 2003; Hardie et al., 2006), which can thus function both in oxygen-and energy-dependent signaling.

3 Primary Oxygen Sensors

3.1 Heme-Based Molecules

Proteins containing a heme moiety and capable of binding molecular oxygen have been identified in diverse taxa from bacteria to vertebrates. Heme-based proteins sense oxygen by binding it reversibly and thereafter initiating a number of signaling cascades with several second messenger molecules, which can ultimately lead to altered gene expression via the activation of specific transcription factors (Wenger, 2000). The blockade of the oxygen-dependent responses by CO, which binds with very high affinity to the oxygen-binding site of many heme proteins, is taken as confirmation that heme proteins act as oxygen sensors (Zhu and Bunn, 1999). Similarly, if the ferrous iron in the heme group is replaced by cations, such as cobalt, which do not bind oxygen, the effect of such replacement on second messenger systems and transcription mimics that of exposure to hypoxia (reviewed by Wenger, 2000). Also, the suggestion that oxygen sensing is heme based is supported by the observation that treatments of cells with iron chelators such as desferroxamine results in a hypoxic response (Wang and Semenza, 1993; Ho and Bunn, 1996).

Simple heme-containing oxygen sensing molecules that respond to environmental changes are present in microbial symbionts (of plants), regulating genes associated with nitrogen fixation. The rhizobial FixL/FixJ system consists of a protein, histidine kinase, and its response regulator, FixJ, which act as an oxygen-sensitive switch. Oxygen binding to the heme moiety of FixL inactivates its kinase activity. This prevents the phosphorylation of FixJ and inhibits its downstream signal transduction pathway (Nakamura et al., 2004).

Thermodynamic studies have shown that both ferrous and ferric forms of the oxygen sensor FixL have a significantly lower oxygen affinity than myoglobin (Rodgers and Lukat-Rodgers, 2005). The evolution of high affinity extracellular and intracellular oxygen sensors in higher organisms may have facilitated formation of more complex body plans which include a nervous system and the occupation of a wider variety of niches.

3.2 The Globin Family of Heme Proteins

Circulating oxygen carriers, hemerythrin, hemocyanin, and hemoglobin, are used by many invertebrates to facilitate oxygen transport from environment to tissues. Members of the globin family serve this function in vertebrates. Oxygen causes a direct change in the structure of these proteins and, thus, they can be considered oxygen sensors. The globins bind oxygen to a Fe-containing porphyrin ring. The binding is often cooperative, if the molecule containing globin is composed of subunits. Of the different globin molecules, myoglobin, cytoglobin, and neuroglobin have been detected in the brain and other neural systems, and thus may have different neuroprotective functions.

3.2.1 Myoglobin

Myoglobin is not only found in muscle, but its isoforms are also expressed in other tissues which have a high metabolic rate and correspondingly high oxygen demand such as liver, gill, and brain (Fraser et al., 2006). Interestingly, expression of the unique myoglobin isoform in neural tissue did not change in response to environmental hypoxia (Fraser et al., 2006) unlike the hypoxia-sensitive upregulation of myoglobin in the heart (Roesner et al., 2006). It has been suggested that constitutive levels of myoglobin isoforms may have other functions in nonmuscle tissues. For example, myoglobin may act as a cytoprotective agent to forestall injury during hypoxia/ischemia and reoxygenation/reperfusion by scavenging free radicals (Mammen et al., 2006). Importantly, myoglobin also plays a role in intracellular oxygen diffusion from the cell surface to mitochondria (Wittenberg and Wittenberg, 2003; Roesner et al., 2006).

3.2.2 Cytoglobin

Cytoglobin has been localized to the nucleus in the cells of several tissues (Geuens et al., 2003). The cytoglobin gene contains both hypoxia response elements (HREs) and mRNA stabilization sites characteristic of an oxygen-regulated gene. Furthermore, real-time quantitative PCR has confirmed that cytoglobin is regulated by hypoxia-inducible factor (HIF)-1α (Fordel et al., 2004).

While cytoglobin is upregulated in response to hypoxia in hippocampal cells in vitro (Fordel et al., 2004), it is not uniformly upregulated throughout the brain in vivo and it is expressed in different brain regions from those in which neuroglobin is found (Mammen et al., 2006). There is no increased expression of cytoglobin mRNA or protein in the neocortex after either chronic or intermittent hypoxia (Li et al., 2006). The brain regions expressing significantly elevated levels of cytoglobin in response to hypoxia are the areas of the archicortex that are sensitive to hypoxia and oxidative stress, namely the hippocampus, thalamus, and hypothalamus. This provides evidence that cytoglobin is an oxygen-responsive globin in vivo (Mammen et al., 2006). Cytoglobin is strongly expressed in the developing mammalian central nervous system (CNS) and it has been suggested that its localization in brain areas sensitive to oxidative stress could be related to a myoglobin-like role in scavenging free radicals associated with the metabolism of oxygen and nitrogen (Mammen et al., 2006).

3.2.3 Neuroglobin

Another recently discovered and characterized intracellular globin, neuroglobin, is expressed in the mammalian central and peripheral nervous systems and has recently been detected in cultured astrocytes from newborn mouse brain (Chen et al., 2005). In ischemic astrocytes, apoptosis increased when cultured cortical astrocytes were treated with neuroglobin antisense (Chen et al., 2005). It has been suggested that neuroglobin is not only associated with areas of high metabolic rate (as indicated by their elevated mitochondrial density) in the brain but also in other neuronal compartments (Hankeln et al., 2005). Neuroglobin mRNA and protein can be detected in neuronal perikarya, axons, and synapses (Hankeln et al., 2005). Neuroglobin makes up less than 0.01% of the protein in the brain (Mammen et al., 2002).

Mammen and coworkers (2002) suggested that neuroglobin expression is correlated with regions of the CNS involved in adaptive stress response pathways. Neuroglobin is found in brain areas that have high levels of nitric oxide (Mammen et al., 2002). Since one of the cytoprotective actions of myoglobin is to detoxify nitric oxide in the heart, it has been suggested that neuroglobin may act similarly in the brain (Mammen et al., 2002; Burmester and Hankeln, 2004). While neuroglobin is strongly expressed in the subthalamic nucleus, only low levels are found in the hypoxia-sensitive cerebellum and hippocampus (Pesce et al., 2004). Data from radiolabeled RNA probes and in situ hybridization suggest that neuroglobin expression is related to oxygen-dependent functions. There is a constitutive level of neuroglobin present in brain areas that respond to changes in oxygen levels, for example: the locus coeruleus, the parabranchial complex, and the periaqueductal grey (reviewed by Mammen et al., 2002). The highest levels of neuroglobin have been reported in the mitochondria-rich photoreceptors of the retina (Hankeln et al., 2005).

Neuroglobin has an oxygen affinity characterized by a P_{50} value of 2 torr, which is twofold higher than that of myoglobin but much lower than that of most hemoglobins. Neuroglobin has been implicated in the storage and intracellular transport of oxygen in highly metabolically active tissues (Burmester et al., 2000; Couture et al., 2001; Trent et al., 2001) and in facilitating O_2 diffusion into mitochondria (Burmester et al., 2000). It seems likely that it could function as an intracellular oxygen sensor as well as participate in the metabolism of reactive species such as NO and ROS. Burmester and Hankeln (2004) suggested that neuroglobin could be involved in the destruction of ROS especially in hypoxic conditions or resulting from oxidative stress that follows hypoxia and reperfusion. However, this suggestion appears not to hold for retina, since there was not a significant increase in neuroglobin mRNA and protein in zebrafish (*Danio rerio*) retina following hypoxia, even though the brain neuroglobin level increased fivefold (Roesner et al., 2006). The lack of neuroglobin upregulation in the retina following hypoxia may be species specific since some hypoxia-tolerant animals such as crucian carp can turn off visual processing at low oxygen levels (Johansson et al., 1997). It has also been suggested that neuroglobin could function as a terminal oxidase to regenerate NAD^+ (Milton et al., 2006), and thereby maintain ATP production when oxygen levels are diminished. Neuroglobin has been shown to have neuroprotective properties demonstrated by reduced neuronal damage following stroke (Sun et al., 2003). Furthermore, it is upregulated in the brain of the anoxia- and hypoxia-tolerant turtle, and it has been suggested that it mediates neuronal survival under anoxia (Milton et al., 2006). This protective effect was diminished by the inhibition of neuroglobin expression with an antisense oligodeoxynucleotide, and enhanced by neuroglobin overexpression (Sun et al., 2003). Fago and coworkers. (2006) suggest that since neuroglobin reacts with ferric cytochrome *c* with rapid kinetics it may well have a role in preventing apoptosis following periods of neuronal stress that cause a surge in the release of cytochrome *c* from mitochondria.

Neuroglobin induction may also be a part of the stress response initiated by heme-based transcription factors or their second messengers, because a transient increase in neuroglobin level was observed in in vitro cultures of cortical neurons after they were exposed to priming conditions that usually result in HIF-1α upregulation such as severely diminished oxygen levels or the addition of cobalt chloride or the iron chelator, desferroxamine (Sun et al., 2003). However, neuroglobin may be regulated by response elements other than HREs (see Hankeln et al., 2005 for a review). Neuroglobin mRNA can be induced directly by hemin in a time- and concentration-dependent manner via a nonhypoxia-dependent second signal transduction pathway that can be blocked by the protein kinase G inhibitor KT5823 (Zhu et al., 2002). This suggests that neuroglobin may be a multifunctional protein that is upregulated by hypoxia and downregulated by protein kinase G. The upregulation of neuroglobin by hemin may represent a HIF-1α-independent pathway and further research is needed to clarify the interaction of HIF with hemin-induced neuroglobin upregulation. Furthermore, Zhu and coworkers (2002) demonstrated that the hypoxic induction of neuroglobin could be prevented by the mitogen-activated protein kinase inhibitor, PD98059, revealing that neuroglobin expression is regulated by more than one signal transduction pathway. Notably, mitogen-activated protein kinases are involved in redox-dependent signaling.

3.3 Cytochromes

The involvement of cytochromes in oxygen sensing has been indicated in many studies (Duranteau et al., 1998; Ehleben et al., 1998; Porwol et al., 2001; Guzy et al., 2005; Guzy and Schumacker, 2006). The cytochromes involved may be mitochondrial (Guzy et al., 2005, 2006) and nonmitochondrial, e.g., cytochrome aa_3 (Porwol et al., 2001), in origin. In both cases, it appears that ROS are the actual transducing molecules for the cytochrome signal. It also appears that CO may significantly regulate the oxygen-dependent cytochrome function (Porwol et al., 2001).

3.4 NADPH Oxidase

Some of the cytochromes suggested to take part in oxygen sensing are parts of the NADPH oxidase enzyme. NADPH oxidase is a heterodimeric flavocytochrome of g22-phox and gp91-phox, which is capable of recruiting the polypeptides p67-phox, p47-phox, and p40-phox to form a membrane bound, multi-subunit structure (Dahan et al., 2002). Recent findings indicate that there are several isoforms of NADPH oxidase in many cell types and that components of some isoforms can act as putative oxygen sensors (reviewed by Acker, 2005). The signaling is mediated via oxygen-regulated ROS formation. NADPH oxidase inhibitors blocked the response of carotid chemoreceptor discharge to hypoxia (Cross et al., 1990) and blocked K^+ and Ca^{2+} currents in type 1 cells of the carotid body (Wyatt et al., 1995). Similarly, mRNA for NADPH oxidase and voltage-gated K^+ channels have been colocalized to hypoxia-responsive pulmonary neuroepithelial bodies and it has been suggested that this tight association represents an oxidase-linked K^+ channel, which acts as an oxygen sensor (Wang et al., 1996). Furthermore, gene knockout studies have shown that gp91-phox null mice had a significantly impaired hypoxic ventilatory response because of the decreased sensitivity of pulmonary neuroepithelial bodies (Kazemian et al., 2001). However, the gp91-phox knockouts show no impairment of carotid body function (Roy et al., 2000) or pulmonary vasoconstriction (Archer et al., 2000) indicating that this single polypeptide is not pivotal in all effector tissues and it is likely that effectors may contain multiple oxygen sensors. If one sensor malfunctions then compensatory changes can be made by other oxygen responsive sensors. Since NADPH oxidase appears to be involved in oxygen-sensitive responses both in plants and in animals, it may be a widespread evolutionarily conserved oxygen sensor.

3.5 Prolyl and Asparagine Hydroxylases

Transcriptional regulation by oxygen is mainly achieved via the function of HIF. To a large extent, its function is regulated either by affecting the stability of the protein by hydroxylation of conserved prolines (proline 402 and 564 in the human protein) and consecutive degradation of the molecule or by affecting the interaction of the molecule with p300 and consecutive DNA binding as a result of hydroxylation of a conserved asparagine residue (Asp803). However, it now appears that in hypoxia-tolerant animals the transcription of HIF may also play a role in the regulation of the HIF pathway (see Shams et al., 2004; Law et al., 2006; Rissanen et al., 2006b). Both prolyl and asparagine hydroxylases use molecular oxygen as a substrate. Their function is oxygen dependent, and thus they function as oxygen sensors affecting hypoxia-inducible gene expression as first demonstrated by Ivan and coworkers (2001) and Jaakkola and coworkers

(2001). The function of prolyl hydroxylases has recently been reviewed by Fandrey et al. (2006). Three types of oxygen-dependent proline hydroxylases (PHD1–3) have been described; the presence of a fourth (PHD4) has been deduced on the basis of genomic information (Oehme et al., 2002). While hydroxylation of conserved prolines (in places of the protein which have the LXXLAP sequence) is achieved by the prolyl hydroxylases, it appears that several other residues are important for the proper functioning of the PHDs (Fandrey et al., 2006). This indicates that in addition to the properties of the enzymes themselves, the three-dimensional structure of HIF-1α affects the hydroxylation. Notably, hydrophobicity plots of various vertebrate HIF-1αs show that the conserved proline residues are in a highly hydrophobic environment, suggesting that the residues are folded within the protein (unpublished results, Rytkönen K, Vuori KAM, Primmer CR and Nikinmaa M., 2007). Thus it is possible that the effect of ROS and calcium on HIF-1α function is mediated by their affect on the three dimensional structure of the protein near the conserved prolines and their consequent accessibility to prolyl hydroxylases. This mechanism would add another way for regulating HIF-1α function by oxygen, since ROS levels can be oxygen dependent. It would also explain, why HIF-1 function has been shown to be ROS dependent in several studies (reviewed by Haddad, 2002; Kietzmann and Gorlach, 2005; Acker et al., 2006), although the enzymatic hydroxylation and consecutive proteasomal breakdown of the protein do not require ROS (Fandrey et al., 2006).

Oxygen-dependent regulation of the DNA binding of HIF is achieved via the function of asparagine hydroxylase (FIH; factor inhibiting hypoxia-inducible factor) (Kaelin, 2005). Because the prolyl and asparagine hydroxylases have different oxygen affinities, it is possible that different genes are regulated by the two enzymes (Dayan et al., 2006). Also, it is possible that the two enzymes regulate HIF function at different oxygen levels.

3.6 AMP Kinase

The function of AMP-activated protein kinase has been reviewed recently by Hardie (2003). The enzyme is composed of three subunits: the catalytic α subunit and the regulatory β and γ subunits. While the system is activated by AMP, it remains inactive even in the presence of this allosteric effector, if not phosphorylated at a critical threonine residue (Hardie, 2003). Because the enzyme is activated by decreasing energy charge (as occurs in hypoxia and anoxia), it is activated by any form of stress that affects energy production/consumption. Thus, much research on AMPK has been directed toward glucose/glycogen/diabetes and fat metabolism studies (Kim et al., 2005; Yun et al., 2005). AMPK function is also affected by ROS (Choi et al., 2001) and NO (Lei et al., 2005), adding to the potential interactions between different regulatory pathways. Since AMPK is involved in regulating cellular energy balance, its activation switches off energy-consuming and switches on energy-producing pathways. One of the major oxygen-consuming processes in cells involves mRNA translation to proteins. Notably, it is inhibited by AMPK in hypoxia also independently from HIF regulation (Liu et al., 2006), showing the importance of energy sensing in hypoxia regulation.

4 Effector Systems

There are basically two types of effector systems involved in oxygen-dependent phenomena: those that exert their oxygen-dependent effect immediately and those that mediate slower oxygen-dependent changes, e.g., at the transcriptional level. The former are often coupled to ion channels that immediately alter neural functions to exert acute effects, for example, ion channels regulate the responsiveness of respiratory neurons involved in peripheral oxygen sensing associated with ventilatory regulation. On the other hand, slower, chronic effects can regulate neural function in such a way that either anaerobic energy production is facilitated or energy is spared usually via oxygen-dependent gene regulation. Often such gene regulation involves HIF. Notably, oxygen-dependent effects may rely on signals in the form of ROS and energy metabolites in addition to molecular oxygen, so that interactions between oxygen, oxidative stresses, and energy metabolism may occur both in acute and chronic oxygen responses.

4.1 Oxygen-Dependent Ion Transport Systems

4.1.1 Exchangers and Cotransporters

Oxygen-dependent cotransporters and exchangers have been studied in most detail in erythrocytes (Gibson et al., 2000) with only sporadic information on other cell types (Tuominen et al., 2003). In erythrocytes, it appears that a major factor regulating oxygen-dependent ion transport is the hydroxyl radical, which has been suggested to increase the activity of KCl cotransport (Bogdanova et al., 2003) and decrease the activity of Na^+/H^+ exchange (Nikinmaa et al., 2003). While data in mammalian and bird erythrocytes have generally been compatible with hemoglobin being the proximal oxygen sensor (Honess et al., 1996; Muzyamba et al., 1999; Muzyamba et al., 2000; Drew et al., 2004; Flatman, 2005), work on hypoxia-intolerant teleost fish species, such as rainbow trout, indicated that in this species bulk hemoglobin could not be the oxygen sensor (Berenbrink et al., 2000). Interestingly, studies on the erythrocytes of the hypoxia-tolerant crucian carp indicate the presence of two different oxygen sensors, one of which has an oxygen affinity similar to (bulk) hemoglobin, and the other has a much lower affinity (Berenbrink et al., 2006). Thus, the data are compatible with minimally two different oxygen sensors of which one may respond to molecular oxygen as the sensed molecule and the other may respond to oxygen radicals (Bogdanova et al., 2001; Berenbrink et al., 2006).

4.1.2 Ion Channels

Most work on the oxygen-dependent ion channels has been done on excitable cells. Some voltage-gated ion channels, such as K^+, Ca^{2+}, and Na^+ channels, have conductances that are regulated by oxygen-dependent factors with the result that the excitability of a cell is oxygen regulated. Fast adaptive changes that compensate for diminished O_2 levels are mediated by oxygen-responsive ion channels and are manifested as changes in excitability, secretion or contractility depending on the cell type (Lopez-Barneo et al., 2001). Cysteine-rich voltage gated ion channels respond to changes in reduced/oxidized redox pairs such as $NADH/NAD^+$ so that they close when oxygen levels drop, favoring the formation of more NAD^+ (Archer, 2000).

A decrease below the threshold P_{O_2}, normally close to 50 Torr, in glomus cells of the carotid body or in the neonatal ductus arteriosus results in an inhibition of the tonic K^+ current. Such oxygen-regulated inhibition of K^+ channels, which may be mediated by mitochondria-derived hydrogen peroxide (Archer et al., 2004), results in an increase in cellular excitability, increased Ca^{2+} influx, and a resultant increase in the level of Ca^{2+} in the cytosol (reviewed by Lopez-Barneo et al., 1999).

Neuroepithelial bodies found in mammalian airways have oxygen-sensitive K^+ currents and are putative airway chemoreceptors that detect changes in the level of O_2 in the airway lumen (Wang et al., 1996). Exposure to acute hypoxia in vivo (Lauweryns et al., 1987) or in vitro (Fu et al., 2002) results in serotonin release, which may have a role in controlling pulmonary vascular tone. The oxygen sensitivity of these channels may result from their close localization to membrane-bound cytochromes (Youngson et al., 1997). It is clear that the closure of voltage-gated K^+ channels by hypoxic exposure has implications for neuroprotection as demonstrated by K^+ channel arrest in hypoxia-tolerant turtles (Pek and Lutz, 1997; Bickler and Buck, 1998; Hochachka and Lutz, 2001; Bickler and Donohoe, 2002), which reduces Ca^{2+} influx (Bickler and Buck, 1998). In some anoxia-tolerant species, neuronal energy is not only conserved by ion channel arrest but also by ATP-sensitive mitochondrial K^+ channel arrest (reviewed by Buck and Pamenter, 2006). This may have implications for clinical interventions.

In the carotid body, hypoxia causes increased Ca^{2+} influx through voltage-gated K^+ channels. This results in an increased afferent input to the brain stem and consequent neurosecretion of catecholamines. Conversely, in the ductus arteriosus a rise in P_{O_2} triggers increased H_2O_2 production. This inhibits voltage-gated K^+ channels and leads in turn to vasoconstriction and the closure of the ductus arteriosus (Archer et al., 2004). Calcium entry also occurs through L-type Ca^{2+} channels involving PKC and/or phosphatase-sensitive pathways (Summers et al., 2000). It is this Ca^{2+} entry through voltage-gated Ca^{2+} channels that mediates hypoxia-responsive neurosecretion by the carotid body. The hypoxic response in the carotid body can be blocked by Ca^{2+} channel blockers (reviewed by Lopez-Barneo et al., 2001). Chronic exposure to hypoxia upregulates T-type voltage-gated Ca^{2+} channels as one of the actions of hypoxia-inducible transcription factors (Del Toro et al., 2003).

Neurons in the caudal hypothalamus appear to be oxygen responsive. They alter their firing rates in hypoxia to contribute to the control of cardiorespiratory responses to hypoxia (Horn and Waldrop, 1997). Oxygen-sensitive Na^+ channels appear to regulate neuronal excitability in the caudal hypothalamus, since, as oxygen levels drop below threshold, the Na^+ current is inhibited by a PKC-dependent mechanism (O'Reilly et al., 1997). Prolonged hypoxia can result in Na^+ channel excitability (Xia et al., 2000), depending on the frequency, intensity, and duration of the hypoxic episodes (Zhao et al., 2005). Furthermore, a subset of voltage-gated Na^+ channels, which are tetrodotoxin sensitive, are not inhibited by hypoxia and generate "persistent" sodium currents, which could cause irreversible neuronal damage as Na^+ homeostasis is lost (Hammarstrom and Gage, 2002). Under experimental conditions the "persistent" sodium current can be inhibited by reducing agents such as dithiothreitol and reduced glutathione (Hammarstrom and Gage, 2002) suggesting that the current is, in fact, redox sensitive.

4.2 Hypoxia-Inducible Factor-Dependent Pathway

In the 1990s, oxygen-dependent gene expression was conclusively demonstrated. The early studies concentrated on the erythropoietin pathway (Semenza and Wang, 1992), but later studies have indicated that oxygen directly affects the expression of many, maybe a hundred, genes (Lahiri et al., 2006). The master regulator of oxygen-dependent gene expression is the hypoxia-inducible transcription factor (HIF), which regulates transcription in the following fashion (Wenger, 2000; Bracken et al., 2003) (see ➋ *Figure 11-1*). The active factor is a nuclear dimer of HIFα and ARNT (HIFβ). While both ARNT and HIFα are continually produced at all oxygen tensions, HIFα is rapidly broken down in normoxia. Consequently, the short half-life of HIFα confers hypoxia sensitivity to the function of the protein dimer. As discussed above, HIFα degradation involves prolyl hydroxylases (Ivan et al., 2001; Jaakkola et al., 2001) that bind at conserved proline residues in the oxygen-dependent degradation domain. It appears that the actual core of the oxygen-dependent degradation domain (including proline564 and proline402 (human nomenclature) which undergo oxygen-dependent hydroxylation by prolyl hydroxylases (Semenza, 2001) is invariable in HIF-1αs across vertebrates (Rytkönen K, Vuori K.A.M., Primmer C.R. and Nikinmaa M.; unpublished observations). The regulation of oxygen-dependent degradation may, however, also requires residues in the vicinity of ODD, which affect the tertiary structure of ODD core. Prolyl hydroxylation enables the interaction of HIFα and the von Hippel–Lindau protein, subsequent ubiquitylation, and proteasomal degradation. Recent results suggest that although prolyl hydroxylation does not require ROS, it can be under redox control by hydroxyl radicals (Liu et al., 2004). In hypoxia, prolyl hydroxylation does not occur, whereby HIFα protein is stable and is transported from cytoplasm to nucleus, where it forms a dimer with ARNT and recruits the general transcriptional activator, CBP/p300. Recruitment of transcriptional activators depends on hydroxylation reactions of conserved asparagine (Mahon et al., 2001). Thereafter, HIF (HIFα+ARNT) binds to HREs present in the promoter/enhancer region of the hypoxia-inducible genes, and gene transcription is stimulated. Since both the stability of HIF and its transcriptional activity are affected by oxygen-dependent enzymes, the oxygen affinity of the gene expression depends on the oxygen affinities of both enzymes. Oxygen availability may further affect the three dimensional structure of HIF protein, since both the DNA binding and the transcriptional activation by HIF appear to be under redox control (Lando et al., 2000; Bracken et al., 2003): serine-to-cysteine mutation at a specific residue in the DNA-binding domain confers redox sensitivity of DNA binding, and nuclear redox regulation by Ref-1 potentiates the hypoxic induction of a reporter gene (Lando et al., 2000). At present, while the principles of how HIF regulates gene expression are clear, the fact that oxygen can affect HIF function at several different places and the lack of measurements means that there is currently no decisive information on the possible differences of oxygen affinities for oxygen-dependent gene expression between tissues and species. Also, the

Figure 11-1
A representation of hypoxia-inducible factor (HIF) function. In normoxia, HIF-1α is broken down after prolyl hydroxylation and its interaction with p300 is diminished after aspargyl hydroxylation. In hypoxia, hydroxylase enzymes are inhibited, whereby the DNA binding of HIF may occur and oxygen-sensitive genes can be induced

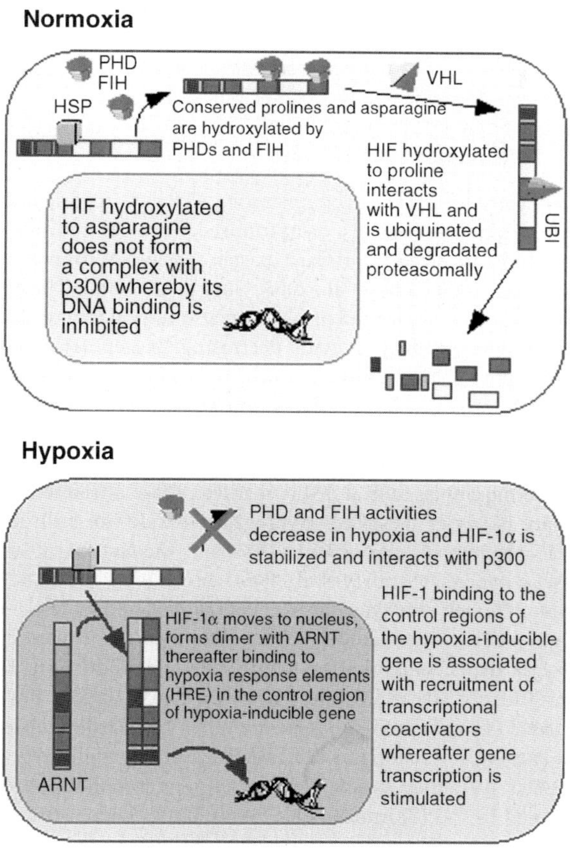

HIF pathway can be stimulated in normoxic conditions by various growth factors, hormones, and cytokines (Richard et al., 2000; Page et al., 2002; Kodama et al., 2003; Ma et al., 2004). In most cases the basis of the stimulation is not known.

4.3 Hypoxia Response Element

In addition to the properties of HIF itself, the HREs especially in the promoter/enhancer regions of the transcribed gene, affect gene expression. HREs may also be present in the introns of oxygen-dependent genes (Rees et al., 2001). The minimal consensus HRE is A/GCGTG (Camenisch et al., 2002). In some cases the presence of HREs alone is not sufficient for hypoxic induction of the genes (Firth et al., 1995), but additional elements such as binding sites for AP1, ATF1/CREB1, HNF4, or Smad3 may be required (Bracken et al., 2003).

4.4 Redox-Responsive Transcription Factors

While HIF has been studied most with regard to hypoxia-inducible gene expression, several other transcription factors also appear to be hypoxia sensitive (Cummins and Taylor, 2005). These include NF-κB, which is a redox-sensitive transcription factor (Fan et al., 2003; Fratelli et al., 2005) that consequently responds to variations in ROS, the level of which is dependent on oxygen level (Bogdanova et al., 2003). Observations suggest that while HIF induces gene expression especially in hypoxia, NF-κB exerts its major influence in hyperoxia (Michiels et al., 2002). Other redox-sensitive transcription factors include PPARγ, Nrf2, AP-1, STAT, and p53 (Kim and Surh, 2006). Many of these show interaction with HIF (Pan et al., 2004) and can be affected by prostaglandins (Kim and Surh, 2006).

4.5 Mitochondrial Function as an Oxygen-Sensitive Effector System

Mitochondria provide the fuel for life processes by generating ATP but can also provide the signals for death via apoptosis if their membrane potential is compromised. Also, as indicated above, mitochondrial cytochromes are suggested to be involved in primary oxygen sensing (Wilson et al., 1994; Zhu and Bunn, 1999; Chandel and Schumacker, 2000; Guzy et al., 2005; Guzy et al., 2006). The oxygen sensor function of mitochondrial cytochromes results in their ROS production also responding to changes in oxygen. Both an increase and a decrease in ROS production with decreasing oxygen tension are feasible. When the mitochondrial electron transport chain is slowed down or arrested, there is a decreased generation of ROS coupled with an increase in reducing equivalents, so either or both could act as signals generated by mitochondria in response to low oxygen. An increase in ROS production in hypoxia can occur if oxygen affects the lifetime of the ubisemiquinone radical in complex III in the mitochondrial inner membrane and its ability to access the ubisemiquinone radical or the relative release of mitochondrial ROS to the matrix versus intermembrane space (Guzy et al., 2006). Hydrogen peroxide can both mimic hypoxia (Canbolat et al., 1998) and counteract the hypoxia responses as shown by the fact that it inhibited the induction of erythropoietin in response to hypoxia (Fandrey et al., 1994) and that the blockade of Epo expression could be reversed by the addition of cobalt chloride and iron chelation (Fandrey et al., 1997). It is possible that mitochondria from different cells (and conditions) differ from each other (Michelakis et al., 2002).

With regard to the oxygen dependence of mitochondrial energy production, isolated mitochondria are capable of producing energy aerobically at much lower oxygen concentrations in vitro than they do in vivo resulting in energy levels lethal to cells in vivo (Gnaiger et al., 1998; Krumschnabel et al., 2000). In intact cells, the (mitochondrial) energy production was decreased at relatively high oxygen tensions (more than 30 mmHg) (Rissanen et al., 2006a). Notably, the oxygen affinity of mitochondrial function is affected both by NO (Koivisto et al., 1997) and by CO (D'Amico et al., 2006), both of which are molecules associated with oxygen sensing. In addition to mitochondrial energy production, apoptosis (programmed cell death) is also oxygen sensitive and affected by mitochondrial function (Araya et al., 1998; Banasiak et al., 2000). The hypoxic induction of apoptosis often involves release of cytochrome *c* from mitochondria, which initiates consequent caspase activation (Chae et al., 2001). It appears that ROS are involved in the hypoxic apoptosis signaling (Kim and Park, 2003). Also, the dependency of apoptosis on oxygen levels shows interaction with NO, CO, and glucose, indicating interdependence of different cellular effector pathways (Madesh et al., 1999; Malhotra and Brosius, 1999; Tofighi et al., 2006).

4.6 Interactions Between Oxygen-Dependent and Other Effector Pathways

Considering the interconnected nature of biochemical pathways, it is not surprising that antagonistic and synergistic effects of other effector pathways on oxygen sensors have been reported. In many cases, a number of stressors act simultaneously so the integrated response may differ in direction or intensity from that caused by a single stressor, either diminishing the oxygen-sensitive response or enhancing it. Furthermore, the effect of cross-protection by stressors needs to be taken into account when considering physiological responses to a specified stressor because oxygen sensors may be affected by more than one type of stress. For example, physiological stress caused by exercise can increase the generation of both ROS and NO, which are key signaling molecules in the oxygen-dependent pathways as discussed above. It has recently been demonstrated that exercise at physiologically relevant levels could result in significant elevations in HIF-1α protein levels coupled with decreased protein levels of the von Hippel–Lindau tumour suppressor, which is responsible for its turnover (Ameln et al., 2005). So it is conceivable that the neurogenerative and neuroprotective effects of exercise may be mediated by activation of the oxygen-sensing pathway.

Both NO and CO may affect (and be involved in) oxygen-dependent phenomena. Accordingly, in vitro experiments have demonstrated that elevated levels of NO or CO can diminish effects of hypoxia on gene transcription. These molecules both activated the internal oxygen-dependent degradation domain and repressed the C-terminal transactivation domain of HIF resulting in less HIF-1α that can bind to DNA (Huang et al., 1999). On the other hand, in vivo experiments have demonstrated that NO is one of the modulators of the hypoxic ventilatory response mediated by the rostral mediolateral medulla (de Paula and Branco, 2003). It is not clear at present, whether different cell types respond in a different manner.

Activated microglia in the CNS express high levels of inducible nitric oxide synthase (iNOS) and occur in high numbers in brain ageing and inflammatory pathologies such as stroke and neurodegenerative diseases (Mander et al., 2005). In addition, exposure to hypoxia and oxidative stress can further increase the number of activated microglia in vivo. Cerebellar granule cells in culture were more susceptible to hypoxia-induced neuronal death when there were increased levels of NO present (Mander et al., 2005). Thus, NO may potentiate the deleterious effects of hypoxia by blocking the action of HIF-1α on the induction of neuroprotective genes via the diminished levels available to bind to DNA, as discussed above (Huang et al., 1999).

The brain, approximately 2% of the body mass, has a high metabolic demand which accounts for approximately 50% of total body glucose utilization. Cross-talk between diminished oxygen and glucose levels occurs for at least two reasons: first, glucose sensing is activated by a signaling pathway that is common with oxygen sensors—the ROS system. The neurons of the arcuate nucleus that modulate insulin release show increased activity in response to glucose or in response to increased ROS as a result of treatment with mitochondrial complex blockers such as rotenone and antimycin, which can be reversed by treatment with antioxidants (Leloup et al., 2006). Second, glucose sensing occurs in some cells which also sense oxygen levels, for example, in the carotid body. While peripheral glucose levels are monitored and responded to by the liver, low glucose levels also inhibit voltage-gated K^+ channels in the glomus cells, as occurs when oxygen levels fall below a threshold (Pardal and Lopez-Barneo, 2002). Since glomus cells respond to both low glucose and low oxygen, it has been suggested that this strategically placed multifunctional sensor regulates oxygen and glucose homeostasis to protect the brain (Pardal and Lopez-Barneo, 2002). Also, interaction between two transcription factors, HIF and glucose-dependent transcription factor, has been described in which the respective response elements in DNA can function both as glucose response elements and HREs (Kietzmann et al., 2002).

The interaction between oxygen-dependent pathways and more general stress-sensing pathways can be seen at an intracellular level, for example in the upregulation of molecular chaperones such as heat-shock proteins (Hsps), which assist in refolding damaged proteins (Burston and Clarke, 1995; Lund, 1995) and maintaining the tertiary structure of proteins during metabolic stress (Gonzalez et al., 1991). Hsps are upregulated from their constitutive level in response to a diverse array of physiological stressors (Basu et al., 2002). These stressors include exposure to psychoactive drugs (Miller et al., 1991), neurodegenerative disease (Harrison et al., 1993), cellular injury (Liang and MacRae, 1997), hypoxia, ischemia (Welch, 1992; Locke and Noble, 1995), and acute temperature change (Airaksinen et al., 1998). In fact, many studies have shown the accumulation of Hsps in hypoxia/anoxia (Patel et al., 1995; Hammerer-Lercher et al., 2001). In many of the studies that show hypoxic/anoxic accumulation of Hsps, it is difficult to say if the response is a general response to stressful conditions or specific to oxygen limitation. The theory of parsimony predicts that a broad range of triggers probably converge on a single molecular target, which once activated serves to upregulate Hsp production and while this remains to be fully tested, there is evidence that induction of Hsp70 is linked to both an oxygen sensor and/or an energy sensor since Hsp70 promoter activation responds to low oxygen (Madamanchi et al., 2001) as well as to decreased cellular energy charge (Kiang and Tsokos, 1998). Furthermore, energy sensors and oxygen sensors appear to act synergistically to regulate Hsp70 levels and it was suggested that a metabolic sensor may be involved in further upregulating the level of Hsp70 above the level that could be induced by anoxia alone (Renshaw et al., 2004). Similarly, cross-protection occurs: when animals are exposed to a sublethal physiological stressor, they are protected from a subsequent stressor of the same or a different modality. In the mammalian heart, heat treatment activates HIF-1α and its target genes including EPO, which may then be responsible for the observed cross-protection to infarction after ischemia reperfusion (Maloyan et al., 2005). In the brain, exposure to brief ischemia made the hypoxia-sensitive CA1 neurons of the hippocampus tolerant to levels of ischemia that were normally lethal (Kitagawa et al., 1990; Kirino, 2002). In addition, nonischemic insults, which resulted in the induction of Hsps in the brain and retina, have been associated with resistance to a variety of insults including ischemia (Franklin et al., 2005). Taken together these data illustrate the multifunctional nature of oxygen-sensitive molecules and suggests that many may be part of the repertoire of responses to stress in general rather than to a specific stressor.

Interaction between oxygen-dependent and other responses occur both in rapid and slower responses. For example, the rapid oxygen-dependent regulation of membrane transport appears quite often to be regulated by ROS, and thereby redox state. Thus, any redox disturbance will affect oxygen-dependent ion transport. As indicated above, gene expression—a slower oxygen-dependent system—is also affected by redox-state dependent transcriptional regulation.

4.7 Oxygen-Dependent and Xenobiotically Induced Gene Expression Pathways

With regard to the function of HIF, its dimerization partner ARNT is also a dimerization partner for many other transcription factors, among them the aryl hydrocarbon receptor (AhR; dioxin receptor). When it was observed that both the xenobiotically induced and hypoxic gene expression used the same dimerization partner, several studies investigated the possibility that one affected the other. While some studies have not been able to show interaction, several others have indicated interaction which may be cell type specific (Gradin et al., 1999; Gassmann et al., 1997; Chan et al., 1999; Nie et al., 2001). There may also be differences in which pathway is preferred (Hofer et al., 2004). In addition to the interaction between aryl hydrocarbon receptor and HIF pathways, studies have indicated an interaction with HIF-pathway and the pathway involved in the generation of diel rhythmicity (Chilov et al., 2001).

4.8 Oxygen- and Temperature-Dependent Responses

Interactions between temperature responses and hypoxia-induced responses may also occur. The possibility of this interaction is of minor importance for the CNS of homeothermic animals, since CNS temperature is usually tightly regulated. If effects in homeotherms are observed, they will be apparent in the peripheral nerves of animals living in cold climates, since there the temperature may be decreased. In contrast, the temperature of all the tissues of poikilothermic animals may show large fluctuations. However, even in mammals temperature affects HIF-1α expression. Increased amounts of HIF-1α protein with concomitant induction of Hsp90 and Hsp70 have been observed in mice exposed to heat (Katschinski et al., 2002), upon heat acclimation in rat (Maloyan et al., 2005) and in human hepatoma cell lines exposed to heat (Katschinski et al., 2002). In poikilotherms, increased expression of HIF occurs during the heat acclimation of *Caenorhabditis elegans* (Treinin et al., 2003). Association of HIF-α with the Hsp heterocomplex Hsp90–Hsp70 stabilizes HIF-1α by protecting it from degradation in both normoxia and hypoxia (Minet et al., 1999; Katschinski et al., 2002; Katschinski et al., 2004). It appears that Hsp90 binds to PAS-B domain of HIF-α and exerts a stabilizing influence on the protein (Katschinski et al., 2004). Whereas the above studies have indicated interaction of the heat-shock response and the HIF response, recent studies (Rissanen et al., 2006b) have indicated that acclimation to a reduced temperature is associated with an increased level of Hsps of the 70 and 90 classes, and increased HIF function, as shown by increased DNA binding of the transcription factor, in a poikilothermic vertebrate, crucian carp. The close association between increased Hsp expression and increased HIF reveal the interaction of temperature and stress-sensitive pathways with oxygen-sensitive pathways.

5 Cellular Systems Described for Oxygen-Dependent Phenomena

5.1 Nonexcitable Cells: Erythrocyte as an Example

Among nonexcitable cells, erythrocytes have been studied in most detail with regard to oxygen-dependent ion transport, and the function of the ion transport pathways has been reviewed by Gibson et al. (2000) (see also ➋ *Sect. 4.1*). As to the physiological role played by the oxygen-dependent pathways, they regulate oxygen transport in hypoxia, and possibly in hyperoxia (Nikinmaa, 2003). While data do not allow firm conclusions to be made about the role that erythrocytes may play in oxygen sensing, the interaction between NO and hemoglobin, and the possibility of consequent blood flow regulation (Stamler et al., 1997), open up the possibility that regulation of erythrocytic oxygen transport is utilized in the regulation of oxygen-dependent responses.

5.2 Excitable Cells: Carotid Body Glomus Cell as a Primary Example

Carotid body glomus cells are cells that sense oxygen in the arterial chemosensory organ. Because of their behavior as oxygen sensors of the peripheral nervous system, they are involved in the control of breathing and have become probably the most important single excitable cell type in which oxygen sensing has been studied. Oxygen sensing by carotid body glomus cells and consecutive mechanisms behind oxygen-dependent nervous signaling to CNS have been the subject of several reviews (Acker and Xue, 1995; Gonzalez et al., 1995a, b; Bisgard, 2000; Prabhakar and Overholt, 2000; Lahiri et al., 2001, 2006; Lopez-Barneo, 2003). Basically, in acute hypoxia, changes in oxygen level are sensed, the activity of potassium channels depends on the oxygen level sensed, leading to an increased efflux of potassium and membrane depolarization in hypoxia. Consequently, calcium influx occurs via calcium channels, neurotransmitters are released, and increased neural discharge occurs in hypoxic conditions (Lahiri et al., 2006). The sequence of oxygen-dependent responses of carotid glomus cells is schematically shown in ➋ *Figure 11-2*. However, although the sequence of events is well characterized and although a Nobel prize was awarded in 1938 to Heymans, a Belgian physiologist, for showing that the carotid body function is responsible for hypoxic hyperventilation, the actual mechanism of oxygen sensing is not clear. Both mitochondrial and membrane models for oxygen sensing have been presented, and several intracellular messenger systems of carotid body glomus cells respond to hypoxia. In the mitochondrial model, the flux of electrons to the final electron acceptor oxygen is reduced, whereby the mitochondrial membrane is depolarized leading to calcium efflux, which stimulates the secretion of excitatory neurotransmitters, and leads to increased activity of afferent nerve fibers (Lahiri et al., 2006). In the membrane model, hypoxia decreases the flux of ions via plasma membrane potassium channels thereby depolarizing the cells, and leading to the influx of calcium from extracellular space (Lahiri et al., 2006). Experimental evidence supporting both models is available. Several inhibitors of mitochondrial function also inhibit the hypoxia response of glomus cells (Lahiri et al., 2001, 2006). The mitochondrial cytochrome a_3 has been suggested as a primary oxygen sensor in the glomus cell (Wilson et al., 1994). On the other hand, extramitochondrial or cell membrane-associated cytochromes have also been implied as primary oxygen sensors (Porwol et al., 2001). In both the mitochondrial and membrane models, redox changes and ROS are involved (Lahiri et al., 2001; Porwol et al., 2001; Lopez-Barneo, 2003), but whereas hypoxia is associated with a decrease in ROS in the case of extramitochondrial control (Porwol et al., 2001), an increase occurs at low oxygen levels in mitochondria (Guzy and Schumaker, 2006). Also, on the basis of present data it appears that both NO and CO can influence the effect of hypoxia on neural discharge from the carotid body glomus cells (Lahiri et al., 2006). Notably, NO influences the oxygen dependency of mitochondrial respiration (Koivisto et al., 1997). In addition, interaction between oxygen-dependent and pH-dependent phenomena occurs in glomus and other excitable cells (Miller et al., 2004; Peers, 2004). It is, furthermore, possible that a globin, neuroglobin, is involved in oxygen sensing, since it has been characterized also in the carotid body (Di Giulio et al., 2006). In sustained hypoxia, it appears that gene expression in the carotid body cells is modified via the HIF-dependent pathway as in other cell types (Fung and Tipoe, 2003; Lopez-Barneo, 2003).

Figure 11-2
A schematic representation of how a decrease in oxygen tension (hypoxia) may affect carotid body glomus cell function. In the mitochondrial model, hypoxia affects either reactive oxygen species (ROS) production or ATP production of mitochondria. Both of these may affect the outward flux of potassium via the potassium channel with the downstream effects shown in the diagram. In the membrane model, the ROS production by membrane-bound molecules (cytochromes) is oxygen sensitive, and thereby affected by hypoxia. Thus, these membrane-bound molecules function as proximal oxygen sensors and cause effects on potassium channels with the downstream effects described in the figure and in the text

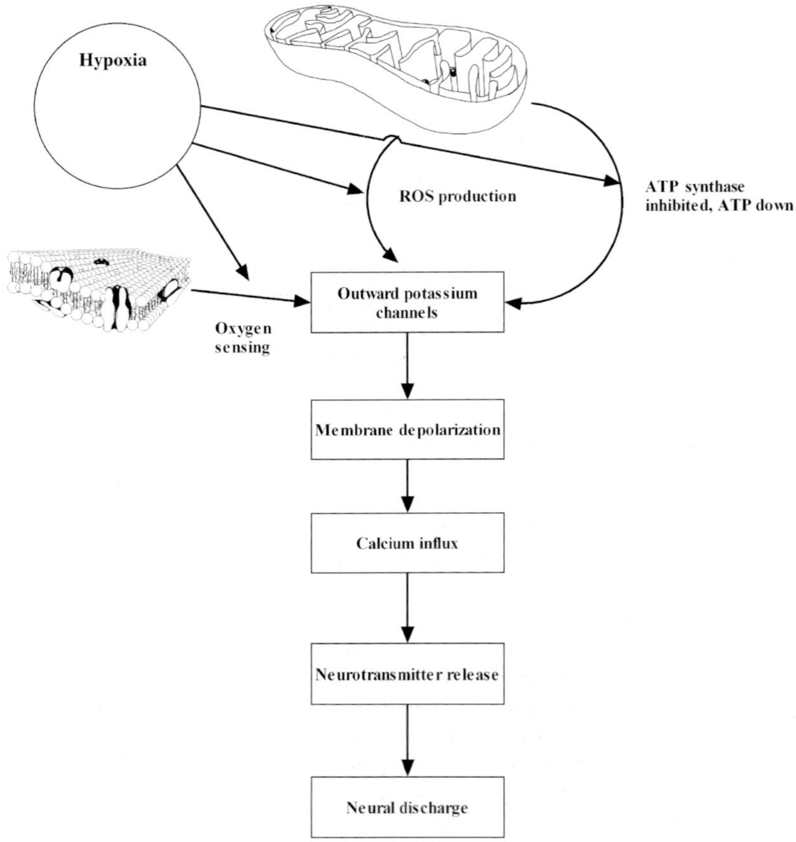

5.3 An Example of Other Excitable Cellular Systems: Gill Neuroepithelial Cells

Cells that were classified as neuroepithelial cells were microscopically observed in fish gills in early 1980s (Dunel-Erb et al., 1982). Since breathing rate in fishes is mainly regulated by oxygen, several studies have investigated the possible effects of oxygen on these cells. The neuroepithelial cells share characteristics with both the carotid body glomus cells and lung neuroepithelial cells (Burleson et al., 2006) and have high levels of biogenic amines (Dunel-Erb et al., 1982; Burleson et al., 2006). The branchial neuroepithelial cells are situated in the primary lamellae between water and blood, but appear to have no direct contact with environmental water (Dunel-Erb et al., 1982). Burleson and coworkers (2006) cultured the putative neuroepithelial cells to evaluate whether they had oxygen-sensitive potassium channels, which are a characteristic of oxygen-sensing cells of the carotid body (Donnelly, 1997; 1999; Lopez-Barneo et al., 1999).

The results indicated this to be the case. Notably, the function of oxygen-sensing cells in the gills of fishes have so far been studied only in a couple of species—notably the zebrafish (Jonz et al., 2004) and the channel catfish (Burleson et al., 2006).

6 The Importance of Oxygen Sensing in Neural Function

When the P_{O_2} in the medulla falls below threshold, putative oxygen sensing neurons increase sympathetic activity (Solomon, 2000) triggered by the release of ATP to stimulate an "adaptive increase" in breathing (reviewed by Gourine, 2005). The oxygen-chemosensitive cell of the central nervous system is encompassed in a distributed network of neurons in the brain stem (reviewed by Neubauer and Sunderram, 2004). Central oxygen responsive neurons mediate short- or long-term adaptations to hypoxia, which provide both retaliatory and pre-emptive neuroprotective responses. The increased ventilatory response provides a neuroprotective function by pre-empting energy failure due to a mismatch of oxidative metabolism with energy expenditure. While many of the short-term rapid responses to P_{O_2} levels below threshold can be mediated by local peripheral oxygen-sensitive sensors including the carotid body and effectors, long-term acclimatization to hypoxia requires alterations in sympathetic outflow in the associated brain stem nuclei (reviewed by Guyenet, 2000). Using a chemodenervated rat model to remove afferent input from the carotid body, Roux and coworkers (2000) investigated the direct effects of hypoxia on neurons of the tractus solitarius and ventrolateral medulla and provided evidence that tyrosine hydroxylase (TH) mRNA levels increased in a similar manner to the sham-operated group and some degree of ventilatory acclimatization still occurred, revealing that neurons in these two nuclei acted as oxygen sensors in their own right. Furthermore, neurons in the rostral ventrolateral medulla also demonstrated intrinsic oxygen sensitivity: since cultured neurons demonstrated hypoxia-mediated excitation they have been proposed as putative oxygen sensors (Mazza et al., 2000).

It has been known for some time that TH protein expression is oxygen sensitive and shows a graded response to the duration of hypoxic exposure in the rat medulla. A short 3-day exposure resulted in a 26–50% increase in TH depending on the subpopulation of medullary nuclei examined. In contrast, 14 days of hypoxic exposure resulted in 31–41% increases in TH, after which the level of TH returned to baseline (Schmitt et al., 1993). While oxygen sensors in the medulla appear to regulate the level of TH without input from the carotid body, the hypoxia-induced increase in turnover rate of noradrenaline (NE) requires input from the carotid body (Soulier et al., 1992).

Microdialysate collected from the phrenic nerve to measure oxygen-sensitive neurotransmitter release from the network of brain stem nuclei affecting phrenic nerve output demonstrated that there was a time-dependent change in neurotransmitter release in response to hypoxia. This response was biphasic with an initial decrease in taurine followed by a sustained increase and the initial increase in phrenic nerve activity was consistent with a rapid elevation in the excitatory neurotransmitter glutamate accompanied by a more gradual increase in the inhibitory neurotransmitter GABA (Hoop et al., 1999). Such hypoxia-induced respiratory depression parallels neuronal hypometabolism in brain stem regions involved in respiratory and cardiovascular control in hypoxia-intolerant mammals (LaManna et al., 1996) and in the hypoxia and anoxia-tolerant tropical reef shark, *Hemiscyllium ocellatum* (Mulvey and Renshaw, 2000). However, neuronal hypometabolism in the anoxia-tolerant reef shark is not related to brain failure since the brain energy charge was maintained above critical levels even after a 50 min exposure to anoxia (Renshaw et al., 2002).

Hypoxia induced increases in cerebral blood flow provides a second neuroprotective preemptive defence strategy. Putative oxygen sensors that regulate cerebral blood flow were investigated in rats with bilateral lesions of the rostral ventrolateral medulla. Exposure to a Pa_{O_2} of 36 ± 1 mmHg significantly increased cerebral blood flow in control animals up to 204%, but not in animals with bilateral lesions (Underwood et al., 1994). However, in animals with bilateral lesions there was a significantly lower cerebral blood flow response to hypoxia because of a 50–69% decrease in cerebral vasodilation, demonstrating that neurons of the rostral ventrolateral medulla have a specific oxygen-sensitive role in controlling vascular tone in response to hypoxia. These changes were not detected in response to hypercapnia (Underwood et al., 1994).

7 Clinical Implications

Asphyxia, sleep apnea, vascular cognitive impairment, stroke, some neurodegenerative diseases, and cardiovascular disease are just a few of the conditions that can result in brain damage as a consequence of reduced oxygen levels. A few minutes of oxygen deprivation is enough to cause neuronal death in the mammalian brain (reviewed by Lutz et al., 2003; Acker and Acker, 2004). This is not so for all vertebrates. The challenge of how to protect vulnerable organs, such as the energetically expensive heart and brain, from hypoxia-induced damage depends on a number of adaptations, including those of oxygen sensing. Many adaptations conferring pronounced hypoxia and/or anoxia tolerance have appeared during vertebrate evolution but they appear to have been retained only by a few species of fish and turtles. Examination of hypoxia-tolerant species has revealed that they can reversibly reprogram gene expression to achieve a "protected phenotype" displaying a suite of retaliatory and pre-emptive mechanisms to forestall cell death (reviewed by Lutz et al., 2003). An understanding of the mechanisms involved in the reversible switch to a protected phenotype may provide an insight into potential intervention strategies that can be used in clinical settings to minimize ischemia–reperfusion injury following stroke and heart attack. Comparative physiology is the nursery ground of several potential treatment strategies for use in clinical settings.

Oxygen levels regulate the pattern of gene expression in health and disease via a master switch, HIF-1α (reviewed by Nikinmaa and Rees, 2005). While HIF-1α acts as a ubiquitous transcription factor to increase cell survival during hypoxia and all animals studied so far express HIF-1α, there are only a few vertebrates that can survive prolonged periods of hypoxia or anoxia. Thus, the ability to change the level of HIF-1α expression per se does not automatically trigger an expression of a hypoxia- or anoxia-tolerant phenotype. An understanding of why HIF-1α prolongs survival in some vertebrates and not in others could lead to new treatments, which involve manipulating HIF-1α levels and its half-life.

Oxygen-regulated gene expression via HIF-1α as a master regulator results in the upregulation of a number of neuroprotective proteins such as EPO, which can protect neurons from apoptotic cell death via its action on the PI-3-k/AKT pathway (Weishaupt et al., 2004). It has been shown that administration of EPO or its induction by hypoxia significantly reduces infarct size in the mammalian brain (Gassmann et al., 2003) and EPO has recently been used successfully in a number of clinical trials (see reviews by Ehrenreich et al., 2004; Gradin et al., 2004; Ren and Finklestein, 2005). As discussed above, the oxygen sensor pathway is closely interlinked with several other stress pathways; so, preconditioning with a stimulus that causes mild cell injury including exposure to low oxygen levels or exercise (Ameln et al., 2005) activates a suite of protective physiological mechanisms (reviewed by Dirnagl et al., 2003), which confer a naturally protected phenotype. For example, stressors that activate the Hsp family of molecular chaperones not only protect cells from subsequent ischemia and block cell death by apoptosis but may also affect a number of neurodegenerative diseases such as Alzheimer's (reviewed by Franklin et al., 2005). The interaction of the oxygen-sensing pathway with other stress pathways opens up the possibility of cross-protection by targeting key neuroprotective chaperones and transcription factors. This is illustrated by the effect of physiological stressors such as heat (Christians et al., 2002) or exercise (Ameln et al., 2005) on elevating the level of Hsps and/or HIF-1α, respectively. Understanding how to manipulate these neuroprotective pathways is expected to lead to strategies designed to preempt injury and facilitate functional recovery.

Future therapies may focus on the activation of HIF-1α and its downstream gene products long enough to gain the neuroprotective benefits but not long enough for HIF to ultimately initiate neuronal death via apoptosis (reviewed by Acker and Acker, 2004). It is also likely that selected gene products, such as neuroprotective globins, that are turned on by intermittent hypoxia (Di Giulio et al., 2006) could be administered to reduce neuronal loss and improve recovery after acute insults due to head trauma, ischemic events, and during chronic diseases such as neurodegenerative diseases of the brain, including the retina.

References

Acker H. 2005. The oxygen sensing signal cascade under the influence of reactive oxygen species. Philos Trans R Soc Lond B Biol Sci 360: 2201-2210.

Acker H, Xue D. 1995. Mechanisms of O_2 sensing in the carotid body in comparison with other O_2-sensing cells. News Physiol Sci 10: 211-216.

Acker T, Acker H. 2004. Cellular oxygen sensing need in CNS function: Physiological and pathological implications. J Exp Biol 207: 3171-3188.

Acker T, Fandrey J, Acker H. 2006. The good, the bad and the ugly in oxygen-sensing: ROS, cytochromes and prolyl-hydroxylases. Cardiovasc Res 71: 195-207.

Airaksinen S, Rabergh CM, Sistonen L, Nikinmaa M. 1998. Effects of heat shock and hypoxia on protein synthesis in rainbow trout (*Oncorhynchus mykiss*) cells. J Exp Biol 201: 2543-2551.

Ameln H, Gustafsson T, Sundberg CJ, Okamoto K, Jansson E, et al. 2005. Physiological activation of hypoxia inducible factor-1 in human skeletal muscle. FASEB J 19: 1009-1011.

Araya R, Uehara T, Nomura Y. 1998. Hypoxia induces apoptosis in human neuroblastoma SK-N-MC cells by caspase activation accompanying cytochrome *c* release from mitochondria. FEBS Lett 439: 168-172.

Archer SL, Weir EK, Reeve HL, Michelakis E. 2000. Molecular identification of O_2 sensors and O_2-sensitive potassium channels in the pulmonary circulation. Adv Exp Med Biol 475: 219-240.

Archer SL, Wu XC, Thebaud B, Moudgil R, Hashimoto K, et al. 2004. O_2 sensing in the human ductus arteriosus: Redox-sensitive K^+ channels are regulated by mitochondria-derived hydrogen peroxide. Biol Chem 385: 205-216.

Banasiak KJ, Xia Y, Haddad GG. 2000. Mechanisms underlying hypoxia-induced neuronal apoptosis. Prog Neurobiol 62: 215-249.

Basu N, Todgham AE, Ackerman PA, Bibeau MR, Nakano K, et al. 2002. Heat shock protein genes and their functional significance in fish. Gene 295: 173-183.

Berenbrink M, Volkel S, Heisler N, Nikinmaa M. 2000. O_2-dependent K^+ fluxes in trout red blood cells: The nature of O_2 sensing revealed by the O_2 affinity, cooperativity and pH dependence of transport. J Physiol (Lond) 526: 69-80.

Berenbrink M, Volkel S, Koldkjaer P, Heisler N, Nikinmaa M. 2006. Two different oxygen sensors regulate oxygen-sensitive K^+ transport in crucian carp red blood cells. J Physiol (Lond) 575: 37-48.

Berra E, Ginouves A, Pouyssegur J. 2006. The hypoxia-inducible-factor hydroxylases bring fresh air into hypoxia signalling. EMBO Rep 7: 41-45.

Bickler PE, Buck LT. 1998. Adaptations of vertebrate neurons to hypoxia and anoxia: Maintaining critical Ca^{2+} concentrations. J Exp Biol 201: 1141-1152.

Bickler PE, Donohoe PH. 2002. Adaptive responses of vertebrate neurons to hypoxia. J Exp Biol 205: 3579-3586.

Bisgard GE. 2000. Carotid body mechanisms in acclimatization to hypoxia. Resp Physiol 121: 237-246.

Bogdanova A, Nikinmaa M. 2001. Reactive oxygen species regulate oxygen-sensitive potassium flux in rainbow trout erythrocytes. J Gen Physiol 117: 181-190.

Bogdanova A, Ogunshola OO, Bauer C, Nikinmaa M, Gassmann M. 2003. Molecular mechanisms of oxygen-induced regulation of Na^+/K^+ pump. Adv Exp Med Biol 536: 231-238.

Bracken CP, Whitelaw ML, Peet DJ. 2003. The hypoxia-inducible factors: Key transcriptional regulators of hypoxic responses. Cell Mol Life Sci 60: 1376-1393.

Buck LT, Pamenter ME. 2006. Adaptive responses of vertebrate neurons to anoxia—Matching supply to demand. Respir Physiol Neurobiol 154: 226-240.

Bunn HF, Poyton RO. 1996. Oxygen sensing and molecular adaptation to hypoxia. Physiol Rev 76: 839-885.

Burleson ML, Mercer SE, Wilk-Blaszczak MA. 2006. Isolation and characterization of putative O_2 chemoreceptor cells from the gills of channel catfish (*Ictalurus punctatus*). Brain Res 1092: 100-107.

Burmester T, Hankeln T. 2004. Neuroglobin: A respiratory protein of the nervous system. News Physiol Sci 19: 110-113.

Burmester T, Weich B, Reinhardt S, Hankeln T. 2000. A vertebrate globin expressed in the brain. Nature 407: 520-523.

Burston SG, Clarke AR. 1995. Molecular chaperones: Physical and mechanistic properties. Essays Biochem 29: 125-136.

Camenisch G, Wenger RH, Gassmann M. 2002. DNA-binding activity of hypoxia-inducible factors (HIFs). Methods Mol Biol 196: 117-129.

Canbolat O, Fandrey J, Jelkmann W. 1998. Effects of modulators of the production and degradation of hydrogen peroxide on erythropoietin synthesis. Resp Physiol 114: 175-183.

Chae HJ, Kim SC, Han KS, Chae SW, An NH, et al. 2001. Hypoxia induces apoptosis by caspase activation accompanying cytochrome *c* release from mitochondria in MC3T3E1 osteoblasts, p38 MAPK is related in hypoxia-induced apoptosis. Immunopharm Immunotox 23: 133-152.

Chan WK, Yao G, Gu YZ, Bradfield CA. 1999. Cross-talk between the aryl hydrocarbon receptor and hypoxia inducible factor signaling pathways—Demonstration of competition and compensation. J Biol Chem 274: 12115-12123.

Chandel NS, Schumacker PT. 2000. Cellular oxygen sensing by mitochondria: Old questions, new insight. J Appl Physiol 88: 1880-1889.

Chen XQ, Qin LY, Zhang CG, Yang LT, Gao Z, et al. 2005. Presence of neuroglobin in cultured astrocytes. Glia 50: 182-286.

Chilov D, Hofer T, Bauer C, Wenger RH, Gassmann M. 2001. Hypoxia affects expression of circadian genes PER1 and CLOCK in mouse brain. FASEB J 15: 2613-2622.

Choi SL, Kim SJ, Lee KT, Kim J, Mu J, et al. 2001. The regulation of AMP-activated protein kinase by H_2O_2. Biochem Biophys Res Comm 287: 92-97.

Christians ES, Yan LJ, Benjamin IJ. 2002. Heat shock factor 1 and heat shock proteins: Critical partners in protection against acute cell injury. Crit Care Med 30: S43-S50.

Couture M, Burmester T, Hankeln T, Rousseau DL. 2001. The heme environment of mouse neuroglobin. Evidence for the presence of two conformations of the heme pocket. J Biol Chem 276: 36377-36382.

Cross AR, Henderson L, Jones OT, Delpiano MA, Hentschel J, et al. 1990. Involvement of an NAD(P)H oxidase as a P_{O_2} sensor protein in the rat carotid body. Biochem J 272: 743-747.

Cummins EP, Taylor CT. 2005. Hypoxia-responsive transcription factors. Pflugers Arch Eur J Physiol 450: 363-371.

Dahan I, Issaeva I, Gorzalczany Y, Sigal N, Hirshberg M, et al. 2002. Mapping of functional domains in the p22 (phox) subunit of flavocytochrome b_{559} participating in the assembly of the NADPH oxidase complex by "peptide walking". J Biol Chem 277: 8421-8432.

D'Amico G, Lam F, Hagen T, Moncada S. 2006. Inhibition of cellular respiration by endogenously produced carbon monoxide. J Cell Sci 119: 2291-2298.

Dayan F, Roux D, Brahimi-Horn MC, Pouyssegur J, Mazure NM. 2006. The oxygen sensor factor-inhibiting hypoxia-inducible factor-1 controls expression of distinct genes through the bifunctional transcriptional character of hypoxia-inducible factor-α. Cancer Res 66: 3688-3698.

Del Toro R, Levitsky KL, Lopez-Barneo J, Chiara MD. 2003. Induction of T-type calcium channel gene expression by chronic hypoxia. J Biol Chem 278: 22316-22324.

de Paula PM, Branco LG. 2003. Nitric oxide in the rostral ventrolateral medulla modulates hyperpnea but not anapyrexia induced by hypoxia. Brain Res 977: 231-238.

Di Giulio C, Bianchi G, Cacchio M, Artese L, Piccirilli M, et al. 2006. Neuroglobin, a new oxygen binding protein is present in the carotid body and increases after chronic intermittent hypoxia. Berlin: Springer-Verlag; pp. 15-19.

Dirnagl U, Simon RP, Hallenbeck JM. 2003. Ischemic tolerance and endogenous neuroprotection. Trends Neurosci 26: 248-254.

Donnelly DF. 1997. Are oxygen dependent K^+ channels essential for carotid body chemo-transduction? Resp Physiol 110: 211-218.

Donnelly DF. 1999. K^+ currents of glomus cells and chemosensory functions of carotid body. Resp Physiol 115: 151-160.

Drew C, Ball V, Robinson H, Ellory JC, Gibson JS. 2004. Oxygen sensitivity of red cell membrane transporters revisited. Bioelectrochemistry 62: 153-158.

Dunel-Erb S, Bailly Y, Laurent P. 1982. Neuroepithelial cells in fish gill primary lamellae. J Appl Physiol 53: 1342-1353.

Duranteau J, Chandel NS, Kulisz A, Shao ZH, Schumacker PT. 1998. Intracellular signaling by reactive oxygen species during hypoxia in cardiomyocytes. J Biol Chem 273: 11619-11624.

Ehleben W, Bolling B, Merten E, Porwol T, Strohmaier AR, et al. 1998. Cytochromes and oxygen radicals as putative members of the oxygen sensing pathway. Resp Physiol 114: 25-36.

Ehrenreich H, Aust C, Krampe H, Jahn H, Jacob S, et al. 2004. Erythropoietin: Novel approaches to neuroprotection in human brain disease. Metab Brain Dis 19: 195-206.

Fago A, Mathews AJ, Moens L, Dewilde S, Brittain T. 2006. The reaction of neuroglobin with potential redox protein partners cytochrome b_5 and cytochrome *c*. FEBS Lett 580: 4884-4888.

Fan CG, Li Q, Ross D, Engelhardt JF. 2003. Tyrosine phosphorylation of I kappa B α activates NF kappa B through a redox-regulated and c-Src-dependent mechanism following hypoxia/reoxygenation. J Biol Chem 278: 2072-2080.

Fandrey J, Frede S, Ehleben W, Porwol T, Acker H, et al. 1997. Cobalt chloride and desferrioxamine antagonize the inhibition of erythropoietin production by reactive oxygen species. Kidney Int 51: 492-496.

Fandrey J, Frede S, Jelkmann W. 1994. Role of hydrogen peroxide in hypoxia-induced erythropoietin production. Biochem J 303: 507-510.

Fandrey J, Gorr TA, Gassmann M. 2006. Regulating cellular oxygen sensing by hydroxylation. Cardiovasc Res 71: 642-651.

Finkel T. 1998. Oxygen radicals and signaling. Curr Opin Cell Biol 10: 248-253.

Firth JD, Ebert BL, Ratcliffe PJ. 1995. Hypoxic regulation of lactate dehydrogenase A: Interaction between hypoxia-inducible factor 1 and cAMP response elements. J Biol Chem 270: 21021-21027.

Flatman PW. 2005. Activation of ferret erythrocyte Na^+-K^+-$2Cl^-$ cotransport by deoxygenation. J Physiol (Lond) 563: 421-431.

Fordel E, Geuens E, Dewilde S, De Coen W, Moens L. 2004. Hypoxia/ischemia and the regulation of neuroglobin and cytoglobin expression. IUBMB Life 56: 681-687.

Franklin TB, Krueger-Naug AM, Clarke DB, Arrigo AP, Currie RW. 2005. The role of heat shock proteins Hsp70 and Hsp27 in cellular protection of the central nervous system. Int J Hyperthermia 21: 379-392.

Fraser J, de Mello LV, Ward D, Rees HH, Williams DR, et al. 2006. Hypoxia-inducible myoglobin expression in nonmuscle tissues. Proc Natl Acad Sci USA 103: 2977-2981.

Fratelli M, Goodwin LO, Orom UA, Lombardi S, Tonelli R, et al. 2005. Gene expression profiling reveals a signaling role of glutathione in redox regulation. Proc Natl Acad Sci USA 102: 13998-14003.

Fridovich I. 1986. Biological effects of the superoxide radical. Arch Biochem Biophys 247: 1-11.

Fu XW, Nurse CA, Wong V, Cutz E. 2002. Hypoxia-induced secretion of serotonin from intact pulmonary neuroepithelial bodies in neonatal rabbit. J Physiol 539: 503-510.

Fung ML, Tipoe GL. 2003. Role of HIF-1 in physiological adaptation of the carotid body during chronic hypoxia. Adv Exp Med Biol 536: 593-601.

Gassmann M, Heinicke K, Soliz J, Ogunshola OO. 2003. Non-erythroid functions of erythropoietin. Adv Exp Med Biol 543: 323-330.

Gassmann M, Kvietikova I, Rolfs A, Wenger RH. 1997. Oxygen- and dioxin-regulated gene expression in mouse hepatoma cells. Kidney Int 51: 567-574.

Geuens E, Brouns I, Flamez D, Dewilde S, Timmermans JP, et al. 2003. A globin in the nucleus! J Biol Chem 278: 30417-30420.

Gibson JS, Cossins AR, Ellory JC. 2000. Oxygen-sensitive membrane transporters in vertebrate red cells. J Exp Biol 203: 1395-1407.

Gloire G, Legrand-Poels S, Piette J. 2006. NF-κB activation by reactive oxygen species: Fifteen years later. Biochem Pharmacol 72: 1493-1505.

Gnaiger E, Lassnig B, Kuznetsov A, Rieger G, Margreiter R. 1998. Mitochondrial oxygen affinity, respiratory flux control and excess capacity of cytochrome *c* oxidase. J Exp Biol 201: 1129-1139.

Gonzalez C, Lopez-Lopez JR, Obeso A, Perez-Garcia MT, Rocher A. 1995a. Cellular mechanisms of oxygen chemoreception in the carotid body. Resp Physiol 102: 137-147.

Gonzalez C, Vicario I, Almaraz L, Rigual R 1995b. Oxygen sensing in the carotid body. Biol Signals 4: 245-256.

Gonzalez MF, Lowenstein D, Fernyak S, Hisanaga K, Simon R, et al. 1991. Induction of heat shock protein 72 like immunoreactivity in the hippocampal formation following transient global ischemia. Brain Res Bull 26: 241-250.

Gourine AV. 2005. On the peripheral and central chemoreception and control of breathing: An emerging role of ATP. J Physiol 568: 715-724.

Gradin K, Mcguire J, Wenger RH, Kvietikova I, Whitelaw ML, et al. 2004. Erythropoietin as a tissue-protective cytokine in brain injury: What do we know and where do we go? Neuroscientist 10: 93-98.

Gradin K, Toftgard R, Poellinger L, Berghard A. 1999. Repression of Dioxin signal transduction in fibroblasts. Identification of a putative repressor associated with Arnt. J Bioc Chem 274: 13511-13518.

Guyenet PG. 2000. Neural structures that mediate sympathoexcitation during hypoxia. Resp Physiol 121: 147-162.

Guzy RD, Hoyos B, Robin E, Chen H, Liu LP, et al. 2005. Mitochondrial complex III is required for hypoxia-induced ROS production and cellular oxygen sensing. Cell Metab 1: 401-408.

Guzy RD, Schumacker PT. 2006. Oxygen sensing by mitochondria at complex III: The paradox of increased reactive oxygen species during hypoxia. Exp Physiol 91: 807-819.

Haddad JJ. 2002. Oxygen-sensing mechanisms and the regulation of redox-responsive transcription factors in development and pathophysiology. Respir Res 3: 26.

Hammarstrom AK, Gage PW. 2002. Hypoxia and persistent sodium current. Eur Biophys J 31: 323-330.

Hammerer-Lercher A, Mair J, Bonatti J, Watzka SBC, Puschendorf B, et al. 2001. Hypoxia induces heat shock protein expression in human coronary artery bypass grafts. Cardiovasc Res 50: 115-124.

Hankeln T, Ebner B, Fuchs C, Gerlach F, Haberkamp M, et al. 2005. Neuroglobin and cytoglobin in search of their role in the vertebrate globin family. J Inorg Biochem 99: 110-119.

Hardie DG. 2003. Minireview: The AMP-activated protein kinase cascade: The key sensor of cellular energy status. Endocrinology 144: 5179-5183.

Hardie DG, Hawley SA, Scott J. 2006. AMP-activated protein kinase—development of the energy sensor concept. J Physiol (Lond) 574: 7-15.

Harrison PJ, Procter AW, Exworthy T, Robert GW, Najlerahim A, et al. 1993. Heat shock protein (hsp70) mRNA expression in human brain: Effects of neurodegenerative disease and agonal state. Neuropathol Appl Neurobiol 19: 10-21.

Ho VT, Bunn HF. 1996. Effects of transition metals on the expression of the erythropoietin gene: Further evidence that the oxygen sensor is a heme protein. Biochem Biophys Res Comm 223: 175-180.

Hochachka PW, Lutz PL. 2001. Mechanism, origin, and evolution of anoxia tolerance in animals. Comp Biochem Physiol B Biochem Mol Biol 130: 435-459.

Hofer T, Pohjanvirta R, Spielmann P, Viluksela M, Buchmann DP, et al. 2004. Simultaneous exposure of rats to dioxin and carbon monoxide reduces the xenobiotic but not the hypoxic response. Biol Chem 385: 291-294.

Honess NA, Gibson JS, Cossins AR. 1996. The effects of oxygenation upon the Cl dependent K flux pathway in equine red cells. Pflugers Arch Eur J Physiol 432: 270-277.

Hoop B, Beagle JL, Maher TJ, Kazemi H. 1999. Brainstem amino acid neurotransmitters and hypoxic ventilatory response. Resp Physiol 118: 117-1129.

Horn EM, Waldrop TG. 1997. Oxygen-sensing neurons in the caudal hypothalamus and their role in cardiorespiratory control. Resp Physiol 110: 219-228.

Huang LE, Willmore WG, Gu J, Goldberg MA, Bunn HF. 1999. Inhibition of hypoxia-inducible factor 1 activation by carbon monoxide and nitric oxide. Implications for oxygen sensing and signaling. J Biol Chem 274: 9038-9044.

Ivan M, Kondo K, Yang HF, Kim W, Valiando J, et al. 2001. HIFα targeted for VHL-mediated destruction by proline hydroxylation: Implications for O_2 sensing. Science 292: 464-468.

Jaakkola P, Mole DR, Tian YM, Wilson MI, Gielbert J, et al. 2001. Targeting of HIFα to the von Hippel-Lindau ubiquitylation complex by O_2-regulated prolyl hydroxylation. Science 292: 468-472.

Johansson D, Nilsson GE, Doving KB. 1997 Anoxic depression of light-evoked potentials in retina and optic tectum of crucian carp. Neurosci Lett 237: 73-76.

Jonz MG, Fearon IM, Nurse CA. 2004. Neuroepithelial oxygen chemoreceptors of the zebrafish gill. J Physiol 560: 737-752.

KaelinWG. 2005. Proline hydroxylation and gene expression. Ann Rev Biochem 74: 115-128.

Katschinski DM, Le L, Heinrich D, Wagner KF, Hofer T, et al. 2002. Heat induction of the unphosphorylated form of hypoxia-inducible factor-1α is dependent on heat shock protein-90 activity. J Biol Chem 277: 9262-9267.

Katschinski DM, Le L, Schindler SG, Thomas T, Voss AK, et al. 2004. Interaction of the PAS B domain with HSP90 accelerates hypoxia-inducible factor-1α stabilization. Cell Physiol Biochem 14: 351-360.

Kazemian P, Stephenson R, Yeger H, Cutz E. 2001. Respiratory control in neonatal mice with NADPH oxidase deficiency. Resp Physiol 126: 89-101.

Kiang JG, Tsokos GC. 1998. Heat shock protein 70 kDa: Molecular biology, biochemistry, and physiology. Pharmacol Ther 80: 183-201.

Kietzmann T, Gorlach A. 2005. Reactive oxygen species in the control of hypoxia-inducible factor-mediated gene expression. Sem Cell Devel Biol 16: 474-486.

Kietzmann T, Krones-Herzig A, Jungermann K. 2002. Signaling cross-talk between hypoxia and glucose via hypoxia-inducible factor 1 and glucose response elements. Biochem Pharmacol 64: 903-911.

Kim EH, Surh YJ. 2006. 15-Deoxy-δ-12,14-prostaglandin J2 as a potential endogenous regulator of redox-sensitive transcription factors. Biochem Pharmacol 72: 1516-1528.

Kim JY, Park JH. 2003. ROS-dependent caspase-9 activation in hypoxic cell death. FEBS Lett 549: 94-98.

Kim KH, Song MJ, Chung J, Park H, Kim JB. 2005. Hypoxia inhibits adipocyte differentiation in a HDAC-independent manner. Biochem Biophys Res Comm 333: 1178-1184.

Kirino T. 2002. Ischemic tolerance. J Cereb Blood Flow Metab 22: 1283-1296.

Kitagawa K, Matsumoto M, Tagaya M, Hata R, Ueda H, et al. 1990. 'Ischemic tolerance' phenomenon found in the brain. Brain Res 528: 21-24.

Kodama T, Shimizu N, Yoshikawa N, Makino Y, Ouchida R, et al. 2003. Role of the glucocorticoid receptor for regulation of hypoxia-dependent gene expression. J Biol Chem 278: 33384-33391.

Koivisto A, Matthias A, Bronnikov G, Nedergaard J. 1997. Kinetics of the inhibition of mitochondrial respiration by NO. FEBS Lett 417: 75-80.

Krumschnabel G, Schwarzbaum PJ, Lisch J, Biasi C, Wieser W. 2000. Oxygen-dependent energetics of anoxia-tolerant and anoxia-intolerant hepatocytes. J Exp Biol 203: 951-959.

Lahiri S, Acker H. 1999. Redox-dependent binding of CO to heme protein controls P_{O_2}-sensitive chemoreceptor discharge of the rat carotid body. Resp Physiol 115: 169-177.

Lahiri S, Roy A, Baby SM, Hoshi T, Semenza GL, et al. 2006. Oxygen sensing in the body. Prog Biophys Mol Biol 91: 249-286.

Lahiri S, Rozanov C, Roy A, Storey B, Buerk DG. 2001. Regulation of oxygen sensing in peripheral arterial chemoreceptors. Int J Biochem Cell Biol 33: 755-774.

LaManna JC, Haxhiu MA, Kutina-Nelson KL, Pundik S, Erokwu B, et al. 1996. Decreased energy metabolism in brain stem during central respiratory depression in response to hypoxia. J Appl Physiol 81: 1772-1777.

Lando D, Pongratz I, Poellinger L, Whitelaw ML. 2000. A redox mechanism controls differential DNA binding activities of hypoxia-inducible factor (HIF) 1α and the HIF-like factor. J Biol Chem 275: 4618-4627.

Lauweryns JM, de Bock V, Decramer M. 1987. Effects of unilateral vagal stimulation on intrapulmonary neuroepithelial bodies. J Appl Physiol 63: 1781-1787.

Law SH, Wu RS, Ng PK, Yu RM, Kong RY. 2006. Cloning and expression analysis of two distinct HIF-α isoforms–gcHIF-1α and gcHIF-4α–from the hypoxia-tolerant grass carp, *Ctenopharyngodon idellus*. BMC Mol Biol 7: 15.

Lee PJ, Jiang BH, Chin BY, Iyer NV, Alam J, et al. 1997. Hypoxia-inducible factor-1 mediates transcriptional activation of the heme oxygenase-1 gene in response to hypoxia. J Biol Chem 272: 5375-5381.

Lei BA, Matsuo K, Labinskyy V, Sharma N, Chandler MP, et al. 2005. Exogenous nitric oxide reduces glucose transporters translocation and lactate production in ischemic myocardium in vivo. Proc Natl Acad Sci USA 102: 6966-6971.

Leloup C, Magnan C, Benani A, Bonnet E, Alquier T, et al. 2006. Mitochondrial reactive oxygen species are required for hypothalamic glucose sensing. Diabetes 55: 2084-2090.

Lesser MP. 2006. Oxidative stress in marine environments. Annu Rev Physiol 68: 253-278.

Li RC, Lee SK, Pouranfar F, Brittian KR, Clair HB, et al. 2006. Hypoxia differentially regulates the expression of neuroglobin and cytoglobin in rat brain. Brain Res 1096: 173-179.

Liang P, MacRae TH. 1997. Molecular chaperones and the cytoskeleton. J Cell Sci 110: 1431-1440.

Lipton P. 1999. Ischemic cell death in brain neurons. Physiol Rev 79: 1431-1568.

Liu LP, Cash TP, Jones RG, Keith B, Thompson CB, et al. 2006. Hypoxia-induced energy stress regulates mRNA translation and cell growth. Mol Cell 21: 521-531.

Liu Q, Berchner-Pfannschmidt U, Moller U, Brecht M, Wotzlaw C, et al. 2004. A Fenton reaction at the endoplasmic reticulum is involved in the redox control of hypoxia-inducible gene expression. Proc Natl Acad Sci USA 101: 4302-4307.

Locke M, Noble EG. 1995. Stress proteins: The exercise response. Can J Appl Physiol 20: 155-167.

Lopez-Barneo J. 2003. Oxygen and glucose sensing by carotid body glomus cells. Curr Opin Neurobiol 13: 493-499.

Lopez-Barneo J, Pardal R, Montoro RJ, Smani T, Garcia-Hirschfeld J, et al. 1999. K^+ and Ca^{2+} channel activity and cytosolic $[Ca^{2+}]$ in oxygen-sensing tissues. Resp Physiol 115: 215-227.

Lopez-Barneo J, Pardal R, Ortega-Saenz P. 2001. Cellular mechanism of oxygen sensing. Annu Rev Physiol 63: 259-287.

Lund PA. 1995. The roles of molecular chaperones in vivo. Essays Biochem 29: 113-123.

Lutz PL, Nilsson GE. 2004. Vertebrate brains at the pilot light. Resp Physiol Neurobiol 141: 285-296.

Lutz PL, Nilsson GE, Prentice HM. 2003. The brain without oxygen: Causes of failure—physiological and molecular mechanisms for survival. Kluwer Academic Publishers. Dordrecht, Boston, London: Third edition.

Lutz PL, Prentice HM. 2002. Sensing and responding to hypoxia, molecular and physiological mechanisms. Integ Comp Biol 42: 463-468.

Ma Y, Freitag P, Zhou J, Brune B, Frede S, et al. 2004. Thyroid hormone induces erythropoietin gene expression through augmented accumulation of hypoxia-inducible factor-1. Am J Physiol Regul Integr Comp Physiol 287: R600-R607.

Madamanchi NR, Li S, Patterson C, Runge MS. 2001. Reactive oxygen species regulate heat-shock protein 70 via the JAK/STAT pathway. Arterioscler Thromb Vasc Biol 21: 321-326.

Madesh M, Ramachandran A, Balasubramanian KA. 1999. Nitric oxide prevents anoxia-induced apoptosis in colonic HT29 cells. Arch Biochem Biophys 366: 240-248.

Mahon PC, Hirota K, Semenza GL. 2001. FIH-1: A novel protein that interacts with HIF-1 α and VHL to mediate repression of HIF-1 transcriptional activity. Genes Dev 15: 2675-2686.

Malhotra R, Brosius FC. 1999. Glucose uptake and glycolysis reduce hypoxia-induced apoptosis in cultured neonatal rat cardiac myocytes. J Biol Chem 274: 12567-12575.

Maloyan A, Eli-Berchoer L, Semenza GL, Gerstenblith G, Stern MD, et al. 2005. HIF-1α-targeted pathways are activated by heat acclimation and contribute to acclimation-ischemic cross-tolerance in the heart. Physiol Genomics 23: 79-88.

Mammen PP, Shelton JM, Goetsch SC, Williams SC, Richardson JA, et al. 2002. Neuroglobin, a novel member of the globin family, is expressed in focal regions of the brain. J Histochem Cytochem 50: 1591-1598.

Mammen PP, Shelton JM, Ye Q, Kanatous SB, McGrath AJ, et al. 2006. Cytoglobin is a stress-responsive hemoprotein expressed in the developing and adult brain. J Histochem Cytochem 54: 1349-1361.

Mander P, Borutaite V, Moncada S, Brown GC. 2005. Nitric oxide from inflammatory-activated glia synergizes with hypoxia to induce neuronal death. J Neurosci Res 79: 208-215.

Marla SS, Lee J, Groves JT. 1997. Peroxynitrite rapidly permeates phospholipid membranes. Proc Natl Acad Sci USA 94: 14243-14248.

Mazza E Jr, Edelman NH, Neubauer JA. 2000. Hypoxic excitation in neurons cultured from the rostral ventrolateral medulla of the neonatal rat. J Appl Physiol 88: 2319-2329.

Michelakis ED, Hampl V, Nsair A, Wu XC, Harry G, et al. 2002. Diversity in mitochondrial function explains differences in vascular oxygen sensing. Circ Res 90: 1307-1315.

Michiels C, Minet E, Mottet D, Raes M. 2002. Regulation of gene expression by oxygen: NF-kappa B and HIF-1, two extremes. Free Radic Biol Med 33: 1231-1242.

Miller EK, Raese JD, Morrison-Bogorad M. 1991. Expression of heat shock protein 70 and heat shock cognate 70 messenger RNAs in rat cortex and cerebellum after heat shock or amphetamine treatment. J Neurochem 56: 2060-2071.

Miller P, Peers C, Kemp PJ. 2004. Polymodal regulation of hTREK1 by pH, arachidonic acid, and hypoxia: Physiological impact in acidosis and alkalosis. Am J Physiol Cell Physiol 286: C272-C282.

Milton SL, Nayak G, Lutz PL, Prentice HM. 2006. Gene transcription of neuroglobin is upregulated by hypoxia and anoxia in the brain of the anoxia-tolerant turtle *Trachemys scripta*. J Biomed Sci 13: 509-514.

Minet E, Mottet D, Michel G, Roland I, Raes M, et al. 1999. Hypoxia-induced activation of HIF-1, role of HIF-1α-Hsp90 interaction. FEBS Lett 460: 251-256.

Mulvey JM, Renshaw GM. 2000. Neuronal oxidative hypometabolism in the brainstem of the epaulette shark

(*Hemiscyllium ocellatum*) in response to hypoxic preconditioning. Neurosci Lett 290: 1-4.

Muzyamba MC, Cossins AR, Gibson JS. 1999. Regulation of Na^+-K^+-$2Cl^-$ cotransport in turkey red cells: The role of oxygen tension and protein phosphorylation. J Physiol 517: 421-429.

Muzyamba MC, Speake PF, Gibson JS. 2000. Oxidants and regulation of K^+-Cl^- cotransport in equine red blood cells. Am J Physiol Cell Physiol 279: C981-C989.

Nakamura H, Kumita H, Imai K, Iizuka T, Shiro Y. 2004. ADP reduces the oxygen-binding affinity of a sensory histidine kinase, FixL: The possibility of an enhanced reciprocating kinase reaction. Proc Natl Acad Sci USA 101: 2742-2746.

Neubauer JA, Sunderram J. 2004. Oxygen-sensing neurons in the central nervous system. J Appl Physiol 96: 367-374.

Nie MH, Blankenship AL, Giesy JP. 2001. Interactions between aryl hydrocarbon receptor (AhR) and hypoxia signaling pathways. Envir Toxicol Pharmacol 10: 17-27.

Nikinmaa M. 2003. Gas transport. Red Cell Membrane Transport in Health and Disease. Bernhardt I, Ellory JC, editors. Berlin: Springer; pp. 489-509.

Nikinmaa M, Bogdanova A, Lecklin T. 2003. Oxygen dependency of the adrenergic Na/H exchange in rainbow trout erythrocytes is diminished by a hydroxyl radical scavenger. Acta Physiologica Scandinavica 178: 149-154.

Nikinmaa M, Rees BB. 2005. Oxygen-dependent gene expression in fishes. Am J Physiol Reg Integr Comp Physiol 288: R1079-R1090.

Oehme F, Ellinghaus P, Kolkhof P, Smith TJ, Ramakrishnan S, et al. 2002. Overexpression of PH-4, a novel putative proline 4-hydroxylase, modulates activity of hypoxia-inducible transcription factors. Biochem Biophys Res Com 296: 343-349.

Olson KR. 2005. Vascular actions of hydrogen sulfide in non-mammalian vertebrates. Antiox Redox Signal 7: 804-812.

O'Reilly JP, Cummins TR, Haddad GG. 1997. Oxygen deprivation inhibits Na1 current in rat hippocampal neurones via protein kinase C. J Physiol (Lond) 503: 479-488.

Page EL, Robitaille GA, Pouyssegur J, Richard DE. 2002. Induction of hypoxia-inducible factor-1α by transcriptional and translational mechanisms. J Biol Chem 277: 48403-48409.

Pan Y, Oprysko PR, Asham AM, Koch CJ, Simon MC. 2004. p53 cannot be induced by hypoxia alone but responds to the hypoxic microenvironment. Oncogene 23: 4975-4983.

Pardal R, Lopez-Barneo J. 2002. Carotid body thin slices: Responses of glomus cells to hypoxia and K^+-channel blockers. Respir Physiol Neurobiol 132: 69-79.

Patel B, Khaliq A, Jarvis-Evans J, Boulton M, Arrol S, et al. 1995. Hypoxia induces HSP 70 gene expression in human hepatoma (HEP G2) cells. Biochem Mol Biol Int 36: 907-912.

Peers C. 2004. Interactions of chemostimuli at the single cell level: Studies in a model system. Exp Physiol 89: 60-65.

Pek M, Lutz PL. 1997. Role for adenosine in channel arrest in the anoxic turtle brain. J Exp Biol 200: 1913-1917.

Pesce A, De Sanctis D, Nardini M, Dewilde S, Moen L, et al. 2004. Reversible hexa- to penta-coordination of the heme Fe atom modulates ligand binding properties of neuroglobin and cytoglobin. IUBMB Life 56: 657-664.

Porwol T, Ehleben W, Brand V, Acker H. 2001. Tissue oxygen sensor function of NADPH oxidase isoforms, an unusual cytochrome aa_3 and reactive oxygen species. Respir Physiol 128: 331-348.

Prabhakar NR, Overholt JL. 2000. Cellular mechanisms of oxygen sensing at the carotid body: Heme proteins and ion channels. Respir Physiol 122: 209-221.

Rees BB, Bowman JA, Schulte PM. 2001. Structure and sequence conservation of a putative hypoxia response element in the lactate dehydrogenase-B gene of Fundulus. Biol Bull 200: 247-251.

Ren JM, Finklestein SP. 2005. Growth factor treatment of stroke. Curr Drug Targets CNS Neurol Disord 4: 121-125.

Renshaw GM, Dyson SE. 1999. Increased nitric oxide synthase in the vasculature of the epaulette shark brain following hypoxia. Neuroreport 10: 1707-1712.

Renshaw GM, Kerrisk CB, Nilsson GE. 2002. The role of adenosine in the anoxic survival of the epaulette shark, *Hemiscyllium ocellatum*. Comp Biochem Physiol B Biochem Mol Biol 131: 133-141.

Renshaw GM, Warburton J, Girjes A. 2004. Oxygen sensors and energy sensors act synergistically to achieve a graded alteration in gene expression: Consequences for assessing the level of neuroprotection in response to stressors. Front Biosci 9: 110-116.

Richard DE, Berra E, Pouyssegur J. 2000. Nonhypoxic pathway mediates the induction of hypoxia-inducible factor 1α in vascular smooth muscle cells. J Biol Chem 275: 26765-26771.

Richmond KN, Burnite S, Lynch RM. 1997. Oxygen sensitivity of mitochondrial metabolic state in isolated skeletal and cardiac myocytes. Am J Physiol 273: C1613-1622.

Rissanen E, Tranberg HK, Nikinmaa M. 2006a. Oxygen availability regulates metabolism and gene expression in trout hepatocyte cultures. Am J Physiol Regul Integr Comp Physiol 291: R1507-R1515.

Rissanen E, Tranberg HK, Sollid J, Nilsson GE, Nikinmaa M. 2006b. Temperature regulates hypoxia-inducible factor-1 (HIF-1) in a poikilothermic vertebrate, crucian carp (*Carassius carassius*). J Exp Biol 209: 994-1003.

Rodgers KR, Lukat-Rodgers GS. 2005. Insights into heme-based O_2 sensing from structure-function relationships in the FixL proteins. J Inorg Biochem 99: 963-977.

Roesner A, Hankeln T, Burmester T. 2006. Hypoxia induces a complex response of globin expression in zebrafish (*Danio rerio*). J Exp Biol 209: 2129-2137.

Rosenthal M, LaManna JC, Jobsis FF, Levasseur JE, Kontos HA, et al. 1976. Effects of respiratory gases on cytochrome *a* in intact cerebral cortex: Is there a critical P_{O_2}? Brain Res 108: 143-154.

Roux JC, Pequignot JM, Dumas S, Pascual O, Ghilini G, et al. 2000. O_2-sensing after carotid chemodenervation: Hypoxic ventilatory responsiveness and upregulation of tyrosine hydroxylase mRNA in brainstem catecholaminergic cells. Eur J Neurosci 12: 3181-3190.

Roy A, Rozanov C, Mokashi A, Daudu P, Al-mehdi AB, et al. 2000. Mice lacking in gp91 phox subunit of NAD(P)H oxidase showed glomus cell $[Ca^{2+}]^i$ and respiratory responses to hypoxia. Brain Res 872: 188-193.

Schmitt P, Pequignot J, Garcia C, Pujol JF, Pequignot JM. 1993. Regional specificity of the long-term regulation of tyrosine hydroxylase in some catecholaminergic rat brainstem areas. I. Influence of long-term hypoxia. Brain Res 611: 53-60.

Semenza GL. 2001. HIF-1, O_2, and the 3 PHDs: How animal cells signal hypoxia to the nucleus. Cell 107: 1-3.

Semenza GL, Wang GL. 1992. A nuclear factor induced by hypoxia via de novo protein synthesis binds to the human erythropoietin gene enhancer at a site required for transcriptional activation. Mol Cell Biol 12: 5447-5454.

Shams I, Avivi A, Nevo E. 2004. Hypoxic stress tolerance of the blind subterranean mole rat: Expression of erythropoietin and hypoxia-inducible factor 1α. Proc Natl Acad Sci USA 101: 9698-9703.

Solomon IC. 2000. Excitation of phrenic and sympathetic output during acute hypoxia: Contribution of medullary oxygen detectors. Respir Physiol 121: 101-117.

Soulier V, Cottet-Emard JM, Pequignot J, Hanchin F, Peyrin L, et al. 1992. Differential effects of long-term hypoxia on norepinephrine turnover in brain stem cell groups. J Appl Physiol 73: 1810-1814.

Stamler JS, Jia L, Eu JP, McMahon TJ, Demchenko IT, et al. 1997. Blood flow regulation by *S*-nitrosohemoglobin in the physiological oxygen gradient. Science 276: 2034-2037.

Summers BA, Overholt JL, Prabhakar NR. 2000. Augmentation of calcium current by hypoxia in carotid body glomus cells. Adv Exp Med Biol 475: 589-599.

Sun Y, Jin K, Peel A, Mao XO, Xie L, et al. 2003. Neuroglobin protects the brain from experimental stroke in vivo. Proc Natl Acad Sci USA 100: 3497-3500.

Tofighi R, Tillmark N, Dare E, Aberg AM, Larsson JE, et al. 2006. Hypoxia-independent apoptosis in neural cells exposed to carbon monoxide in vitro. Brain Res 1098: 1-8.

Treinin M, Shliar J, Jiang H, Powell-Coffman JA, Bromberg Z, et al. 2003. HIF-1 is required for heat acclimation in the nematode *Caenorhabditis elegans*. Physiol Genomics 14: 17-24.

Trent JT 3rd, Watts RA, Hargrove MS. 2001. Human neuroglobin, a hexacoordinate hemoglobin that reversibly binds oxygen. J Biol Chem 276: 30106-30110.

Tuominen A, Rissanen E, Bogdanova A, Nikinmaa M. 2003. Intracellular pH regulation in rainbow trout (*Oncorhynchus mykiss*): Hypoxia stimulates sodium/proton exchange. J Comp Physiol B Biochem Syst Env Physiol 173: 301-308.

Underwood MD, Iadecola C, Reis DJ. 1994. Lesions of the rostral ventrolateral medulla reduce the cerebrovascular response to hypoxia. Brain Res 635: 217-223.

Wang D, Youngson C, Wong V, Yeger H, Dinauer MC, et al. 1996. NADPH-oxidase and a hydrogen peroxide-sensitive K^+ channel may function as an oxygen sensor complex in airway chemoreceptors and small cell lung carcinoma cell lines. Proc Natl Acad Sci USA 93: 13182-13187.

Wang GL, Semenza GL. 1993. Desferrioxamine induces erythropoietin gene expression and hypoxia-inducible factor 1 DNA-binding activity: Implications for models of hypoxia signal transduction. Blood 82: 3610-3615.

Weir EK, Hong ZG, Porter VA, Reeve HL. 2002. Redox signaling in oxygen sensing by vessels. Respir Physiol Neurobiol 132: 121-130.

Weishaupt JH, Rohde G, Polking E, Siren AL, Ehrenreich H, et al. 2004. Effect of erythropoietin axotomy-induced apoptosis in rat retinal ganglion cells. Invest Ophthalmol Vis Sci 45: 1514-1522.

Welch WJ. 1992. Mammalian stress response: Cell physiology, structure/function of stress proteins, and implications for medicine and disease. Physiol Rev 72: 1063-1081.

Wenger RH. 2000. Mammalian oxygen sensing, signalling and gene regulation. J Exp Biol 203: 1253-1263.

Wenger RH, Gassmann M. 1996. Little difference. Nature 380: 100.

Werner E. 2004. GTPases and reactive oxygen species: Switches for killing and signaling. J Cell Sci 117: 143-153.

Wilson DF, Mokashi A, Chugh D, Vinogradov S, Osanai S, et al. 1994. The primary oxygen sensor of the cat carotid body is cytochrome a_3 of the mitochondrial respiratory chain. FEBS Lett 351: 370-374.

Wittenberg JB, Wittenberg BA. 2003. Myoglobin function reassessed. J Exp Biol 206: 2011-2020.

Wolin MS, Ahmad M, Gupte SA. 2005. Oxidant and redox signaling in vascular oxygen sensing mechanisms: Basic concepts, current controversies, and potential importance of cytosolic NADPH. Am J Physiol Lung Cell Mol Physiol 289: L159-L173.

Wyatt CN, Peers C. 1995. Ca^{2+}-activated K^+ channels in isolated type I cells of the neonatal rat carotid body. J Physiol 483: 559-565.

Xia Y, Fung ML, O'Reilly JP, Haddad GG. 2000. Increased neuronal excitability after long-term O_2 deprivation is mediated mainly by sodium channels. Brain Res Mol Brain Res 76: 211-219.

Youngson C, Nurse C, Yeger H, Curnutte JT, Vollmer C, et al. 1997. Immunocytochemical localization on O_2-sensing protein (NADPH oxidase) in chemoreceptor cells. Microsc Res Tech 37: 101-106.

Yun H, Lee M, Kim SS, Ha J. 2005. Glucose deprivation increases mRNA stability of vascular endothelial growth factor through activation of AMP-activated protein kinase in DU145 prostate carcinoma. J Biol Chem 280: 9963-9972.

Zhao P, Xue J, Gu XQ, Haddad GG, Xia Y. 2005. Intermittent hypoxia modulates Na^+ channel expression in developing mouse brain. Int J Dev Neurosci 23: 327-333.

Zhu H, Bunn HF. 1999. Oxygen sensing and signaling: Impact on the regulation of physiologically important genes. Respir Physiol 115: 239-247.

Zhu Y, Sun Y, Jin K, Greenberg DA. 2002. Hemin induces neuroglobin expression in neural cells. Blood 100: 2494-2498.

Index